Health Aspects of Lipoproteins

Health Aspects of Lipoproteins

Edited by **Caroline Gardner**

R CALLISTO
REFERENCE

New York

Published by Callisto Reference,
106 Park Avenue, Suite 200,
New York, NY 10016, USA
www.callistoreference.com

Health Aspects of Lipoproteins
Edited by Caroline Gardner

International Standard Book Number: 978-1-63239-418-7 (Hardback)

Contents

Permissions

List of Contributors

Preface

Searching for the word "lipoprotein" in the databases of Medline or PubMed, one gets more than 100,000 results, this shows the prevailing interest in the topic. This book will be valuable to new investigators in the field to get familiar with the general topic of lipoprotein research and will guide scientists interested in this domain. The sections cover topics like important issues of lipoprotein structure, clinical chemical methods, characterization of dys- and hyper lipoproteinemias and endoscopic treatments on lipoprotein.

Various studies have approached the subject by analyzing it with a single perspective, but the present book provides diverse methodologies and techniques to address this field. This book contains theories and applications needed for understanding the subject from different perspectives. The aim is to keep the readers informed about the progresses in the field; therefore, the contributions were carefully examined to compile novel researches by specialists from across the globe.

Indeed, the job of the editor is the most crucial and challenging in compiling all chapters into a single book. In the end, I would extend my sincere thanks to the chapter authors for their profound work. I am also thankful for the support provided by my family and colleagues during the compilation of this book.

Editor

Lipoprotein Structure and Assembly

Lipoprotein Structure and Dynamics: Low Density Lipoprotein Viewed as a Highly Dynamic and Flexible Nanoparticle

Ruth Prassl and Peter Laggner

Additional information is available at the end of the chapter

1. Introduction

Low density lipoproteins (LDLs) are the principal transporter of cholesterol and fat in human blood. Circulating LDLs guarantee a constant supply of cholesterol for tissues and cells, whereas cholesterol is required for membrane synthesis, modulation of membrane fluidity and the regulation of cell signaling pathways. The function of LDL in metabolism is mediating by cellular uptake via receptor-mediated endocytosis followed by lysosomal degradation [1,2], and is strongly dependent on the lipid distribution, the structure of LDL particles and on the proper conformational orientation of apolipoprotein B100 (apo-B100). Apo-B100 is the sole protein component of LDL being mainly located on the surface of LDL. Apart from their well established role as lipid transporter, LDL particles are intimately involved in the progression of cardiovascular diseases such as atherosclerosis or stroke, which are among the most prevalent causes of death in developed countries [3]. In particular, raised plasma levels of LDL are linked to an increased risk for disease. Moreover, dysregulations of LDL due to abnormalities in LDL structure have been identified as independent predictors of risk for coronary heart diseases [4,5]. LDL particles by themselves are highly heterogeneous in nature, varying in their buoyant densities, size, surface charge and chemical composition [6,7], and these biochemical characteristics determine the fate of LDL in the subendothelial space [8,9]. For example, small, dense LDL subclasses are more atherogenic than their light counterparts, which are more susceptible to modifications [5,10]. Modifications of LDL, primarily through oxidation, enzymatic degradation or lipolysis are the initiating factors in early atherosclerosis. In that case, LDL particles accumulate in the intima of the arterial wall where apo-B100 binds to proteoglycans of the extracellular matrix through ionic interactions. As a consequence, LDL becomes trapped in the subendothelium, where it is prone to oxidation processes, aggregation and fusion. Bioactive lipids, such as oxidized phospholipids, lysolipids or oxidized cholesterylester, are released from LDL particles, which are simultaneously non-specifically altered. A broad spectrum of diverse LDL particles with non-defined physicochemical properties is generated that, in turn, promotes a rapid uptake of these particles by macrophages to form foam cells [11]. This is one of the key steps in the progression of atherosclerosis. Today, atherosclerosis is known to be a chronic inflammatory disorder of the blood vessels and recognized as a prevailing cause of cardiovascular disorders, the leading causes of morbidity and mortality worldwide [12]. Since the early initiation of atherosclerosis strongly depends on the metabolism of LDL, which is predominantly triggered by molecular characteristics of LDL, it is of paramount biomedical importance to explore structural features of LDL particles in great detail. However, mostly due to the complex nature of LDL particles many questions concerning molecular details are still unanswered.

This article will review our current knowledge on the structure and dynamics of LDL particles. In fact, several recent studies revealed that the molecular organisation and dynamics of LDL core lipids, in close relationship to the intrinsic dynamics of LDL surface components, control not only the metabolism of lipids in humans, but determine the role of LDL in the pathogenesis of cardiovascular diseases. In this article, we will give a short historical review on LDL structure and then present prevailing concepts on the self-organisation of LDL. Special emphasis will be paid to dynamic features of LDL particles. In particular, we will discuss the interplay between structure and dynamics in more detail. Finally, we will give an outlook to promising future strategies to clarify the molecular structural details of LDL and how to exploit LDL nanoparticles for medical needs.

2. Molecular architecture of LDL

LDLs are composed of lipids and protein, which assemble to form a supramolecular complex with a molecular mass exceeding 2.5 - 3.0 million Da and involving 2000 to 3000 lipid molecules. Thus, LDL particles are commonly described as micellar complexes, macromolecular assemblies, self-organized nanoparticles or microemulsions. Regardless of diverse definitions, it is generally accepted that assembled LDL particles are organized into two major compartments, namely an apolar core, comprised primarily of cholesteryl esters (CE), minor amounts of triglycerides (TG) and some free unesterified cholesterol (FC). The core is surrounded by an amphipathic outer shell. This shell is composed of a phospholipid (PL) monolayer containing the larger part (>2/3) of the FC molecules and one single copy of apo-B100, which is one of the largest known monomeric glycoproteins [13]. Figure 1 provides an overview on characteristic properties of LDL together with a schematic presentation of an LDL particle. Since molecular interactions between different kinds of lipids have turned out to be highly complex, it is almost impossible to separate the surface and core regions exactly from each other. Accordingly, in some recent reports an additional hydrophobic interfacial layer composed of phospholipid acyl chains, FC, some CE molecules and hydrophobic protein domains is defined. This description takes account for the interplay between neutral core lipids and the surface layer [14].

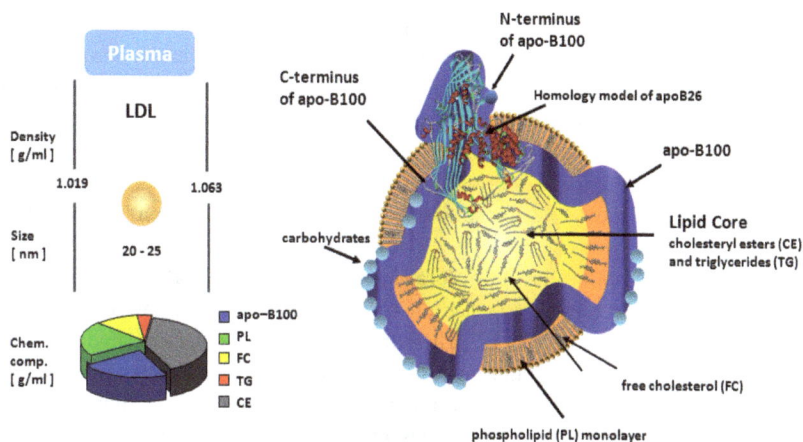

Figure 1. Molecular organisation of LDL. LDL particles are isolated from human plasma within a defined density range. Their particle size varies between 20 to 25 nm. LDLs are built up by a hydrophobic lipid core of cholesterylester (CE) and triglyceride (TG) molecules, which make up more than 40% of particle mass surrounded by a phospholipid (PL) monolayer corresponding to about 20% of particle mass. Varying amounts of free cholesterol (FC) are incorporated in the shell and the core regions. One single copy of apo-B100 (550 kDa) is embedded in the surface monolayer, partially penetrating the core and covering about 40 to 60% of the surface area. The carbohydrate moieties are distributed along the protein chain and are surface exposed. The N-terminal end of apo-B100 (about 26% of total) is hydrophilic and shows a high homology to lamprey lipovitellin. The C-terminal end was shown to be located close to the N-terminus.

Since LDL particles are highly heterogeneous, especially with respect to the chemical composition of the core lipids, the actual size of LDL particles varies between 20 to 25 nm, with an average particle diameter of about 22 nm. This intrinsic heterogeneity allows a subdivision of LDLs into distinct highly homogeneous LDL subspecies, which are identified on the basis of their hydrated densities, which normally lies between the extremes of d, 1.019 and 1.063 g ml^{-1} [15]. These subspecies also differ in their physico-chemical characteristics, receptor binding affinity [16], susceptibility to oxidative modifications [17,18], and in their atherogenic behaviour. Following these lines, it is important to consider LDL as a flexible construct, which needs to respond to changing environmental conditions during lipid exchange. Hence, during particle remodelling, apo-B100 and the surface PL molecules have to rearrange to compensate for changes in the surface area and surface pressure [6]. It is known, that apo-B100 predominantly resides on the surface of LDL and displaces PL molecules, concomitantly changing the diffusion and order parameter of lipids as shown in a recent near atomistic simulation study [19]. Based on simple geometrical considerations taking into account the surface PL monolayer (about 700 PL molecules) with an average area per lipid of 0.65 nm² and an LDL particle diameter of 22 nm, large parts of the surface layer must be covered by the protein to avoid unfavourable hydrophobic contacts. In support of these considerations, a loose surface packing of PL molecules was derived from molecular dynamics simulations [19]. This low surface pressure enables hydrophobic amino acid regions of apo-B100 to penetrate into the interfacial regions, predominantly formed by the acyl chains of PLs. Consequently, apo-B100 might interact more readily with the neutral core lipids, and indeed it was shown that some of the CE molecules align along the β-sheet structures of apo-B100 [20], thereby driving CE molecules to the surface, where they become part of the interfacial layer. Particularly noteworthy is the fact that the lipids within the interfacial layer are not homogeneously distributed but form local microenvironments [14]. More precisely, two nanodomains were identified, one rich on sphingomyelin and FC, the other one rich in phosphatidylcholine and poor in FC. The latter was shown to be associated more closely with apo-B100 [21]. Even though, one has to keep in mind that these domains are not static or confined in size and number and co-determine the intrinsic dynamics of LDL. Based on these types of findings, it seems reasonable to suggest that variations in the molecular organisation of lipid/apo-B100 impact the structure of LDL, and have to be considered to act as physiological determinants of LDL function.

3. Structural models of LDL

Our present understanding of the structure of LDL particles has emerged from the concerted application of different physico-chemical techniques with early ground-breaking findings derived from neutron- or X-ray small angle scattering data [22-25] complemented by results from negative staining electron micoscopic (e.m.) [26,27] and spectroscopic techniques [28,29]. For comprehensive reviews on different biophysical studies applied on LDL species see refs. [30,31]. In recent years structural investigations using cryo-e.m. reconstruction techniques have become prevalent and with time 3-dimensional models with improved resolution were presented [32-37]. While in earlier studies LDLs are described as quasi-spherical particles, later studies presented a new view of the overall particle structure displaying an oblate elliptical particle shape. Moreover, recent 3D-images show convincing data that LDL can be considered as discoidal-shaped particle with two flat surfaces on opposite sides. In this model, apo B100 encircles LDL at the edge of the particle, while the PL monolayer is rather located at the flat surfaces which are parallel to the CE layers in the core [36,37]. To get a better impression of what LDL looks like in a structure map obtained by 3D-reconstruction from cryo-e.m, we show some images in Figure 2 revealing the surface density distribution on LDL. It has to be stated that this model strictly holds true for LDL particles with the core lipids being in a frozen liquid-crystalline state.

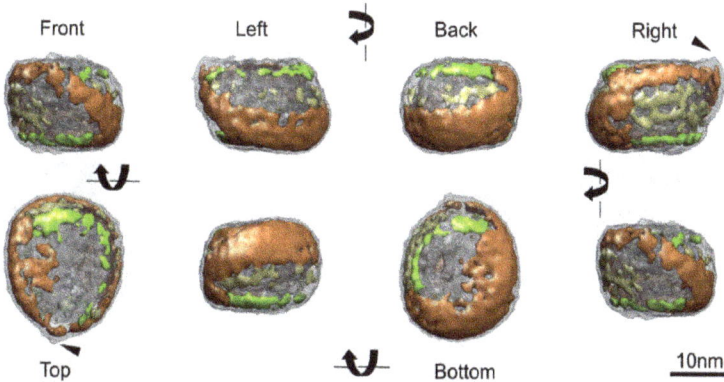

Figure 2. Density distribution at the surface of LDL. The 3D-density map derived from cryo e.m. images by reconstitution reveals the oblate overall particle shape of LDL shown in gray. The overlaid high density regions represent the backbone of apo-B100, colored in orange. The belt surrounds the particle to form an enclosed circle. The second group of high density regions (green) contours the rims and complements the backbone enclosing lower-density regions. The high density regions on the sidewall (yellow) are structures extending from the backbone. A knob-like protrusion is visible at the pointed end (indicated by triangles in the right and top views). The 3D-map is turned 90° in each frame. Reprinted with permission from ref. [37].

4. Core lipid packing and lipid phase transition

Despite of compositional heterogeneity, LDL particles share one common feature: the CE molecules in the core undergo a structural transition from an ordered liquid-crystalline phase to a fluid oil-like state as function of temperature and chemical composition [38]. More precisely, the actual transition temperature, which is close to body temperature, is inversely correlated to the content of triglycerides within the lipid core [22,39]. Based on these characteristics, several models for CE packing have been suggested including a spherical concentric layer model derived primarily from X-ray and neutron scattering data [40,41]. More recently, the concept of a flat lamellar structure came up. This model is derived from single-particle reconstructions from cryo-e.m. images of LDL in vitreous ice [32,34]. An ordered three-layer internal lamellar structure with a distance of about 3.6. nm between the single lamellae was reported [32], in agreement with repeat distances derived from X-ray scattering patterns for LDL below the transition temperature. While these images were observed for LDL particles being in the liquid crystalline phase before snap-freezing, diverse results were reported for LDL particles frozen from a state above the phase transition temperature [42,43]. One plausible explanation for these discrepancies might be that the melting rate of the core lipids proceeds extremely fast. It has been shown that the physical state of core lipids changes within milliseconds [44]. This fast kinetics has caused experimental difficulties for long time, however, a recent experimental approach by speeding up freezing allowed to trap the lipids in the molten state [45]. The authors report on a co-existing phase of layered and broken shells for LDL particles, which are shock-frozen in a state above the phase transition. This is the first time to visualize the nucleation process of CEs within LDL. Most interesting, the images indicate intermediate states between the order/disorder phase transition. Figure 3 shows the dynamic model of the core CE packing during the phase transition and gives a comparison of the internal features of reconstructed 3D-volumes of LDL.

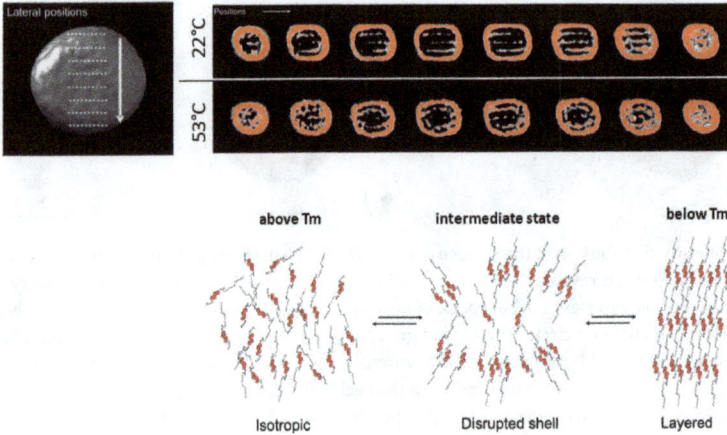

Figure 3. Schematic picture of the dynamic model of LDL core lipid packing during the phase transition. Comparision of the internal features of the reconstructed 3D-volume of LDL snap-frozen from below (22°C) and above (53°C) the phase transition temperature (Tm). Samples prepared from 22°C show a layered organisation while samples prepared from 53°C reveal a disorded shell like structure, which is concentric to the surface. Note, the overall shape of LDL has also changed slightly. The lower panel shows a hypothetical model for the core lipid packing depicting the dynamic process of the core lipid phase transition upon cooling from isotropic to layered passing through an intermediate state. Modified with permission from ref. [45].

In summary, it seems reasonable to argue that both the overall shape and core lipid packing of LDL particles are highly sensitive to changes in temperature and lipid composition. Indeed, this newly proposed patch nucleation behavior permits the temporary formation of local molecular microenvironments as suggested previously by our group in terms of trigylceride segregation [46]. In the next paragraph we will address some interesting questions in support of above hypotheses.

Does a lipid microphase separation occur in LDL particles as a function of the relative core content of CE and TG ?

As already mentioned, the transition temperature correlates with the lipid composition, however, a discontinuity in the concentration dependence was observed [46]. A break in the concentration dependence of a transition temperature in a mixed lipid system constitutes an index for the existence of a phase separation at the break point. In isolated triglyceride - cholesteryl ester systems no indication of a phase separation at similar compositions was found [39,47]. It appears therefore, that structural constraints within the LDL particle determine this effect. Experimental data provide evidence that at low TG content (below 12%) the TG molecules separate into distinct hydrophobic nanoenvironments while the CEs form a smectic liquid crystalline layer. With increasing TG content the thermal stability of the CE layer is decreased by intermixing with TG [46]. This hypothesis implies that the TG-rich fluid nanodomains can serve as a reservoir for lipophilic minor constituents, such as vitamins (tocopherol, carotenoids etc.) below the phase transition. The local concentration of these antioxidants and hence their efficiency in scavenging lipophilic free radicals is higher than if they were dissolved in the bulk volume of total apolar lipids. At the same conditions the CE molecules are strongly immobilized and the intracellular degradation of LDL is decelerated [48], equally the activity of lipid transfer proteins is diminished [49,50]. Based on these considerations it is tempting to speculate that circulating LDL, as a consequence of the variation in blood temperature, periodically undergoes a thermal transition resulting in a transient increase in the local core concentration of minor constituents [46]. Here, it should be emphasized that a periodic redistribution of lipophilic solutes, and also for example of drugs, into the confined LDL core volume could represent an attractive approach to the modulation of biochemical reactions, which would not occur at sufficient rates under the normal conditions of relative concentration. Studies along these lines could indeed verify the long missing physiological role of the thermal LDL transition.

Can LDL structure follow quasi-isothermal changes in blood temperature during its circulation, or does it remain adiabatically metastable in the molten-lipid state?

In order to provide evidence to answer this question we have applied time resolved X-ray scattering experiments using a high flux synchrotron generated X-ray beam. Thus, we have been able to trigger the thermal transition in either direction (heating and cooling) simultaneously monitoring associated structural changes in sub-second time intervals. With our special instrumental setup we managed to evaluate the kinetics of core-transition by T-jump and T-drop experiments [44]. We found that the melting transition proceeds faster than 10 milliseconds indicating that thermal-induced lipid reorganisation takes place at the time scale of blood circulation. As the velocity of blood-flow can be as low as 0.3 mm/s in peripheral blood capillaries the residence time for LDL particles in cooler regions of the body can be several seconds. Consequently, LDL can easily follow periodic temperature changes during blood circulation and assist the redistribution of lipophilic constituents within its core nanodomains forming fluid defect zones. For biomedicine, this strengthens the hypothesis that the core lipids of LDL not only act as passive chemical substrates in metabolism, but that their physical state within the LDL nanoparticles has the potential to control their metabolic fate in normal and atherosclerotic cholesterol transport.

Does the core lipid transition have a physiological meaning ?

Despite its occurrence conspicuously close to blood temperature and the variation of the transition temperature of LDL among different subjects, no clear evidence for a physiological or patho-biochemical role of this transition has so far been found. It is now generally accepted in literature that the rearrangement of the core lipids also affects the overall structure and shape of the LDL particle. Morphological changes in turn can impact receptor-binding activity as well as the action of lipid hydrolyzing enzymes. Equally, the susceptibility of LDL particles to oxidative modifications and lipid peroxidation might be correlated to temperature [18]. As oxidized LDL play a crucial role in the pathogenesis of atherosclerosis, any contribution to the comprehension of antioxidant efficiency may be of therapeutic potential [2,51], further pointing to the physiological relevance of the lipid core organisation. However, this vital question still remains unanswered.

5. Apo-B100 is a flexible string wrapped around the surface of LDL

As already indicated above, the physicochemical state of the core lipids is intimately related to the structure and dynamics of the particle surface, which consists of about 700 phospholipid molecules and one single copy of apo-B100. Apo-B100 is a huge glycoprotein and its polypeptide chain consists of 4536 amino acid residues with an estimated molecular mass of about 550 kDa for the glycosilated form [52,53]. Apo-B100 is a single chain protein with a total contour length of about 70 nm [54] and can be viewed as a highly flexible molecular string composed of single domains [20]. Five consecutive domains were identified based on secondary structure elements representing the main conformational motifs of apo-B100. The single domains are connected by flexible interdomain linker regions, which allow relative movements of domains to each other. The feasibility of such motions was shown in a low resolution model of detergent solubilized apo-B100, which was derived from small angle neutron scattering data [55]. In this model, compact rigid domains are visible being connected by flexible interdomain linkages, which possess a substantial degree of freedom in their spatial orientation. A hypothetical spatial arrangement of the apo-B100 molecule on a spherical LDL particle was created after assigning the secondary structure elements, which were deduced from a secondary structure prediction, to the surface of apo-B100 (Figure 4). Likewise, the averaged surface shape of the 3D-model would allow for variations in the thickness of the apo-B100 molecule by about 1 nm. Such variations are most likely required to compensate for changes in the surface area upon lipid exchange and particle shrinking during endogeneous lipoprotein conversion from very low density lipoprotein (VLDL) to LDL.

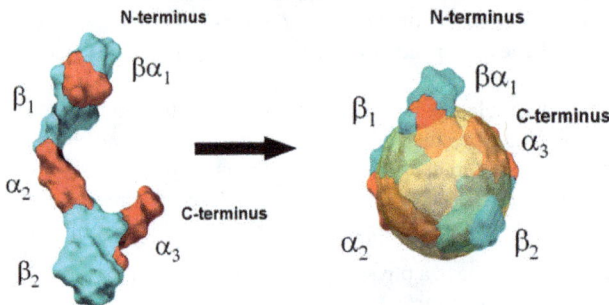

Figure 4. Reconstituted low resolution model of lipid-free apo-B100 derived from small angle neutron scattering data. Apo-B100 shows an elongated arch-like morphology indicating single domains and mobile less defined linker regions. A hypothetical model of a spherical LDL particle after superposition of the structural model of apo-B100 is shown (adapted from ref. [55]). Secondary structure modules are assigned to the surface after a secondary structure prediction was performed. The results nicely correspond to the pentapartite model suggested by [20].

Concerning the topology of apo-B100 on the surface of LDL the most detailed information is obtained from cryo-e.m. images (see also Figure 2). Chatterton et al. were among the first to visualize apo-B100 as string circumventing LDL, and to report on mapped epitopes of apo-B100 distributed over one hemisphere of the LDL particle [56,57]. Recent single particle 3D-reconstructions from immuno cryo e.m. images delineated a more accurate picture of apo-B100 revealing a looped topology of the protein backbone with distinct epitopes identified along the protein chain. According to this model, epitopes in the LDL receptor binding domain are located on one side of LDL, whereas epitopes located in the N-terminal and C-terminal domains are in close vicinity to each other on the opposite side of LDL [36]. In addition, a prominent protrusion is visible in the images at the pointed end of the particle. A similar knob-like region was apparent in the low resolution model of lipid-free apo-B100 shown in Figure 4. This protrusion most probable represents the non-lipid associated globular N-terminal domain of apo-B100, which shows a high homology to lamprey lipovitellin [58]. Except for the N-terminal domain, little is known about the molecular organisation of the structural motifs, whose amphipathic nature determine lipid association. However, to evaluate lipid-protein interactions physical parameter like interfacial elasticity or molecular dynamics have to be considered. In this context, it was suggested that the hydrophobic β-sheet domains of apo-B100 act as elastic lipid anchors, whereas the amphipathic α-helical domains respond rapidly to changes in surface pressure [59,60]. In any case, it can be assumed that alterations in the adsorption and penetration depth of apo-B100 in the phospholipid monolayer and in the lipid core are accompanied by structural rearrangements of the domains and changes in the orientation of the domains relative to each other. In the course of such elastic motions, intramolecular rearrangements are likely to alter the overall hydrophobicity and surface activity of single protein domains. These modifications not only affect lipid-protein interaction, but are equally important for molecular and cellular recognition of apo-B100.

6. Apo-B100 containing lipoproteins are very soft and flexible

LDL particles are formed in the circulation by lipolytic conversion of TG-rich VLDL particles. This enzyme mediated endogenous transformation is accompanied by an extensive shrinking in particle size from about 50-80 nm for VLDL to ~20 nm for LDL. In the course of remodelling, apo-B100 remains bound to its nanocarrier stabilizing the lipid assembly by maintaining structural integrity. To accomplish this, apoB100 has to become more condensed or relaxed depending on the lipid packing density. Likewise, this dynamic process is modulated by the molecular mobility of the surrounding microenvironment. To test for this hypothesis we have recorded temperature dependent molecular motions in VLDL and LDL particles using elastic incoherent neutron scattering [61]. With this technique, motions in the nano- to picosecond time scale can be recorded. The calculated dynamic force constants are a direct measure for the resilience of the particles. The results show that at physiological temperatures VLDL particles are very soft, elastic and mobile as compared to LDL, which is more rigid (see Figure 5). This observation supports the notion that apo-B100 in VLDL is loosely packed at the interface covering a large surface area with low interfacial surface tension [59]. During particle conversion from VLDL to LDL, however, the relative number of surface molecules increases and a higher molecular packing density leads to a compression of the lipid anchored protein regions and an overall stiffening of the LDL particle [60].

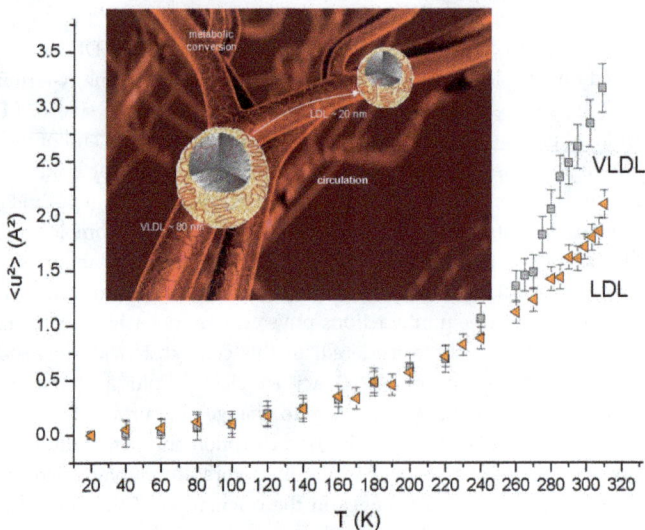

Figure 5. Molecular motions in LDL and VLDL. Elastic temperature scans are recorded with elastic incoherent neutron scattering. The mean square thermal fluctuations ($<u^2>$) are shown as function of temperature. The molecular resiliences are derived from the slopes in the curves. It is seen that VLDL has an increased motion at elevated temperatures compared to LDL. Parts of this figure are reproduced, with permission, from ref. [60].

To conclude, the intrinsic conformational flexibility and elasticity of apo-B100 containing lipoprotein particles is most likely critical for specific affinities of lipoproteins to receptors, antibodies or enzymes. Moreover, it would seem that the susceptibility of lipoproteins to oxidative modifications and hence their atherogenicity is influenced by their dynamic nature.

7. LDLs are flexible nanotransporters circulating in blood

In the search for new and improved therapeutics, the field of nanomedicine dealing with functionalized nanoparticles for molecular imaging and therapy is rapidly emerging. Nanoparticles offer new opportunities to transfer active substances directly to the diseased site in the body. By additional surface coatings or functionalizations, the properties of nanoparticles can be tuned to specific needs. Within the last two decades, a variety of artificial nanoparticles have been designed for targeted delivery of drugs or contrast agents. Many of these nanoconstructs are developed for cancer therapy taking advantage of the leaky vasculature of tumours. Apart from tumour targeting, increasing efforts are devoted to the treatment and imaging of atherosclerotic plaques (for a recent review see ref. [62]). Over time, a broad and versatile nanoparticle platform was created in which liposomes and biodegradable polymers have turned out to be the most promising candidates. It is important to mention that several nanomedicine products have already been established on the market and numerous products are successfully applied in clinical trials [63]. However, inherent problems of nanoparticles are biocompatibility and low stability in vivo, since most nanoparticles become rapidly cleared by the reticuloendothelial system. In contrast to artificial systems, lipoproteins are naturally occurring nanoparticles evading recognition by the body's immune system. Hence, lipoproteins are excellent candidates with attractive properties to be considered as molecular transporters. A great advantage of LDL over other nanoparticles is the fact that LDL particles stay in circulation for several days, and are not cleared immediately by the mononuclear phagocyte system of the liver and spleen. The average lifetime of an LDL particle is 2-3 days and this time span is about three times longer as reported for long-circulating liposomes, currently applied in chemotherapy [64]. It was recognized that certain tumor cells overexpress LDL receptor, however, the targeting specificity is limited as the LDL receptor is ubiquitously expressed throughout the body, most prominent in the liver. However, using apo-B100 as inherent targeting sequence the enhanced circulation times in blood enable drug-loaded LDL particles to bind to specific receptors exposed on the surface of e.g. tumor or atherosclerotic plaque. Once recognized by the receptor, the functionalized LDL particles become internalized, accumulate in the tissue and exert an enhanced effect (reviewed by [65]). The intrinsic targeting properties of LDL to atherosclerotic plaques are already utilized for early diagnosis and detection of atherosclerotic lesions by different imaging modalities (for reviews see refs. [66,67]). However, to modify lipoprotein particles for medical purposes, care has to be taken not to compromise essential biophysical and structural features of LDL with the goal to preserve the biological activity. In general, there are several possibilities to create multi-functionalized lipoprotein particles. Some representative examples are shown in the scheme in Figure 7. One possibility is to load hydrophobic drugs (e.g. chemotherapeutics, antibiotics, vitamins, signal emitting molecules or small nanocrystals) in the lipophilic inner core of LDL. This can be accomplished by different techniques including lyophilisation, solvent evaporation and reconstitution procedures [68,69]. However, LDL particles can not be reconstituted so easily and remote drug/contrast agent loading into native lipoprotein particles is still a tedious approach currently not being standardized. Amphiphilic substances (drugs or marker molecules) or fatty acid modified chelator complexes can be incorporated in the PL monolayer [70,71]. This has successfully been done in numerous biophysical studies and for diagnostic purposes. Finally, the surface of LDL can be modified by protein labeling. This is done by covalent attachment of substances to the lysine and cysteine amino acid residues of apo-B100. Such substances include fluorophores, radionuclides or metal ions for molecular imaging [65]. Alternatively, targeting sequences (e.g. folic acid) can be coupled to apo-B100 with the purpose to reroute LDL to alternate receptors, which, in case of folate, are more specifically expressed in tumor cells [72].

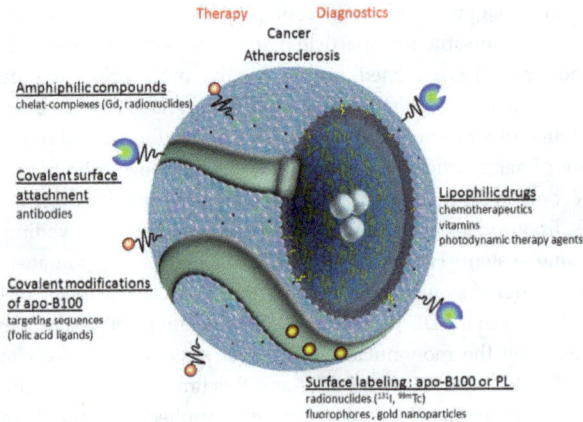

Figure 6. Scheme giving some examples of how LDL particles can be modified to act as natural endogenous nanoparticles for targeted drug delivery or multifunctional molecular imaging.

To construct lipoprotein mimetic particles, also referred to as lipoprotein related particles, artificial lipoprotein particles have to be reassembled from individual lipid and protein entities. This approach was highly successful for high density lipoproteins using apo-AI mimetic peptide sequences [73]. For LDL, this approach was not pursued yet and will be much more complicated considering the complex dynamic nature of apo-B100.

Over the last few years, a promising nanoparticle platform was established, which exploits the endogenous properties of natural lipoproteins being non-toxic, non-immunogenic and biodegradable. Although this platform still offers vast potential for improvements, first promising results in enhanced multimodal imaging of tumors and atherosclerotic plaques are achieved giving hope that further endeavors to combine diagnostics and personalized therapeutics will also be successful.

8. Conclusions and future directions

The intrinsic flexibility and dynamics of LDL lipids and protein in conjunction with the inherent compositional heterogeneity of LDL particles has hitherto hampered successful structure determinations at atomic level. Recent technological developments, however, allowed to restore characteristic structural features of individual LDL particles at low resolution. In particular, using cryo e.m. 3D-reconstruction techniques several groups have succeeded in imaging morphological and topological details of LDL to a resolution limit of approximately 2 nm [34-36]. Now, new concepts will be needed to make further progress in the development of high resolution models of LDL. One promising way is to put stronger emphasis on protein crystallography in combination with computational modelling and molecular dynamics simulations. X-ray crystallography apprears to be a hopeless pursuit with heterogeneous and flexible particles like LDL. Nevertheless, our earlier attempts of crystallisation have been partially successful [74]. Additional efforts, however, have to be focussed on the stabilization of apo-B100 in a more rigid state, perhaps by co-crystallisation with monoclonal antibodies. An alternative way ahead would be to work with lipid-free apo-B100 stabilized by detergent-mimetic systems, e.g. amphipathic designer peptides, or to proceed with truncated fragments of apo-B100.

At present there is still a deficit in our knowledge concerning the molecular lipid trafficking mechanisms of LDL. To know the atomic structure of LDL, in particular of apo-B100, may well contribute to a better understanding of biologic aspects of cardiovascular diseases, especially with respect to future strategies towards rational pharmaceutical interventions.

Author details

Ruth Prassl and Peter Laggner
Institute of Biophysics and Nanosystems Research,
Austrian Academy of Sciences, Graz, Austria

Acknowledgement

This manuscript is based in part upon work supported by the Austrian Science Fund under grant number P-20455.

9. References

[1] Brown MS and Goldstein JL (1976) Receptor-mediated control of cholesterol metabolism. Science 191: 150-154.

[2] Steinberg D, Parthasarathy S, Carew S, Khoo JC, Witztum JL (1989) Beyond cholesterol. Modifications of low-density lipoprotein that increase its atherogenicity. New Engl.J.Med. 320: 915-924.

[3] Lusis AJ (2000) Atherosclerosis. Nature 407: 233-241.

[4] Packard C, Caslake M, Shepherd J (2000) The role of small, dense low density lipoprotein (LDL): a new look. Int.J.Cardiol. 74 Suppl 1: S17-S22.

[5] Packard CJ (2006) Small dense low-density lipoprotein and its role as an independent predictor of cardiovascular disease. Curr.Opin.Lipidol. 17: 412-417.

[6] McNamara JR, Small DM, Li ZL, Schaefer EJ (1996) Differences in LDL subspecies involve alterations in lipid composition and conformational changes in apolipoprotein B. J.Lipid Res. 37: 1924-1935.

[7] Chapman MJ, Guerin M, Bruckert E (1998) Atherogenic, dense low-density lipoproteins. Pathophysiology and new therapeutic approaches. Eur.Heart J. 19 Suppl A: A24-A30.

[8] Pentikainen MO, Oksjoki R, Oorni K, Kovanen PT (2002) Lipoprotein lipase in the arterial wall: linking LDL to the arterial extracellular matrix and much more. Arterioscler.Thromb.Vasc.Biol. 22: 211-217.

[9] Skalen K, Gustafsson M, Rydberg EK, Hulten LM, Wiklund O, Innerarity TL, Boren J (2002) Subendothelial retention of atherogenic lipoproteins in early atherosclerosis. Nature 417: 750-754.

[10] Hurt-Camejo E, Camejo G, Sartipy P (2000) Phospholipase A2 and small, dense low-density lipoprotein. Curr.Opin.Lipidol. 11: 465-471.

[11] Williams KJ and Tabas I (2005) Lipoprotein retention--and clues for atheroma regression. Arterioscler.Thromb.Vasc.Biol. 25: 1536-1540.

[12] Hansson GK and Hermansson A (2011) The immune system in atherosclerosis. Nature Immunology 12: 204-212.

[13] Kostner, G. M. and Laggner, P. (1989) in Human Plasma Lipoproteins - Clinical Biochemistry, Principles, Methods, Applications 3 (Fruchart, J. C. and Shepherd, J., eds.), pp. 23-54, Walter de Gruyter, Berlin - New York.

[14] Hevonoja T, Pentikainen MO, Hyvonen MT, Kovanen PT, Ala-Korpela M (2000) Structure of low density lipoprotein (LDL) particles: basis for understanding molecular changes in modified LDL [In Process Citation]. Biochim.Biophys.Acta 1488: 189-210.

[15] Chapman MJ, Laplaud PM, Luc G, Forgez P, Bruckert E, Goulinet S, Lagrange D (1988) Further resolution of the low density lipoprotein spectrum in normal human plasma: physicochemical characteristics of discrete subspecies separated by density gradient ultracentrifugation. J.Lipid Res. 29: 442-458.

[16] Nigon F, Lesnik P, Rouis M, Chapman MJ (1991) Discrete subspecies of human low density lipoproteins are heterogeneous in their interaction with the cellular LDL receptor. J.Lipid Res. 32, 1741-1753.

[17] Dejager S, Bruckert E, Chapman MJ (1993) Dense low lipoprotein subspecies with diminished oxidative resistance predominate in combined hyperlipidemia. J.Lipid Res. 34, 295-308.

[18] Schuster B, Prassl R, Nigon F, Chapman MJ, Laggner P (1995) Core lipid structure is a major determinant of the oxidative resistance of low density lipoprotein. Proc.Natl.Acad.Sci.USA 92: 2509-2513.

[19] Murtola T, Vuorela TA, Hyvonen MT, Marrink SJ, Karttunen M, Vattulainen I (2011) Low density lipoprotein: structure, dynamics, and interactions of apoB-100 with lipids. Soft Matter 7: 8135-8141.

[20] Segrest JP, Jones MK, De Loof H, Dashti N (2001) Structure of apolipoprotein B-100 in low density lipoproteins. J.Lipid Res. 42: 1346-1367.

[21] Sommer A, Prenner E, Gorges R, St³tz H, Grillhofer H, Kostner GM, Paltauf F, Hermetter A (1992) Organization of phosphatidylcholine and sphingomyelin in the surface monolayer of low density lipoprotein and lipoprotein(a) as determined by time-resolved fluorometry. J.Biol.Chem. 267: 24217-24222.

[22] Atkinson D, Deckelbaum RJ, Small DM, Shipley GG (1977) Structure of human plasma low-density lipoproteins: Molecular organization of the central core. Proc.Natl.Acad.Sci.USA 74: 1042-1046.

[23] Laggner P, Degovics G, Müller KW, Glatter O, Kostner GM, Holasek A (1977) Molecular packing and fluidity of lipids in human serum low density lipoproteins. Hoppe-Seyler's Z.Physiol.Chem. 358: 771-778.

[24] Laggner P and Kostner GM (1978) Thermotropic changes in the surface structure of lipoprotein B from human-plasma low-density lipoproteins. A spin-label study. Eur.J.Biochem. 84: 227-232.

[25] Laggner P, Kostner GM, Rakusch U, Worcester DL (1981) Neutron small-angle scattering on selectively deuterated human plasma low density lipoproteins. The location of polar phospholipid headgroups. J.Biol.Chem. 256, 11832-11839.

[26] Gulik-Krzywicki T, Yates M, Aggerbeck LP (1979) Structure of serum low-density lipoprotein. J.Mol.Biol. 131: 475-484.

[27] Spin JM and Atkinson D (1995) Cryoelectron microscopy of low density lipoprotein in vitreous ice. Biophys.J. 68: 2115-2123.

[28] Laggner P, Chapman MJ, Goldstein S (1978) An X-Ray Small-Angle Scattering Study of Trypsin Treated Low Density Lipoprotein from Human Serum. Biochem.Biophys.Res.Commun. 82: 1332-1339.

[29] Lund-Katz S, Ibdah JA, Letizia JY, Thomas MT, Phillips MC (1988) A ^{13}C NMR characterization of lysine residues in apolipoprotein B and their role in binding to the low density lipoprotein receptor. J.Biol.Chem. 263: 13831-13838.

[30] Prassl, R., Schuster, B., and Laggner, P. (1997) in Supramolecular Structure and Function 5 (Pifat, G., ed.), pp. 47-73, Balaban Publishers.

[31] Prassl R and Laggner P (2009) Molecular structure of low density lipoprotein: current status and future challenges. Eur.Biophys.J.Biophys.Lett. 38: 145-158.

[32] Orlova EV, Sherman MB, Chiu W, Mowri H, Smith LC, Gotto AM (1999) Three-dimensional structure of low density lipoproteins by electron cryomicroscopy. Proc.Natl.Acad.Sci.U.S.A 96: 8420-8425.

[33] Van Antwerpen R (2004) Preferred orientations of LDL in vitreous ice indicate a discoid shape of the lipoprotein particle. Arch.Biochem.Biophys. 432: 122-127.

[34] Ren G, Rudenko G, Ludtke SJ, Deisenhofer J, Chiu W, Pownall HJ (2010) Model of human low-density lipoprotein and bound receptor based on cryoEM. Proc Natl Acad Sci U S A 107: 1059-1064.

[35] Kumar V, Butcher SJ, Oorni K, Engelhardt P, Heikkonen J, Kaski K, Ala-Korpela M, Kovanen PT (2011) Three-Dimensional cryoEM Reconstruction of Native LDL Particles to 16 angstrom Resolution at Physiological Body Temperature. PLoS ONE 6.

[36] Liu YH and Atkinson D (2011) Enhancing the Contrast of ApoB to Locate the Surface Components in the 3D Density Map of Human LDL. Journal of Molecular Biology 405: 274-283.

[37] Liu YH and Atkinson D (2011) Immuno-electron cryo-microscopy imaging reveals a looped topology of apoB at the surface of human LDL. J.Lipid Res. 52: 1111-1116.

[38] Deckelbaum RJ, Shipley GG, Small DM, Lees RS, George PK (1975) Thermal transitions in human plasma low density lipoproteins. Science 190, 392-394.

[39] Deckelbaum RJ, Shipley GG, Small DM (1977) Structure and interactions of lipids in human plasma low density lipoproteins. J.Biol.Chem. 252: 744-754.

[40] Laggner P and Müller K (1978) The structure of serum lipoproteins as analysed by X-ray small-angle scattering. Q.Rev.Biophys. 11: 371-425.

[41] Laggner P, Kostner GM, Degovics G, Worcester DL (1984) Structure of the cholesteryl ester core of human plasma low density lipoproteins: Selective deuteration and neutron small- angle scattering. Proc.Natl.Acad.Sci.USA 81: 4389-4393.

[42] Sherman MB, Orlova EV, Decker GL, Chiu W, Pownall HJ (2003) Structure of triglyceride-rich human low-density lipoproteins according to cryoelectron microscopy. Biochemistry 42: 14988-14993.

[43] Coronado-Gray A and Van Antwerpen R (2005) Lipid composition influences the shape of human low density lipoprotein in vitreous ice. Lipids 40: 495-500.

[44] Prassl R, Pregetter M, Amenitsch H, Kriechbaum M, Schwarzenbacher R, Chapman JM, Laggner P (2008) Low density lipoproteins as circulating fast temperature sensors. PLoS ONE 3: e4079 .

[45] Liu Y, luo D, Atkinson D (2010) Human LDL core cholesterol ester packing: 3D image reconstruction and SAXS simulation studies. J Lipid Res 51.

[46] Pregetter M, Prassl R, Schuster B, Kriechbaum M, Nigon F, Chapman J, Laggner P (1999) Microphase separation in low density lipoproteins. Evidence for a fluid triglyceride core below the lipid melting transition. J.Biol.Chem. 274: 1334-1341.

[47] Small, D. M. (1986) in The Physical Chemistry of Lipids - From Alkanes to Phospholipids pp. 395-473, Plenum Press, New York and London.

[48] Lusa S and Somerharju P (1998) Degradation of low-density-lipoprotein cholesterol esters by lysosomal lipase in-vitro - effect of core physical state and basis of species selectivity. Bba-Lipid Lipid Metab 1389: 112-122.

[49] Morton RE and Parks JS (1996) Plasma cholesteryl ester transfer activity is modulated by the phase transition of the lipoprotein core. J.Lipid Res. 37: 1915-1923.

[50] Zechner R, Kostner GM, Dieplinger H, Degovics G, Laggner P (1984) In vitro modification of the chemical composition of human plasma low-density lipoproteins: Effects on morphology and thermal properties. Chem.Phys.Lipids 36: 111-119.

[51] Esterbauer H, Dieber-Rotheneder M, Waeg G, Striegl G, Jürgens G (1990) Biochemical, Structural, and Functional Properties of Oxidized Low-Density Lipoprotein. Chem.Res.Toxicol. 3: 77-92.

[52] Chen S-H, Yang C-Y, Chen PF, Setzer D, Tanimura M, Li W-H, Gotto AM, Jr., Chan L (1986) The complete cDNA and amino acid sequence of human apolipoprotein B-100. J.Biol.Chem. 261: 2918-2921.

[53] Knott TJ, Pease RJ, Powell LM, Wallis SC, Rall SC, Innerarity TL, Blackhart B, Taylor WH, Marcel Y, Milne R, Johnson D, Fuller M, Lusis AJ, McCarthy BJ, Mahley RW, Levy-Wilson B, Scott J (1986) Complete protein sequence and identification of structural domains of human apolipoprotein B. Nature 323: 734-738.

[54] Phillips ML and Schumaker VN (1989) Conformation of apolipoprotein B after lipid extraction of low-density lipoproteins attached to an electron microscope grid. J.Lipid Res. 30: 415-422.

[55] Johs A, Hammel M, Waldner I, May RP, Laggner P, Prassl R (2006) Modular structure of solubilized human apolipoprotein B-100. Low resolution model revealed by small angle neutron scattering. J.Biol.Chem. 281: 19732-19739.

[56] Chatterton JE, Phillips ML, Curtiss LK, Milne RW, Marcel YL, Schumaker VN (1991) Mapping apolipoprotein B on the low density lipoprotein surface by immunoelectron microscopy. J.Biol.Chem. 266: 5955-5962.

[57] Chatterton JE, Schlapfer P, Bütler E, Gutierrez MM, Puppione DL, Pullinger CR, Kane JP, Curtiss LK, Schumaker VN (1995) Identification of apolipoprotein B 100 Polymorphisms that affect low-density lipoprotein metabolism: Description of a new approach involving monoclonal antibodies and dynamic light scattering. Biochemistry 34: 9571-9580.

[58] Mann CJ, Anderson TA, Read J, Chester SA, Harrison GB, Kochl S, Ritchie PJ, Bradbury P, Hussain FS, Amey J, Vanloo B, Rosseneu M, Infante R, Hancock JM, Levitt DG, Banaszak LJ, Scott J, Shoulders CC (1999) The structure of vitellogenin provides a molecular model for the assembly and secretion of atherogenic lipoproteins. J.Mol.Biol 285: 391-408.

[59] Wang L, Walsh MT, Small DM (2006) Apolipoprotein B is conformationally flexible but anchored at a triolein/water interface: a possible model for lipoprotein surfaces. Proc.Natl.Acad.Sci.U.S.A 103: 6871-6876.

[60] Wang L, Martin DD, Genter E, Wang J, McLeod RS, Small DM (2009) Surface study of apoB1694-1880, a sequence that can anchor apoB to lipoproteins and make it nonexchangeable. J Lipid Res 50: 1340-1352.

[61] Mikl C, Peters J, Trapp M, Kornmueller K, Schneider WJ, Prassl R (2011) Softness of atherogenic lipoproteins: a comparison of very low density lipoprotein (VLDL) and low density lipoprotein (LDL) using elastic incoherent neutron scattering (EINS). J Am Chem Soc 133: 13213-13215.

[62] Lobatto ME, Fuster V, Fayad ZA, Mulder WJM (2011) Perspectives and opportunities for nanomedicine in the management of atherosclerosis. Nature Reviews Drug Discovery 10: 835-852.

[63] Duncan R and Gaspar R (2011) Nanomedicine(s) under the Microscope. Molecular Pharmaceutics 8: 2101-2141.

[64] Allen TM and Cullis PR (2004) Drug delivery systems: entering the mainstream. Science 303: 1818-1822.

[65] Ng KK, Lovell JF, Zheng G (2011) Lipoprotein-Inspired Nanoparticles for Cancer Theranostics. Accounts of chemical research 44: 1105-1113.

[66] Frias JC, Lipinski MJ, Lipinski SE, Albelda MT (2007) Modified lipoproteins as contrast agents for imaging of atherosclerosis. Contrast.Media Mol.Imaging 2: 16-23.

[67] Cormode DP, Skajaa T, Fayad ZA, Mulder WJ (2009) Nanotechnology in medical imaging: probe design and applications. Arterioscler Thromb Vasc Biol 29: 992-1000.

[68] Hammel M, Laggner P, Prassl R (2003) Structural characterisation of nucleoside loaded low density lipoprotein as a main criterion for the applicability as drug delivery system. Chem.Phys.Lipids 123: 193-207.

[69] Song LP, Li H, Sunar U, Chen J, Corbin I, Yodh AG, Zheng G (2007) Naphthalocyanine-reconstituted LDL nanoparticles for in vivo cancer imaging and treatment. International Journal of Nanomedicine 2: 767-774.

[70] Corbin IR, Li H, Chen J, Lund-Katz S, Zhou R, Glickson JD, Zheng G (2006) Low-density lipoprotein nanoparticles as magnetic resonance imaging contrast agents. Neoplasia 8: 488-498.

[71] Chen LC, Chang CH, Yu CY, Chang YJ, Hsu WC, Ho CL, Yeh CH, Luo TY, Lee TW, Ting G (2007) Biodistribution, pharmacokinetics and imaging of Re-188-BMEDA-labeled pegylated liposomes after intraperitoneal injection in a C26 colon carcinoma ascites mouse model. Nuclear Medicine and Biology 34: 415-423.

[72] Zheng G, Chen J, Li H, Glickson JD (2005) Rerouting lipoprotein nanoparticles to selected alternate receptors for the targeted delivery of cancer diagnostic and therapeutic agents. Proc.Natl.Acad.Sci.U.S.A 102: 17757-17762.

[73] Zhang ZH, Chen J, Ding LL, Jin HL, Lovell JF, Corbin IR, Cao WG, Lo PC, Yang M, Tsao MS, Luo QM, Zheng G (2010) HDL-Mimicking Peptide-Lipid Nanoparticles with Improved Tumor Targeting. Small 6: 430-437.

[74] Prassl R, Chapman JM, Nigon F, Sara M, Eschenburg S, Betzel C, Saxena A, Laggner P (1996) Crystallization and preliminary X-ray analysis of a low density lipoprotein from human plasma. J.Biol.Chem. 271: 28731-28733.

New Insights into the Assembly and Metabolism of ApoB-Containing Lipoproteins from *in vivo* Kinetic Studies: Results on Healthy Subjects and Patients with Chronic Kidney Disease

Benjamin Dieplinger and Hans Dieplinger

Additional information is available at the end of the chapter

1. Introduction

Lipoproteins are complexes consisting of a lipid core of mainly triglycerides and cholesterol esters surrounded by a surface monolayer of phospholipids, free cholesterol and specific protein components named apolipoproteins [1]. Most apolipoproteins undergo complex exchange reactions and serve many metabolic functions including transport, enzyme cofactors and receptor ligands. Except for the covalently linked apolipoprotein(a)-apolipoproteinB-100 (apo(a)-apoB) complex in Lipoprotein(a) [Lp(a)], apolipoproteins are non-covalently associated with each other and the lipid core.

Lipoprotein disorders are often associated with cardiovascular disease (CVD), atherosclerosis and other organ dysfunctions [2, 3]. To prevent and treat these diseases and to fully understand their cause, it is necessary to characterise the underlying metabolic disorders [1]. The conventional initial approach to do this is by measuring concentrations of plasma lipids or apolipoproteins. However, abnormal concentrations of lipids and apolipoproteins can result from changes in the production, conversion or catabolism of lipoprotein particles. Therefore, although static measurements and functional assays are important techniques to gain first in vivo functional insights, it is necessary to study their metabolic pathway to understand the complexity of lipoprotein function and pathophysiology [4, 5].

Animal models cannot sufficiently replace human studies to explore lipoprotein metabolism due to substantial species specificity. This holds particularly true for conventional laboratory animals such as mice and rats which – unless genetically modified or induced by special diet - do not develop atherosclerosis (see review [6]). The same argument is valid for investigations using cellular model systems. Since the liver is the central organ responsible for lipoprotein metabolism and primary human hepatocytes are of only limited use in research, most cellular studies in lipoprotein metabolism have been conducted in human hepatoma cells lines. These lines express, secrete and assemble a lipoprotein pattern which is substantially different from the respective human counterpart [7].

For all these reasons, the in vivo investigation of metabolic pathways in human subjects is the ultimate approach to elucidate physiological or pathological functions of metabolites in the human body. Historically, such human kinetic studies were performed using radioactive tracers; this methodology is, however, nowadays of only restricted use. Therefore, stable-isotope tracer kinetic studies in human subjects with clear advantages regarding safety and technical issues have replaced the radiotracer methods to become an important research tool for achieving a quantitative understanding of the dynamics of metabolic processes in vivo.

The aim of this review is to shortly describe the methodology and illustrate how the approach has expanded our understanding of physiological mechanisms as well as the pathogenesis of disorders of human lipoprotein metabolism. We will then specifically address the assembly mechanism of the atherogenic Lp(a) complex and focus on the kinetics of apoB-containing lipoproteins in patients with chronic kidney disease. This patient group is well-known for its high risk for atherosclerotic complications and a 10- to 20-fold increased cardiovascular mortality compared to the general population [8].

2. Principles of tracer technology

Exogenous and endogenous labelling techniques have been used to study the in vivo metabolism of an endogenous molecule, the tracee (see review [4]). In the exogenous method, the same molecule, in form of a usually radioactively labelled tracer, is introduced into the bloodstream [9]. In lipoprotein studies, this methodology first requires purification of the target molecule or particle and ex-vivo radiolabelling followed by reinfusion into the circulation. The physological integrity of the target molecule might, however, suffer from such procedure. Furthermore, in case of multiprotein complexes (which most lipoproteins are), the kinetics of individual protein components cannot be investigated by this approach. As an example, the investigation of in vivo kinetics of both protein components of Lp(a), as described in this article, to study its assembly mechanism would not be possible with the exogenous labelling approach.

In contrast, in endogenous labelling, a labelled precursor of the molecule of interest, in case of proteins usually a labelled amino acid, is used to label the target molecule by infusion into the circulation of a suitable proband. Ideally, the tracer can easily be detected and quantified, has the same kinetic behaviour as the tracee, and does not perturb the system. Usually, kinetic studies are performed in steady state, where the rates of input and output for a given unlabelled tracee substance are equal and time invariant. Thus, the information provided by the tracer reflects the behaviour of the tracee [10, 11]. At various times, the target protein or particle has to be purified from the blood of human probands and the amount of tracer is quantified to provide a kinetic curve. A mathematical model is then constructed to extract all the information contained in the kinetic curve. By fitting a model to the data, it is possible to calculate the parameters of the model that characterize the flux of molecules between kinetically homogeneous pools. For example, it is thus possible to investigate the whole pathway including production, conversion or catabolism of lipoprotein particles, information that cannot be obtained by static measurements alone.

The term stable isotope refers to a non-radioactive isotope of a given atom that is less abundant in a molecule within a biological system than the lightest naturally occurring isotope. The most common stable isotope used as metabolic tracer for apolipoprotein kinetic studies is [2H3]-leucine. Stable isotope tracers are much safer than radioactive tracers for both the study subject and the investigator. Furthermore, the duration of stable isotope experiments is normally less than 24 hours which is much shorter compared to radiotracer techniques which may need up to 14 days of examination [9].

2.1. Tracer administration

A tracer can be administered intravenously as either a single bolus injection, a primed constant infusion (i.e., a constant infusion given immediately after a priming bolus), or as a combination of both. The tracer bolus administration offers superior dynamics compared with the primed constant infusion, because the enrichment curves (the tracer/tracee ratios) after a bolus injection correspond to the impulse response of the system. It is therefore suitable to study components of lipoprotein metabolism with a slow rate of turnover. Another advantage of bolus administration is that it facilitates the determination of newly synthesized particles, as the intracellular precursor enrichment is greater at the start of the study. This argument therefore counts particularly when investigating kinetics of particle assembly, as described in 3.1.1. Practically, the bolus infusion is also most convenient for both subjects and investigators.

2.2. Multicompartment models for data analysis

Multicompartment modelling is a superior method to dissect the complexities of lipoprotein metabolism, and has been widely applied to systems in which material is transferred over time between compartments connected in a specific structure to permit the movement of material amongst the compartments [12].

Each compartment is assumed to be a homogenous entity within which the entities being modelled are equivalent. For instance, the compartments may represent different types of lipoprotein particles that are kinetically homogeneous and distinct from other material in the system. Very often, the data can be described by more than one model. To ensure that the best model is selected, it is necessary to carefully examine the fitting of the kinetic curve, to determine the precision of the parameter estimates, and to perform statistical tests to compare results obtained with different models. However, the complexity of a multicompartment model is usually a compromise for what is practically possible. A very simple model may not adequately describe the kinetic heterogeneity present within the system. A model that is too complex, on the other hand, will not be supported by experimental data and, hence, will have little predictive value. Furthermore, even if the development of models is based on experimental data, several assumptions are required in order to derive the model that is to be used. Thus, mathematical models do not determine the kinetics of lipids directly; rather, they derive an indirect approximation.

The software SAAM (Epsilon Group, Charlottesville, VA, USA) has become the first choice for modelling lipoprotein kinetic studies. The SAAM II program was recently developed by SAAM Inst., Inc., Seattle, WA, USA, and is frequently used to analyse lipoprotein tracer data using compartmental models [13, 14]. The primary kinetic parameter resulting after modelling with SAAM II is the fractional synthesis rate (FSR) which, under steady state conditions, is identical to the fractional catabolic rate (FCR) and has the dimension of pools/day. The reciprocal value of FSR/FCR is called retention time (RT, given in days) and indicates the residence time of the investigated tracee (the target apolipoprotein in our cases) in the circulation. The product of FSR multiplied by the concentration of tracee is called production rate (PR) and is usually expressed as mg/kg body weight/day.

3. Metabolism of apoB-containing lipoproteins

Dietary lipids are absorbed in the intestine and packaged into large, triglyceride-rich chylomicrons which undergo lipolysis to form chylomicron remnants. In the last step of the so-called exogenous lipoprotein pathway, these particles are finally taken up by the liver. The liver then secretes triglyceride-rich lipoproteins known as very low-density lipoproteins (VLDLs) representing the first step oft the endogenous lipoprotein pathway (Figure 1). Lipoprotein kinetic studies have shown that VLDLs are metabolically heterogeneous. Following lipolysis by endothelium-bound lipoprotein lipase (LPL) and hepatic lipase (HL), these particles are converted via intermediate-density lipoproteins (IDL, also called VLDL remnants) to low-density lipoprotein (LDL) or taken up by the liver. LDL is catabolized mainly by the liver or peripheral tissues via the LDL receptor. Increased plasma concentrations of LDL are a major risk factor for CVD. ApoB-100 is the major apolipoprotein of chylomicrons, VLDL, IDL and LDL.

Lipoprotein(a) [Lp(a)] consists of an LDL-like particle which is covalently bound to the glycoprotein apolipoprotein(a) [apo(a)] by disulfide linkage and derives from the liver [15] (Figure 2). Among individuals, Lp(a) plasma concentrations vary more than 1000-fold, ranging from less than 0.1 mg/dl to more than 300 mg/dl. Depending on the investigated population and the used genetic approach, it has been shown that between 30% and 90% of this variation in plasma concentrations of Lp(a) is determined by the apo(a) gene locus, encoding proteins from <300 to >800 kDa [16-18]. Apo(a) size is negatively correlated with Lp(a) concentrations, such that low-molecular-weight (LMW) apo(a) isoforms express on average high Lp(a) plasma concentrations, while high-molecular-weight (HMW) isoforms are usually associated with lower concentrations (reviewed in reference [15]). Elevated plasma concentrations of Lp(a) have been found associated with an increased risk of developing CVD in many studies which was confirmed by recent large meta-analyses [19, 20]. In vivo kinetic studies using radio-labeled Lp(a) indicated that the large differences in Lp(a) concentrations seen among individuals are determined by synthesis and not degradation [9, 21].

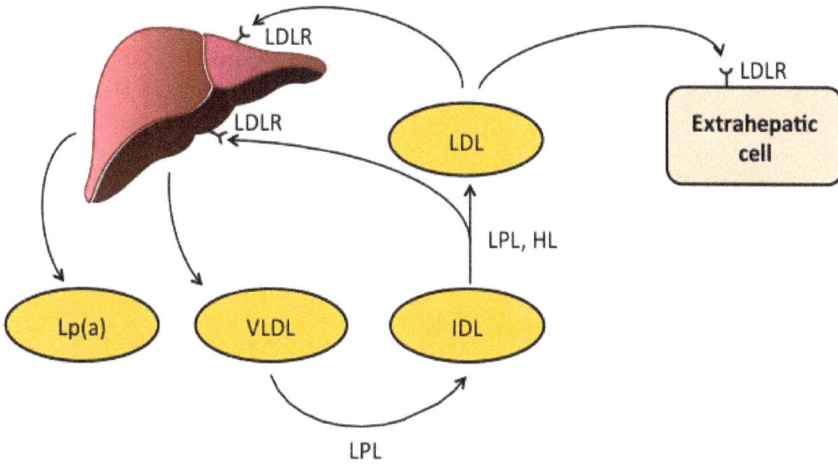

Figure 1. Endogenous metabolic pathway of apolipoprotein B (apoB)-containing lipoproteins. Triglyceride-rich very-low-density lipoproteins (VLDL) are synthesized and secreted by the liver into the blood stream and their triglycerides catabolized by the endothel-bound enzyme lipoprotein lipase (LPL) resulting in intermediate-densitly lipoproteins (IDL). LPL and hepatic lipase (HL) further convert IDL to low-density lipoproteins (LDL) which are removed from the circulation by the liver and extrahepatic tissue cells via LDL-receptor (LDLR)-mediated endocytosis. Lipoprotein(a) [Lp(a)] is synthesized and secreted by the human liver into circulation.

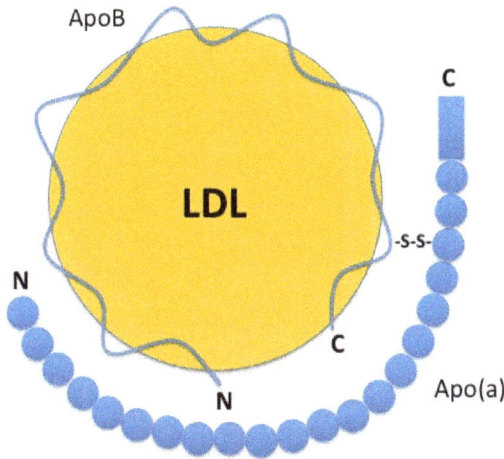

Figure 2. Structure of lipoprotein(a) [Lp(a)]. Lp(a) consists of an LDL-like particle and the disulfide-bridge-linked glycoprotein apolipoprotein(a) [apo(a)] which exerts high sequence homology to plasminogen. Apo(a) consists of an inactive protease domain (blue rectangle) and identical as well as non-identical repeats of kringle domains (blue circles). The number of identical kringles vary among individuals and gives rise to a genetically determined molecular size polymorphism of apo(a).

3.1. Biosynthesis of Lp(a)

Lp(a) has been the target of extensive and successful research particularly with respect to the unusually high degree of genetic control of its expression. In contrast, metabolism and physiological roles as well as pathogenicities of Lp(a) are still poorly understood, as recently reviewed by Dubé et al. [22]. The mechanisms that control Lp(a) secretion and assembly were investigated mostly by means of cellular hepatocyte model systems, yielded contrasting results and thus remain highly controversially discussed (see review [23]). Assembly of apo(a) and apoB to Lp(a) is generally viewed as a two-step procedure [24, 25]. In a first step, distinct domains within the apoB molecule initially associate with apo(a) in a non-covalent interaction to bring the two molecules into close proximity. In a second step, a disulfide bond is formed between apo(a) cysteine 4057 and apoB cysteine 4326 residues [24, 26]. Whether this disulfide bond is formed through a spontaneous oxidation reaction or through a specific enzymatic reaction is unclear [27, 28].

The location of this assembly process is the subject of controversial discussion as well. Intracellular, extracellular and/or plasma membrane-associated assembly procedures have been reported to occur in various cell systems [23]. Lp(a), like many other oligomeric protein complexes, may assemble in the endoplasmic reticulum of the hepatocyte and be secreted as a whole particle [29, 30]. Alternatively, newly synthesized apo(a) could bind extracellularly to preexisting LDL or VLDL circulating in the plasma. Most authors postulate an extracellular assembly of Lp(a) based on studies conducted in various cellular model systems. White *et al.* could not detect an intracellular apo(a)-apoB complex by adding anti-apo(a) antiserum to the culture medium of primary baboon hepatocytes, but found such complexes attached to the plasma membrane. The authors therefore concluded that, in that cellular system, Lp(a) is primarily assembled after secretion and to some extent also on the plasma membrane [31]. This conclusion has to be, however, critically evaluated since baboon hepatocytes secrete most of their apoB as VLDL, which does not associate with apo(a) [32]. Similar studies in apo(a)-transfected HepG2 cells could not demonstrate an intracellular apo(a)-apoB assembly for this human hepatocyte model and thus confirmed the results from the baboon studies [24, 33, 34]. Nevertheless, there is also evidence for intracellular assembly of Lp(a) in cell culture systems. Bonen *et al.* were able to detect an intracellular apo(a)-apoB complex in HepG2 cells transfected with an apo(a) minigene [35]. HepG2 cells have been reported to secrete a triglyceride-rich lipoprotein particle with an LDL density that does not exist at all in human plasma [36]. Taken together, the extracellular Lp(a) assembly proposed by numerous *in vitro* studies needs to be reviewed with caution, because these studies used cellular models that do not reflect the physiological lipoprotein metabolism.

3.1.1. In vivo metabolism of Lp(a) and LDL in healthy subjects

Kinetic *in vivo* studies in humans have unfortunately also produced controversial results. Krempler *et al.* injected radiolabeled VLDL in Lp(a)-positive healthy probands and found no metabolic relationship between apoB in VLDL or LDL and apoB in Lp(a). The authors therefore concluded that Lp(a) seems to be synthesized as a separate lipoprotein

New Insights into the Assembly and Metabolism of ApoB-Containing Lipoproteins from In vivo Kinetic
Studies: Results on Healthy Subjects and Patients with Chronic Kidney Disease

27

independently of other apoB-containing lipoproteins [37, 38]. Two *in vivo* turnover studies using stable-isotope labeling techniques came to the same conclusion: Morrisett *et al.* and Su *et al.* observed similar synthesis rates of Lp(a)-apo(a) and Lp(a)-apoB [39, 40]. While these findings are compatible with an intracellular assembly of nascent apo(a) and apoB to Lp(a), two other kinetic studies concluded that Lp(a) originates from *de novo* hepatic LDL as well as from plasma LDL [41, 42].

We investigated by stable-isotope technology the metabolism of apo(a) and apoB-100, the two major Lp(a) protein components, in comparison to apoB of LDL in nine healthy probands. The metabolic data accumulating in this study after appropriate modeling present a scenario of virtually complete intracellular assembly of Lp(a) [43].

Mean FSR, RT and PR values of apo(a) from Lp(a) were similar to those of apoB from Lp(a) but significantly different from the kinetic parameters of LDL-apoB. The differences were particularly large between the PR values of LDL and Lp(a) since this parameter takes into account plasma concentrations that are much higher for LDL than for Lp(a).

Tracer/tracee data from Lp(a)-apo(a), Lp(a)-apoB, LDL-apoB and VLDL-apoB were analyzed based on the multicompartment model shown in Figure 3 in order to investigate whether Lp(a) assembles from circulating LDL or from *de novo* produced "hepatic" LDL. 92% of leucine in Lp(a)-apoB originated from the hepatic apoB pool. The remaining 8% derived from plasma LDL-apoB. LDL-apoB stemmed from two sources, namely from VLDL-apoB (54%) and from de novo synthesis (46%).

The kinetic parameters obtained from this *in vivo* turnover study of Lp(a) metabolism in healthy men allow three major conclusions: i) Since FSRs of both protein components of Lp(a) is very similar and different from those of LDL, an almost exclusive intracellular hepatic Lp(a) assembly can be assumed. This analysis, however, does not allow any conclusions to be drawn on where (inside the hepatocyte, at its plasma membrane or, eventually, in the space of Dissé) this assembly takes place. ii) Apo(a) FSR/FCR is positively related to the number of apo(a) kringle 4 repeats (e.g. apo(a) molecular size), suggesting that plasma Lp(a) concentrations are controlled not only by synthesis but also to some smaller extent by degradation. iii) Longer plasma RT of apo(a) from probands with LMW apo(a) isoforms compared to those with HMW apo(a) isoforms help to explain the potential atherogenicity of higher concentrations in carriers with LMW apo(a) isoforms.

The *de novo* synthesis of LDL is an absolute prerequisite for the postulated (intra)cellular hepatic assembly of Lp(a). Such a "direct" LDL production has been questioned by some investigators who presume that it may instead be the consequence of a very fast lipolytic pathway [45]. However, metabolic studies of apoB metabolism using stable-isotope technology fitted by multicompartmental modeling support a significant "direct" LDL production by the liver [46, 47]. A substantial amount of nascent LDL production was also detected in cultured primary human hepatocytes [48] but not in HepG2 cells [36]. Lp(a) secretion was previously demonstrated in such cells, thus additionally supporting the view of "direct" LDL synthesis by human hepatocytes [49].

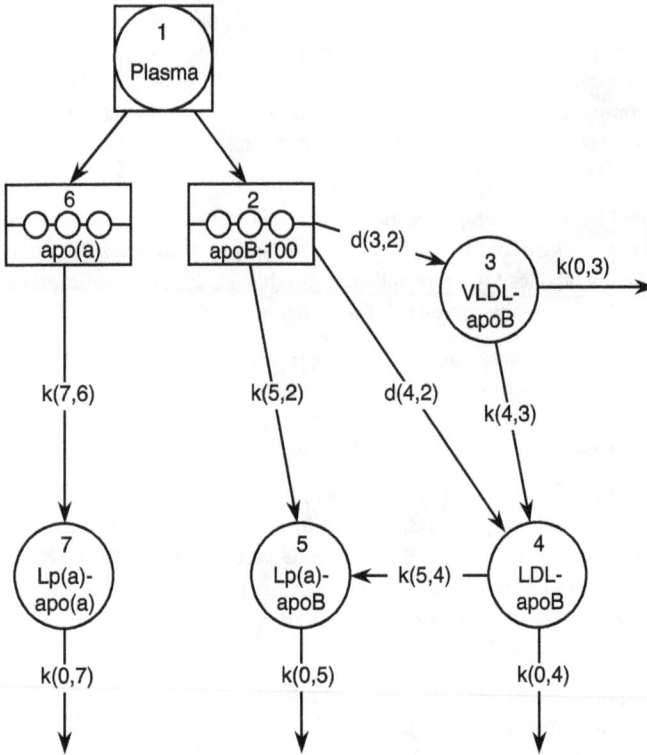

Figure 3. Multicompartmental model for apoB-100 and apo(a) metabolism. A plasma leucine pool (compartment 1) was used as a forcing function, and delay compartments that account for assembly and subsequent secretion of apoB-100 (compartment 2) and apo(a) (compartment 6), respectively. ApoB in VLDL, LDL, Lp(a), and apo(a) in Lp(a) consist all of single compartments. The input of apoB in Lp(a) is twofold: one via *de novo* synthesis from the liver and one from LDL-apoB. d(i,j) denotes the distribution of transfer from the delay compartment j to compartment i and k(i,j) represents the rate constant from compartment j to compartment i. In this model, tracer/tracee data for VLDL, LDL, Lp(a) apoB, and Lp(a) apo(a) as well as leucine masses (nmol/L) in these compartments were fitted simultaneously. The fractional catabolic rate (FCR) of apo(a)-Lp(a) and apoB-Lp(a) was equal to k(0,7) and k(0,5), respectively (taken from Frischmann et al. [44], with permission).

3.2. Lipoprotein metabolism in HD patients suffering from chronic kidney disease (CKD)

Dyslipidemia in patients with CKD and hemodialysis (HD) patients is distinct from other organ-specific diseases with far-reaching therapeutic consequences (see review [2]). It involves all lipoprotein classes, shows considerable variations depending on the stage of CKD [50, 51] and is further modified by concurrent diseases such as diabetes [52] and nephrotic syndrome [53]. In addition, major qualitative compositional changes in lipoprotein particles, such as oxidation, glycation, carbamylation and formation of small dense LDL (sdLDL – see below) which render the particles more atherogenic, have been observed [54]. Reduced activities of plasma cholesterol esterification and cholesterol ester transfer between lipoproteins – key factors for the so-called „reversed cholesterol transport" – result in substantially abnormal lipid composition of virtually all lipoprotein classes in HD patients [55].

Plasma triglycerides start to increase in early stages of CKD and show the highest concentrations in nephrotic syndrome and in patients treated with peritoneal dialysis (PD). In pre-dialysis CKD patients, the accumulation of triglycerides is the consequence of both an increased PR and a decreased FCR of triglyceride-rich lipoproteins [56]. The increased production of triglyceride-rich lipoproteins is possibly a consequence of impaired carbohydrate tolerance and enhanced hepatic VLDL synthesis [57]. The reduced catabolism is likely due to decreased activities of LPL and HL [58, 59], two endothelium-associated lipases that cleave triglycerides into free fatty acids for energy production or storage.

Diminished catabolism results in the accumulation of IDL particles contributing to compositional and size heterogeneity of triglyceride-rich lipoproteins in plasma of CKD patients. IDL are rich in apoE, a ligand that is important for removal from the circulation by binding to the LDL receptor [60]. The arterial wall therefore is exposed to high plasma concentrations of IDL which may predispose to atherosclerosis [54].

Elevated plasma concentrations of LDL cholesterol and –apoB are common in nephrotic syndrome and PD but do not occur in patients with advanced CKD, treated with HD. There are, however, qualitative changes in LDL in patients with CKD and dialysis patients. The fraction of sdLDL, which is considered to be highly atherogenic, is increased in HD patients. sdLDL is a subtype of LDL which penetrates the vessel wall more efficiently than normal LDL, becomes oxidized, and triggers atherosclerotic processes. In addition, sdLDL exert a high affinity for macrophages promoting their entry into the vascular wall to participate in the formation of foam cells and atherosclerotic plaques [61].

In kidney disease, elevated plasma Lp(a) concentrations are not only genetically determined but also a consequence of kidney failure [62]. In predialysis CKD patients, Lp(a) concentrations are influenced by the glomerular filtration rate (GFR). In patients with HMW apo(a) isoforms but not in those with LMW apo(a) isoforms, plasma Lp(a) concentrations begin to increase in stage 1 CKD before GFR starts to decrease [50]. This isoform-specific increase in plasma Lp(a) concentrations was observed in several but not all studies in CKD and HD patients [50, 62-66]. In contrast, in patients with nephrotic syndrome [67, 68] and in PD patients [63], increases in plasma Lp(a) concentrations occur in all apo(a) isoform groups, probably as a consequence of the pronounced protein loss and a subsequently increased production in the liver [69]. After successful kidney transplantation, a decrease in plasma Lp(a) can be observed in HD patients with HMW apo(a) isoforms [70, 71] and in PD patients with all apo(a) isoform groups [72]. Thus, the elevation of Lp(a) in CKD is due to non-genetic causes, mostly influenced by the degree of proteinuria [50, 67] and less by the cause of kidney disease [63].

In summary, the hallmarks of uremic dyslipidemia include hypertriglyceridemia and increased circulating concentrations of IDL, sdLDL and Lp(a). HD patients are characterised by normal LDL concentrations, whereas patients with nephrotic syndrome and CKD patients treated by PD are diagnosed with elevated LDL concentrations.

3.2.1. Dyslipidemia and CVD in CKD

Forty years ago, Lindner and colleagues described in their seminal report the excessive risk of CVD in HD patients for the first time [73]. Later, Foley et al. extended these observations by reporting a 10 to 20 times higher mortality rate in HD patients compared to the general population [8]. While in the general population high plasma concentrations of apoB-containing lipoproteins, low concentrations of HDL cholesterol and high total triglyceride concentrations are associated with an increased atherosclerotic cardiovascular risk [74], most investigations, including cross-sectional [75-78] and longitudinal [66, 79-87] studies, do not support the association between dyslipidemia and CVD in hemodialysed CKD populations or even observe opposite associations. Indeed, a worse survival among HD patients has been observed with low rather than high BMI [88], blood pressure [89] and serum/plasma concentrations of cholesterol [90]. This seemingly paradoxical phenomenon is often called „reverse epidemiology" [91] and exemplified in crossing curves when relating BMI with mortality in HD patients and the general population [92].

While the BMI-associated death risk shows an almost linear negative gradient in HD patients [92], the relationship between plasma total cholesterol and mortality has been found to be U-shaped [93]. The group with total cholesterol between 200 and 250 mg/dl had the lowest risk for death, whereas those with levels >350 mg/dl had a relative risk of 1.3-fold and those with levels <100 mg/dl had a relative risk of 4.2-fold. The association between low total cholesterol and increased mortality, however, was reduced after statistical adjustment for plasma albumin levels. This dichotomous relationship was confirmed in the Choices for Healthy Outcomes in Caring for ESRD (CHOICE) study [94], which showed a nonsignificant negative association of cardiovascular mortality with plasma total as well as non–HDL cholesterol levels in the presence of inflammation and/or malnutrition; in contrast, there was a positive association between total and non–HDL cholesterol and mortality in the absence of inflammation or malnutrition. These observations are compatible with the hypothesis that the inverse association of total cholesterol levels with mortality in dialysis patients is mediated by the cholesterol-lowering effect of malnutrition and/or systemic inflammation and not due to a protective effect of high cholesterol concentrations.

The association of Lp(a) with atherosclerotic complications and CVD has been investigated in numerous studies in dialysis patients. Like other atherogenic lipoproteins, Lp(a) has been found to contribute to the high cardiovascular burden [66, 79, 84, 95-97]. When apo(a) phenotyping was performed along with plasma Lp(a) concentrations, an association between the apo(a) K-IV repeat polymorphism and CV complications was consistently observed.

Two final considerations regarding the impact of classical risk factors for the development of CVD in CKD patients are, however, worth mentioning: the cardiovascular risk for an individual CKD patient at a given time point is the sum (or combination) of risk exposure before and after developing CKD. When taking Lp(a) concentrations and apo(a) isoforms as an example, a previously healthy subject with low Lp(a) concentrations and a HMW apo(a) isoform develops CKD with subsequently rising Lp(a) concentrations covering a relatively short period of his lifespan. A subject with LMW apo(a) isoform, on the other hand, has genetically caused elevated Lp(a) concentrations for his whole life which do not substantially increase after developing CKD. Since the HMW apo(a) carrier is exposed to elevated atherogenic Lp(a) for a much shorter period of his life, this condition has to be considered less CVD-prone than having LMW apo(a). This example demonstrates the importance of the „longitudinal" factor when considering risk factors for CVD in CKD patients.

Finally, as already discussed in the introduction, the quantification of a target parameter deemed to be associated with or predictive for a disease can only provide a static picture and hardly reflects the true in vivo metabolism. Seemingly normal blood concentrations of suspected marker candidates can only be validated by kinetic studies in humans and have been therefore performed also in CKD patients. They have provided novel and unexpected information regarding the physiology and pathology of atherogenic apoB-containing lipoproteins (see review [98]).

3.2.2. Delayed in vivo catabolism of LDL and IDL in HD patients as potential cause of premature atherosclerosis

For better understanding the atherogeneity of apoB-containing lipoproteins in HD patients and to resolve the apparent discrepancy between their obviously impaired lipoprotein metabolism and e.g. normal LDL plasma concentrations, we studied the in vivo kinetics of VLDL, IDL and LDL by stable isotope technology in HD patients and compared them to those of healthy controls [12].

This study demonstrated for the first time severely decreased FCRs of IDL- and LDL-apoB in HD patients as compared to controls (Figure 4), whereas the in vivo kinetics of VLDL did not change significantly. A decreased FCR of IDL- and LDL-apoB is identical to a prolonged RT of these highly atherogenic particles. The longer RT of these lipoproteins results in an extended exposure to oxidation for IDL and LDL in a highly oxidative environment. This is in line with experimental data showing a highly significant correlation of 5-hydroxy-2-aminovaleric acid (HAVA) in LDL, an oxidation product of apoB, with LDL RT in normolipidemic controls [99]. In accordance with these results, two previously conducted randomized placebo-controlled studies revealed a significant reduction in composite cardiovascular disease endpoints when HD patients were treated for two years with supplementation of antioxidants such as vitamin E [100] or acetylcysteine [101].

Most remarkably, the observed impaired metabolism of apoB-containing lipoproteins is accompanied by normal concentrations of LDL-apoB and elevated levels of IDL-apoB (Figure 4), in line with previous reports which found increased concentrations of IDL as an independent risk factor for atherosclerosis in HD patients [102]. A closer look at the kinetic data reveals that the normal concentrations of LDL are the result of a combination of decreased FCR and PR. This pattern therefore demonstrates convincingly the strength of kinetic studies in contrast to simply quantifying blood concentrations of a target marker such as LDL concentrations. Its normal concentrations are masked by two metabolic disorders which neutralise each other and result in normal values such as observed in the general population. The altered lipoprotein metabolism therefore puts HD patients at high risk for developing atherosclerotic disease despite their normal total and LDL cholesterol concentrations. Since most lipid-lowering drugs act by "normalising" the RT of the major atherogenic lipoproteins IDL and LDL [103], these drugs are expected to correct some of the basic defects of the severely disturbed lipoprotein metabolism in HD patients. Therefore, kinetic studies on the impact of lipid-lowering medication on the lipoprotein metabolism in CKD patients were a logic consequence of the observed, above-described findings (see chapter 3.2.5.).

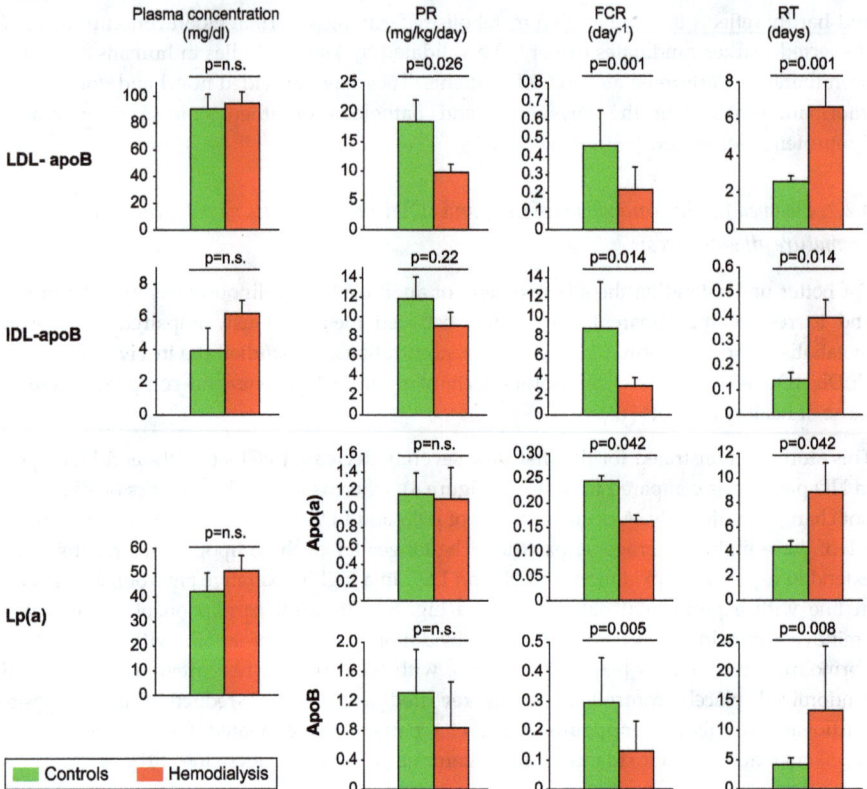

Figure 4. Kinetic parameters of apoB in LDL, IDL and Lp(a) and apo(a) in Lp(a). Plasma concentrations, production rates (PR), fractional catabolic rates (FCR) and residence times (RT) are given for healthy controls (green columns) and HD patients (red columns). Columns represent mean values ± SD. Results for LDL and IDL are taken from Ikewaki et al. [12], those for Lp(a) from Frischmann et al. [44].

Due to the laborious nature of these studies and the complexity of the metabolic modeling, only few studies have been performed so far in CKD patients either by radiotracer or stable isotope technology. Our kinetic data seem to contrast with a previously published turnover study in Finnish HD patients performed with conventional radiotracer techniques. While the authors found decreased LDL clearance rates in predialysis CKD patients [104] they could not find a significant difference in LDL-apoB FCR between HD patients and controls [105]. More recently, Prinsen et al., by using stable isotopes, found unchanged FCRs for LDL-apoB in CKD patients treated with peritoneal dialysis [106]. Chan et al. injected radio-labeled VLDL into HD patients with or without hyperlipidemia and found decreased FCRs of VLDL-apoB and IDL-apoB (the latter only in hyperlipidemic patients) [107]. LDL kinetics were not investigated in this study. The reason for these discrepancies is not clear. There might be ethnic differences in the lipoprotein metabolism between the investigated patient populations of different ethnic origin. One major difference between our and the Finnish study is an age difference between patients and controls in our but not in the Finnish study. Our control subjects were considerably younger than the HD patients (35 vs. 51 years). At first glance, this age difference might explain to some extent the dramatic differences found in our study, since LDL clearance rates have been repeatedly described to decrease with age presumably due to down-regulated hepatic LDL receptor expression in the elderly [108, 109]. Based on the results of these studies, an age difference of 15 years (as observed in our work) would result in an approximately 10% change in FCR values and could therefore not explain the more than two-fold difference in our study. The observed differences in kinetic parameters can therefore not be explained by age differences between study groups.

Several mechanisms may contribute to our observations. First, the diminished LDL catabolism in HD patients might be explained by a possible contribution of LDL uptake by the healthy human kidney which does not function appropriately (or at all) in chronic kidney failure. In fact, glomerular cells like mesangial or epithelial cells have been shown *in vitro* to express lipoprotein receptors and to take up LDL comparably to fibroblasts and hepatocytes [110]. It is, however, completely unclear whether the kidney plays a significant role in LDL catabolism *in vivo*. Perfusion studies in rat kidneys indicate that virtually no intact LDL is cleared from the circulation by the kidney [111]. Second, an impaired lipolytic cascade in HD patients most likely also contributes to our results. The relatively normal VLDL concentrations and kinetic parameters and the correspondingly impaired IDL parameters are in good accordance with previous findings of normal lipoprotein lipase (LPL) but significantly decreased activities of hepatic triglyceride lipase (HL) in HD patients [59]. Since HL promotes the conversion of IDL to LDL, a decrease in HL activity might contribute to the accumulation of IDL and reduced production rates of LDL (without accumulating small, dense LDL) in HD patients.

3.2.3. Kinetics of Lp(a) in hemodialysis patients

We previously performed *in vivo* kinetic studies using stable-isotope techniques to elucidate the mechanism for increased plasma Lp(a) concentrations in HD patients [44]. PRs of apo(a) and apoB, the two apolipoproteins contained in Lp(a), were normal, when compared to control subjects with similar plasma Lp(a) concentrations (Figure 4). The FCR of these apolipoproteins was, however, significantly reduced compared to controls resulting in a much longer plasma RT for apo(a) of almost 9 days, compared to only 4.4 days in controls. Since the PR of Lp(a) did not differ between HD patients and controls, its decreased clearance in HD patients leads to increased Lp(a) plasma concentrations and is likely the result of loss in kidney function [44]. A role of the kidney in the catabolism has been previously supported by the observation of renovascular arteriovenous differences in Lp(a) concentrations [112] as well as apo(a) fragments in urine [113, 114].

Comparing kinetic data in HD patients [44] with those in patients with nephrotic syndrome [69] points to fundamental differences in the metabolism of Lp(a) and other proteins between these two patient groups. Patients with nephrotic syndrome do not differ with respect to the FCR of Lp(a) compared to controls but have increased Lp(a) PRs [69]. It is well known that nephrotic patients show a generally increased lipoprotein synthesis of lipoproteins [115]. Since kidney function is relatively well preserved in nephrotic syndrome, a decreased clearance of Lp(a) in these patients is not likely to be expected. Metabolic differences between nephrotic and dialysis patients are not only evident for Lp(a) but also for albumin. Whereas the FCR of albumin in HD patients is similar or even reduced compared to controls, the FCR in patients with nephrotic syndrome is increased [116, 117].

3.2.4. Consequences of the impaired metabolism of atherogenic lipoproteins in HD patients

The observation of markedly decreased FCRs of apoB of LDL and IDL as well as apo(a) and apoB in Lp(a) causes a prolonged RT of these highly atherogenic lipoproteins. Due to the long retention period, "aged" lipoprotein complexes are thus more susceptible for alterations such as oxidation damage, which was shown to be associated with accelerated atherogenesis in HD patients [118]. Previous kinetic studies investigated the metabolism of the two LDL subclasses, "buoyant" LDL1 and the smaller cholesterol-poor "dense" LDL2, in subjects with familial defective apoB-100 (FDB). The authors found a more than four-fold longer RT for small dense LDL2 in those patients as compared to normolipidemic controls [99]. It was therefore suggested that oxidative damage of an "aged" LDL2, which is present in large concentrations in both blood and the subendothelial space, may be an important mechanism for the development of premature atherosclerosis in patients with familial defective apoB-100. Since the LDL-like particle of Lp(a) is compositionally similar to LDL2 [41], it is tempting to speculate that the increased RT of circulating Lp(a) might pose an additional risk factor for the increased incidence of cardiovascular disease in HD patients.

3.2.5. Influence of statin treatment on kinetic parameters in hemodialysed patients

In the general population, therapy with HMG-CoA-reductase inhibitors (statins) which inhibit endogenous cholesterol biosynthesis has shown to improve outcome in several atherosclerotic diseases [119, 120]. The inhibition of cholesterol biosynthesis subsequently leads to up-regulation of LDL receptors and therefore increased clearance and thus reduced RT of circulating LDL [103]. Statins also have a beneficial role as anti-inflammatory agents, which is independent of their lipid-lowering effect. Inflammation is highly prevalent in patients with CKD and is consistently associated with cardiovascular morbidity and mortality. In line with this metabolic background, the first studies in HD patients demonstrated a substantial normalisation of the dyslipidemic plasma profile and reduced progression of renal disease [121, 122] and in one study also reduced mortality [123] in these patients.

In contrast and quite surprisingly, three previously conducted large, randomized, placebo-controlled trials on statin treatment in CKD patients had not led to significant benefits regarding their primary cardiovascular outcome. Two of those studies, the German Diabetes Dialysis (4D) Study and the Study to Evaluate the Use of Rusovastatin in Subjects on Regular Hemodialysis (AURORA) were performed on HD patients, one study, the Assessment of Lescol in Renal Transplantation (ALERT) Study on patients who had undergone kidney transplantation. Their primary endpoints (death of cardiovascular cause, nonfatal myocardial infarction or nonfatal stroke) were virtually unchanged [124-126]. However, there has been a promising risk reduction in the secondary endpoint 'all cardiac events combined' in one study [126]. A simulated study of exactly the same trial using a large historical database with more than 10.000 patients also demonstrated that statin use was associated with some benefit [127]. A comprehensive review of outcome data from the 4D and AURORA trials found no benefit of statin therapy in either the whole study group of HD patients or after stratification for inflammatory marker levels [128]. More recently, another, much larger trial including 9270 patients with chronic kidney disease, the Study of Heart and Renal Protection (SHARP) could show a significant risk reduction in cardiovascular events in a mixed population of patients with kidney disease including 2/3 predialysis and 1/3 HD patients treated with a combination of a statin and ezitimibe. This effect did not differ between HD and predialysis patients [129].

Based on our previous studies on lipoprotein kinetics in HD patients and the above-described conflicting results regarding their cardiovascular risk profile after statin treatment, we examined by stable-isotope technology the in vivo kinetics of apoB-containing particles in HD patients before and after treatment with atorvastatin (Schwaiger et al., unpublished).

In this study we described for the first time effects of HMG-CoA reductase inhibition on apoB metabolism in CKD patients treated with HD. Low-dose atorvastatin, given for three months to six male patients, lowered, as expected, concentrations of VLDL- and LDL-apoB, both accompanied by a significant increase of their FCR, while hepatic production of both apolipoproteins was not altered. This led, as expected, to a lower RT of these atherogenic apoB-containing particles comparable to RT values of healthy subjects with normal kidney function. The observed findings therefore argue for a beneficial effect of statin therapy regarding cardiovascular events in HD patients similar to described for the general population.

To understand why statins have surprisingly failed to reduce cardiovascular events in HD patients, the basic mechanisms underlying the pathophysiology of CVD in CKD must be critically considered. In contrast to the general population, CKD patients suffer, in addition to dyslipidemia, from several further complex comorbid conditions including diabetes mellitus, hypertension, oxidative stress, inflammation, insulin resistance, anemia and disturbances in mineral metabolism. Lipid lowering therapy by statins have the potential to ameliorate only some but no all of those conditions (see review [130]). Taken together, statin therapy in CKD maybe recommended based on our kinetic studies on apoB-containing lipoproteins, optimally combined with medication to treat atherogenic non-lipid factors in HD patients.

4. Conclusion

Kinetic in vivo studies in human subjects are superior to many methodological approaches including animal and cell culture models and thus represent the ultimate approach to understand basic metabolic pathways in humans. They have clearly revolutionized human lipoprotein research and have particularly resulted in novel insights into the metabolism of atherogenic apoB-containing lipoproteins some of which have been the subject of our previous investigations and object of this review.

Author details

Benjamin Dieplinger*
Department of Laboratory Medicine, Konventhospital Barmherzige Brüder Linz, Austria

Hans Dieplinger**
Division of Genetic Epidemiology, Department of Human Genetics and Molecular Pharmacology, Medical University of Innsbruck, Austria

Acknowledgement

We would like to thank Ramona Berberich, Linda Fineder, Michael E. Frischmann, Tatsuo Hosoya, Katsunori Ikewaki, Paul König, Florian Kronenberg, Seibu Mochizuki, Yoshinobu Nakada, Ulrich Neyer, Keio Okubo, Hermann Salmhofer Jürgen R. Schäfer, Johannes P. Schwaiger, Horst Schweer, Alex Starke, Evi Trenkwalder and Emanuel Zitt for invaluable contributions over many years within the various projects covered in this chapter. Funding support from the Austrian Science Fund (P10090-MED, P12358-MED), the Austrian National Bank (6721/4, 9331) and Pfizer-Austria (ATV-A-02-007G) is greatly appreciated.

5. References

[1] Mahley RW, Innerarity TL, Rall SCJ, Weisgraber KH. Plasma lipoproteins: Apolipoprotein structure and function. J Lipid Res 1984;25:1277-1294.

[2] Kwan BC, Kronenberg F, Beddhu S, Cheung AK. Lipoprotein metabolism and lipid management in chronic kidney disease. J Am Soc Nephrol 2007;18:1246-1261.

[3] Williams KJ, Tabas I. The response-to-retention hypothesis of early atherogenesis. Arterioscler Thromb Vasc Biol 1995;15:551-561.

---

* benjamin.dieplinger@bs-lab.at
** Corresponding Author, hans.dieplinger@i-med.ac.at

[4]   Boren J, Taskinen MR, Adiels M. Kinetic studies to investigate lipoprotein metabolism. J Intern Med 2012;271:166-173.

[5]   Adiels M, Olofsson SO, Taskinen MR, Boren J. Overproduction of very low-density lipoproteins is the hallmark of the dyslipidemia in the metabolic syndrome. Arterioscler Thromb Vasc Biol 2008;28:1225-1236.

[6]   Jawien J, Nastalek P, Korbut R. Mouse models of experimental atherosclerosis. J Physiol Pharmacol 2004;55:503-517.

[7]   Javitt NB. Hep G2 cells as a resource for metabolic studies: Lipoprotein, cholesterol, and bile acids. FASEB J 1990;4:161-168.

[8]   Foley RN, Parfrey PS, Sarnak MJ. Clinical epidemiology of cardiovascular disease in chronic renal disease. Am J Kidney Dis 1998;32:S112-119.

[9]   Krempler F, Kostner GM, Bolzano K, Sandhofer F. Turnover of lipoprotein(a) in man. J Clin Invest 1980;65:1483-1490.

[10] Barrett PH, Chan DC, Watts GF. Thematic review series: patient-oriented research. Design and analysis of lipoprotein tracer kinetics studies in humans. J Lipid Res 2006;47:1607-1619.

[11] Ikewaki K, Rader DJ, Sakamoto T, et al. Delayed catabolism of high density lipoprotein apolipoproteins A-I and A-II in human cholesteryl ester transfer protein deficiency. J Clin Invest 1993;92:1650-1658.

[12] Ikewaki K, Schaefer JR, Frischmann ME, et al. Delayed in vivo catabolism of intermediate-density lipoprotein and low-density lipoprotein in hemodialysis patients as potential cause of premature atherosclerosis. Arterioscler Thromb Vasc Biol 2005;25:2615-2622.

[13] Barrett PH, Bell BM, Cobelli C, et al. SAAM II: Simulation, analysis, and modeling software for tracer and pharmacokinetic studies. Metabolism 1998;47:484-492.

[14] Cobelli C, Foster DM. Compartmental models: theory and practice using the SAAM II software system. Adv Exp Med Biol 1998;445:79-101.

[15] Utermann G. Lipoprotein(a). In: Scriver C. R., Beaudet A. L., Sly W. S.,Valle D. (eds), The metabolic bases of inherited disease, New York, McGraw Hill Inc., 2001:2753-2787.

[16] Clarke R, Peden JF, Hopewell JC, et al. Genetic variants associated with Lp(a) lipoprotein level and coronary disease. N Engl J Med 2009;361:2518-2528.

[17] Kraft HG, Köchl S, Menzel HJ, Sandholzer C, Utermann G. The apolipoprotein(a) gene: a transcribed hypervariable locus controlling plasma lipoprotein(a) concentration. Hum Genet 1992;90:220-230.

[18] Schmidt K, Kraft HG, Parson W, Utermann G. Genetics of the Lp(a)/apo(a) system in an autochthonous Black African population from the Gabon. Eur J Hum Genet 2006;14:190-201.

[19] Bostom AG, Gagnon DR, Cupples LA, et al. A prospective investigation of elevated lipoprotein (a) detected by electrophoresis and cardiovascular disease in women: The Framingham Heart Study. Circulation 1994;90:1688-1695.

[20] Erqou S, Kaptoge S, Perry PL, et al. Lipoprotein(a) concentration and the risk of coronary heart disease, stroke, and nonvascular mortality. JAMA 2009;302:412-423.

[21] Rader DJ, Cain W, Ikewaki K, et al. The inverse association of plasma lipoprotein(a) concentrations with apolipoprotein(a) isoform size is not due to differences in Lp(a) catabolism but to differences in production rate. J Clin Invest 1994;93:2758-2763.

[22] Dube JB, Boffa MB, Hegele RA, Koschinsky ML. Lipoprotein(a): more interesting than ever after 50 years. Curr Opin Lipidol 2012;23:133-140.

[23] Dieplinger H, Utermann G. The seventh myth of lipoprotein(a): where and how is it assembled? Curr Opin Lipidol 1999;10:275-283.

[24] Brunner C, Kraft HG, Utermann G, Müller HJ. Cys4057 of apolipoprotein(a) is essential for lipoprotein(a) assembly. Proc Natl Acad Sci U.S.A. 1993;90:11643-11647.

[25] Trieu VN, McConathy WJ. A two-step model for lipoprotein(a) formation. J Biol Chem 1995;270:15471-15474.

[26] McCormick SP, Ng JK, Taylor S, Flynn LM, Hammer RE, Young SG. Mutagenesis of the human apolipoprotein B gene in a yeast artificial chromosome reveals the site of attachment for apolipoprotein(a). Proc Natl Acad Sci U.S.A. 1995;92:10147-10151.

[27] Becker L, Cook PM, Koschinsky ML. Identification of sequences in apolipoprotein(a) that maintain its closed conformation: a novel role for apo(a) isoform size in determining the efficiency of covalent Lp(a) formation. Biochemistry 2004;43:9978-9988.

[28] Becker L, Nesheim ME, Koschinsky ML. Catalysis of covalent Lp(a) assembly: evidence for an extracellular enzyme activity that enhances disulfide bond formation. Biochemistry 2006;45:9919-9928.

[29] Gething MJ, Sambrook J. Protein folding in the cell. Nature 1992;355:33-45.

[30] Hurtley SM, Helenius A. Protein oligomerization in the endoplasmic reticulum. Annu Rev Cell Biol 1989;5:277-307.

[31] White AL, Lanford RE. Biosynthesis and metabolism of lipoprotein(a). Curr Opin Lipidol 1995;6:75-80.

[32] White AL, Rainwater DL, Lanford RE. Intracellular maturation of apolipoprotein[a] and assembly of lipoprotein[a] in primary baboon hepatocytes. J Lipid Res 1993;34:509-517.

[33] Lobentanz EM, Krasznai K, Gruber A, et al. Intracellular metabolism of human apolipoprotein(a) in stably transfected Hep G2 cells. Biochemistry 1998;37:5417-5425.

[34] Koschinsky ML, Côté GP, Gabel B, Van der Hoek YY. Identification of the cysteine residue in apolipoprotein(a) that mediates extracellular coupling with apolipoprotein B-100. J Biol Chem 1993;268:19819-19825.

[35] Bonen DK, Hausman AML, Hadjiagapiou C, Skarosi SF, Davidson NO. Expression of a recombinant apolipoprotein(a) in HepG2 cells. Evidence for intracellular assembly of lipoprotein(a). J Biol Chem 1997;272:5659-5667.

[36] Dashti N, Alaupovic P, Knight-Gibson C, Koren E. Identification and partial characterization of discrete Apolipoprotein B containing lipoprotein particles produced by human hepatoma cell line HepG2. Biochemistry 1987;26:4837-4846.

[37] Krempler F, Kostner GM, Bolzano K, Sandhofer F. Lipoprotein(a) is not a metabolic product of other lipoproteins containing apolipoprotein B. Biochim Biophys Acta 1979;575:63-70.

[38] Krempler F, Kostner G, Bolzano K, Sandhofer F. Studies on the metabolism of the lipoprotein Lp(a) in man. Atherosclerosis 1978;30:57-65.

[39] Morrisett J, Gaubatz J, Nava L, et al. Metabolism of apo(a) and apoB-100 in human lipoprotein(a). In: Catapano A., Gotto Jr. A. M., Smith L. C.,Paoletti R. (eds), Drugs affecting lipid metabolism, Dordrecht, The Netherlands, Kluwer Academic Publishers and Fondazione Giovanni Lorenzini, 1993:161-167.

[40] Su W, Campos H, Judge H, Walsh BW, Sacks FM. Metabolism of Apo(a) and ApoB100 of lipoprotein(a) in women: effect of postmenopausal estrogen replacement. J Clin Endocrinol Metab 1998;83:3267-3276.

[41] Demant T, Seeberg K, Bedynek A, Seidel D. The metabolism of lipoprotein(a) and other apolipoprotein B-containing lipoproteins: a kinetic study in humans. Atherosclerosis 2001;157:325-339.

[42] Jenner JL, Seman LJ, Millar JS, et al. The metabolism of apolipoproteins (a) and B-100 within plasma lipoprotein (a) in human beings. Metabolism 2005;54:361-369.

[43] Frischmann ME, Ikewaki K, Trenkwalder E, et al. In vivo stable-isotope kinetic study suggests intracellular assembly of lipoprotein(a). Atherosclerosis 2012;(in press).

[44] Frischmann ME, Kronenberg F, Trenkwalder E, et al. In vivo turnover study demonstrates diminished clearance of lipoprotein(a) in hemodialysis patients. Kidney Int 2007;71:1036-1043.

[45] Shames DM, Havel RJ. De novo production of low density lipoproteins: Fact or fancy. J Lipid Res 1991;32:1099-1112.

[46] Pietzsch J, Wiedemann B, Julius U, et al. Increased clearance of low density lipoprotein precursors in patients with heterozygous familial defective apolipoprotein B-100: a stable isotope approach. J Lipid Res 1996;37:2074-2087.

[47] Packard CJ, Demant T, Stewart JP, et al. Apolipoprotein B metabolism and the distribution of VLDL and LDL subfractions. J Lipid Res 2000;41:305-318.

[48] Bouma ME, Pessah M, Renaud G, Amit N, Catala D, Infante R. Synthesis and secretion of lipoproteins by human hepatocytes in culture. In Vitro Cell Dev Biol 1988;24:85-90.

[49] Kagawa A, Azuma H, Akaike M, Kanagawa Y, Matsumoto T. Aspirin reduces apolipoprotein(a) (apo(a)) production in human hepatocytes by suppression of apo(a) gene transcription. J Biol Chem 1999;274:34111-34115.

[50] Kronenberg F, Kuen E, Ritz E, et al. Lipoprotein(a) serum concentrations and apolipoprotein(a) phenotypes in mild and moderate renal failure. J Am Soc Nephrol 2000;11:105-115.

[51] Kronenberg F, Kuen E, Ritz E, et al. Apolipoprotein A-IV serum concentrations are elevated in patients with mild and moderate renal failure. J Am Soc Nephrol 2002;13:461-469.

[52] Krentz AJ. Lipoprotein abnormalities and their consequences for patients with type 2 diabetes. Diabetes Obes Metab 2003;5 Suppl 1:S19-27.

[53] Kronenberg F. Dyslipidemia and nephrotic syndrome: recent advances. J Ren Nutr 2005;15:195-203.

[54] Shoji T, Ishimura E, Inaba M, Tabata T, Nishizawa Y. Atherogenic lipoproteins in end-stage renal disease. Am J Kidney Dis 2001;38:S30-33.

[55] Dieplinger H, Schoenfeld PY, Fielding CJ. Plasma cholesterol metabolism in end-stage renal disease. Difference between treatment by hemodialysis or peritoneal dialysis. J Clin Invest 1986;77:1071-1083.

[56] Batista MC, Welty FK, Diffenderfer MR, et al. Apolipoprotein A-I, B-100, and B-48 metabolism in subjects with chronic kidney disease, obesity, and the metabolic syndrome. Metabolism 2004;53:1255-1261.

[57] Appel G. Lipid abnormalities in renal disease. Kidney Int 1991;39:169-183.

[58] Chan MK, Persaud J, Varghese Z, Moorhead JF. Pathogenic roles of post-heparin lipases in lipid abnormalities in hemodialysis patients. Kidney Int 1984;25:812-818.

[59] Oi K, Hirano T, Sakai S, Kawaguchi Y, Hosoya T. Role of hepatic lipase in intermediate-density lipoprotein and small, dense low-density lipoprotein formation in hemodialysis patients. Kidney Int Suppl 1999;71:S227-228.

[60] Brown MS, Goldstein JL. A receptor-mediated pathway for cholesterol homeostasis. Science 1986;232:34-47.

[61] Littlewood TD, Bennett MR. Apoptotic cell death in atherosclerosis. Curr Opin Lipidol 2003;14:469-475.

[62] Dieplinger H, Lackner C, Kronenberg F, et al. Elevated plasma concentrations of lipoprotein(a) in patients with end-stage renal disease are not related to the size polymorphism of apolipoprotein(a). J Clin Invest 1993;91:397-401.

[63] Kronenberg F, König P, Neyer U, et al. Multicenter study of lipoprotein(a) and apolipoprotein(a) phenotypes in patients with end-stage renal disease treated by hemodialysis or continous ambulatory peritoneal dialysis. J Am Soc Nephrol 1995;6:110-120.

[64] Milionis HJ, Elisaf MS, Tselepis A, Bairaktari E, Karabina SA, Siamopoulos KC. Apolipoprotein(a) phenotypes and lipoprotein(a) concentrations in patients with renal failure. Am J Kidney Dis 1999;33:1100-1106.

[65] Stenvinkel P, Heimburger O, Tuck CH, Berglund L. Apo(a)-isoform size, nutritional status and inflammatory markers in chronic renal failure. Kidney Int 1998;53:1336-1342.

[66] Zimmermann J, Herrlinger S, Pruy A, Metzger T, Wanner C. Inflammation enhances cardiovascular risk and mortality in hemodialysis patients. Kidney Int 1999;55:648-658.

[67] Kronenberg F, Lingenhel A, Lhotta K, et al. The apolipoprotein(a) size polymorphism is associated with nephrotic syndrome. Kidney Int 2004;65:606-612.

[68] Wanner C, Rader D, Bartens W, et al. Elevated plasma lipoprotein(a) in patients with the nephrotic syndrome. Ann Intern Med 1993;119:263-269.

[69] de Sain-van der Velden MG, Reijngoud DJ, Kaysen GA, et al. Evidence for increased synthesis of lipoprotein(a) in the nephrotic syndrome. J Am Soc Nephrol 1998;9:1474-1481.

[70] Kronenberg F, König P, Lhotta K, et al. Apolipoprotein(a) phenotype-associated decrease in lipoprotein(a) plasma concentrations after renal transplantation. Arterioscler Thromb 1994;14:1399-1404.

[71] Kronenberg F, Lhotta K, Konig P, Margreiter R, Dieplinger H, Utermann G. Apolipoprotein(a) isoform-specific changes of lipoprotein(a) after kidney transplantation. Eur J Hum Genet 2003;11:693-699.

[72] Kerschdorfer L, Konig P, Neyer U, et al. Lipoprotein(a) plasma concentrations after renal transplantation: a prospective evaluation after 4 years of follow-up. Atherosclerosis 1999;144:381-391.

[73] Lindner A, Charra B, Sherrard DJ, Scribner BH. Accelerated atherosclerosis in prolonged maintenance hemodialysis. N Engl J Med 1974;290:697-701.

[74] Wilson PW, D'Agostino RB, Levy D, Belanger AM, Silbershatz H, Kannel WB. Prediction of coronary heart disease using risk factor categories. Circulation 1998;97:1837-1847.

[75] Cheung AK, Sarnak MJ, Yan G, et al. Atherosclerotic cardiovascular disease risks in chronic hemodialysis patients. Kidney Int 2000;58:353-362.

[76] Guz G, Nurhan Ozdemir F, Sezer S, et al. Effect of apolipoprotein E polymorphism on serum lipid, lipoproteins, and atherosclerosis in hemodialysis patients. Am J Kidney Dis 2000;36:826-836.

[77] Koch M, Kutkuhn B, Trenkwalder E, et al. Apolipoprotein B, fibrinogen, HDL cholesterol, and apolipoprotein(a) phenotypes predict coronary artery disease in hemodialysis patients. J Am Soc Nephrol 1997;8:1889-1898.

[78] Stack AG, Bloembergen WE. Prevalence and clinical correlates of coronary artery disease among new dialysis patients in the United States: a cross-sectional study. J Am Soc Nephrol 2001;12:1516-1523.

[79] Cressman MD, Heyka RJ, Paganini EP, O'Neil J, Skibinski CI, Hoff HF. Lipoprotein(a) is an independent risk factor for cardiovascular disease in hemodialysis patients. Circulation 1992;86:475-482.

[80] Degoulet P, Legrain M, Reach I, et al. Mortality risk factors in patients treated by chronic hemodialysis. Report of the Diaphane collaborative study. Nephron 1982;31:103-110.

[81] Hocher B, Ziebig R, Altermann C, et al. Different impact of biomarkers as mortality predictors among diabetic and nondiabetic patients undergoing hemodialysis. J Am Soc Nephrol 2003;14:2329-2337.

[82] Iseki K, Fukiyama K. Predictors of stroke in patients receiving chronic hemodialysis. Kidney Int 1996;50:1672-1675.

[83] Koda Y, Nishi S, Suzuki M, Hirasawa Y. Lipoprotein(a) is a predictor for cardiovascular mortality of hemodialysis patients. Kidney Int Suppl 1999;71:S251-S253.

[84] Kronenberg F, Neyer U, Lhotta K, et al. The low molecular weight apo(a) phenotype is an independent predictor for coronary artery disease in hemodialysis patients: a prospective follow-up. J Am Soc Nephrol 1999;10:1027-1036.

[85] Ohashi H, Oda H, Ohno M, Watanabe S, Sakata S. Lipoprotein(a) as a risk factor for coronary artery disease in hemodialysis patients. Kidney Int Suppl 1999;71:S242-S244.

[86] Schwaiger JP, Lamina C, Neyer U, et al. Carotid plaques and their predictive value for cardiovascular disease and all-cause mortality in hemodialysis patients considering renal transplantation: a decade follow-up. Am J Kidney Dis 2006;47:888-897.

[87] Shoji T, Emoto M, Shinohara K, et al. Diabetes mellitus, aortic stiffness, and cardiovascular mortality in end-stage renal disease. J Am Soc Nephrol 2001;12:2117-2124.

[88] Kalantar-Zadeh K, Kopple JD, Kilpatrick RD, et al. Association of morbid obesity and weight change over time with cardiovascular survival in hemodialysis population. Am J Kidney Dis 2005;46:489-500.

[89] Kalantar-Zadeh K, Kilpatrick RD, McAllister CJ, Greenland S, Kopple JD. Reverse epidemiology of hypertension and cardiovascular death in the hemodialysis population: the 58th annual fall conference and scientific sessions. Hypertension 2005;45:811-817.

[90] Nishizawa Y, Shoji T, Ishimura E, Inaba M, Morii H. Paradox of risk factors for cardiovascular mortality in uremia: is a higher cholesterol level better for atherosclerosis in uremia? Am J Kidney Dis 2001;38:S4-7.

[91] Kalantar-Zadeh K. What is so bad about reverse epidemiology anyway? Semin Dial 2007;20:593-601.

[92] Kalantar-Zadeh K, Block G, Humphreys MH, Kopple JD. Reverse epidemiology of cardiovascular risk factors in maintenance dialysis patients. Kidney Int 2003;63:793-808.

[93] Lowrie EG, Lew NL. Death risk in hemodialysis patients: the predictive value of commonly measured variables and an evaluation of death rate differences between facilities. Am J Kidney Dis 1990;15:458-482.

[94] Longenecker JC, Coresh J, Powe NR, et al. Traditional cardiovascular disease risk factors in dialysis patients compared with the general population: the CHOICE Study. J Am Soc Nephrol 2002;13:1918-1927.

[95] Longenecker JC, Klag MJ, Marcovina SM, et al. Small apolipoprotein(a) size predicts mortality in end-stage renal disease: The CHOICE study. Circulation 2002;106:2812-2818.

[96] Kronenberg F, Kathrein H, König P, et al. Apolipoprotein(a) phenotypes predict the risk for carotid atherosclerosis in patients with end-stage renal disease. Arterioscler Thromb 1994;14:1405-1411.

[97] Longenecker JC, Klag MJ, Marcovina SM, et al. High lipoprotein(a) levels and small apolipoprotein(a) size prospectively predict cardiovascular events in dialysis patients. J Am Soc Nephrol 2005;16:1794-1802.

[98] Kronenberg F, Ikewaki K, Schaefer JR, Konig P, Dieplinger H. Kinetic studies of atherogenic lipoproteins in hemodialysis patients: do they tell us more about their pathology? Semin Dial 2007;20:554-560.

[99] Pietzsch J, Lattke P, Julius U. Oxidation of apolipoprotein B-100 in circulating LDL is related to LDL residence time. In vivo insights from stable-isotope studies. Arterioscler Thromb Vasc Biol 2000;20:E63-67.

[100] Boaz M, Smetana S, Weinstein T, et al. Secondary prevention with antioxidants of cardiovascular disease in endstage renal disease (SPACE): randomised placebo-controlled trial. Lancet 2000;356:1213-1218.

[101] Tepel M, van der Giet M, Statz M, Jankowski J, Zidek W. The antioxidant acetylcysteine reduces cardiovascular events in patients with end-stage renal failure: a randomized, controlled trial. Circulation 2003;107:992-995.

[102] Shoji T, Nishizawa Y, Kawagishi T, et al. Intermediate-density lipoprotein as an independent risk factor for aortic atherosclerosis in hemodialysis patients. J Am Soc Nephrol 1998;9:1277-1284.

[103] Vega GL, Grundy SM. Influence of lovastatin therapy on metabolism of low density lipoproteins in mixed hyperlipidaemia. J Intern Med 1991;230:341-350.

[104] Hörkkö S, Huttunen K, Korhonen T, Kesäniemi YA. Decreased clearance of low-density lipoprotein in patients with chronic renal failure. Kidney Int 1994;45:561-570.

[105] Hörkkö S, Huttunen K, Kesäniemi YA. Decreased clearance of low-density lipoprotein in uremic patients under dialysis treatment. Kidney Int 1995;47:1732-1740.

[106] Prinsen BH, Rabelink TJ, Romijn JA, et al. A broad-based metabolic approach to study VLDL apoB100 metabolism in patients with ESRD and patients treated with peritoneal dialysis. Kidney Int 2004;65:1064-1075.

[107] Chan PC, Persaud J, Varghese Z, Kingstone D, Baillod RA, Moorhead JF. Apolipoprotein B turnover in dialysis patients: its relationship to pathogenesis of hyperlipidemia. Clin Nephrol 1989;31:88-95.

[108] Ericsson S, Eriksson M, Vitols S, Einarsson K, Berglund L, Angelin B. Influence of age on the metabolism of plasma low density lipoproteins in healthy males. J Clin Invest 1991;87:591-596.

[109] Millar JS, Lichtenstein AH, Cuchel M, et al. Impact of age on the metabolism of VLDL, IDL, and LDL apolipoprotein B-100 in men. J Lipid Res 1995;36:1155-1167.

[110] Quaschning T, Koniger M, Kramer-Guth A, et al. Receptor-mediated lipoprotein uptake by human glomerular cells: comparison with skin fibroblasts and HepG2 cells. Nephrol Dial Transplant 1997;12:2528-2536.

[111] Pegoraro AA, Gudehithlu KP, Cabrera E, et al. Handling of low-density lipoprotein by the renal tubule: release of fragments due to incomplete degradation. J Lab Clin Med 2002;139:372-378.

[112] Kronenberg F, Trenkwalder E, Lingenhel A, et al. Renovascular arteriovenous differences in Lp[a] plasma concentrations suggest removal of Lp[a] from the renal circulation. J Lipid Res 1997;38:1755-1763.

[113] Kostner KM, Maurer G, Huber K, et al. Urinary excretion of apo(a) fragments. Role in apo(a) catabolism. Arterioscler Thromb Vasc Biol 1996;16:905-911.

[114] Mooser V, Seabra MC, Abedin M, Landschulz KT, Marcovina S, Hobbs HH. Apolipoprotein(a) kringle 4-containing fragments in human urine. Relationship to plasma levels of lipoprotein(a). J Clin Invest 1996;97:858-864.

[115] Kaysen GA, de Sain-van der Velden MG. New insights into lipid metabolism in the nephrotic syndrome. Kidney Int Suppl 1999;71:S18-21.

[116] Giordano M, De Feo P, Lucidi P, et al. Increased albumin and fibrinogen synthesis in hemodialysis patients with normal nutritional status. J Am Soc Nephrol 2001;12:349-354.

[117] Kaysen GA. Albumin turnover in renal disease. Miner Electrolyte Metab 1998;24:55-63.

[118] Shoji T, Fukumoto M, Kimoto E, et al. Antibody to oxidized low-density lipoprotein and cardiovascular mortality in end-stage renal disease. Kidney Int 2002;62:2230-2237.

[119] Randomised trial of cholesterol lowering in 4444 patients with coronary heart disease: the Scandinavian Simvastatin Survival Study (4S). Lancet 1994;344:1383-1389.

[120] Cannon CP, Braunwald E, McCabe CH, et al. Intensive versus moderate lipid lowering with statins after acute coronary syndromes. N Engl J Med 2004;350:1495-1504.

[121] Fried LF, Orchard TJ, Kasiske BL. Effect of lipid reduction on the progression of renal disease: a meta-analysis. Kidney Int 2001;59:260-269.

[122] Nishizawa Y, Shoji T, Tabata T, Inoue T, Morii H. Effects of lipid-lowering drugs on intermediate-density lipoprotein in uremic patients. Kidney Int Suppl 1999;71:S134-136.

[123] Seliger SL, Weiss NS, Gillen DL, et al. HMG-CoA reductase inhibitors are associated with reduced mortality in ESRD patients. Kidney Int 2002;61:297-304.

[124] Fellstrom BC, Jardine AG, Schmieder RE, et al. Rosuvastatin and cardiovascular events in patients undergoing hemodialysis. N Engl J Med 2009;360:1395-1407.

[125] Holdaas H, Holme I, Schmieder RE, et al. Rosuvastatin in diabetic hemodialysis patients. J Am Soc Nephrol 2011;22:1335-1341.

[126] Wanner C, Krane V, Marz W, et al. Atorvastatin in patients with type 2 diabetes mellitus undergoing hemodialysis. N Engl J Med 2005;353:238-248.

[127] Chan KE, Thadhani R, Lazarus JM, Hakim RM. Modeling the 4D Study: statins and cardiovascular outcomes in long-term hemodialysis patients with diabetes. Clin J Am Soc Nephrol 2010;5:856-866.

[128] Krane V, Wanner C. Statins, inflammation and kidney disease. Nat Rev Nephrol 2011;7:385-397.

[129] Baigent C, Landray MJ, Reith C, et al. The effects of lowering LDL cholesterol with simvastatin plus ezetimibe in patients with chronic kidney disease (Study of Heart and Renal Protection): a randomised placebo-controlled trial. Lancet 2011;377:2181-2192.

[130] Epstein M, Vaziri ND. Statins in the management of dyslipidemia associated with chronic kidney disease. Nat Rev Nephrol 2012;8:214-223.

Diagnosis of Lipoprotein Disorders

A Non-Atherogenic and Atherogenic Lipoprotein Profile in Individuals with Dyslipoproteinemia

Stanislav Oravec, Johannes Mikl, Kristina Gruber and Elisabeth Dostal

Additional information is available at the end of the chapter

1. Introduction

The 1985 Nobel Prize in Medicine was awarded to American lipidologists Goldstein and Brown for their work in identifying the role of the LDL receptor pathway in lipoprotein metabolism and in maintaining the homeostasis of blood cholesterol (Goldstein & Brown 1985).

The discovery of the LDL receptor and an understanding of its role in lipid metabolism in health and illness were a milestone in research into metabolic disorders in lipids. At the same time, some other successes in lipoprotein research were also reported: a new understanding of the role of oxidized LDL in atherosclerosis pathogenesis (Steinberg 1987, Witztum & Steinberg 1991); an update of the Ross theory on atherosclerosis genesis (Ross 1986); studies with hypolipidemics; a cholestyramine study, the Coronary Drug Project with niacin, and the Helsinki Heart Study with gemfibrosil. The next two decades was devoted to the effort to create sophisticated criteria for determining risk groups in populations, developing a consensus about cholesterol, and adopting pharmacological uniformity to achieve so-called target lipid values in at-risk individuals with dyslipidemia. The well-defined criteria as a result of these efforts gave hope to at-risk individuals for longer-term survival without ischemic vascular accidents (Canner *et al.* 1986, Frick *et al.* 1987, Expert panel 2001).

Generally, it was confirmed that hypercholesterolemia represents a risk factor for the development of cardiovascular diseases. In addition to arterial hypertension and nicotine abuse, hypercholesterolemia is considered one of three cardinal risk factors.

Cholesterol in plasma is transported by a sophisticated lipoprotein complex system and is also an active part of this lipoprotein system. Different parts of the lipoprotein system are called lipoprotein families. Every lipoprotein family transports different concentrations of cholesterol in blood plasma, but the major conveyor of cholesterol in plasma is the family of Low Density Lipoproteins, i.e., the LDL family. LDL is considered an atherogenic part of the lipoprotein system (Kwiterovich 2002a, 2002b).

LDL transports a major cholesterol load from the liver to the peripheral cells of the body. Under conditions of impaired LDL catabolism in the periphery, LDL particles persist in the circulation, their physical-chemical characteristics are modified, and the physiological pathway of LDL degradation - via LDL receptors - fails. The consequence of this sequence of events is the formation of an alternative metabolic pathway of LDL degradation through scavenger receptors and the formation of cholesterol deposits in the subendothelial space of the arterial wall. In this way, the process of atherogenesis and atherothrombosis begins; and LDL particles play a crucial role at the beginning and in the development of this injury process in the vessel walls (Berneis & Krauss 2002, Haffner 2006, Fruchart et al. 2008).

LDL-cholesterol became a criterion for the degree of atherogenic risk for the development of atherothrombosis. A high LDL-cholesterol concentration in plasma correlates positively with the premature onset of cardiovascular diseases, and is considered a strong cardiovascular risk factor. From this point of view, the aim of treatment of hypercholesterolemia, in secondary as well as in primary prevention, is the reduction of LDL concentration in plasma and a lowering of the cholesterol level to the "target reference values" (Expert panel 2001, Backers 2005).

However, in the last few decades, lipoprotein research has focused on the phenomenon of atherogenic and non-atherogenic lipoproteins, atherogenic and non-atherogenic lipoprotein profiles, and on the phenotype A vs. phenotype B characterization (Austin et al. 1990, Chait et al. 1993, Van et al. 2007). The traditional approach to hypercholesterolemia as an atherogenic risk factor for the development of degenerative diseases of the cardiovascular system became a target of criticism. Castelli published evidence that more than 75 percent of patients with an acute coronary syndrome or a myocardial infarction had normal plasma values of cholesterol, LDL cholesterol and/or HDL cholesterol (Castelli 1988, 1992, 1998). Thus, it was necessary to look for other risk factors in plasma, at levels that could cause an acute coronary event. An increased cholesterol level, as an universal explanation for the origin of atherogenesis, was no longer valid.

A reasonable explanation was found in atherogenic lipoprotein subpopulations, the presence of which in plasma, even in very low concentrations, could impair the integrity of the vessel wall and lead to endothelial dysfunction with its fatal consequences: formation of atherothrombotic plaques, acute myocardial infarction, stroke, and sudden death (Nichols & Lundmann 2004, Rizzo & Berneis 2006, Shoji et al. 2009, Zhao et al. 2009).

Those laboratory analysis methods became an essential contribution to the identification of atherogenic lipoprotein entities, which simplified the analysis and quantification of the atherogenic lipoprotein subfractions. Gradient gel elecrophoretic separation of LDL and HDL subclasses or proton nuclear magnetic resonance spectroscopy were the methods of choice for the analysis of these entities (Rainwater et al. 1997, Alabakovska et al. 2002, Otvos et al. 2003).

Recently, electrophoresis of plasma lipoproteins on the polyacrylamide gel (PAG) Lipoprint LDL System is one of several diagnostic analytical methods for the identification and quantitative evaluation of lipoprotein subfractions, i.e., the atherogenic and non-atherogenic lipoproteins (Hoefner et al. 2001).

The LDL System has become a staple in routine laboratory analysis and in the diagnosis of lipoprotein metabolism disorders, and has also been recommended by the FDA for human medicine. Lipoprint LDL enables the analysis of 12 lipoprotein subfractions: VLDL; IDL1; IDL2; IDL3; LDL1; LDL2; LDL 3-7; HDL; and determines an atherogenic lipoprotein profile phenotype B versus a non-atherogenic lipoprotein profile phenotype A.

Atherogenic lipoprotein profiles are characterized by a predominance of atherogenic lipoproteins, namely very low density (VLDL), intermediate density IDL1, and IDL2, and particularly by the presence of small dense lipoproteins with low density (LDL). Profiles identify highly atherogenic LDL subfractions that form the LDL 3-7 fractions (Tab.1). These subfractions are smaller, with a diameter < 26.5 nm (265 Angström) and they float within a density range of 1.048 – 1.065 g/ml, i.e., a higher density than LDL1 and LDL2. On the PAG they are detected as subtle bands on the anodic end of the gel right behind HDL that migrate to the head of separated lipoproteins.

Small dense LDL are highly atherogenic for ((Berneis&Krauss 2002, Lamarche et al. 1999, Packard 2003, Carmena et al. 2004):	
*low recognition by LDL-receptors (configuration alterations Apo B) →	
*enhanced aptitude for oxidation and acetylation →	
*Oxid-LDL	→ release of pro-inflammatory cytokines → muscle cell apoptosis
*Oxid-LDL	→ release of metalloproteinase → collagen degradation
*Oxid-LDL	→ enhanced aptitude for trapping by macrophages (scavenger-receptors) → stimulation of foam cell formation
*easier penetration into the subendothelial space and formation of cholesterol deposits	

Table 1.

On the basis of lipoprotein separation by the Lipoprint LDL System, a non-atherogenic normolipidemia, an atherogenic normolipidemia, a non-atherogenic hyperbetalipoproteinemia and an atherogenic hyperlipoproteinemia can be characterized (Oravec 2006a, 2006b, 2007a, 2007b).

Two of these are identified as new lipoprotein profiles with high clinical significance: an **atherogenic normolipidemia** and a **non-atherogenic hyper-betalipoproteinemia LDL1,2.**

A non-atherogenic hyperbetalipoproteinemia LDL1,2 involves individuals with a high concentration of plasma cholesterol, predominantly transported by LDL1 and LDL2 subfractions. However, these individuals are at low risk for a cardiovascular event based on cardiologic and angiologic examimation results, and have familial history negative for cardiovascular diseases.

Conversely, an atherogenic normolipidemia was identified in a group of individuals with normal cholesterol and triglyceride concentrations in plasma, who had a high concentration of strongly atherogenic small dense LDL in the lipoprotein profile. These individuals could be at higher risk for a cardiovascular event despite normolipidemia.

In our clinical study, we characterized hypercholesterolemic individuals with untreated hypercholesterolemia, who had a non-atherogenic hyperbetalipoproteinemia, as well as normolipemic individuals who were currently without clinical or laboratory signs of damage to the cardiovascular system, but who, nevertheless, had an atherogenic lipoprotein profile. All these subjects underwent a medical examination to identify the extent of the arterial vessel damages caused by hypercholesterolemia, or dyslipidemia.

2. Patients and methods

The hypercholesterolemic individuals with untreated hypercholesterolemia were tested by Lipoprint LDL analysis. In this group of hypercholesterolemic subjects, 145 individuals with a non-atherogenic lipoprotein profile were identified.

Of the total number, 15 individuals were under 40 years of age without clinically apparent impairment and no laboratory signs of cardiovascular disease. These subjects formed one subgroup of younger people (34 years +- 5 years). The subgroup of younger subjects was separated from the older individuals with hypercholesterolemia because a separate analysis of the older subjects with hypercholesterolemia was performed to confirm that undamaged vessels in older individuals persist even into old age, and that diagnosed hypercholesterolemia does not cause an atherogenic impairment in the vessels. The subgroup of older subjects consisted of 130 individuals (32 males, 57 +-11 years of age; and 98 females, 62 +- 9 years of age).

The medical examination, which included a physical examination, blood pressure, and ECG examination, bicycle stress test, echocardiography, and duplex ultrasound examination of the carotid arteries, confirmed that there was no impairment of the cardiovascular system. Only mild signs of clinically irrelevant aortic valve sclerosis were found in the subgroup of older subjects.

Individuals with hyperglycemia, diabetics, and those individuals who were being treated with lipid-lowering drugs were excluded from the study.

The control group consisted of 165 normolipidemic volunteers, all nonsmokers, who had no clinically apparent impairment, or laboratory signs of cardiovascular disease. Volunteers were recruited from medical students at the medical facility. The average age of the subjects was 21.5 ± 2.5 years, and the group involved 65 males and 100 females. All subjects gave written, informed consent, and the study was approved by the local ethics committee.

A blood sample from the antecubital vein was collected in the morning after a 12-hour fasting period. EDTA-K$_2$ plasma was obtained and the concentration of total cholesterol and triglycerides in plasma was analyzed, using an enzymatic CHOD PAP method (Roche Diagnostics, Germany).

The quantitative analysis of lipoprotein families and lipoprotein subfractions included: VLDL; IDL1; IDL2; IDL3; LDL1; LDL2; LDL3-7; and HDL. A non-atherogenic lipoprotein profile, phenotype A, versus an atherogenic lipoprotein profile, phenotype B, was determined using the Lipoprint LDL System (Quantimetrix Corp., USA; (Hoefner *et al.* 2001). The analysis of HDL subclasses, with their subpopulations, including large HDL-, intermediate HDL-, and small HDL- subclasses in plasma, was also performed using the Lipoprint HDL System (Morais *et al.* 2003).

The Score of the Anti-Atherogenic Risk (SAAR) was calculated as the ratio between non-atherogenic and atherogenic lipoproteins in plasma (Oravec 2007a). SAAR values over 10.8 represented a non-atherogenic lipoprotein profile, whereas values under 9.8 represented an atherogenic lipoprotein profile. The cut-off values for a non-atherogenic lipoprotein profile and an atherogenic lipoprotein profile were calculated from the results of 940 Lipoprint LDL analyses. Using the Quantimetrix Lipoprint LDL system interpretation, all 940 individuals were examined (general group of subjects) and tested for the occurrence of atherogenic and non-atherogenic lipoprotein profiles, and were then divided into the two subgroups of subjects with an LDL profile:

• Indicative of Type A, i.e., a non-atherogenic lipoprotein profile phenotype A
• Not indicative of Type A, i.e., an atherogenic lipoprotein profile, phenotype B (Hoefner *et al.* 2001).

For practical interpretation of the analysed lipoprotein profiles using the Lipoprint LDL System, for the non-atherogenic lipoprotein phenotype A, the subtypes 1a, 1b, 2a, 2b, 3, and 4 were introduced, because of the large profile heterogeneity in the non-atherogenic lipoprotein profile phenotype A. With regard to the atherogenic lipoprotein phenotype B, only subtype 5 and subtype 6 were introduced. (Oravec 2007b). (Tab.2)

Statistical evaluation of obtained values was performed by an unpaired student's t-test. The level of significance was set at $p < 0.05$.

3. Results

The subjects with a non-atherogenic hypercholesterolemia had a significantly increased concentration of total cholesterol and lipoprotein parameters ($p<0.0001$), except for LDL 3-7 subfractions (small dense LDL), which were significantly lower ($p<0.0001$), compared to the control group (Tab.3). The highest increase of concentrations was found for total cholesterol, LDL cholesterol, HDL cholesterol, IDL3, and LDL1 subfractions. The concentration of LDL1 exceeded the LDL1 concentration in the control group by more than 88 percent. The LDL1 concentration in the younger hypercholesterolemic subjects reached 1.84 mmol/l, i.e., more than twice, comparing to 0.89 mmol/l in the control group. (Tab.3, Tab.5). The rise of LDL2 concentration (32 percent in younger hypercholesterolemic subjects), did not match the increase in LDL1 concentrations (Tab.3 - Tab.6).

A. Non-atherogenic lipoprotein profile, phenotype A ......   **59 %**

1a. Subtype: Non-atherogenic lipoprotein profile phenotype A.....   11 %
    Atherogenic lipoproteins absent
    LDL cholesterol normal

1b. Subtype: Non-atherogenic lipoprotein profile phenotype A.....   10 %
    Atherogenic lipoproteins absent
    LDL cholesterol elevated

2a. Subtype: Non-atherogenic lipoprotein profile phenotype A......   12%
    Atherogenic lipoproteins present in traces
    LDL cholesterol normal

2b. Subtype: Non-atherogenic lipoprotein profile phenotype A...   11%
    Atherogenic lipoproteins present in traces
    LDL cholesterol elevated

3. Subtype: Non-atherogenic lipoprotein profile phenotype A.......   3%
    Atherogenic lipoproteins present
    LDL cholesterol normal

4. Subtype: Non-atherogenic lipoprotein profile phenotype A 12%
    Atherogenic lipoproteins present
    LDL cholesterol elevated

B. Atherogenic lipoprotein profile phenotype B ....   **41 %**

5. Subtype: Atherogenic lipoprotein profile phenotype B...............   12%
    Atherogenic lipoproteins present
    LDL cholesterol normal

6. Subtype: Atherogenic lipoprotein profile phenotype B..................   29%
    Atherogenic lipoproteins present
    LDL cholesterol elevated

An atherogenic lipoprotein profile was identified in 41% of examined individuals in a general group of subjects (n = 940), (Oravec 2007b).

Table 2. Incidence rate of non-atherogenic vs. atherogenic lipoprotein subtypes in a general group of subjects (n = 940)

T-Chol (mmol/l ±SD)	TAG	VLDL	IDL1	IDL2	IDL3	LDL1	LDL2	LDL3-7	T-LDL	HDL	SAAR
Control 4.31	1.16	0.62	0.39	0.28	0.33	0.89	0.41	0.04	2.34	1.33	36.1
n = 165 ±0.62	±0.39	±0.16	±0.16	±0.09	±0.12	±0.28	±0.21	± 0.04	±0.54	±0.32	±20.6
H-βLP 6.71	1.29	0.74	0.55	0.51	0.82	1.68	0.52	0.01	4.09	1.88	76.0
n= 145 ±0.90	±0.49	±0.21	±0.16	±0.12	±0.23	±0.36	±0.21	±0.01	±0.69	±0.46	±17.0
Control vs. HLP p< 0.0001	n.s.	<..p< 0.0001..>									

Legend: T-cholesterol: total cholesterol, T-LDL: total LDL-cholesterol, H-βLP: hyperbetalipoproteinemia

Table 3. Plasma concentration of lipids, lipoproteins, and SAAR score in the group of hypercholesterolemic subjects vs. control normolipidemic subjects

T-Chol (mmol/l ±SD)	TAG	VLDL	IDL1	IDL2	IDL3	LDL1	LDL2	LDL3-7	T-LDL	HDL	SAAR
Control 4.31	1.16	0.62	0.39	0.28	0.33	0.89	0.41	0.04	2.34	1.33	36.1
n = 165 36.1	±0.39	±0.16	±0.16	±0.09	±0.12	±0.28	±0.21	± 0.04	±0.54	±0.32	±20.6
H-βLPs 6.73	1.30	0.73	0.55	0.52	0.80	1.67	0.52	0.01	4.08	1.93	76.5
n= 130 ±0.91	±0.48	±0.19	±0.16	±0.13	±0.23	±0.35	±0.22	±0.01	±0.69	±0.45	±18.1
Control vs. HLP p< 0.0001	n.s.	<..p< 0.0001..>									

Legend: H-βLPs : hyperbetalipoproteinemia subgroup of seniors

Table 4. Plasma concentration of lipids, lipoproteins, and SAAR score in the subgroup of older hypercholesterolemic subjects and controls

T-Chol (mmol/l ±SD)	TAG	VLDL	IDL1	IDL2	IDL3	LDL1	LDL2	LDL3-7	T-LDL	HDL	SAAR
Control 4.31	1.16	0.62	0.39	0.28	0.33	0.89	0.41	0.04	2.34	1.33	36.1
n = 165 ±0.62	±0.39	±0.16	±0.16	±0.09	±0.12	±0.28	±0.21	± 0.04	±0.54	±0.32	±20.6
H-βLPjr 6.62	1.20	0.84	0.58	0.44	0.80	1.84	0.54	0.01	4.20	1.46	71.1
n= 15 ±0.80	±0.59	±0.28	±0.18	±0.01	±0.25	±0.42	±0.18	±0.01	±0.64	±0.23	±13.2
Control vs. HLP p< 0.0001	n.s.	<..................p< 0.0001..................>			n.s.	n.s.		p< 0.0001	n.s.	p< 0.0001

Legend: H-βLP jr : hyperbetalipoproteinemia subgroup of younger hypercholesterolemic subjects

Table 5. Plasma concentration of lipids, lipoproteins, and SAAR-score in the subgroup of younger hypercholesterolemic subjects and controls

T-Chol (mmol/l ±SD)	TAG	VLDL	IDL1	IDL2	IDL3	LDL1	LDL2	LDL3-7	T-LDL	HDL	SAAR
H-βLP jr 6.62	1.20	1.20	0.58	0.44	0.80	1.84	0.54	0.01	4.20	1.46	71.1
n= 15 ±0.80	±0.59	±0.28	±0.18	±0.01	±0.25	±0.42	±0.18	±0.01	±0.64	±0.23	±13.2
H-βLPs 6.73	1.30	0.73	0.55	0.52	0.80	1.67	0.52	0.01	4.08	1.93	76.5
n= 130 ±0.91	±0.48	±0.19	±0.16	±0.13	±0.23	±0.35	±0.22	±0.01	±0.69	±0.45	±18.1

<.. n.s. ....> p< 0.001 n.s.

juniors v.s. seniors

Legend: H-βLP jr.: Hyperlipoproteinemia subgroup of younger subjects
H-βLP s.: Hyperlipoproteinemia subgroup of older subjects

Table 6. Plasma concentration of lipids, lipoproteins, and SAAR-score in the subgroup of younger (n=15) versus older (n=130) hypercholesterolemic subjects

The lipid and lipoprotein parameters in younger and older hypercholesterolemic subjects were very similar, and the results were not statistically significantly different between the groups, except that HDL cholesterol in the older hypercholesterolemic individuals was statistically significant higher ($p < 0.001$) compared to the control group (Tab.6). Results similar to those in older hypercholesterolemic subjects were obtained when the group of younger hypercholesterolemic subjects was compared to the control group (Tab.5), except for LDL2, LDL 3-7, and HDL lipoproteins, where the changes in the cholesterol concentrations - increased in LDL2- and decreased in LDL3-7 subfractions were not significant.

	T-HDL mmol/l ± SD	HDL large	HDL intermediate	HDL small
Control	1.31	0.59	0.56	0.15
(n=103)	± 0.29	± 0.23	± 0.10	± 0.09
H-βLP	1.51	0.70	0.65	0.15
(n=110)	± 0.34	± 0.46	± 0.42	± 0.12
	p< 0.0001	p< 0.005	p< 0.005	n.s.

Legend: T-HDL: total HDL

Table 7. Plasma concentration of HDL lipoprotein subclasses

Tab.7 shows the HDL-cholesterol concentration and HDL subclasses, analysed by the Lipoprint HDL System. The concentration of total HDL cholesterol (T-HDL) in the group of hypercholesterolemic subjects was significantly higher ($p < 0.0001$), compared to the control group. There was an increased concentration of both HDL subclasses, i.e. the HDL large subclass ($p < 0.005$) and the HDL intermediate subclass ($p < 0.005$) in the hypercholesterolemia subjects. The difference in the concentration of the HDL small subclass between hypercholesterolemic subjects and the control group was not confirmed.

	T-Chol (mmol/l ±SD)	TAG	VLDL	IDL1	IDL2	IDL3	LDL1	LDL2	LDL3-7	T-LDL	HDL	SAAR
Subjects with a non atherogenic profile, n = 155	4.31 ±0.62	1.12 ±0.38	0.62 ±0.16	0.39 ±0.17	0.28 ±0.09	0.33 ±0.12	0.91 ±0.27	0.40 ±0.21	0.03 ±0.03	2.33 ±0.54	1.33 ±0.32	38.1 ±19.6
Subjects with an atherogenic profile, n = 10	4.37 ±0.50	1.63 ±0.30	0.72 ±0.14	0.36 ±0.08	0.28 ±0.06	0.27 ±0.08	0.67 ±0.17	0.55 ±0.14	0.25 ±0.06	2.37 ±0.34	1.27 ±0.36	5.3 ±2.0
All subjects n = 165	**4.31** ±0.62	**1.16** ±0.39	**0.62** ±0.16	**0.39** ±0.16	**0.28** ±0.09	**0.33** ±0.12	**0.89** ±0.28	**0.41** ±0.21	**0.04** ± 0.04	**2.34** ±0.54	**1.33** ±0.32	**36.1** ±20.6
nonath.vs.athero		$p<0.001$					$p< 0.01$	$p< 0.02$	$p< 0.0001$			$p< 0.0001$

Table 8. Plasma concentration of lipids, lipoproteins, and the SAAR score in the subgroups of normolipemic control volunteers

Tab.8 shows the lipid and lipoprotein values obtained and the Score for Anti-Atherogenic Risk (SAAR) in the examined group of 165 control subjects.

In a subgroup of 155 subjects, a non-atherogenic lipoprotein profile phenotype A was identified. In a subgroup of 10 subjects, an atherogenic lipoprotein profile phenotype B was identified. Both lipoprotein phenotypes were confirmed by the Lipoprint LDL method. All examined subjects had normal values of cholesterol and triglycerides. The highest significant difference ($p<0.0001$) between the subgroup with an atherogenic lipoprotein profile phenotype B and a non-atherogenic lipoprotein profile phenotype A was found in the subfractions LDL 3-7, i.e., small dense LDL ($p< 0.0001$), which represent strongly atherogenic lipoproteins. The SAAR score also showed highly significant differences in the values between the atherogenic and the non-atherogenic subgroup ($p<0.0001$). There was a higher concentration of triglycerides ($p<0.001$) in the atherogenic subgroup. LDL1 was higher in the non-atherogenic subgroup ($p<0.01$) and LDL2 was higher in the atherogenic subgroup.

4. Discussion

The identification of atherogenic and non-atherogenic lipoproteins in the plasma lipoprotein spectrum represents a deeper analysis of lipoprotein parameters than a routine analysis of plasma cholesterol, triglycerides, or lipoproteins like LDL, HDL, and VLDL. These lipid parameters only provide limited information about the percentage of subjects in the general population (general group of subjects) who are at-risk for a sudden attack for cardiovascular or cerebral-vascular event. The 41 percent of the subjects from our large population of 950 individuals, who were identified by this analytical method, would not otherwise have been identified, confirming the value of this information for physicians (Tab.2) know that, based on mortality statistics, approximately 50 percent of deaths are caused by cardiovascular events. It may be that this 41 percent represents a major part of the 50 percent of deaths attributable to a cardiovascular cause, and the individuals with atherogenic dyslipidemia are surely at risk for a sudden cardiovascular event. Thus these individuals could be target for close monitoring, have a follow-up examination, and the optimal treatment could be recommended.

In addition, the identification of six percent of normolipidemic young healthy individuals with an atherogenic lipoprotein profile among clinically healthy volunteers questions our knowledge and generally accepted belief that normolipidemia, 'per se', represents an optimal health lipid constellation (Tab.8). An atherogenic normolipidemia in the lipoprotein profile of our clinically healthy subjects represents a new phenomenon. These individuals are also at risk for the development of premature cardiovascular ischemic disease and should undergo close medical follow-up. If these individuals receive no preventive anti-atherothrombotic measures, the manifestation of cardiovascular ischemic diseases is certain later in life.

The findings of hypercholesterolemia in clinically healthy subjects, without clinically apparent signs of cardiovascular disease or laboratory confirmation of cardiovascular disease, and with a negative history for the occurence of cardiovascular events, stimulated an active search for hypercholesterolemic indviduals and the initiation of a medical examination of these subjects.

For the identification of the hypercholesterolemic individuals with a non-atherogenic lipoprotein profile, a new innovative electrophoretic method for the analysis of plasma lipoproteins on polyacrylamide gel (PAG) was used (Hoefner *et.al* 2001). The method can analyze the total lipoprotein spectrum of examined subjects, identify an atherogenic/non-atherogenic lipoprotein profile, and quantify the atherogenic lipoprotein subpopulations in plasma, including strongly atherogenic LDL subpopulations, i.e., the small dense LDL, which form the subfractions LDL 3-7. In the absence of atherogenic lipoproteins, or when the atherogenic lipoproteins form a minor part of the whole lipoprotein spectrum, a non-atherogenic lipoprotein profile exists.

The identification of a non-atherogenic hypercholesterolemia offers new information, which suggests a re-evaluation of the belief that the whole LDL family is an atherogenic lipoprotein part of the plasma lipoprotein spectrum. Our results confirme the results of several previous research studies. They show that only a part of the LDL is atherogenic. Atherogenic are small dense LDL, subfraction of LDL, which are associated with the premature development of ischemic cardiovascular diseases. In contrast, LDL1 and LDL2, even in higher concentrations in plasma, do not represent a high cardiovascular risk. Also negative cardiological examination with normal results: only milde signs of clinically irrelevant aortic valve sclerosis, support and confirm the non-atherogenicity of large 'buoyant' LDL subfractions in the individuals with hyperbetalipoproteinemia LDL1,2. Fig.1 - 4. Based on these laboratory results and medical findings, the medical approach to these hypercholesterolemic individuals needs to be revised. The intensive hypolipidemic treatment should not be recommended, and the question also remains, whether any treatment at all, in cases of non-atherogenic hypercholesterolemia, in general, is a reasonable clinical decision. The reduction of total LDL-cholesterol as a target for hypolipidemic treatment for prevention of atherogenesis and atherothrombosis seems to be no longer necessary.

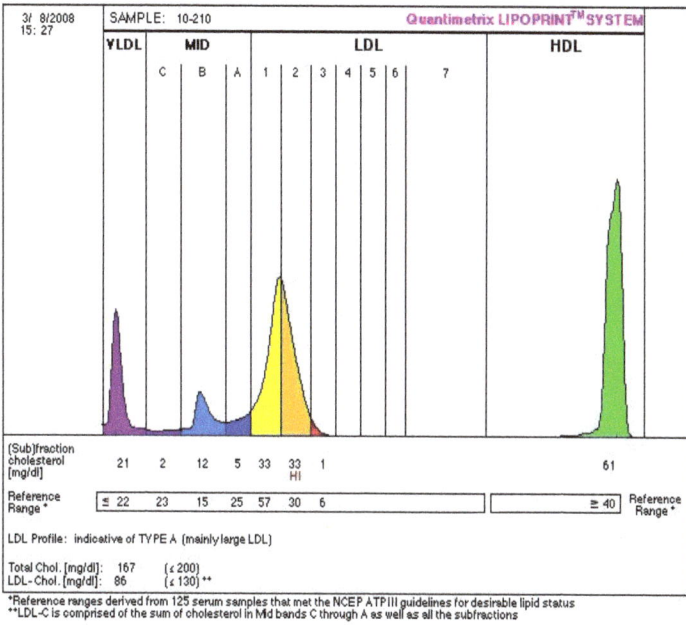

Figure 1. Non-atherogenic normolipidemia

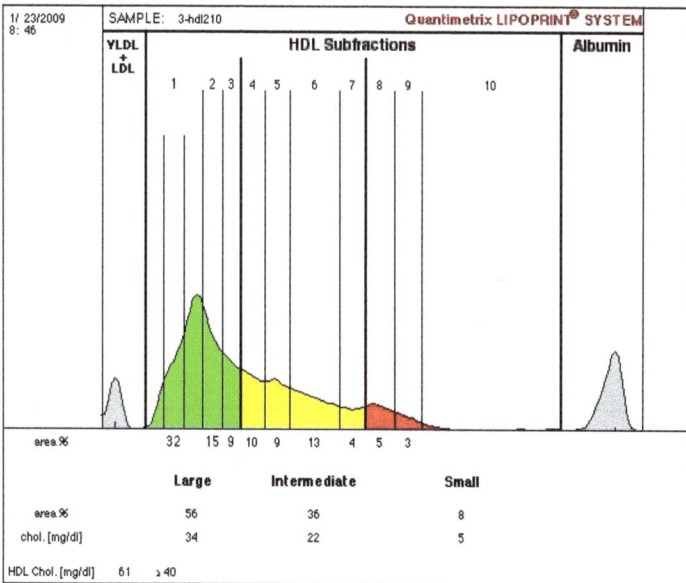

Figure 2. Non-atherogenic normolipidemia HDL subfractions

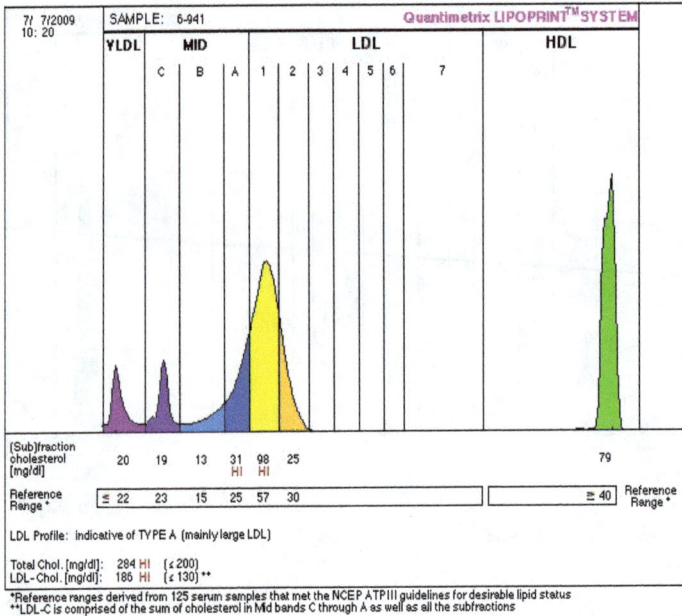

Figure 3. Non-atherogenic hyperbetalipoproteinemia LDL1,2

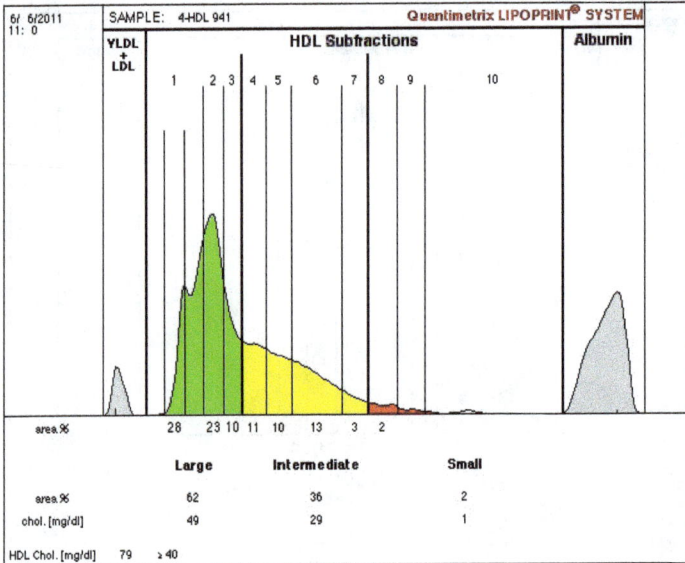

Figure 4. Non-atherogenic hyperbetalipoproteinemiaLDL1,2 HDL subfractions

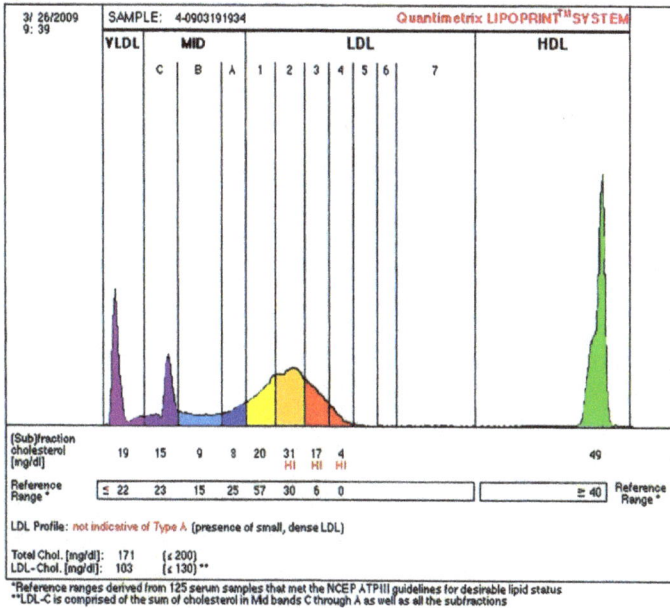

Figure 5. Atherogenic normolipidemia

LDL represent a lipoprotein family created by several LDL subfractions with different characteristics and different role in the intermediary metabolism and in the atherothrombogenesis. LDL1 and LDL2 subfractions are important physiological major conveyors of cholesterol in plasma. These subfractions are an important source for the biosynthesis of highly physiologically effective drugs and structures in the body (steroid hormones, bile acids, vitamin D3, membranes of cells and of subcellular structures). Lowering of the concentration of LDL1 and LDL2 by using a non-specific hypolipidemic treatment has a negative effect on several physiological processes, which create the optimal maintenance of healthy equilibrium in the body. LDL1 and LDL2 seem to be a not atherogenic part of LDL. The non-specific lowering of total cholesterol reduces in the first step the concentration of cholesterol in LDL1, LDL2 subfractions. A protective part of LDL (LDL1, LDL2) is removed and the strong atherogenic small dense LDL persist.

The non-specific hypolipidemic treatment does not form a non-atherogenic lipoprotein constellation. On the contrary, along with the impairment of endocrine steroid synthesis in the body, with an unjustified hypolipidemic treatment approach, the atherogenicity of the plasma will be increased. Figure 6 - 8 shows a Lipoprint LDL picture of atherogenic normolipidemia obtained frequently after hypolipidemic treatment of atherogenic hypercholesterolemia.

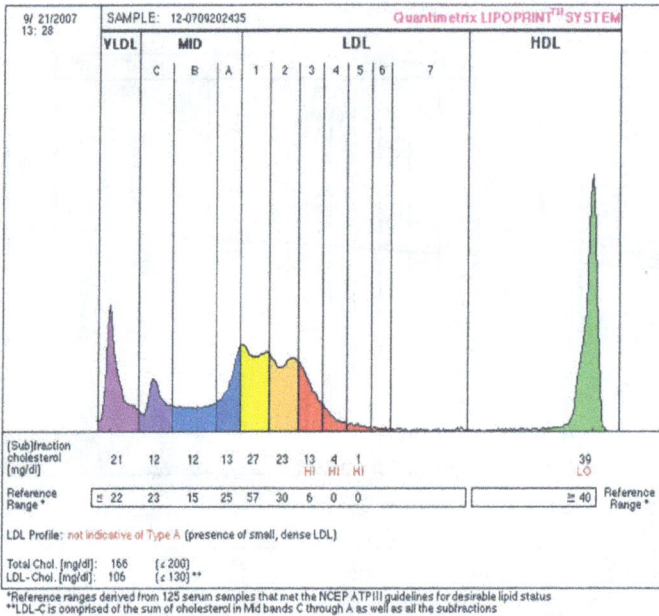

Figure 6. Atherogenic normolipidemia obtained frequently after hypolipidemic treatment of atherogenic hypercholesterolemia

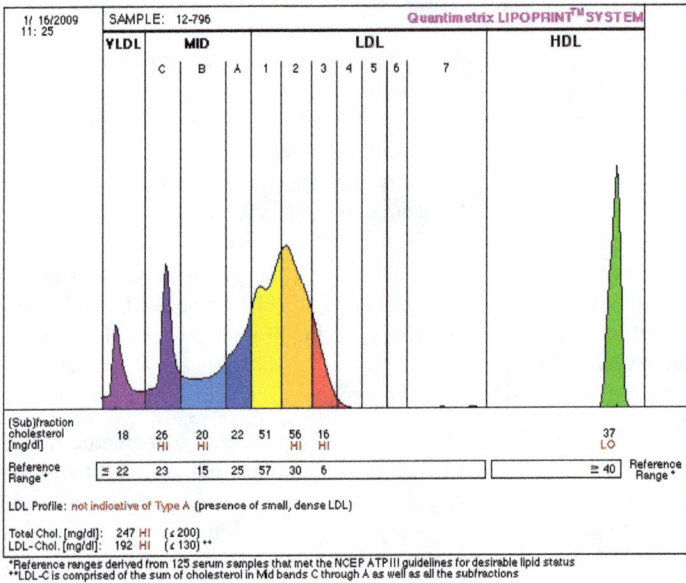

Figure 7. Atherogenic hypercholesterolemia

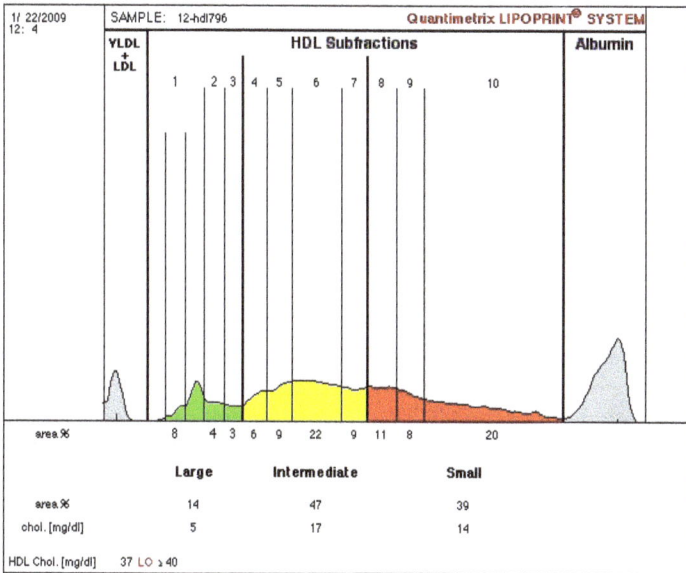

Figure 8. Atherogenic hypercholesterolemia HDL subfractions

In our study a group of individuals with hypercholesterolemia was divided into two subgoups: younger and older subjects (Tab.3-6). The reason was to differentiate the influence of the age factor on the lipoprotein constellation and on the quality of the vascular wall, especially in the group with older subjects. The quality of the arteries was evaluated by medical examination. Tested individuals were examined, including physical examination, blood pressure, and ECG examination, a bicycle stress test, echocardiography, and duplex ultrasound examination of the carotid arteries. The medical results confirmed that the vessel wall was not seriously impaired, not even in older subjects with hypercholesterolemia, which is why a hyper-betalipoproteinemiaLDL1,2 does not represent a serious cardiovascular risk for individuals with this type of hypercholesterolemia.

The results of HDL subclass analysis (Lipoprint HDL System (Morais *et al.* 2003) in individuals with a non-atherogenic hyperbetalipoproteinemia LDL1 confirm a supposition of low atherogenicity in hyperberalipoproteinemia LDL1,2 (Tab.7), The lipoprotein profile of HDL typically contains a predominance of HDL large and HDL intermediate subclasses, which confer a protective, anti-atherogenic effect on the vessel wall (Morais 2005, Muniz & Morais 2005 Oravec *et al.* 2011c). The small HDL subclass with atherogenic characteristics was present in the lipoprotein profile in low concentrations only, compared to the control group of healthy volunteers. Fig.3 – 4.

The major findings can be summarized as follows:

1. In examined subjects with hypercholesterolemia, a non-atherogenic lipoprotein profile, phenotype A was confirmed with a high concentration of LDL1 and LDL2 subfractions. In particular, the LDL1 subfraction was nearly double that of the LDL1 of the control group, and, in some individual cases, three times that of the control group average (Oravec et al. 2011b).

2. The lipoprotein electrophoresis confirmed only a trace concentration of LDL3-7 subpopulations (1mg LDL 3-7 cholesterol/dl, i.e., 0.0256 mmol/l). In the overwhelming majority of subjects (60%) indeed, there was an absence of the atherogenic LDL 3-7 in the lipoprotein profile of these subjects. (Plasma lipoprotein profiles for patients with confirmed cardiovascular disease are generally characterized by a high concentration of small dense LDL) (Kwiterovich 2000, Maslowska 2005, Oravec 2010, Oravec et al. 2010a, 2010b, Oravec et al. 2011a).

3. The concentration of HDL was significantly increased (p<0.0001) compared to the control group, with an overwhelming majority of the non-atherogenic HDL subpopulations, HDL large and HDL intermediate. The concentration of small dense HDL was not increased (Tab.7), Fig 1-4. Small dense HDL form an atherogenic part of the HDL lipoprotein spectrum, and their higher plasma concentration corelates with the development of cardiovascular diseases (Luc et al. 2002, St Pierre et al. 2005, Morais 2005, Muniz & Morais 2005, Oravec et al. 2011d), Fig.7,8. The structural representation of HDL subpopulations confirmed a non-atherogenic type of lipoprotein profiles in our examined group of hypercholesterolemic subjects.

4. The examined individuals, despite increased total cholesterol and LDL cholesterol values, were healthy, without apparent clinical signs of cardiovascular disease (angina pectoris, cardiac insufficiency, myocardial infarction, or other survived cardiovascular events). There is evidence that an optimal anti-atherogenic LDL profile (see the lipoprotein results) could actually have a vasoprotective effect in tested hypercholesterolemic individuals. Based on the present results, a further, more extensive study will continue to evaluate the Lipoprint electrophoretic method as a standard method for the diagnosis of cardiovascular risk, along with the standard tests now used (ECG examination, bicycle stress test, echocardiography, and duplex ultrasound examination of the carotid arteries).

5. The newly introduced SAAR, a ratio of non-atherogenic/atherogenic lipoproteins, also confirmed a non-atherogenic lipoprotein constellation in the plasma of hypercholesterolemic individuals (Oravec 2007a).

Based on the results of examined individuals with hypercholesterolemia, these conclusions can be drawn:

1. LDL1 and LDL2 do not fulfill the criteria of atherogenicity for lipoprotein entities that is usually ascribed to LDL lipoproteins.

2. LDL1 and LDL2 subfractons in hypercholesterolemic indidviduals, in our study group, created a non-atherogenic hypercholesterolemia - a non-atherogenic hyperbetalipo-

proteinemia LDL1,2 without the presence of atherogenic small dense LDL (or with traces only) that are typically associated with a high concentration of cardiovascular protective HDL subfractions in the plasma lipoprotein spectrum.

We report the existence of a newly described type of hypercholesterolemia, **a non-atherogenic hyperbetalipoproteinemia LDL 1,2**, characterized by a minimal onset of cardiovascular complications, even in those individuals who are not treated with hypolipidemic therapy.

The hypercholesterolemic subjects of the study group are still undergoing follow-up examinations.

4.1. Atherogenic normolipidemia

Generally, a normolipidemia is interpreted as an equilibrated state of lipoprotein metabolism, characterized by total cholesterol and triglyceride values within reference ranges. We know from clinical experience that patients with normolipidemia are better protected from development of cardiovascular diseases and degenerative vessel changes, a source of cardio-vascular disease.

In normolipidemia, of the goal is to create a non-atherogenic lipoprotein profile and to lower or eliminate the risk of atherosclerosis development and prevent the rise of an acute cardiovascular event. However, the existence of an atherogenic normolipidemia disproves the theory that normolipidemia provides protection against the development of atherosclerotic vessel impairment. A premature atherosclerosis development can be found even in young people, adolescents with the high risk (Backers 2005; Rizzo & Berneis 2006).

An atherogenic lipoprotein profile is characterized by the rich presence of atherogenic lipoproteins, very low density lipoprotein (VLDL), intermediate density lipoproteins (IDL1, IDL2), and especially, by the presence of small dense low-density lipoproteins (sdLDL), which form LDL 3-7 subfractions, and which are strongly atherogenic (Lamarche *et al.* 1997; Gardner et al. 1996; Rajman *et al.* 1996; Halle *et al.* 1998, Austin *et al.* 1994).

An analysis of the lipoprotein profile by the Lipoprint LDL system reveals a new lipoprotein composition in lipoprotein profile and focuses authors on a new clinical-diagnostic phenomenon: an **atherogenic normolipidemia**. Compared to the well known atherogenic dyslipidemia, or atherogenic hyperlipoproteinemia, this new **atherogenic normolipidemia** (Oravec *et al.* 2010; Oravec *et al.* 2011d) is not identifiable by common biochemical diagnostic analysis.

This phenomenon represents a serious cardiovascular risk for individuals with this profile, and these individuals at high cardiovascular risk are not currently identified, diagnosed, medically registered, or treated. The presence of an **atherogenic normolipidemia** enlarges the portion of the population at increased risk for a cardiovascular event, however these individuals at risk do not participate on the protective measures of primary cardiovascular prevention. Fig.5. Medical community does not know till now, that the individuals with an atherogenic normolipimia are at-risk individuals for the development of premature ischemic cardiovascular diseases. Identification of the type of lipoprotein profile (atherogenic vs. non-atherogenic) by this innovative electrophoretic method for lipoprotein analysis in plasma represents a beneficial contribution to actual lipid diagnostics. This system provides the analysis of lipoprotein parameters but also offers new interpretation for lipoprotein profiles, including an actual framework of the practising scheme for diagnostics and treatment of dyslipidemias.

The Score of Anti-Atherogenic risk SAAR, newly introduced parameter, a ratio of non-atherogenic/atherogenic lipoproteins, also confirmes atherogenic lipoprotein constellation and determines the degree of the atherogenic risk of subjects with atherogenic normolipidemia (Oravec 2007a; Oravec 2007b; Oravec 2010).

5. Summary

A new method of electrophoretic lipoprotein separation on polyacrylamide gel (PAG)using the Lipoprint LDL System can quantify non-atherogenic and atherogenic plasma lipoproteins, including small dense LDL, i.e. strong atherogenic lipoprotein subpopulations.

With respect to the predominance of a non-atherogenic or atherogenic lipoproteins in thewhole lipoprotein profile, this method distinguishes a non-atherogenic lipoprotein profilephenotype A from an atherogenic lipoprotein profile phenotype B.

The contribution of this method is to confirm the existence of a non-atherogenic type of hyper-betalipoproteinaemia and the existence of normolipidemia with atherogenic lipoprotein profile, along with the common and well-known atherogenic hyperlipoproteinemia and non-atherogenic normolipidemia.

According to our preliminary analysis of a normolipidemic population, an atherogenic lipoprotein profile was revealed in 6% of normolipidemic young healthy individuals.

More than 40% of the examined individuals in the general group of subjects had an atherogenic lipoprotein profile phenotype B. These people represent an at-risk population.

However, the tools by which is possible to identify these individuals at risk for a cardiovascular event are limited.

A non-atherogenic hyperbetalipoproteinemiaLDL1,2 can be identified, which represents approxmately 20% of examined individuals with hypercholesterolemia and 10% of individuals in a general group of subjects. HyperbetalipoproteinemiaLDL1,2 is not associated with the premature development of arterial vascular impairment.

Author details

Stanislav Oravec
2nd Department of Internal Medicine,
Faculty of Medicine, Comenius University, Bratislava, Slovak Republic

Johannes Mikl
Department of Cardiology, Hietzing Hospital, Austria

Kristina Gruber
Department of Internal Medicine, Landesklinikum Thermenregion Baden, Austria

Elisabeth Dostal
Krankenanstalten Dr. Dostal, Vienna, Austria

Acknowledgement

This study was supported by an EU structural research fund Interreg III AT-SR, project code: 1414-02-000-28 in years 2006-2008.

We would like to acknowledge the excellent technical assistance of MTA Barbara Reif, MTA Judith Trettler and MTA Karin Waitz, Krankenanstalten Dr.Dostal, Vienna, Austria and also to acknowledge the excellent technical assistance of MTA Olga Reinoldova, 2nd Department of Internal Medicine, Faculty of Medicine, Comenius University, Bratislava, Slovakia

6. References

[1] Goldstein JL, Brown MS. Receptor mediated endocytosis:concepts emerging from the LDL-receptor system. Ann Rew Cell Biol 1985; 1: 1-39

[2] Steinberg D. Lipoproteins and the pathogenesis of atherosclerosis. Circulation 1987; 76: 504 - 7

[3] Witztum JL, Steinberg D. Role of oxidized low density lipoprotein in atherosclerosis. J Clin Invest 1991; 84: 1086 - 95

[4] Ross R. The pathogenesis of atherosclerosis – an update. N Engl J Med 1986; 314: 365 - 374

[5] Canner PL, Berge KG, Wenger NK, Stamler J, Friedman L, Prineas RJ et al. Fifteen year mortality in Coronary Drug Project patients, long term benefit with niacin. J Amer Coll Cardiol 1986; 8: 1245 - 55

[6] Frick MH, Elo O, Haapa K, Heinonen OP, Heinsalmi P, Helo P, Huttunen JK, Kaitaniemi P, Koskinen P, Manninen V et al. Helsinki Heart Study: primary prevention trial with gemfibrozil in middle aged men with dyslipidemia. N Engl J Med 1987; 317: 1237 - 45

[7] Kwiterovich PO. Clinical Relevance of the Biochemical, Metabolic and Genetic Factors that influence Low density Lipoprotein Heterogeneity. Am J Card 2002; 90 (Suppl 8A): 30i-48i

[8] Kwiterovich PO. Lipoprotein Heterogeneity: Diagnostic and Therapeutic Implications. Am J Card 2002; 90 (Suppl 8A): 1i-10i

[9] Berneis KK, Krauss RM. Metabolic origins and clinical significance of LDL heterogeneity. J Lipid Res. 2002; 43: 1363-79.

[10] Expert Panel on Detection Evaluation and Treatment of High Blood Cholesterol in Adults. Executive summary of the third report of the National Cholesterol Education Program (NCEP) expert panel of detection, evaluation and treatment of high blood cholesterol in adults (Adult Treatment Panel III). JAMA 2001; 285: 2488 - 97

[11] Backers JM. Effect of Lipid-Lowering Drug Therapy on Small-dense Low-Dense Lipoprotein. Ann Pharmacol 2005; 39: 523 - 26.

[12] Austin MA, King MC, Vranizan KM, Krauss RM. Atherogenic lipoprotein phenotype. A proposed genetic marker for coronary heart disease risk. Circulation 1990; 82: 495-506

[13] Chait A, Brazo RL, Tribble DL, Krauss RM. Susceptibility of small, low- density lipoproteins to oxidative modification in subjects with the atherogenic lipoprotein phenotype, pattern B. Amer J Med 1993; 94: 350-6

[14] Van J, Pan J, Charles MA, Krauss R, Wong N, Wu X. Atherogenic lipid phenotype in a general group of subjects. Arch Pathol Lab Medicine 2007; 131: 1679 – 85

[15] Castelli WP. Cholesterol and lipids in the risk of coronary artery disease – The Framingham Heart Study. Can J Cardiol 1988; (Suppl A): 5A-10A.

[16] Castelli WP. Epidemiology of triglycerides; a view from Framingham. Am J Cardiol 1992; 70: 43-49

[17] Castelli WP. The new pathophysiology of coronary artery disease. Am J Cardiol 1998; 82: (Suppl 2): 60-85

[18] Nicholls S, Lundmann P (2004). The emerging role of lipoproteins in atherogenesis: beyond LDL cholesterol. Semin Vasc Med 2004; 4: 187-195

[19] Rizzo M, Berneis K. Low density lipoprotein size and cardiovascular prevention. Europ J Int Med 2006; 17: 77 - 80.

[20] Shoji T, Hatsuda S, Tsuchikura S, Shinohara K, Komoto E, Kovama H, Emoto M, Nishizhawa Y. Small dense low-density lipoprotein cholesterol concentration and carotid atherosclerosis. Atherosclerosis 2009; 202: 582 - 588.

[21] Zhao ChX, Cui YH, Fan Q, Wang PH, Hui R, Cianflone K, Wang DW. Small Dense Low-Density Lipoproteins and Associated Risk Factors in Patients with Stroke. Cerebrovasc Dis 2009; 27: 99-104.

[22] Rainwater DL, Moore PH jr, Shelledy WR, Dyer TD, Slifer SH. Characterization of a composite gradient gel for the electrophoretic separation of lipoproteins. J Lipid Res 1997; 38: 1261-1266

[23] Alabakovska SB, Todorova BB, Labudovic DD, Tosheska KN. Gradient gel electrophoretic separation of LDL and HDL subclasses on BioRad Mini Protean II and size phenotyping in healthy Macedonians. Clin Chim Acta 2002; 317: 119-123.

[24] Otvos JD, Jeyarajah EJ, Bennet SW, Krauss RM.Development of a proton nuclear magnetic resonance spectroscopic method for determining plasma protein concentrations and subspecies distribution from a single, rapid measurement. Clin Chem 1992; 38: 1632- 38

[25] Hoefner DM, Hodel SD, O'Brien JF, Branum EL, Sun D, Meissner I, McConnell JP. Development of a rapid quantitative method for LDL subfraction with use of the Quantimetrix Lipoprint LDL system. Clin Chem 2001; 472: 266-274.

[26] Lamarche B, Lemieux I, Despres JP. The small, dense LDL phenotype and the risk of coronary heart disease : epidemiology, patho-physiology and therapeutic aspects. Diabetes Metab 1999; 25: 199-211

[27] Packard CJ. Triacylglycerol-rich lipoproteins and the generation of small dense low-density lipoprotein. Biochem Soc Transactions 2003; 31: 1066 - 69

[28] Carmena R, Duriez P, Fruchart JC. Atherogenic lipoprotein particles in atherosclerosis. Circulation 2004; 109: III2-III7

[29] Oravec S. Nová laboratórno-medicínska pomoc v diagnostike dyslipoproteinemií a kardiovaskulárnych ochorení: Identifikácia LDL podskupín. Med Milit Slov 2006a; 8: 28-32.

[30] Oravec S. Identifikácia subpopulácií LDL triedy –Aktuálny prínos v diagnostike porúch metabolizmu lipoproteínov a ochorení kardiovaskulárneho systému. Med Milit Slov 2006b; 8: 32-34.

[31] Oravec S. Nové perspektívy v diagnostike porúch metabolizmu lipoproteínov - prínos v interpretácii výsledkov. Med Milit Slov 2007a; 9: 42-45

[32] Oravec S. Nové možnosti posúdenia kardiovaskulárneho rizika u pacientov s obezitou a metabolickými ochoreniami. Med Milit Slov 2007b; 9: 46-49.

[33] Morais J, Neyer G, Muniz N. Measurement and Distribution of HDL subclasses with the new Lipoprint® HDL Method (pdf format). Presented at AACC, Philadelphia, PA , June 2003

[34] Morais J. Quantimetrix shows that all HDL subfractions may not protect against heart disease. AACC international congress of Clinical Chemistry, Orlando, FL, June 2005

[35] Muniz N, Morais J. Coronary heart disease. High density lipoprotein subclasses associated with heart disease. Medical Letter on the CDL and FDA, July 31st, 2005

[36] Maslowska M, Wang HW, Cianflone K. Novel roles for acylation stimulatory protein/C3ades Atg: a review of recent in vitro and in vivo evidence. Vitam Horm 2005; 70: 309-32

[37] Kwiterovich PO jr. The metabolic pathways of HDL,LDL and triglycerides. A current review. Am J Card 2000; 86 (Suppl 1): 5-10

[38] Oravec S. Den drohenden Herztod erkennen- und vermeiden. Der Mediziner 2010; 4: 6-7

[39] Oravec S, Dukát A, Gavorník P, Caprnda M, Kucera M. Lipoproteínový profil séra pri novozistenej arteriálnej hypertenzii. Úloha aterogénnych lipoproteínov v patogenéze ochorenia. Vnitr Lek 2010a; 56: 967-971.

[40] Oravec S., Dukát A., Gavorník P., Čaprnda M, Reinoldová O.Zmeny v lipoproteínovom spektre pri končatinovo-cievnej ischemickej chorobe.Vnitř. Lék 2010b; 56(6): 620-623

[41] Oravec S, Dukat A, Gavornok P, Caprnda M, Kucera M, Ocadlik I. Contribution of the atherogenic lipoprotein profile to the development of arterial hypertension. Brat Lek Listy 2011a; 112: 4-7

[42] 42) Luc G, Bard J-M, Ferriéres J, Evans A, Amouyel P, Arveiler D, Fruchart J-Ch, Ducimetière P, Prime Study Group. Value of HDL-cholesterol, apolipoprotein A-I, Lipoprotein A-I, Lipoprotein A-I/A-II in prediction of coronary heart disease . The Prime Study. Arterioscler Thromb Vasc Biol. 2002; 22: 1155- 61

[43] St-Pierre AC, Cantin B, Daganais GR, Mauriege P, Bernard PM, Despres JP, Lamarche B. Low density lipoprotein subfractions and the long-term risk of ischemic heart disease in

men : 13-year follow-up data from the Quebec Cardiovascular Study. Arterioscler Thromb Vasc Biol. 2005; 25: 553-559

[44] Fruchart JC, Sacks FM, Hermans MP et al. The residual risk reduction initiative: a call to action to reduce residual vascular risk in dyslipidaemic patients. Diabetes Vasc Res 2008; 5: 319-335

[45] Chun Xia Zhao, Ying Hua Cui, Qiao Fan, Pei Hua Wang, Ruitai Hui, Cianflone K, Dao Wen Wang. Small Dense Low-Density Lipoproteins and Associated Risk Factors in Patients with Stroke. Cerebrovasc Dis 2009; 27: 99-104

[46] Haffner SM. The metabolic syndrome: inflammation, diabetes mellitus and cardiovascular disease. Am J Cardiol 2006; 97: 3A-11A

[47] Lamarche B, Tchernof A, Moorjani S, Cantin B, Dagenais GR, Lupien PJ, Despres JP. Small dense LDL lipoprotein particles as a predictor of the risk of ischemic heart disease in men. Prospective results from the Quebec Cardiovascular Study. Circulation 1997; 95: 69-75

[48] Gardner CD, Fortman SP, Krauss RM. Association of small low-density lipoprotein particles with the incidence of coronary artery disease in men and women. JAMA 1996; 276: 875-881

[49] Rajman I, Kendall MJ, Cramb R, Holder RL, Salih M, Gammage MD. Investigation of low density lipoprotein subfractions as a coronary risk factor in normotriglyceridaemic men. Atherosclerosis 1996; 125: 231-42

[50] Halle M, Berg A, Baumstark MW, Keul L. LDL-Subfraktionen und koronare Herzerkrankung – Eine Übersicht. Zeitschrift Kardiol 1998; 87: 317-30

[51] Austin MA, Hokanson JE, Brunzell JD. Characterization of low-density lipoprotein subclasses: methodologic approaches and clinical relevance. Curr.Opinion Lipidol 1994; 5: 395-403

[52] Oravec S, Gruber K, Dostal E, Mikl J. Hyper-betalipoproteinenmia LDL1,2: a newly identified non-atherogenic hypercholesterolemia in a group of hypercholesterolemic subjects. Neureoendocrinol Lett 2011b; 32: 322-327

[53] Oravec S, Dostal E, Dukat A, Gavorník P, Kucera M, Gruber K. HDL subfractions analysis: A new laboratory diagnostic assay for patients with cardiovascular diseases and dyslipoproteinemia. Neuroendocrinol Lett 2011c; 32: 502-509

[54] Oravec S, Dukat A, Gavorník P, Lovásová Z, Gruber K. Atherogenic normolipidemia – a new phenomenon in the lipoprotein of clinically healthy subjects. Neuroendocrinol Lett 2011d; 32:317-321

The Importance of Lipid and Lipoprotein Ratios in Interpretetions of Hyperlipidaemia of Pregnancy

D.S. Mshelia and A.A. Kullima

Additional information is available at the end of the chapter

1. Introduction

In spite of the fact that the hyperlipidaemia of pregnancy is usually considered physiological [1-12], all pregnant women develop hypertriglyceridaemia with subsequent formation of small, dense low-density lipoprotein(LDL) particles, both of which are an independent risk factor of coronary heart disease(CHD) [13]. By 3rd trimester most women have a lipid profile which could be considered highly atherogenic in the nonpregnant state [14]. Similarly, animal model studies showed that maternal hypercholesterolaemia during pregnancy even when temporary and limited to pregnancy triggers pathogenic events in the fetal aorta, greatly enhanced atherogenesis later in life[14, 15]. On the other hand, intrauterine growth retardation (IUGR) has been associated with pre-eclampsia [16], as a result of decreased maternal lipid transfer to the fetus secondary to placental abnormalities. IUGR has also been associated with failure of development of hyperlipidaemia during pregnancy with subsequent reduction in maternal lipid reaching the fetus in a normal placenta [17, 18]. Generally, serum lipid and lipoprotein levels in pregnancy are modulated by complex interactions between genetics, medical complications of pregnancy, co-existing medical conditions, and other maternal factors [9, 19]. This underscores the need to take a meticulous and decisive approach in interpreting hyperlipidaemia of pregnancy. In searching for an emergent or new cardiovascular risk factor, concerning lipid and lipoprotein in adult males and nonpregnant women, several lipoprotein ratios or atherogenic indices have been defined[20]. These ratios were found to provide information on risk factors difficult to quantify by routine analyses and could be a better mirror of the metabolic and clinical interactions between lipid fractions[21]. Despite findings of [22] in a registry study of heterozygous familial hypercholesterolaemia(FH) mothers, who observed no significant untoward effect of lipid-lowering drugs during pregnancy, the current trend is that Statins, classified by FDA as category X, should be avoided in pregnancy[23, 24]. The use of lipid and lipoprotein ratios in interpreting pregnancy associated hyperlipidaemia may provide a balanced hyperlipidaemia not only in normal pregnancy but also in the other modulators of lipid metabolism in pregnancy.

2. Pathophysiology of hyperlipidaemia of pregnancy

Pregnancy is a dynamic state consequent of the fact that normal fetal development needs the availability of essential nutrients such as glucose, free fatty acids(FFAs), long-chain polyunsaturated fatty acids(LCPUFAs), amino acids, minerals, vitamins, to be continuously supplied to the growing fetus despite intermittent maternal food intake[10,25]. The dynamism of the gestational period support fetal growth and development while maintaining maternal homeostasis and preparation for lactation. This is achieved by complex and continuously evolving adjustments in maternal nutrient metabolism occurring throughout gestation.

Many of these maternal adjustments occur in the early stages of pregnancy when the fetus is too small to make considerable metabolic demands of the mother, resulting in the maternal metabolism working from a different baseline compared with the nonpregnant state. This period is called the anabolic phase. In late pregnancy, however, the maternal metabolic processes become more complicated because of the two-way interaction between the mother and the developing fetus. This is caked the catabolic phase.

The changes in nutrient metabolism can be described by several general concepts[8]: (a) adjustments in nutrient metabolism are driven by hormonal changes, fetal demands and maternal nutrient supply; (b) more than one potential adjustment exists for each nutrient; (c) maternal behavioural changes augment physiologic adjustment; and (d) a limit exists in the physiologic capacity to adjust nutrient metabolism to meet pregnancy needs, which when exceeded, fetal growth and development are impaired. Subsequently, metabolic adaptations, during pregnancy are essential [26]: 1, To ensure adequate growth and development of the fetus; 2, to provide the fetus with adequate energy stores and substrates that are needed following birth; 3, and, to provide the mother with sufficient energy stores and substrates to cope with the demands of pregnancy as well as those of labour and lactation. One of the maternal metabolic adjustments during pregnancy includes accumulation of fat depots in maternal tissues[26]. During this anabolic phase, the number of insulin receptors on the adipocytes increases, culminating into increased insulin sensitivity, increase lipoprotein lipase(LPL) activity which hydrolyses circulating triglycerides for tissue uptake, enhanced lipogenesis and marked maternal fat deposition(about 3.5 to 6.0kg) which is used as energy sources for the mother so that glucose is spared for the developing fetus in the catabolic part of the pregnancy[27, 28]. Lean women increase their fat stores more than obese women per kg body weight, likely due to higher insulin sensitivity in them, in early pregnancy, promoting lipid uptake and de novo synthesis.

The important attributes of fat deposits during the anabolic phase in pregnant women are :(1) Hyperphagia, present in pregnant women and increases as gestational time advances. This progressive increase in the availability of exogenous substrates actively contributes to maternal accumulation of fat depots [29]; (2) Promotion of lipogenesis and suppression of lipolysis mediated by progressive increase in insulin and its sensitivity and enhanced by progesterone and cortisol [30]; (3) The proportional increase in adipose tissue lipoprotein lipase (LPL) activity [1,12,31] which hydrolyzes triglycerides(TGs) in form of TG-rich lipoproteins, chylomicron and very-low density lipoprotein(VLDL), which are respectively converted into remnant particles and intermediate-density lipoprotein (IDL). The hydrolytic products, non-esterified fatty acids(NEFA) and glycerol, are partially taken up by subjacent tissues [11, 12, 32, 33]; (4) the unique capacity of tissue to utilize intracellularly the glycerol released during lipolysis. Under normal circumstances, the negligible glycerol kinase activity in adipose tissues hampers the utilization of glycerol for glycerol-3-phosphate synthesis and its use for the synthesis of TGs [34,35]. However, an increase in glycerol kinase activity and its subsequent capacity to metabolize glycerol has been found in rodents under condition of hyperinsulinaemia and enhanced fat accumulation, such as occurs in obesity [35, 36]. The lower lipolytic activity together with the augmented capacity of the tissues for the synthesis of glycerol-3-phosphate for uses in TG synthesis from both glucose and intracellular released glycerol results in net intracellular accumulations of TGs. Since all these pathways are stimulated by insulin, it is proposed that the enhanced insulin responsiveness [37] in the presence of an augmented response of the pancreatic β-cells to the insulinotropic stimulus of glucose that has been found in early pregnant women [38] would be the principal driving forces for the net fat depot accumulation at this stage of pregnancy. These ultimately lead to maternal fat accumulations in the anabolic phase of gestation.

The anabolic condition of adipose tissue during early pregnancy switches to a net catabolic state during the last 1/3 of gestation. The signals responsible for this switch from lipid storage to lipid mobilization are not well understood; however, placental hormones that increase with advancing gestation, known to induce maternal insulin resistance, may play a major role. Placental growth hormone, human placental lactogen, leptin, and tumour necrosis factor-α(TNF-α) are placental hormones that induce insulin resistance. The presence of high plasma levels of placental hormones, known to have lipolytic effects, human placental lipase (HPL), an augmented production of catecholamine secondary to maternal hypoglycemia [38], and the insulin-resistant condition present at this stage [39, 40], appear to be responsible for the net breakdown of maternal fat depots, consistently causing increments of plasma nonesterified fatty acids(NEFA) and glycerol levels during the 3rd trimester of gestation. The main destination of these lipolytic products released from maternal adipose tissue is the maternal liver. They are converted in the liver into their respective active forms, acyl-CoA and glycerol-3-phosphate, to become partially re-esterified for the synthesis of triglycerides, which are transferred to native VLDL particles and released into the circulation. Acyl-CoA can also be converted throughout the β-oxidation pathway to acetyl-CoA for energy production and ketone body synthesis, whereas glycerol may also be used for glucose synthesis. Fetal-placental glucose and amino acids utilization rates are highest at 22 to 26weeks decreasing near term. In contrast lipid transport is maximal in the 3rd trimester coincident with rapid fetal fat accretion, this spares the mother to utilize glucose during this period. Humans are born with the highest percentage of fat (12 to 15%) compared to any species and 90% deposition occurs in the last 10weeks of gestation, exponentially increasing to 7g/day near term. The preferential use of glycerol released from maternal adipose tissue for gluconeogenesis acquire greater importance during maternal fasting period, when circulating glucose levels are lower than under nonpregnant conditions[34]. Under fed condition in early gestation, plasma ketone body values are even lower in pregnant than in nonpregnant condition [41], indicating an enhanced use of these fuels by maternal tissues as alternative substrate to glucose. However, during fasting period maternal ketogenesis become highly accelerated, as indicated by the exaggerated increase in plasma ketone bodies that occur [41]. This benefit the fetus in two ways: (1) ketone bodies are used by maternal tissues, thus, saving glucose for essential function and delivery to the fetus, (2) placental transfer of ketone bodies is very efficient, attaining the same concentration in fetal plasma as in maternal circulation[42]. In addition, ketone bodies may be used by the fetus as oxidative fuels as well as substrate for brain lipid synthesis [43].

Insulin is well known to inhibits adipose tissue lipolytic activity, hepatic gluconeogenesis and ketogenesis but to increases adipose tissue LPL activity. Thus, it is not surprising that all of these pathways change in the opposite direction which is consistent with insulin resistance occurring in later part of pregnancy. These pathways become even further modified under uncontrolled gestational diabetes mellitus(GDM), where insulin resistance is further enhanced [44].

3. Maternal hyperlipidaemia of pregnancy

The enhanced net breakdown of fat depots during late pregnancy is associated with hyperlipidaemia, which chiefly corresponds to plasma rises in TGs with smaller rise in phospholipids and cholesterol [44]. The greatest increased in plasma TGs corresponds to VLDL values but TGs also accumulates in other lipoprotein fractions, which do not normally transport them, such as LDL and HDL [45]. The high TGs concentration secondary to lipolysis in the presence of increased cholesteryl ester transfer protein(CETP) activity, occurring in midgestation[45], contributes to the accumulation of TGs in the lipoprotein fractions of higher densities, LDL and HDL[44, 45]. CETP facilitates the exchange of TGs by esterified cholesterol between VLDL and either LDL or HDL. Furthermore, during late gestation the activity of hepatic lipase (HL) greatly decreased [45]. HL converts the buoyant HDL-2-TG-rich particles into small HDL-3-TG-poor particle allowing a proportional accumulation of buoyant HDL-2-TG-rich particle.

Other hormonal dynamism occurring during pregnancy contributing to maternal hypertriglyceridaemia are, **table** 1, consistently increasing oestrogen concentration almost throughout the gestation period and oestrogen has been shown to (1) increase endogenous production of VLDL-TGs [46]; (2) reduce adipose tissue LPL activity [33, 45], and (3) inhibition of hepatic TG lipase activity [33, 44]. Thus, the oestrogenic influence over TG metabolism suggests an increased circulating VLDL-TG. Although the role of progesterone in TG metabolism is not certain, its administration in rats had a lipid neutral effect. Thus, the interaction between oestrogen and progesterone would favour hypertriglyceridaemia. Prolactin may inhibit adipose tissue LPL while stimulating breast LPL in late gestation [45]. Thus, the physiologic outcome of increasing concentration of Prolactin with advancing pregnancy would be a shift in storage from the adipocytes to the breast in preparation for lactation.

Enzymes	Activities
Adipose tissue lipoprotein lipase(LPL)	decrease
Diacylglycerol acetyltransferase	decrease
Cholesterol 7-alpha hydroxylase	decrease
Placental lipoprotein lipase (PLPL)	increase
Placental triglycerides hydroxylase	Increase
Phospholipase A2(PLA2)	Increase
Cholesterol ester transfer protein(CETP)	Increase
Hepatic lipase	Decrease
Hormones and cytokines	Concentrations
Estradiol	Increase
Insulin	Sensitivities increase during first trimester but subsequently decreases from second trimester to end of gestation
Human placental lactogen	Increase
Prolactin	Increase
Cortisol	Increase
Glucagon	Increase
Porgesterone	Increase
Leptin	Increase
Tumor Necrosis Factor-alpha(TNF-alpha)	Increase
Human chorionic somatomammotropin(HCS)	Increase

Table 1. Hormone and enzyme changes during the course of pregnancy.

The combined effects of enhanced liver production of VLDL [47, 48], decreased removal of these particles from the circulation due to low LPL activity [45,49], high CETP activity and low HL activity, would not only be responsible for the accumulation of TGs in LDL but also for the proportional accumulation of TG in buoyant TG-rich HDL-2b subfractions at the expense of the cholesterol-rich and TG-poor HDL-2a or HDL-3[45].

The increasing insulin-resistance in late gestation and continuously increasing plasma oestrogen levels occurring during pregnancy are the main hormonal factors responsible for these metabolic changes resulting into the development of maternal hypertriglyceridaemia, see table 2.

Lipid and lipoproteins (mg/dl)	First trimester	Second trimester	Third trimester	Nonpregnant controls
HDL-C	67±12	83±19	81±17	69±10
LDL-C	90±17	130±46	136±33	99±23
TGs	79±27	151±80	245±73	77±34
TC	173±18	243±53	267±30	183±23
ApoA-1	170±27	204±22	196±28	163±24
ApoA-2	49±7	52±6	49±5	47±6
ApoB	70±21	91±25	113±29	61±22
ApoC-11	265±13	299±18	314±21	237±11
ApoC-111	141±3	188±5	217±6	121±19
ApoE	41±12	42±9	49±19	42±20
Lp(a)	60(0-1440)	63(2-1210)	54(0-1230	86(11-473)
VLDL-1	19(12-55)	47(26-110)	109(38-170)	23(5-85)
VLDL-2	17(7-45)	36(20-77)	103946-168)	23(13-44)
IDL	26(13-54)	58(24-100)	124(79-157)	35(18-62)
Total LDL	200(135-323)	292(206-410)	353(244-534)	207(150-363)
LDL-1	33(16-52)	49(37-70)	67(27-96)	50(22-130)
LDL-11	143(95-231)	160(103-287)	201(59-316)	135(72-258
LDL-111	28(15-56)	32(24-165)	123(43-192)	31(5-68)

Table 2. The increasing lipid and lipoproteins during course of gestation(courtesy: Ahmet Basaran, MD)

4. Placental transfer of maternal lipid and lipoproteins and their metabolites to the fetus

The human placenta contains VLDL, LDL, HDL, and scavenger receptors as well as LDL receptor-related proteins. The placenta also has LPL, phospholipase-A2 (PLA-2) and intracellular lipase activities as well as plasma membrane fatty acid-binding protein (FABP/GOT2), fatty acid translocase (CD36), fatty acid transfer protein (FATP) and different cytoplasmic FABPs [29, 42,50, 51]. Thus, lipid and lipoproteins in maternal plasma can be taken up and handled by the placenta, allowing LCPUFAs associated with plasma lipoproteins to be transferred to the fetus. The human placenta is capable of transporting free fatty acids(FFAs) by diffusion and selectively increases the transport of essential fatty acids (EFAs) and their long-chain polyunsaturated fatty acids (LCPUFA) derivatives by fatty acid carrier proteins.

Although lipoprotein TGs does not directly cross the placental barrier, the placenta has mechanisms to release fatty acids(FAs) circulating in maternal plasma lipoproteins into the fetus. In addition, high levels of TGs in maternal circulation may create a steep concentration gradient across the placenta, which accelerates their transport and deposition in fetal tissues. In term human trophoblasts, insulin and fatty acids have been shown to enhance the expression of adipophylin, which is associated with cellular lipid droplets and implicated in cellular fatty acid uptake and storage of neutral lipids

4.1. Fatty acids transfer

The supply of LCPUFA is important for fetal growth and tissue development especially for the development of the nervous system and the considerable requirements of these LCPUFAs in the fetus must be provided by their placental transfer [52]. The plasma membrane fatty acid-binding proteins present in human placental membrane [51,52] are responsible for the preferential uptake of LCPUFAs. A selective cellular membrane of certain FAs may also contributes to the placental transfer process, as would the conversion of a certain proportion of arachidonic acid(AA) to prostaglandins(PGs)[52], the incorporation of some FAs into phospholipids[50-52], the oxidation of placental fatty acids[53] and the synthesis of FAs[52,53]. Even though essential fatty acids(EFA) as well as LCPUFAs are transferred across the placenta, the fetus needs to receive substantial amounts of preformed AA and docosahexaenoic acid(DHA) which can be synthesized to a limited extent from the EFA. The two dietary EFAs are linoeic acid(18:2ω-6) and α-linolenic acid(18:3ω-3), which are precursors of the ω-6 and ω-3 LCPUFA, respectively. The synthesis of AA and DHA do not take place in the fetus or the placenta in substantial amounts, owing to the low activities of the desaturating enzymes. Both AA and DHA are abundant in the brain and the retina and their appropriate supply during pregnancy and the neonatal period is critical for proper function [1,54]. Maternal plasma NEFA, though in smaller proportion than lipoprotein TGs, is an important source of polyunsaturated fatty acids(PUFAs) for the fetus [51,52]. Maternal plasma NEFAs correlates with those in the fetus and maternal adipocytokines have been associated with fetal growth[1]. The combination of these processes determines the actual rate of placental FAs transfer and its selectivity, consequent to the proportional enrichment of certain LCPUFAs, such as AA and DHA in fetal as compared with maternal compartments [52, 54].

4.2. Cholesterols

Cholesterol plays a key role in embryonic and fetal development hence the demands for cholesterol in the embryo and fetus is relatively high. Cholesterol is an essential component of cell membrane influencing the fluidity and passive permeability by interacting with phospholipids and sphingolipids [55]. It's the precursor of bile acids and steroid hormones. It is also required for cell proliferation and development of the growing body, cell differentiation, and cell-to-cell communications, and is the precursor of oxysterol, which regulates key metabolic processes. Available cholesterol in fetus is contributed by: (1) transfer from the mother especially during the first half of the gestation and too little cholesterol due to lack of maternal cholesterol or reduced expressions of placental lipoprotein receptors is correlated with small fetuses and a trends for microcephally; and (2) Fetal synthesis especially during the last half of gestation. Too little cholesterol due to lack of synthesis leads to a spectrum of congenital defects as seen in infants with Smith-Lomli-

Opitz Syndrome(SLOS) who are unable to synthesize cholesterol at normal rate due to null/null mutations in 3β-hydroxysteriod Δ7-reductase, the enzyme that converts 7-dehydrocholesterol to cholesterol. The placental endothelial cells are capable of transporting substantial amounts of cholesterol to the fetal circulation and this mechanism is further enhanced by liver-X receptors and induced up regulation of ATP-binding cassette transporter, ABCA1 and ABCG1[56].

4.3. Glycerol

Maternal Plasma glycerol levels are consistently elevated during late pregnancy, but crosses the placenta less than glucose or L-alanine [1,25, 57] though they all have similar molecular weights. Transfer of maternal glycerol via the placenta is by simple diffusion (2). However, its effective and rapid utilization through other pathways, such as gluconeogenesis and glyceride glycerol synthesis in the mother[10,25] results in its low plasma concentration and this very active kinetics impede the formation of the adequate gradient to create the appropriate driving forces for its placental transfer.

4.4. Ketone bodies

In the 3rd trimester of pregnancy, under fed conditions, plasma ketone body concentrations remain low although are greatly increase compared to nonpregnant condition under fasting [58] consequent to enhanced adipose tissue lipolysis. The lipolysis accelerates delivery of NEFA to the liver and enhanced ketogenesis. Ketone bodies can easily cross the placenta and be used as fuels and lipogenic substrates by the fetus. The transfer of ketone bodies across the placenta occurs either by simple diffusion or by a low-specificity carrier-mediated process [25]. The activities of ketone body metabolizing enzyme are present in fetal tissues (brain, liver and kidneys)[1,25] and can be increased by conditions of maternal ketonaemia such as occurs in starvation, during late pregnancy[39] or high-fat feeding[25]. Ketone bodies are used by the fetus as oxidative fuels as well as substrates for brain lipid synthesis [25]. However, in maternal hyperketonaemia as occurs in poorly controlled diabetes patients associated with transfer of excessive arrival of ketone bodies to the fetus seems to be responsible for the major damages [10], increasing stillbirth rate, incidence of malformations, and impaired neurophysiologic development [10]. Subsequently, it could be recommended that pregnant mothers, if possible, should avoid starvation and high fat diet especially in the 3rd trimester.

5. The importance of lipid and lipoprotein ratios in hyperlipidaemia in adult male and nonpregnant females

While cholesterol is a key component of the development of atherosclerosis, LDL-C concentration has been the prime index of cardiovascular disease(CVD) risk and the main target for therapy[21]. However, currently, there is almost unanimous agreement among epidemiologists and clinicians that coronary risk assessment based exclusively on LDL-C is not optimal[59]. Therefore in the recent past, efforts have been made in seeking emergent or new cardiovascular risk factors to improve cardiovascular disease prediction[20] and in an attempt to optimize the predictive capacity of lipid profile, several lipoprotein ratios or "atherogenic indices" have been defined. In the Framingham study, the TC:HDL-C ratio, a useful summary of the joint contribution of total cholesterol(TC) and HDL-C to coronary heart disease(CHD) risk[60], was also found to be an excellent predictor of CHD risk, with a hazard ratio of 1.21 for a 1.0 increment in ratio[60]. The value of this ratio should be emphasized when lipid profile is within desirable range. It was shown that patients with high-risk LDL-C levels >160mg/dl(4.2mmol/L) and low TC: HDL-C ratio (≤5.0) had an incidence of CHD of 4.9%. This was similar to those with low levels of both LDL-C(≤130mg/dl, 3.4mmol/L) and TC:HDL-C ratios, 4.6%[60]. By contrast, subjects with low-risk LDL-C levels(≤130mg/dl, 3.4mmol/L) and high TC:HDL-C ratio(>5.0) had a 2.5-fold higher incidence of CHD than those with similar LDL-C levels but low TC:HDL-C ratio[60]. For example, TC of 231mg/dl(5.89mmol/L) and HDL-C of 42mg/dl(1.09mmol/L) gives a TC:HDL-C ratio of 5.5, which indicate moderate atherogenic risk[61]. On the other hand, with the same level of TC, if HDL-C were 60mg/dl(1.55mmol/L), the ratio would be 3.8[61]. However, in the Helsinki Heart Study[62], it was demonstrated that the LDL-C:HDL-C ratio that paints the most relevant picture of a person's cardiovascular health risk especially when triglyceridaemia is taken into account and the risk is significantly higher in the presence of hypertriglyceridaemia. When there is no reliable calculation of LDL-C, especially when triglyceridaemia exceeds 300mg/dl(3.36mmol/L), it is preferable to use the TC:HDL-C ratio. Similarly individuals with high concentration of triglycerides, VLDL fraction shows cholesterol enrichment and thus the LDL-C:HDL-C ratio may underestimate the magnitude of the lipoprotein abnormalities in them[21]. Subsequently, both TC:HDL-C, known as the atherogenic or Castelli index, and LDL-C:HDL-C ratios are two important components and indicators of vascular risk, the predictive values of which is greater than isolated parameters used independently, particularly LDL-C. These ratios can provide information on risk factors difficult to quantify by routine analyses and could be a better mirror of the metabolic and clinical interactions between lipid fractions. Their applications therefore in interpreting hyperlipidaemia of pregnancy cannot be over emphasized.

5.1. ApoB:ApoA-1 ratio

Apolipoprotein-B(apoB) represents most of the protein contents in LDL and is also present in IDL and VLDL. ApoA-1 is the principal apolipoprotein in HDL and is believed to be a more reliable parameter for measuring HDL than cholesterol content since it is not subject to variation. Therefore, the apoB:apoA-1 ratio is also highly valuable for detecting atherogenic risk, and there is currently sufficient evidence to demonstrate that it is better for estimating vascular risk than the TC:HDL-C ratio[63-65]. The apoB:apoA-1 ratio was found to be stronger than the TC:HDL and LDL:HDL ratios in predicting risk[63]. ApoB:ApoA-1 ratio reflects the balance between two completely opposite processes. Transport of cholesterol to peripheral tissues, with its subsequent arterial internalization, and reverse transport to the liver[66]. Consequently, a larger ratio will implies higher amount of cholesterol from atherogenic lipoprotein circulating through the plasma compartment and likely to induce endothelial dysfunction and trigger the atherogenic process. On the other hand, a lower ratio will indicate less vascular aggression by plasma cholesterol and increased more effective reverse transport of cholesterol, as well as other beneficial effects, thereby reducing the risk of CVD. However, its use is limited by the fact that apolipoprotein measurement methods are not widely used as lipoprotein methods

5.2. TG:HDL ratio

Known as the atherogenic plasma index shows a positive correlation with HDL-C estimation rate(FER$_{HDL}$) and an inverse correlation with LDL size[67]. Therefore, the phenotype of LDL and HDL particles is clearly synchronized with the FER$_{HDL}$. The simultaneous use of TG and HDL in this ratio reflects the complex interaction of lipoprotein metabolism overall and can be useful for predicting plasma atherogenecity especially in pregnant women who manifesting with hypertriglyceridaemia of pregnancy. An atherogenic plasma index[Log(TGs:HDL)] over 0.5 has been proposed as the cutoff point indicating atherogenic risk[67].

5.3. LDL-C:apoB ratio

Although apoB is not an apolipoprotein exclusive to LDL, since it is present in other atherogenic lipoproteins such as IDL and VLDL, the LDL:apoB ratio provides approximate information on LDL particle size. A ratio of <1.3 indicate the predominance of LDL particle with low cholesterol content, consistent with small, dense LDL particle[68].

Variations in plasma lipid and lipoprotein ratios in adult men and nonpregnant women have been associated with more substantial alterations in metabolic indices predictive of future consequences of hyperlipidaemia than individual components of plasma lipid profile alone[69, 70] and as discussed above. Given the physiological role of gestational hyperlipidaemia in fetal development and the fact that the adaptations in maternal lipid metabolisms taking place throughout gestation is not without consequences, an urgent establishment of reference values for lipid and lipoprotein ratios in normal pregnancy is highly recommended.

5.4. The hyperlipidaemia of pregnancy, a dyslipidaemia? Find out!

5.4.1. The importance of lipid and lipoprotein ratios in interpreting the hyperlipidaemia of pregnancy

In normal nonpregnant adult population, higher concentrations of plasma triglycerides are associated with preferentially higher VLDL-1 concentration [71]. This particle is secreted by the liver to supply tissues with TGs fatty acids in the post absorptive state. The concentration of VLDL-2, the principal precursor in the circulation to IDL and LDL, does not change as dramatically. In addition, in normal nonpregnant adult population, a higher concentration of VLDL-1 is associated with a failure of insulin action and increased risk of CHD. In contrast, in pregnant women, as pregnancy progresses and high TG levels developed, VLDL-1 and VLDL-2 rose together so that the ratio, instead of increasing 2-fold, as would be predicted from population studies in the nonpregnant subjects (VLDL-1 o VLDL-2 ratio at a plasma TGs of 0.5mmol/L is 1.0 compared to 2.0 at plasma TGs of 2.5mmol/L)[71], remain constant. Sattar[33], *et al*, found a parallel increase in the small cholesterol-rich VLDL-2(17 to 103mg/dl) and the larger TG-rich VLDL-1(19 to 109mg/dl) at 35 weeks. Similarly, the relationships of VLDL constituents expressed as ratios were not significantly different comparing antepartum and postpartum observations, however, the TG/C ratio was higher at all of these times compared to controls, but the composition of these fractions was similar to that seen in a recent cross-sectional survey of healthy adults (19). The increase in VLDL-TG during gestation is likely due to an increase in VLDL synthesis rather to a compositional change in the VLDL particle, as a study showed no significant increase in VLDL TG/C ratio over time, and the ratios is similarly lower in all the trimesters compared to nonpregnant period(see table 3)

Ratios	First trimester	Second trimester	Third trimester	Nonpregnant control
TC:HDL-C	2.56	3.37	3.90	3.29
LDL-C:HDL-C	1.44	1.95	2.37	1.79
TGs:HDL-C	0.56	0.79	1.16	0.64
VLDL-TGs:CL	1.64±1.53	2.47±3.91	2.57±3.60	3.69±3.48
LDL-TG:CL	0.46±	1.24±2.68	0.56±0.29	0.14±0.08
HDL-TG:CL	0.58±0.21	0.60±0.19	0.69±0.32	0.21±0.09
HDL-TG:ApoA-1	4.09±1.55	5.24±1.43	6.13±1.28	2.73±0.71
LDL-CL:ApoB	2230±339	2222±228	2113±305	2506±167
LDL-TG:ApoB	217±59	256±41	332±60	157±32
LDL-PL:ApoB	748±123	753±66	727±109	824±64
IDL-TG:ApoB	2026±1085	1666±360	1550±202	1530±371
VLDL-TG:ApoB	6272±1924	6278±1629	5551±1416	7040±2778

Table 3. Lipid and lipoprotein ratios in the three trimesters of normal pregnancy.

Taken together, and as shown in **table 3**, although one of the consequences of pregnancy is that maternal lipid metabolism is specifically altered, using the lipid and lipoprotein ratios, the hyperlipidaemia occurring in the later part of pregnancy appears to be a balanced hyperlipidaemia. These are discussed below

During the course of normal pregnancy, plasma TGs and cholesterol rise by 200-400% and 25-50% respectively. The total LDL mass increased during gestation (median concentration increased by about 70%, 200-353mg/dl) between 10 to 35weeks, see table 4. The lipid become enriched with TGs and depleted in cholesterol. The larger, more buoyant subclasses of LDL (LDL-1 and LDL-2) predominant in healthy pregnant females and may in the reproductive age, whereas smaller, denser LDL-3 often occur after menopause [11, 72]. Several studies showed there to be an association between elevated plasma TG concentrations, small, dense LDL [11, 73] and decreased HDL cholesterol [74], in particular HDL-2 cholesterol [73]

VARIABLES	10 WEEKS OF GESTATION	35 WEEKS OF GESTATION
Triglyceride(mean)	69.65mg/dl	227.69mg/dl
Total cholesterol(mean)	172.57mg/dl	282.38mg/dl
HDL-C(mean)	64.73mg/dl	65.50mg/dl
VLDL-1(mean)	19mg/dl	109mg/dl
VLDL-2(mean)	17mg/dl	103mg/dl
IDL(mean)	26mg/dl	124mg/dl
LDL	200mg/dl	333mg/dl
LDL-1	33mg/dl	67mg/dl
LDL-11	143mg/dl	201mg/dl
LDL-111	28mg/dl	123mg/dl
LDL-1	17% of total LDL	20% of total LDL
LDL-11	69% of total LDL	49% of total LDL
LDL-111	14% of total LDL	32% of total LDL

Table 4. Magnitude of changes in lipid and liporpotein values from first to third trimester.

In men and nonpregnant females, plasma TG is the major determinant of small, dense LDL, occurring for 40-60% of the variability of this fraction in the plasma [71,75,76]. In addition, recent cross-sectional studies [70,74] have prompted the suggestion that, within the relationship between plasma TGs and LDL subfractions profile, there is a threshold effect. At low-normal plasma TG concentrations, there is a positive association between LDL-2(the major LDL species) concentration and plasma TGs. Above a certain plasma value, however(reportedly about 1.5mmol/L in men)[71,75], LDL-2 concentration correlates negatively with plasma TGs, and LDL-3 concentration which had been relatively constant below this TG concentrations, correlates positively with plasma TG. Generally, percent LDL-3(and LDL-3 mass) changed little in early gestation despite increasing TG concentrations. However, there appeared to be considerable variation between individuals in the gestational age and plasma TGs intervals at which change in the LDL profile first manifested—the elevated TG levels already present in the first trimester may be responsible for the increased in dense LDL

In line with the alarming observations in LDL subclasses and total LDL mass, LDL-1 mass increased around 2-fold, from 33 to 67mg/dl; LDL-2 mass increased least by around 40% from a median of 143 to 201mg/dl, reaching a maximum of 218mg/dl at 30weeks gestation, whereas in sharp contrast, LDL-3 mass increased by greater than 4-fold from 23 to 123mg/dl.

However, as concentration of LDL-2 is declining, that of LDL-3 is increasing and implying that the ratio may tend towards a unit.

Towards end of second trimester to end of gestation, the concentrations of VLDL, IDL, and LDL-1/LDL-2 further increased, producing a distribution of lipoproteins dominated by buoyant lipoprotein species, in particular LDL-1. In line with this, Winkler's[11], et al data do not support the idea that the same mechanisms as those described for the atherogenic lipoprotein phenotypes govern lipid metabolism in late pregnancy. Therefore, in uncomplicated pregnancy there appears to be a balance between potentially damaging factors such as altered lipid metabolism and as yet poorly understood protective mechanisms [11,33,75]. However, the clinical significance of gestational lipoprotein metabolisms may arise if this balance is compromised as in hypertensive disorders of pregnancy. It is in these circumstances then when the application of these ratios is very important, for example; Toescu, [77] et al while comparing lipid levels between pregnant diabetic women(types 1 and 2 and GDM) and pregnant nondiabetic counterparts did not demonstrated any significant differences among the groups according to trimesters, implying that the observed hyperlipoproteinaemia during pregnancy is independent of diabetes status[10]

Kilby,[78] et al although observed higher lipid levels and increased in TC, TGs, VLDL/LDL ratio, HDL-C with gestational age in type 1 DM, similarly found no significant difference from gestationally matched controls[78] in their study. Investigations are required to characterize lipid and lipoprotein profile using ratios in the other modulators, particular these will assists clinicians while dealing with hyperlipidaemia of pregnancy considering the limited quantification opportunities.

Currently the applications of lipid and lipoprotein ratios in interpreting the hyperlipidaemia of pregnancy are limited particularly in the poor developing nations. In spite of the fact that the hyperlipidaemia of pregnancy is usually considered physiological, serum lipid and lipoprotein levels in pregnancies are generally modulated by complex interactions between genetics, medical complications of pregnancy, co-existing medical conditions and other maternal factors[19], **table 5**.

Therefore the hyperlipidaemia during pregnancy could be classified according to clinical implications and future prospects as in **fig 1**, particularly where there is limited opportunity of investigations do to poverty.

In our laboratory [79] the ratios were found to be important particularly where measurement of lipid and lipoprotein is not routinely done due to poverty. In addition hyperlipidaemia in pregnancy is confounded by other conditions that may predispose to hyperlipidaemia, such as obesity, diabetes mellitus, chronic renal insufficiency, and pre-eclampsia. Similarly subfractions of lipoproteins are usually not done due to limited methodology. Without the use of lipid and lipoprotein ratios particularly considering these confounding conditions which are also likely to present with hyperlipidaemia, interpreting the hyperlipidaemia of pregnancy is encountered with difficulties.

Medical complications of pregnancy
1. Pre-eclampsia
2. Pregnancy-induced hypertension
3. Gestational diabetes mellitus
4. Intra-uterine growth restriction(retardation)
5. Prelipaemia
Co-existing medical conditions
1. Obesity
2. Types 1 and 2 diabetes mellitus
3. Hypothyroidism
4. Hypertension
5. Renal diseases, particularly nephritic syndrome
6. Alcoholism
7. Medications, eg LMWt-heparin and glucocorticoid
Other maternal factors
1. BMI(Obesity)
2. Maternal weight gain in the index pregnancy
3. Maternal nutrition
4. Pre-pregnancy lipid levels

Table 5. Factors that can also modulate lipid and lipoprotein concentrations in pregnancy (genetic factors not mentioned)

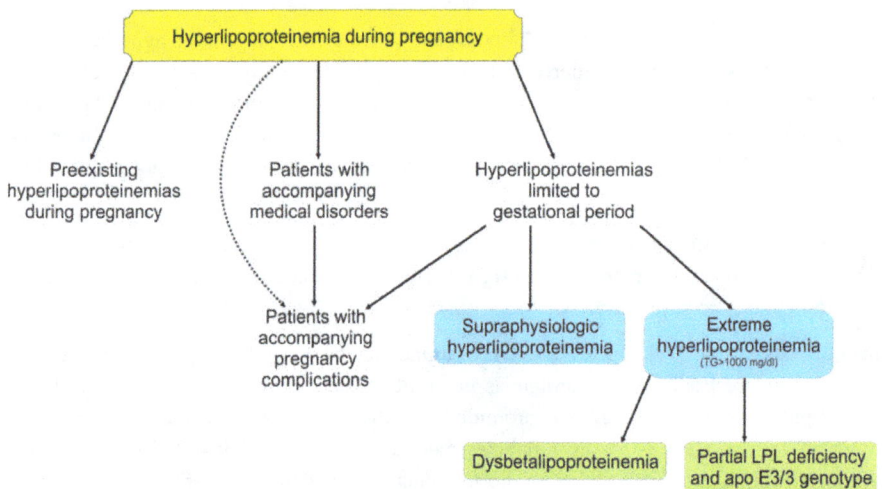

Image from courtesy of Ahmet BASARAN, MD

Figure 1. Classification of hyperlipidaemia of pregnancy

Whilst the hyperlipidaemia of pregnancy is considered physiological, studies have demonstrated that deviations present as a two-edged sword. On one hand, development of the physiological hyperlipidaemia out of proportion could be associated with many consequences and on the other hand failure to development the required proportion of physiological hyperlipidaemia of pregnancy could also be associated with some consequences, **Table 6** and these will be discussed subsequently

Consequences of hyperlipidaemia
Complications associated with hyperlipidaemia of pregnancy:
Cholesterol gallstones
Intrahepatic cholestasis
Acute pancreatitis
Endothelial dysfunction
Unanswered questions concerning hyperlipidaemia of pregnancy
Is hyperlipidaemia of pregnancy atherogenic?
Is hyperlipidaemia of pregnancy a dyslipidaemia?
Future effect of pregnancy-induced Supraphysiologic hyperlipidaemia
Pre-eclapmsia and hyperlipidaemia of pregnancy
Consequences of failure of development of hyperlipidaemia
IUGR—Intra-uterine growth retardation
Future development of metabolic syndrome in affected fetus

Table 6. Consequences of deviations of Hyperlipidaemia of pregnancy.

6. Is normal pregnancy atherogenic?[80]

The change in triglycerides in normal pregnancy is important in relation to lipoprotein subclasses, such as LDL. These lipoproteins contain subfractions of various sizes, densities and compositions, which differ in their ability to initiate atherogenesis [81]. One of the subfractions of LDL (LDL-3) is small, dense LDL particles which do not bind readily to the LDL receptors and therefore remain in the circulation for longer time, penetrate the arterial intima better than do larger ones[82] and are more readily oxidized, probably because they contain less vitamin E and other antioxidants[83]. Finally, their uptake into macrophages to create form cells, and initiate atherogenesis, is facilitated [84]. This may explain their identification as an independent risk factor for coronary heart disease [82-84].

Plasma triglycerides are the major determinant of small, dense LDLs, accounting for 40-60% of the variability of this fraction in the plasma [75]. VLDL represents the major precursor of LDL and reflects plasma TGs levels. Two subclasses of VLDL have been defined: a large and buoyant fraction enriched with TGs (VLDL-1) and a smaller, denser fraction(VLDL-2). It follows from the association between LDL subclasses and raised TGs that VLDL-1 may be important as a vehicle in the process of neutral lipid exchange and generation of small, dense LDLs. Cholesterol esters are transferred from LDL and HDL to VLDL-1 by cholesterol ester transfer protein(CETP) in exchange for TGs and the increased concentration of VLDL-1, due to hypersecretion by the liver promote TG transfer into LDL during pregnancy[33]. TG-enriched LDL particles subsequently undergo a size reduction through the action of hepatic lipase, resulting in the formation of small, dense LDL subfractions. In addition Lippi[2], et al demonstrated in their study that advanced pregnancy is associated with an increased prevalence of undesirable or abnormal values for total cholesterol, LDL-C and TGs in the second trimester, and total cholesterol, LDL-C, TGs TC:HDL ratio in third trimester demonstrating that physiological pregnancy is associated with a substantial modification of lipid and lipoproteins metabolism from the second trimester, providing reference ranges for traditional and emerging cardiovascular risk predictors throughout the gestational period. Therefore, is normal pregnancy atherogenic?

All pregnant women develop a transient hyperlipidaemia associated with hypertriglyceridaemia, and subsequent formation of small, dense LDL particles, both of which are an independent risk factor for CHD, and by 3rd trimester most women have a lipid profile which would be considered highly atherogenic in the non-pregnant state[13]. Increased prevalence of angina, cholesterol gallstone, and obesity in postmenopausal women who have had several pregnancies has been observed [85]. Yet the long-term consequences of multiple pregnancy, gestational diabetes or maternal obesity in LDL subfractions and lipid profile are unknown. Further studies are recommended to determine if certain women are at increased risk of CVD in later life because of effects on their lipid profile during pregnancy. In contrast, increasing suggestions are that maternal hypercholesterolaemia during pregnancy even when temporary and limited to pregnancy triggers pathogenic events in the fetal aorta, greatly enhanced fatty streak formation and that may influence atherogenesis later in life[14,15]. Fetal plasma cholesterol levels are high and are proportional to the maternal cholesterol levels [14] in second trimester, decline with increasing fetal age[14] and are even lower at term birth. This is supported by the fact that lipid levels observed in umbilical cord blood(UCB) from normal pregnancy were significantly lower than those found in maternal blood with exception of HDL-C, and that LDL:HDL ratio in neonate of normal pregnancy are much lower than the value in normal pregnant mothers[16]. The high HDL levels and a lower LDL:HDL ratio in UCB suggest that the fetus of a normal pregnancy is protected against atherogenic lipoprotein[16]. Despite these findings, studies at autopsy demonstrated that atherosclerosis progresses much faster in offsprings of hypercholesterolaemic mother than in offsprings of normocholesterolaemic mothers[86]. Same studies observed that at each time point, offsprings of hypercholesterolaemic mothers had 1.5 to 3-fold larger lesions than offsprings of normocholesterolaemic mothers, and they suggested that, pathogenic programming in utero increases the susceptibility to atherogenic risk factors later in life and maternal intervention with cholesterol-lowering agents reduce postnatal lipid peroxidation and atherosclerosis in their offsprings[87]. A registry study by Toleikyte,[22] et al, of heterozygous familial hypercholesterolaemic(FH) mothers observed that: the serum levels of cholesterol in the nonpregnant, nontreated women were 370mg/dl(9.59mmol/L); no maternal cardiovascular deaths were observed; the children of mothers with FH were no more likely than the general population to be born prematurely, have low birth weight, or have congenital malformations; and that no congenital malformations were observed in the 19 pregnancies associated with the use of lipid-lowering drugs during pregnancy. However, the current trend is that Statins, classified by FDA as category X in pregnancy, should be avoided in pregnancy [23,24]. Although there are observations for and against the maternal hyperlipidaemia being atherogenic to the fetus and increasing tendencies of future atherosclerosis, a long-term follow-up studies of offsprings of mothers with FH who did not inherited the disease is recommended. The result will demonstrate evidence of effects of maternal hyperlipidaemia on fetal atherosclerosis and or predisposition to future atherosclerosis in these offsprings.

6.1. Fetal lipoproteins in pre-eclampsia

Successful placental development is very important for normal fetal growth, and may condition health and well being during childhood and even adulthood [88], because it forms the interface for nutrients, fluids and gas exchange between mother and fetus. Pre-eclampsia (PE), a human pregnancy specific vascular disorder, defines a pathologic condition that affects the mother and can adversely influence the feto-placental unit. PE is associated with placental dysfunction, oxidative stress[1, 89], dyslipidaemia[16,90] and endothelial cell activation, and is a major cause of maternal and fetal morbidity and mortality[88] A pro-atherogenic lipid profile, characterized by increased TG levels with reduced HDL concentration[90, 91] and increased small, dense LDL particles[90] has been described. In contrast other studies demonstrated a dominance of buoyant LDL-1 and a significant decreased of small, dense LDL, namely LDL-5 and LDL-6[92]. Notwithstanding, it has been suggested that an abnormal lipid metabolism may not only be a manifestation of PE, but dyslipidaemia may play an essential role in its pathogenesis

Apart from genetic predisposition, the second group of disorders associated with an increased risk of PE includes a variety of chronic conditions such as dyslipidaemia, diabetes mellitus, hypertension, renal diseases, and various thrombophilias[93]. These disorders can be convincingly grouped together based on their common association with vascular endothelial dysfunction, especially those which have been included in the proposed metabolic syndrome [93]. Ironically also all are associated with dyslipidaemia. Although PE is a multifactorial disorder, it is therefore tempting to ask, could dyslipidaemia be the central focal point linking these disorders into pathogenesis of PE? One of the abnormalities found in the abnormal placental bed is presence of acute atherosis in desidual vessels, characterized by accumulations of foam cells and perivascular mononuclear cell infiltration. Reduced placental perfusion and placental/fetal hypoxia may develop.

Catarino[16], et al, while comparing lipid and lipoproteins in normal pregnant and PE pregnant women found an enhanced physiologic hyperlipidaemia. However, the most striking difference noted in PE women was the rise TGs that almost double the median value compared to normal pregnant women. Higher TGs value has been associated with endothelial dysfunction and may therefore play an important role in the pathogenesis of PE. Considering the placental dysfunction and lipid changes occurring in PE, fetal lipid metabolism can be affected due to an altered placental transfer of lipids. Maternal TGs does not cross the placenta. It has to be hydrolyzed by human placental LPL into FFAs which is then transported to the fetus. Fetal TG levels are therefore dependent on maternal TGs. Moreover, the placenta also contains receptors for VLDL, LDL and HDL which also transport TGs and other esterified lipid to the fetus (23)

Catarino[16],et al observed that lipid levels observed in umbilical cord blood (UCB) from normal pregnancy were significantly lower than those found in maternal blood except for HDL, which was similar. In addition, LDL:HDL ratio in neonates of normal pregnancies are much lower than the values in normal pregnant mothers. In contrast, lower values of HDL and ApoA-1 and higher TG levels were noted in neonates of PE mothers. In addition, higher LDL:HDL ratio, a decreased HDL which is more pronounced than ApoA-1, suggest that fetal loading of ApoA-1 with cholesterol is significantly less in PE. Hence fetal HDL composition is likely to be altered due possibly to enrichment with the enhanced hypertriglyceridaemia. Also observed in the PE is a significantly higher value of TGs in UCB which is parallel with significant increased TGs in maternal blood. Since hypertriglyceridaemia is considered a maternal adaptation in order to assure fetal growth in normal pregnancy, the exaggerated hypertriglycedaemia noticed in PE mothers could be a compensation pathway to face the uteroplacental hypoperfusion in order to enhance FAs transfer to the fetus. In line with this, it seems LPL expression is also enhanced in PE as was observed in IUGR [94]. Taken together, it appears lipid transfer from maternal side in PE mothers to their fetus are altered in both quantity and quality and does not seems to be protective as noticed in neonates of normal pregnancies. In support of this PE has been associated with reduced fetal birth weight [95, 96] and the expression of lipoprotein receptors are decreased in the placenta of women with PE.

PE pregnancies is associated with an enhanced hypertriglyceridaemia, which seems to have a negative impact on fetal lipid profile, as reflected by a higher atherogenic LDL:HDL ratio, decreased HDL, disproportionate decreased in HDL and ApoA-1 and enhanced hypertriglycedaemia, children born in pregnancies associated with PE deserve a closer clinical follow-up later in life.

6.2. Role of lipid metabolism in pathogenesis of intrauterine growth restriction (IUGR)[17]

It was proposed that the abnormal lipid metabolism noted in pre-eclampsia was in an attempt to compensate for the placental insufficiency [97], given the physiological role of gestational hyperlipidaemia in supplying both cholesterol and triglycerides to the rapidly developing fetus [98]. In contrast Dabi[17], et al demonstrated that concentration of total cholesterol (TC), TGs, LDL and VLDL observed to increase with increasing gestational age in normal pregnancies, these lipids and lipoproteins decreased with increasing gestational age in pregnancies with IUGR. HDL did not change significantly. These findings certainly indicated that pregnancies complicated by IUGR are associated with an abnormal lipid profile, particularly decreased levels of TC, TGs, LDL and HDL(Dabi [17],et al Sattar[18], et al), see table 7

Weeks	Group A			Group B		
	28-31	32-36	37-40	28-31	32-36	37-40
TC	216.3+29.2	202.5+26.0	191.5+34.5	238.0+17.0	250.7+27.6	260.3+23.6
TGs	173.83+78.18	168.23+51.73	137.83+18.25	148.25+15.31	160.36+24.43	171.14+41.56
HDL-C	42.0+5.3	43.0+4.4	46.4+3.2	44.4+4.5	43.0+6.0	41.6+8.2
LDL	139.4+23.9	126.3+21.8	117.0+31.8	165.9+26.5	174.8+26.5	184.5+23.4
VLDL	34.78+15.63	33.64+10.34	27.26+3.81	29.77+3.06	32.05+4.91	34.22+8.31

Table 7. Lipid and Lipoprotein changes with advancing gestational age(Group A=Pts with IUGR and Group B=Pt with normal pregnancy)

It is well known that normal fetal development needs the availability of both essential fatty acids and long chain polyunsaturated fatty acids (LCPUFAs), thus making a persuasive case indicating a relationship between nutritional status of mother during gestation reflecting her lipid profile and fetal growth. From observations in study by Dabi[17], et al and similar findings in other studies, it is possible that the decreased concentrations of TC, TGs, VLDL and LDL may have decreased availability of glycerol, LCPUFAs and essential fatty acids to the fetuses of mothers with otherwise normal pregnancy ultimately leading to IUGR. In addition to above findings, Sattar[18], et al observed a decreased in levels of VLDL-2 and IDL in IUGR pregnancies, which are precursors of LDL. Taken together, the decreased cholesterol levels (mainly reflected as decrease LDL) may be due to decreased synthesis of LDL in women with IUGR. The HDL: apoA and apoB:apoA ratios were found to be higher in the IUGR and was suggested that blood lipid modifications in the IUGR group are partly secondary to changes in HDL metabolism and the competitive inhibition of fibrinolysis by apoB which is increased in pregnancy with IUGR. This indicated that apoA: apoB ratio could be a good marker for the early detection of IUGR. Taken together also, these findings definitely generated considerable interest in certain aspects of fetal growth and its relationship to blood lipid levels during pregnancy. However, more study is recommended aiming at analyzing the otherwise normal pregnancies associated with IUGR, particularly to elucidate the hypothesis that the decrease in TGs(and particularly LDL and VLDL-2) compromises the supply of substrates for energy production to the growing fetus resulting in IUGR. The effect of the changes in lipid profile and its translation in changes in blood viscosity needs more extensive research including detailed analyses of apoA and apoB levels in these patients.

6.3. Pregnancy-induced Supraphysiologic hyperlipidaemia

It is well known in literature that hyperlipidaemia is a normal metabolic adjustment in pregnancy benefiting both mother and the fetus. However, some women may not be able to adapt to the hyperlipidaemic stress of pregnancy. In addition, in similarity with other gestational metabolic syndromes such as gestational diabetes mellitus (GDM), pregnancy-induced hypertension (PIH), pre-eclampsia, eclampsia, etc, some of them may develop a state of Supraphysiologic hyperlipidaemia, defined as lipid levels greater than 95th percentile for the corresponding gestational age, because of failed adaptation to requirement of pregnancy. Supraphysiologic hyperlipidaemia may serve as a marker for what is cited by Montes[99], et al, a 'prelipaemia' in the same way that GDM is a marker for pre-diabetes.

The characteristics of Supraphysiologic hyperlipidaemia, as observed by Montes[99], et al, are that, the antepartum hyperlipidaemia may return to normal levels postpartum more slowly than normal, the presence of HDL cholesterol concentrations that are persistently low antepartum and postpartum, and the patients do have hyperlipidaemic family members. In contrast, hypercholesterolemia is not greatly exaggerated in pregnancy among these women. Are there future consequences of the pregnancy-induced Supraphysiologic hyperlipidaemia? Long-term follow-up studies of women with genetically well-characterized disorders of lipoprotein metabolism are required to determine if an abnormal lipoprotein response in pregnancy can identify prelipaemic subjects and distinguish among the major disorders of lipoprotein metabolism. Identification of the prelipaemia will provide an opportunity to study prospectively the natural progression, potential for atherosclerosis, and possible treatment of hyperlipidaemia from early adulthood.

Author details

D.S. Mshelia and A.A. Kullima
University of Maiduguri/University of Maiduguri Teaching Hospital, Nigeria

Acknowledgement

I, D.S. Mshelia, greatly appreciate the support given to me by my lovely wife, Aisha A Buba and my children. Similarly, I, A.A. Kullim, sincerely appreciate the support given to me by my wife, Yagana Umara and my children (Inna and Yaanama).

7. References

[1] Herrera E and Ortega-Senovilla H(2010). Maternal lipid metabolism during normal pregnancy and its implications to fetal development. Clin Lipidol; 5(6):899-911.
[2] Lippi G, Albiero A, Montagnana M, et al (2007). Lipid and lipoprotein profile in physiological pregnancy; 53(3-4):173-177.
[3] Brizzi P, Tonolo G, Esposito F, Puddu L, Dessole S, Maioli M, et al, (1999). Lipoprotein metabolism during normal pregnancy; 181:430-434
[4] Hadden DR, and McLaughlin C (2009). Normal and abnormal maternal metabolism during pregnancy. Seminars in fetal and neonatal medicine; 4:66-71
[5] Hytten F, Chamberlain G (1980). Clinical physiology in obstetrics. Oxford, United Kingdom: Blackwell Scientific Publications;---
[6] Food and Nutrition Board, Committee on Nutrition of mother and preschool child. Laboratory indices of nutritional status in pregnancy. Washington, DC: National Academy of Sciences, 1978

[7] Denne SC, Patel D, Kalhan SC (1991). Leucine kinetics and fuel utilization during a brief fast in human pregnancy. Metabolism 40:1249-1256

[8] King JC (2000). Physiology of pregnancy and nutrient metabolism. Am J Clin Nutr. 719Supll0: 1218S-1225S.

[9] Basaran A(2009). Pregnancy-induced hyperlipoproteinaemia: Review of the literature. Reproductive Sciences. 1695):431-437.

[10] Butte NF (2000). Carbohydrate and lipid metabolism in pregnancy: normal compared with gestational diabetes mellitus. Am J Clin Nutr. 71(suppl): 1256S-1261S

[11] Winkler K, Wetzka B, Hoffmann MM, Fredrich I, Kinner M, Baumstark MW, et al (2000). Low density lipoprotein(LDL) subfractions during pregnancy: Accumulation of buoyant LDL with advancing gestation. Metab 85: 4543-4550

[12] Alvarez JJ, Montelongo A, Iglesias, A, Lasuncion MA, and Herrera E (1996). Longtudinal study on lipoprotein profile, high density lipoprotein subclass, and postheparin lipase during gestation in women. J Lipid Research; 37:299308

[13] Lyall F and   Greer AI(1996). The vascular endothelium in normal pregnancy and pre-eclampsia. Reviews of reproduction 1:107-116.

[14] Napoli C, D'Armiento FP, Mancini FP, Witztum JL, Palumbo G, Palinski W(1997). Fatty streak formation occurs in human fetal aortas and is greatly enhanced by maternal hypercholesterolaemia. Intimal accumulation of low density lipoprotein and its oxidation precede monocyte recruitment into early atherosclerotic lesions. J. Clin. Invest. 100:2680-2690.

[15] Napoli C, Witztum JL, de Nigris F, Palumbo G, D'Armiento FP, Palinski W.(1999). Intracranial ateries of human fetuses are more resistant to hypercholesterolemia-induced fatty streak formation than extracranial arteries. Circulation 99:2003-2010.

[16] Catarino C, Rebelo I, Belo L, Rocha-Pereira P, Rocha S, Casto EB, Patricio B, et al (2008). Fetal lipoprotein changes in pre-eclampsia. Acta Obstetricia et Gynecologica 87:628-634.

[17] Dabi DR, Manish P and Vikas G (2004). A cross-sectional study of lipids and lipoproteins in pregnancies with intrauterine growth retardation. 64(5):467-472.

[18] Sattar N, Greer IA, Galloway PJ, Packard CJ, Stepherd J, Kelly T, et al (1999). Lipid and lipoprotein concentrations in pregnancies complicated by intrauterine growth restriction. J clin Endocrinol Metab. 84:128-130.

[19] Hegele RA (1991). Hyperlipidaemia in pregnancy. Can Med Assoc J. 145(12):1596.

[20] Yusuf S, Hawken S Qunpuu S, et al (2004). Effect of potentially modifiable risk factors associated with myocardial infarction in 52 countries(the INTERHEART study): case-control study. Lancet. 364:937-952.

[21] Millan J, Pinto X, Munoz A, Zuniga M, Rubies-Prat J, Pallardo LF, et al(2009). Lipoprotein ratios: Physiological significance and clinical usefulness in cardiovascular prevention. Vascular health and risk Management 5:757-765

[22] Toleikyte I, Retterstol K, Leren PT, Iversen PO (2011). Pregnancy outcome s in familial hypercholesterolaemia: a registry-based study. Circulation 124:1606-1614.

[23] Ito MK, McGowan MP, Moriarty PM (2011). Management of familial hypercholesterolaemia in adult patients: recommendations from the National Lipid

association Expert Panel on familial Hypercholesterolaemia. I Clin Lipidol. 5(3 suppl): S38-S45.

[24] Edison RJ, Muenke M (2004). Central nervous system and limb anomalies in case reports of first-trimester statin exposure. N. Engl. J Med. 350:1579-1582.

[25] Herrera E. Lipid metabolism in pregnancy and its consequences in the foetus and newborn. Endocrine 2000; 19:43-55.

[26] Blackburn ST, Loper DL (1992). Carbohydrate, fat, and protein metabolism. In: Blacburn ST, Loper DL, Editors, Maternal, fetal, and neonatal physiology: a clinical perspective, Philadelphia:W.B. Saunders: p, 583-613

[27] Clapp j, Seaward BL, Seamaker RH, and Hiser j (1982). Maternal physiologic adaptations to early human pregnancy. Lancet 1:588-592

[28] Hytten FE (1991). Weight gain in pregnancy. In: Hytten F. Chambelain G, editors. Clinical physiology in obstetric. 2nd ed. Oxyford Blacwell: p. 174

[29] Herrera E (2002). Lipid metabolism in pregnancy and its consequences in the fetus and newborn. 19(1): 43-55.

[30] Salameh W and Mastrogianis D (1994). Maternal hyperlipidaemia in pregnancy. Clin Obstet. Gynaecol 37:66

[31] Couch SC, Philipson EH, Bendel RB, Pujda LM, Milvae RA and Lammi-Keefe CJ (1998). Elevated lipoprotein lipids and gestational hormones in women with diet-treated gestational diabetses mellitus compared to healthy pregnant controls. J Diabetes and Its Complications 12(1): 1-9

[32] Mazur A, Ozgo M and Rayssiguier Y (2009). Altered plasma triglyceride-rich lipoproteins and triglyceride secretion in feed-restricted pregnant ewes. Veterinani Medicina, 54(9): 412-418

[33] Sattar N, Greer IA, Louden J, Lindsay G, Mcconnell M, Shepher J and Packard CJ (1997). Lipoprotein subfraction changes in normal pregnancy: threshold effect of plasma triglyceride on appearance of small, dense low density lipoprotein. J Clin Endocrinol Metab. 82(8):2483-2491

[34] Herrera E, Lasuncion MA, Zorzano A (1992): Changes with starvation in the rat of the lipoprotein lipase activity and hydrolysis of triacylglycerols from triacylglycerol-rich lipoproteins in adipose tissue preparations. Biochem J 210(3):639-643

[35] Koschinsky T, Gries FA, and Herberg L (1971): Regulation of glycerol kinase by insulin in isolated fat cells and liver of Bar Harbor obese mice. Diabetologia 7(5): 316-322

[36] Thenen SW and Mayer J (1975): adipose tissue glycerokinase activity in genetic and acquired obesity in rats and mice. Proc Soc. Exp. Boil Med. 148(4):953-957

[37] Ramos MP, Crespo-Solans MD, del Campo S, Cacho J and Herrera E (2003): fat accumulation in the rat during early pregnancy is modulated by enhanced insulin responsiveness. Am J Physiol Endocrinol Metba. 285(2):E318-E328

[38] Buch I, Hornnes PJ, Kuhl C (1986): Glucose tolerance in early pregnancy. Acta Endocrinol (Copenh.) 112(2):263-266

[39] Herrera EM, knopp RH and Freikel N (1969). Urinary excretion of epinephrine and norepinephrine during fasting in late pregnancy in the rat. Endocrinology 84(2): 447-450

[40] Sivan E and Boden g (20030. Free fatty acids, insulin resisitance, and pregnancy. Curr. Diab. Rep. 3(4): 319-322

[41] Lopez-Saldado I, Betancor-Fernandez A and Herrera E (2002). Differential metabolic response to 4hr food deprivation at different periods of pregnancy in the rat. J Physiol Biochem 58(2): 75-83.

[42] Herrera E and Lasuncion MA(2004). Maternal-fetal transfer of lipid metabolites. In: fetal and neonatal physiology (vol 1). Polin RA, fox WW, Abman SH(Eds). Saunders, Philadelhia, PA, USA

[43] Shambaugh GE, Metzger BE, Radosevich JA(1992). Nurient metabolism and fetal brain development. In: Perinatal biochemistry, Herrera E and Knopp RH(Eds). CRC Press, Posca Raton, FL, USA

[44] Monlelango A, Lasuncion MA, Pallardo LF and Herrera E(1992). Longitudinal study of plasma lipoproteins and hormones during pregnancy in normal and diabetic women. Diabetes 42(12): 1651-1659

[45] Alvarez JJ, Montelongo A, Iglesias A, Lasuncion MA and Herrera E(1996): Longitudinal study on lipoprotein profile, high density lipoprotein subclass, and postheparin lipases during gestation in women. J Lipid Res 37(2):299-308

[46] Desoye G, Schweditsch MO, Pfeiffer KP, Zechner R and Kostner GM(1987). Correlation of hormones with lipid and lipoprotein levels during normal pregnancy and postpartum. J Clin Endocrinol Metab. 64:704-712

[47] Wasfi I, Weinstein I and Heimberg M(1980). Increased formation of triglyceride from oleate in perfused livers from pregnant rats. Endocrinology 107(2): 584-590

[48] Wasfi I, Weinstein I and Heimberg M(1980). Hepatic metabolism of [1-14C]oleate in pregnancy. Biochim. Biophys. Acta 619(3):471-481

[49] Ramos P and Herrera E(1995). Reversion of insulin resistance in the rat during late pregnancy by 72h glucose infusion. Am J Physiol. 269(5 Pt 1), E858-E863)

[50] Jones HN, Powell TL, Jansson T(2007). Regulation of placental nutrient transport—a review. Placenta 28(8-9):763-774

[51] Koletzko B, Larque E and Demmelmair H(2007). Placental transfer of long-chain polyunsaturated fatty acids(LU-PUFA). J Perinat Med 35(suppl 1):S5-S11

[52] Dutta-Roy AK(2000). Transport mechanism for long-chain polyunsaturated fatty acids in the human placenta. Am J clin Nutr. 71(Suppl 1):315S-322S

[53] Herrera E, Ortega H, Alvino G, Giovannini N, Amusquivar E, and Cetin I(2004). Relationship between plasma fatty acids profile and antioxidant vitamins during normal pregnancy. Eur J Clin Nutr. 58(9): 1231-1238

[54] Koletzko B and Braun M(1991). Arachidonic acid and early human growth:Is there a relation?. Ann nutr Metab 35(3):128-131

[55] Woollett LA(2005). Maternal cholesterol in fetal development: transport of cholesterol from the maternal to fetal circulation. Am J clin nutr. 82:1155-1161

[56] Stefulj J, Panzenboeck U, Becker T, et al, (2009). Human endothelial cells of the placental barrier efficiently deliver cholesterol to the fetal circulation via ABCA1 and ABCG 1. Circ Res 104950:600-608

[57] Lasuncion MA, Lorenzo J, palacin M and Herrera E(1987). Maternal factors modulating nutrient transfer to fetus. Boil Neonate 51(2):86-93

[58] Martin-Hidalgo A, Holm C, Belfrage P, Schotz MC and Herrera E(1994). Lipoprotein lipase and hormone-sensitive lipase activity and mRNA in rat adipose tissue during pregnancy. Am J Physiol 266(6 Pt 1):E930-E935

[59] Superko HR and King S 111 (2008). Lipid management to reduce cardiovascular risk: a new strategy is required. Circulation 117:560-568

[60] Loshak D. Ratio of Total to LDL Cholesterol is Best Predictor of Coronary heart Disease: A DG Review of: "Efficacy of cholesterol levels and ratios in predicting future coronary heart disease in a Chinese population". Am. J. Cardiol. 2001; 88:737-743.

[61] Gotto AM, Assmann G, Carmena R, et al(2000). The ILIB lipid hand-book for clinical practice: blood lipids and coronary heart disease. 2nd ed New York, NY: International Lipid Information Bureau; p 52, 53, 201

[62] Manninen V, Tenkanen L, Koskinen P, et al(1992). Joint effects of serum triglyceride and LDL cholesterol and HDL cholesterol concentrations on coronary heart disease risk in the Helsinki Heart Study. Implications for treatment. Circulation 85:37-45

[63] Walldius G, Junter I, Aastveit A, Holme I, Furberg CD, Sniderman AD(2004). The ApoB-ApoA-1 ratio is better than the cholesterol ratios to elstimate the balance between the plasma proatherogenic and antiatherogenic lipoproteins and to predict coronary risk. Clin Chem Lab Med 42:1355-1363

[64] Sniderman AD, Junter I, Holme I, Aastveit A, Walldius G(2006). Error that result from using the ApoB/ApoA-1 ratio to identify the lipoprotein-related risk of vascular disease. J Intern Med 259:455-461

[65] Walldius G, Junter I(2006). The ApoB/apoA-1 ratio: a stronger, new risk factor for cardiovascular disease and a target for lipid-lowering therapy---a review of evidence. J Intern Med 259:493-519

[66] Thompson A and Danesh J(2006). Association between apolipoprotein B, apolipoprotein A1, the apolipoprotein B/A-1 ratio and coronary heart disease: a literature-based meta-analysis of prospective studies. J Intern Med 259:481-492

[67] Dobiasova M and Frohlich J(2001). The plasma parameter log(TG/HDL-C) as an atherogenic index: correlation with lipoprotein particle size and esterification rate in apoB-lipoprotein-depleted plasma(FERHDL). Clin Biochem 34:583-588

[68] Vega GL, Beltz WF, Grundy SM(1985) Low density lipoprotein metabolism in hypertriglyceridaemic and normolipidemic patients with coronary heart disease. J Lipid Res 26:115-126

[69] Linn S, Fulwood R, Carroll M, Brook JG, Johnson C, Kalsbeek WD, and Rifkind BM(1991). Serum total cholesterol:HDL cholesterol ratios in US white and Black adults by selected demographic and socioeconomic variables(HANES II). Am J Public Health; 81(8): 1038-1043

[70] American Heart Association. Cholesterol statistics from National Health and Nutrition Examination Survey(NHANES), 1999-2004, National Center for Health Statistics and the NHLB1. Available at:
http://www.americanheart.org/presenter.jhtml?identfier=4506. Accessed on 12-11-07.

[71] [71]Tan CE, Squires L, Caslake MJ, et al(1995). Relationship between very low and low density lipoprotein subfractions in normolipaemic men and women. Arterioscler Thomb Vasc Biol 13:1839-1848

[72] McNamara JR, Campos H, Ordovas JM, Peterson JH, Wilson PWF, Schaefer E(1987). Effect of gender,age, and lipid status on low density lipoprotein subfraction distribution. Arteriosclerosis 7:483-490

[73] Austin MA, King MC, Vraizian KM, Krauss RM(1990). The atherogenic lipoprotein phenotypes: a proposed gentic marker for coronary heart disease risk. Circulation 82:495-506

[74] Krauss RM(1991). The tangled web of coronary risk factors. Am J Med 90:36S-41S

[75] Griffin BA, freeman DJ, Tait GW, et al(1994). Role of plasma triglyceride in the regulation of plasma low density lipoprotein(LDL) subfractions: relative contribution of small, dense LDL to coronary heart disease risk. Atherosclerosis 106:241-253

[76] Austin MA and Edward KL(1996). Small, dense low density lipoprotein, and insulin resistance syndrome and noninsulin-dependent diabetes. Curr Opin Lipidol 7:167-171

[77] Toescu V, Nuttall SL, Martin U, et al(2004). Changes in plasma lipids and markers of oxidative stress in normal pregnancy and pregnancies complicated by diabetes. Clin Sci(Lond). 106:93-98

[78] Kilby MD, Neary RH, Mackness MI, Durrington PN(1998). Fetal and maternal lipoprotein metabolism in human pregnancy complicated by type 1 diabetes mellitus. J Clin Endocrinol Metab. 83:1736-1741

[79] Mshelia DS, Kulima AA, Gali RM, Kawuwa MB, Mamza YP, Habu SA, et al(2010).The use of plasma lipid and lipoprotein ratios in interpreting the hyperlipidaemia of pregnancy. 30(8):804-808

[80] Martin U, Davies C, Hayavi S, Hartland A and Dunne F(1999). Is normal pregnancy atherogenic?. Clin Sci 96: 421-425

[81] Rajman I, Maxwell S, Cramb R and Kendall MJ(1994). Particle size: the key to the atherogenic lipoprotein? Q J Med 87:709-720

[82] Dejager S, Brukert E and Chapman MJ (1993). Dense LDL subspecies with diminished oxidative resistance predominate in combined hyperlipidaemia. J Lipid Res 34:295-308

[83] Tribble DL, Van der Berg JIM, Motchnik PA et al (1994). Oxidative susceptibility of low density lipoprotein subfractions is related to their ubiquinol-10 and alpha-tocopherol content. Proc Natl. Acad. Sci USA 91: 1183-1187

[84] Campos H, Genest JJ, Blijlevens E et al (1992). Low density lipoprotein particles size and coronary artery disease. Arteriosclerosis Thomb 12:187-195

[85] Bengtsson C, Rybo G and Westerberg H (19730. Number of pregnancies, use of contraceptives and menopausal age in women with ischaemic heart disease, compared to a population sample of women. Acta Med Scand Suppl 549: 75-81

[86] Napoli C, Glass CK, Witztum JL, Deutsch R, D'Armiento FP Palinski W(1999). Influence of maternal hypercholesterolaemia during pregnancy on progression of early atherosclerotic lesions in childhood: Fate of early lesions in Children(FELIC) study. Lancet 354:1234-1241

[87] Palinski W, D'armiento FP, Witztum JL, de Nigris F, Casanada F, Condorelli M, et al.(2001). Maternal hypercholesterolaemia and treatment during pregnancy influence the long-term progression of atherosclerosis in offspring of rabbits 89:991-996

[88] Barker D(1998). In utero programming of chronic disease . Clin Sci. 95:115-128

[89] Serdar Z, Gur E, Colakodullary M, Develiodlu O, Saradol E (2003). Lipid and protein oxidation and antioxidant function in women with mild and severe pre-eclampsia. Arch Gynaecol Obstet 268:19-25

[90] Belo L, Caslake M, Gaffney D, Santos-Silva A, Pereira LL, Quintanilha A, et al(2002). Cahnges in LDL size and HDL concentration in normal and pre-eclamptic pregnancies. Atherosclerosis 162:425-432

[91] Williams M, Woelk G, King I, Jenkis I and Mahomed K(2003). Plasma carotenoids, retinol, tocopherols and lipoproteins in pre-eclamptic and normotensive pregnant Zimbabwean women. Am J Hypertens 16:665-672

[92] Winkler M, Wetzka B, Hoffmann M Friedrich I, Kinner M, Baumstark M et al(2003). Triglyceride-rich lipoproteins are associated with hypertension in pre-eclampsia. J Clin Endocrinol Metab 88:1162-1166

[93] Cudihy D and Lee RV(2009). The Pathophysiology of pre-eclampsia: current clinical concepts. 29970: 576-582

[94] Tabano S, Alvino G, Antonazzo P, Grati F, Miozzo M, Cetin I(2006). Placenta LPL gene expression is increased in severe intrauterine growth-restricted pregnancies. Pediatr Res. 59:250-253

[95] Lau T, Pang M, Sabota D, Leung T(2005). Impact of hypertensive disorders of pregnancy at term on infant birth weight. 84:875-877

[96] Sanchez S, Zhang C, Williams M(2003). The influence of maternal triglyceride lvels on infant birth weight in Peruvian women with pre-eclampsia. J Mater Fetal Noenatal Med. 13:328-333

[97] Sitadevi C, Patrudu MB, Kumar Ym et al(1981). Longitudinal study of serum lipids and lipoproteins in normal pregnant and puerperium. Trop Geogr Med 33:319-323

[98] Franz H and Wendler D(1992). A controlled study of maternal serum concentration of lipoproteins in pregnancy-induced hypertension. Arch gynecol Obstet 252:1-6

[99] Montes A, Walden CE, knopp RH, Cheung M, Chapman MB and Albers JJ(1984). Physiologic and Supraphysiologic increases in lipoprotein lipids and apolipoproteins in late pregnancy and postpartum: Possible markers for the diagnosis of 'prelipemia'. Arteriosclerosis 4:407-417.

Chapter 5

The apoB/apoA-I Ratio is a Strong Predictor of Cardiovascular Risk

Göran Walldius

Additional information is available at the end of the chapter

1. Introduction

In the present paper the rationale for including apolipoprotein (apo)B and apoA-I into clinical practice is reviewed. ApoB and apoA-I are the two major apolipoproteins involved in lipid transport and in the processes causing atherosclerosis and its complications. ApoB is the major protein in Very Low Density (VLDL), Intermediate Density (IDL) and Low Density Lipoproteins (LDL), one protein per particle (1). ApoA-I is the major protein in High Density Lipoprotein (HDL) particles **(Figure 1)**. The apoB number indicates the total number of atherogenic particles, the higher the number the higher is the cardiovascular (CV) risk.

ApoA-I reflects the anti-atherogenic potential in HDL particles, the higher the value the better protection of CV risk. The apoB/apoA-I ratio (apo-ratio) indicates the balance between atherogenic and anti-atherogenic particles, the higher the value, the higher is the CV risk. In previous papers we (2-6) and others (7-12) have reviewed the importance of apolipoproteins, mainly apoB and apoA-I, but also other apolipoproteins like apoC-II and apoCIII, apoE, and Lp(a) as markers of atherogenic risk. In this review many new data on apoB, apoA-I and the apo-ratio and their relations to cardiovascular (CV) risk are presented. The majority of these studies were published in the last 6 year period.

The debate today (mid 2012) is about whether LDL-C should remain as the primary variable for CV risk evaluation and target for lipid-lowering therapy. During the last few years non-HDL-C has been found and proposed to be the next primary target for CV risk evaluation and target for treatment (9-11,13,14). Notably, although LDL-C and non-HDL-C are considered the best CV risk markers most large studies of CV risk have shown that the lipid ratios, i.e. the TC/HDL-C, the LDL-C/HDL-C and the non-HDL-C/HDL-C ratios, are stronger than any specific single lipid fraction (2,3,4,6,15). The major aim of this paper is therefore to review papers on apoB, apoA-I and the apo-ratio related to risk of atherosclerosis and various clinical complications like myocardial infarction (MI), stroke and other severe events to find out if there is evidence for using apoB and apoA-I, and especially the apoB/apoA-I ratio (apo-ratio) motivating clinical use of these risk markers/predictors. Both similarities, but mainly differences between apos and conventional lipids to predict CV risk, will be highlighted. Methodological aspects and the role of apoB and apoA-I, the two determinants of the apo-ratio, will first be commented. The major part of the paper describes the role of the apo-ratio as a CV risk marker/predictor. The overall conclusion from this paper will be that apoB, apoA-I and the apo-ratio merit to be included in future guidelines in order to be recognized and used in clinical practice.

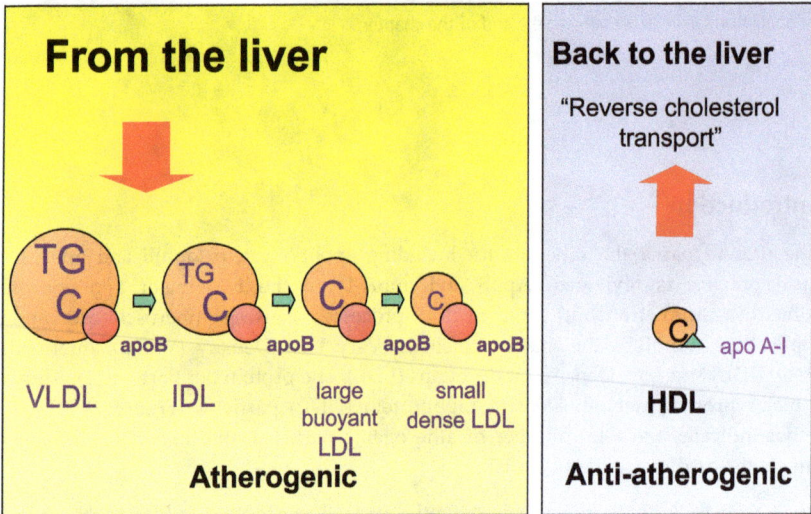

Figure 1. The figure shows the atherogenic particles containing one apoB protein per VLDL, IDL, large buoyant LDL, small dense LDL particles and the anti-atherogenic lipoproteins containing apoA-I. The balance between apoB and apoA-I, i.e., the apoB/apoA-I ratio, reflects the balance between the "bad cholesterol particles and the good cholesterol particles". This apo-ratio is strongly related to cardiovascular risk, the higher the ratio, the higher is the risk. (From reference 3).

2. Methodological pros and cons for using apoB, apoA-I and the apo-ratio versus conventional lipids

2.1. Methodological problems for various lipids

The most commonly used method world-wide to measure LDL-C is based on the Friedewald formula (16). However, errors are common and the methodological problems and shortcomings are not commonly recognized but have been discussed in many papers (17-25). Thus, the formula (LDL-C = TC – HDL-C – TG/5) is not valid for blood samples having triglycerides (TG) above 3.5-4 mmol/L, for patients with type III hyperlipoproteinemia or chylomicronemia or non-fasting specimens (17-19). The errors for

LDL-C can be false positive in the range of 2-17% or false negative between 12-15% if TG levels are very low or closer to 4 mmol/L. This may create large problems for both clinicians and patients since patients may be misclassified as being at risk or not at risk according to guidelines. Similarly, it may be difficult for the clinician to evaluate if a patient has been adequately treated to the target of LDL-C. Newer so called "direct LDL-C methods" have been developed and they are homogeneous methods, that is, assays that do not require a preliminary separation step, such as ultracentrifugation, or manual manipulation of the sample for determining LDL-C (9,18-20). However, these methods, although standardized at a given laboratory, do not always correlate well over the whole range of lipid values, and they are not even internationally standardized like those for apolipoproteins.

The practical problems of measuring HDL-C are also of concern and correlation between various methods are sometimes even worse than those for LDL-C (18,19,26). Consequently, the values for non-HDL-C (TC minus HDL-C) may also be subject to large variations due to the errors mainly for measuring HDL-C. However, there is an advantage for non-HDL-C over LDL-C determined by the Friedewald formula since non-HDL-C is not subject to influence of non-fasting that may distort the TG levels and make it difficult to obtain a correct value for LDL-C (27). Furthermore, non-HDL-C contains C from all atherogenic fractions i.e. VLDL, IDL and various forms of LDL. Thus, non-HDL-C which indicates the total mass of C is more likely to reflect the variation of atherogenic particle set up for many patients with various genotypes and phenotypes. Such patients may have a greater chance to be correctly identified as risk individuals based on non-HDL-C, rather than to an imprecise measure of only LDL-C. For the interested reader of methods and concerns of validity, see further excellent reviews (18,19,26,27).

2.2. Methodological advantages for apolipoproteins

There are methodological advantages of using apoB and apoA-I compared to LDL-C and HDL-C since the apo-methods have been internationally standardized according to WHO-IFCC already in 1990-ies (26,28,29). The standardization initiatives for apo B have proceeded more quickly and more successfully than for LDL-C. The WHO-IFCC collaboration has resulted in the development of secondary reference material to ensure traceability of manufacturer calibrators to an approved standard. The bias and imprecision for 22 immunonephelometric and immunoturbidimetric assays ranged were usually below 5%. These errors are commonly smaller than that for calculated LDL-C and lipid ratios. Costs for measuring apos can be much reduced if apos are introduced as routine methods. However, pedagogical aspects (education of physicians, patients and laymen), and the well documented and cemented LDL-paradigm will make it difficult to convince guideline committees to introduce apoB and apoA-I as CV risk predictors. Importantly, this should not invalidate that apos are accepted as strong risk markers especially since so many other methods determining LDL-C and HDL-C are accepted in guidelines despite rather weak correlations between various methods due to incomplete standardizations.

3. Physiological and pathophysiological aspects of apoB

3.1. ApoB production, circulation and distribution

ApoB-100 is produced in the liver and apoB-48 is synthesized in the gut (3,12). ApoB-100 is the dominating protein in plasma compared with minute amounts of apoB-48 even in the postprandial state. In most conditions, more than 90% of all apoB in blood is found in LDL. There are excellent reviews of how apoB-100 assembles VLDL in the liver, more details on VLDL composition (12), and some comments on the genetics of apoB (30-33). ApoB is present in VLDL, IDL large buoyant LDL, and small dense LDL (sdLDL), with one molecule of apoB in each of these atherogenic particles (1). Importantly, apoB does not occur on HDL particles. Thus, total apoB reflects the total number of potentially atherogenic particles (Figure 1). This is principally different from non-HDL-C which indicates the total mass of C. ApoB produced in the liver stabilizes and allows the transport of C and TG in plasma VLDL, IDL, large buoyant LDL and sdLDL. ApoB also serves as the ligand for the apoB and apoB,E receptors thereby facilitating uptake of C in peripheral tissues and in the liver as reviewed (2,3,12). ApoB may provoke atherogenesis since it can be entrapped in the arterial wall of the coronary arteries and also as exemplified by findings in femoral plaques (12,34,35) where it may be modified, oxidized and glucosylated and therefore also contribute in the process of plaque formation. In this process LDL-C with apoB infiltrates the arterial wall and many factors like adhesion molecules, cytokines, growth factors are involved in oxidation processes leading to inflammation and growth of plaques unless HDL bound apoA-I can neutralize these processes (see elsewhere in this paper). Interestingly, already in 1976 Hoff presented data showing that apoB and apoA-I were found in the arterial wall of the coronary and carotid arteries as well as in the aorta (35). Olofsson et al (12) discuss the intra-arterial metabolism of apoB and apoA-I and also Fogelstrand and Borén (36).

3.2. Plasma levels of apoB and target values for therapy

The levels of apoB in plasma may vary from 0.2 to above 3 g/L, with highest values for those with hereditary hypercholesterolemias. In the "normal case" the values for males and females do not differ much. Reference values have been published by Cantois et al. already in 1996 (37). The values slowly increase from childhood to adult life (2,3). Those who live to ages above 75 years commonly have relatively low apoB values since those with higher values may have died due to various CV events. During lipid-lowering therapy apoB targets have been recommended to be < 0.90 g/L for those at moderate risk and < 0.80 for those at a high risk, see further below (3,9,11,38). Values should be given in two decimals.

3.3. ApoB versus LDL-C and risk for CV events

One of the first publications on clinical risks during the course of myocardial infarction (MI) related to apoB and also to apoA-I was presented by Avogaro already in 1978 (39,40). In 1980 Sniderman et al. presented data indicating that hyperapoB with normal C levels was related to coronary atherosclerosis (41). Since then many reports have been published indicating that apoB is involved in atherogenesis and its complications like MI. In 1996 Lamarche et al. (42) showed that apoB was strongly associated with onset of coronary heart disease in 2,155 men aged 45–76 years followed for 5 years (Quebec Cardiovascular Study). The predictive effect of apoB remained after adjustment for TG, HDL-C and TC/HDL-C. ApoA-I was protective, but not as strong as the harmful apoB in multivariate analysis. In the 10-year follow up of the Atherosclerosis Risk in Communities (ARIC) study, apoB was measured in 12,339 middle-aged participants (43) and had predictive power above that of LDL-C, TG and HDL-C. However, despite strong univariate associations for apoB and LDL-C, apoB did not contribute to risk prediction in subgroups with elevated TG, with lower LDL-C, or with high apoB relative to LDL-C. This may be due to the error for apoB determination which was estimated at 17% which is considerably higher than the approximate 5% that is common in most recent trials.

Importantly, apoB has been found to have a stronger relation with CV risk than LDL-C in several other studies as reported in coming sections. These include the AMORIS study (44), especially at low values of LDL-C (see below), the Thrombo Study (45), the Thrombo Metabolic Syndrome Study (46), the Northwick Park Heart Study (47), the Nurses' Health Study (48) and amongst patients with type 2 diabetes in the Health Professionals Follow-up Study (49).

In the Copenhagen City Heart Study Benn et al. (50) studied 9,231 asymptomatic women and men from the Danish general population followed prospectively for 8 years and observed the following incident events: ischemic heart disease 591, MI 278, ischemic cerebrovascular disease 313, ischemic stroke 229, and any ischemic CV event 807. ApoB, adjusted for multiple common confounding risk factors, had a higher predictive ability than LDL-C in all these various ischemic events (p < 0.03 to < 0.001). They suggested that prediction of future ischemic cardiovascular events could be improved by measuring apoB.

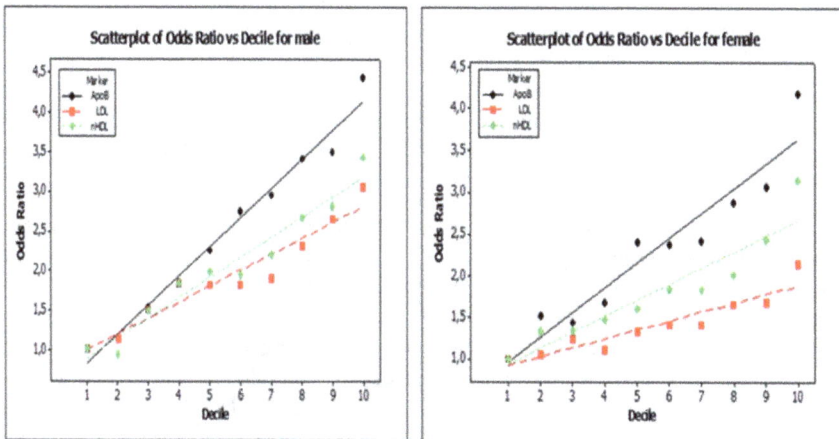

Figure 2. The AMORIS study; apoB, non-HDL-C and LDL-C (x-axis, deciles) versus risk of myocardial infarction (Odds Ratio) (y-axis) in males (left) and females (right).

In addition, in the placebo groups of several major statin clinical trials such as 4S (51), AFCAPS/TexCAPS (52,53) and LIPID (54) apoB was more informative than LDL-C as an index of the risk of CV events. Taken together, this strongly indicates that apo B is superior to

LDL-C in recognizing the risk of CV disease and effects of statin therapy. Additional results (55) also favor apoB over LDL-C, and others are also reported in the section on the apo-ratio below. Such major studies are the AMORIS (3,44,56,57). In our study we found the steepest risk-relationship for MI with increasing values of apoB followed by non-HDL-C and the lowest increase in relation to LDL-C values with similar risk progressions for men and women **(Figure 2 and Figure 3 left)**. Also in the INTERHEART (58,59) and ISIS-studies (60) as well as those summarized in the ERFC-meta-analyses (8,10) apoB was strongly related to risk of MI. In meta-analyses similar strong findings for apoB versus LDL-C are summarized by a large number of international scientists and clinicians in more detail (4,13,61,62).

4. ApoB versus non-HDL-C

ApoB indicates the number of atherogenic particles whereas non-HDL-C indicates the C mass from all atherogenic fractions like VLDL, IDL and the large buoyant LDL and the most atherogenic sdLDL fractions. But is apoB similar to or better than non-HDL-C in predicting risk? Although there is a similarity between apoB and non-HDL-C, they may have different metabolic fate and thus impact on risk. The rationale for using non-HDL-C is based on the fact that there is a close relationship between non-HDL-C and apoB values. Usually the correlation is about 0.80–0.85. However, correlation is not the same as concordance. In fact, two variables can be highly correlated but also be highly discordant, i.e. they do not correspond well. Either they are too high or too low compared with the other variable. Importantly, discordance produces major errors in the middle of the population distribution. Sniderman has frequently presented data with explanations of the advantages of apoB over LDL-C and non-HDL-C (4,13,14,21,62,63). Commonly, the sdLDL-particles contribute much to the large numbers of atherogenic particles, i.e. the apoB number is high and these small particles can easily penetrate into the arterial wall. However, in conditions with high non-HDL-C due to high VLDL-C and high large buoyant LDL-C the sdLDL-particles may be rather low in numbers indicating comparatively low numbers of apoB particles. These larger cholesterol-containing VLDL and IDL particles, although rich in C, do not easily penetrate into the arterial wall.

Of interest, the number of apoB is more closely associated with insulin resistance or markers of the metabolic syndrome than either LDL-C or non-HDL-C (3,62,63). Thus, in patients with hypertriglyceridaemia with normal, or even low LDL-C values, i.e. patients with the metabolic syndrome (MetS), and in patients with overt diabetes, apoB has been shown to be superior to non-HDL-C in predicting vascular risk (3,62,63). Again, even if non-HDL-C and apoB correlate, they are not the same biologically or clinically. In most cases apoB is associated with a higher CV risk than non-HDL-C as well as LDL-C. Furthermore, non-HDL-C may not be that easy to understand or explain for the clinician or the patient once they have learnt that the bad C is LDL-C.

Many studies and clinical trials have been published showing that apoB has a stronger capacity to identify all different phenotypes and to better predict CV risk than both LDL-C and non-HDL-C (3,11,61-63). Sniderman et al. (62) have published a convincing meta-analysis of results from published epidemiological studies that contains estimates of the relative risks of LDL-C, non-HDL-C and apoB of fatal or non-fatal ischemic CV events. Twelve reports including 23,455 subjects and 22,950 events, were analyzed. Whether analyzed individually or in head-to-head comparisons, apoB was the most potent marker of CV risk RR = 1.43; (95% CI, 1.35-1.51), LDL-C was the least RR = 1.25; (1.18-1.33), and non-HDL-C was intermediate RR = 1.34; (1.24-1.44). Only HDL-C accounted for any substantial portion of the variance of the results among the studies. They commented that in patients in whom LDL composition is normal, the cholesterol markers and apoB are equivalent markers of risk, i.e. correlation between the three markers is high. However, when the markers are discordant, that is, when LDL-C is normal but LDL-particles (P)(= apoB) is high or, alternatively, when LDL-C is high but LDL-P are normal, then apoB and non-HDL-C are better markers of risk than LDL-C. They calculated the number of clinical events prevented by a high-risk treatment regimen of all those >70th percentile of the US adult population using each of the 3 markers. Over a 10-year period, a non-HDL-C strategy would prevent 300,000 more events than an LDL-C strategy, whereas an apoB strategy would prevent 500,000 more events than a non-HDL-C strategy. These examples emphasize the greater potential for using apoB rather than non-HDL-C and LDL-C.

However, in another major meta-analysis by Boekholdt et al. (64) they studied 62,154 patients enrolled in 8 statin trials published between 1994 and 2008. Among 38,153 statin treated patients 158 developed fatal MI, 1,678 non-fatal MI, 615 fatal events from other coronary artery disease, 2,806 hospitalizations for unstable angina, and 1,029 fatal or nonfatal strokes occurred during follow-up. The adjusted HRs for major CV events per 1-SD increase were 1.13 (95% CI, 1.10-1.17) for LDL-C, 1.16 (1.12- 1.19) for non-HDL-C, and 1.14 (1.11-1.18) for apoB. These HRs were significantly higher for non-HDL-C than LDL-C (p = 0.002) and apoB (p = 0.02). Thus, from both these meta-analyses non-HDL-C stands out as a stronger predictor of CV diseases than LDL-C. The explanation for the different findings of apoB in these two meta-analyses is unclear but may be explained by the fact that the first study is based on data from a prospective risk studies, whereas the second study reflects effects of statins on lipid and lipoprotein metabolism. Further comments are given in the discussion.

5. Physiological and pathophysiological aspects of apoA-I

There are many subgroups of particles of HDL with different lipid and apo compositions (3,12,29). Beyond apoA-I there are other apos such as apoA-II, apoA-III, apoC-III, apoD and apoM. ApoA-I is the major protein in HDL and this protein is taken to represent HDL metabolism since it occurs almost exclusively in HDL particles. By measuring HDL-C the amount of C transported in blood is indicated to represent the reverse cholesterol transport (RCT), a major protective aspect of HDL metabolism – by laymen named "the good cholesterol". ApoA-I initiates the RCT process in peripheral tissues. ApoA-I has also many other functions beyond RCT since apoA-I is involved in anti-inflammation, anti-oxidation, anti-infectious activity, anti-proteas activity, anti-apoptotic, and anti-thrombotic functions (3). Furthermore, apoA-I can initiate the endothelial production of nitric oxide that is of vital help in producing vasodilation (3). Furthermore apoA-I may help to regulate glucose-insulin homeostasis. Thus, by measuring apoA-I you may get additional "protective" effects above those given only by the HDL-C number. For methodological reasons Warnick and others prefer to use apoA-I rather than HDL-C methods (19). HDL and apoA-I metabolism are reviewed in more detail, see ref. (3,12,29,65-67).

5.1. Plasma levels of apoA-I and target values for therapy

The plasma concentration of apoA-I can vary from 0.1 to over 3 g/L. Reference values have been published by Cantois et al. already in 1996 (37). There is little variation with fasting-non-fasting (68). Normally women have 0.1-0.3 g/L higher apoA-I values than men, similar to the higher HDL-C values for women. After menopause apoA-I values commonly decrease in parallel with HDL-C. However, there have been few published recommendations regarding what should be a "normal apoA-I value". A normal value for any adult should be at least close to 1 g/L or above. So far, there have been few recommendations on valid cut values indicating increased CV risk and target values. Values should be given in two decimals. For further comments, see the section on the apo-ratio.

5.2. Biological variation of apoA-I

ApoA-I and ApoA-II may also enter the cerebrospinal flow via the choroid plexus (69). Reduction in the HDL apoA-I/apoC-III ratio, changes in the HDL subpopulation distribution and an increase in HDL oxidation potential correlated with the development of MI in young patients (70). In a Korean study of 15,154 healthy subjects higher CRP levels were associated with significantly lower HDL-C and apoA-I levels, and also higher apoB values (71). In a US population of 8,708 apparently healthy population apoA-I was strongly positively associated, whereas apoB was significantly reduced with alcohol intake. Similarly the transaminases AST, ALT and gamma-GT increased with higher alcohol consumption (72).

5.3. ApoA-I and risk for CV events

Already in 1978,1979, Avogaro et al. (39,40) showed that apoA-I was as good as lipids in predicting myocardial infarction (MI) in those under 50 years of age but apoA-I was a better predictor in those over 60 years of age. In the Swedish APSIS study (73) Held et al. in 1994 studied patients with angina pectoris. During a median follow-up time of 3.3 years (2,663 patient years), 37 patients suffered a CV death, 30 suffered a non-fatal MI and 100 underwent a revascularization. Apo-I and TG were predictors of CV death or non-fatal MI in univariate analyses, but only apoA-I remained as an independent predictor in multivariate analyses. All lipid variables except LDL-C were related to the risk of revascularization in univariate analyses, but only apoA-I and apoB were independent predictors of such events. They concluded that apolipoprotein levels were better predictors of CV events than other lipid parameters in patients with stable angina pectoris.

Many studies have shown an inverse relationship between apoA-I and MI (3,44,74,52,53,59,60). In a study of Japanese Americans apoA-I predicted coronary heart disease only at low concentrations of HDL (75). High apoA-I values have been found to correlate with low risk for MI in AMORIS as indicated **(Figure 3, right).** Luc et al. also found that apoA-I is the best prospective risk marker of several other apoproteins in HDL (76). In the large INTERHEART case-control study apoA-I had a greater protective effect of MI at a wider range of apoA-I values than HDL-C (58,59). Patel et al. (77) found that ApoA-I levels are a consistent discriminator of atherosclerotic burden among patients with stable CAD.

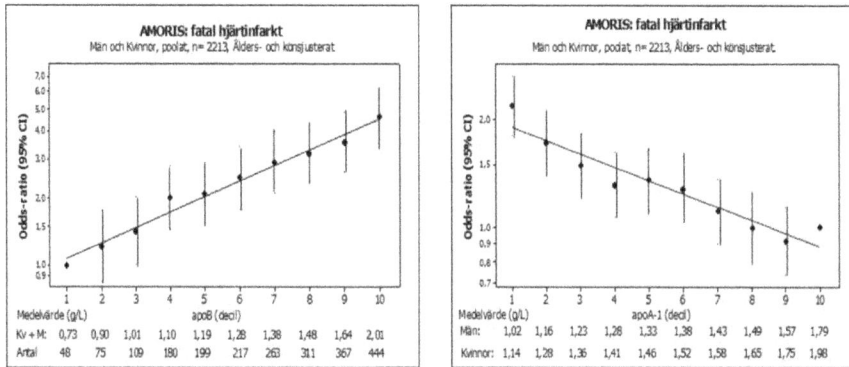

AMORIS: fatal hjärtinfarkt
Män och Kvinnor, poolat, n= 2213, Ålders- och könsjusterat

Medelvärde (g/L)				apoB (decil)					
1	2	3	4	5	6	7	8	9	10
Kv+M: 0,73	0,90	1,01	1,10	1,19	1,28	1,38	1,48	1,64	2,01
Antal 48	75	109	180	199	217	263	311	367	444

AMORIS: fatal hjärtinfarkt
Män och Kvinnor, poolat, n= 2213, Ålders- och könsjusterat

Medelvärde (g/L)				apoA-1 (decil)					
1	2	3	4	5	6	7	8	9	10
Män: 1,02	1,16	1,23	1,28	1,33	1,38	1,43	1,49	1,57	1,79
Kvinnor: 1,14	1,28	1,36	1,41	1,46	1,52	1,58	1,65	1,75	1,98

Figure 3. Left; The AMORIS study: Fatal myocardial infarction (Odds Ratio) is related to increasing values of apoB. The values are adjusted for age, TC and TG. Right; The AMORIS study: Fatal myocardial infarction is related to decreasing values of the apoA-I. The values are adjusted for age, gender, TC and TG. Similar pattern for men (män) and women (kvinnor).

In the CORONA study performed in patients with severe heart failure (placebo versus rosuvastatin) apoA-I, in univariate analysis, was the second best (after apoB plus apoA-I) of all different lipid fractions in predicting total death and MACE (MAjor Coronary Events). Furthermore, in a multivariate stepwise analysis apoA-I ranked fifth, better than high sensitivity CRP (hsCRP), of all 14 predictors of outcome where no conventional lipid fraction was significant. The best predictor was pro-BNP (78).

In a study of risk of stroke in Taiwan it was shown that apoA-I but not apoB levels may serve as an effect modifier of hypertension for the risk of stroke events (79).

In the combined analysis of data from the IDEAL statin trial and the Epic-Norfolk case-control study (80) very high HDL-C due to enlarged HDL-particles values were associated with increased rather than decreased CV risk. However, in contrast, apoA-I appears not to turn into a significant risk factor at high plasma concentrations. They conclude that apoA-I is associated with CHD risk independently from HDL size suggesting that the cardioprotective role of large HDL might be more closely related to its apoA-I content than to HDL size per se. These observations may have important consequences for future CAD risk assessment and novel treatment strategies. Indeed, several experimental studies have pointed to a crucial role for apoA-I in protection against atherosclerosis (3,12,65,81).

In the AFCAPS /TexCAPS statin trial (placebo versus lovastatin) multivariate analysis showed that apoA-I was better than HDL to predict outcome (52,53). In addition, the apo-ratio was the best of all lipids and apo-fractions to explain CV risk reduction, see further below.

6. General comments on the validity of using a ratio as a primary marker of risk

Lipid and lipoprotein ratios like TC/HDL-C and LDL-C/HDL-C have been used in various international guidelines for decades to define CV risk. However, LDL-C has in the vast number of guidelines dominated as the primary risk marker why ratios rarely are used today in clinical practice. One major reason why the lipid ratios are questioned as relevant risk markers is due to the fact that HDL-C is included in the value for TC, so HDL-C occurs both in the nominator and denominator of the ratio. Similarly, since LDL-C most commonly is derived by the Friedewald formula, HDL-C is involved as a factor for calculating LDL-C and therefore also indirectly in the nominator and denominator of that ratio. Therefore physicians are hesitant to the mathematical way of dividing various lipid numbers to obtain a mathematical, but, in their mind, not a biologically relevant ratio. When so called direct methods are used for measuring LDL-C this problem is less. In recent years non-HDL-C has been recommended as the next primary risk variable and the new non-HDL-C/HDL-C ratio has been defined. Interestingly, this ratio gives the same final number of risk as that of the TC/HDL-C ratio.

Most researchers and guidelines recommend the use the TC/HDL-C ratio since calculation of this ratio is not dependent on that blood sampling has been performed in the fasted state. This is the same argument as for using non-HDL-C rather than calculated LDL-C. The challenge now is can the apo-ratio, which summarizes the CV risk related to all atherogenic and all anti-atherogenic variables into one number, be the next rational choice as a primary risk variable? Does the apo-ratio add to information already obtained by lipids and lipid ratios? And are the values for apoB, apoA-I and especially the apo-ratio much influenced by other confounding risk factors? These and many other questions are addressed in the sections below based on a vast number of publications.

7. Prospective cardiovascular risk studies – Relations to apoB, apoA-I and the apoB/apoA-I-ratio

7.1. The AMORIS prospective study and risk of myocardial infarction (MI)

The Swedish AMORIS (Apolipoprotein-related MOrtality RISk) study is the largest of all studies in which apoB and apoA-I have been measured in more than 175.000 individuals followed prospectively for up to 25 years. The participating subjects were recruited from health check-ups during 1985-1996. Their age ranged from below 10 years to above 90 years. In these years health screening was very common in Sweden. Subjects included in the database called AMORIS were mainly healthy, not acutely ill or hospitalized and no subject participated in clinical trials. They were all treated by their general practitioners in the greater Stockholm area and they constitute a valid socio-economical cohort of the greater Stockholm population as indicated in several of our papers presented below. Large blood screening programs were used including some 8,000 determinations of LDL-C according to Friedewald. Simultaneously apoB and apoA-I were analyzed by automated immunoturbidimetric methods in all 175.000 subjects according to the WHO-IFCC protocol and in collaboration with their representatives (82). LDL-C was calculated according to the

Jungner formula (44). The Jungner formula yields the same LDL-C values as those obtained by using the Friedewald formula (16) as confirmed by Talmud et al. (47). Also HDL-C values were determined by the Friedewald formula once LDL-C was calculated by the Jungner formula. All analyses in the AMORIS laboratory database (confounding clinical risk factors like hypertension, diabetes and obesity were available in cohorts) were performed by automated methods at the same CALAB laboratory headed by Ingmar Jungner. Several papers were published describing the apoB, apoA-I and the apo-ratio characteristics of the population and the methods (3,21,44,82-85). In an early AMORIS study we have previously noted that patients with type IIB dyslipidemias, i.e. combined hypercholesterolemia and hypertriglyceridemia, had the highest apo-ratio (86). The subsequent CV manifestations were related to the laboratory variables obtained at the first visit to the physician.

In 2001 we presented the first endpoint paper based on 98,722 men and 76,831 women in the Lancet (44). We found a strong direct relationship between apoB and an indirect inverse relationship between apoA-I and risk of MI (men = 864, women = 359) **(Figure 3)**. Furthermore, apoB was a stronger risk factor than LDL-C especially at low values of LDL-C. The apo-ratio was the strongest lipid-related factor **(Figure 4)**. In the left part of the figure the values for apoB and apoA-I divided into quartiles are displayed in a three dimensional way. Thus, in those with highest values of apoB and in those with lowest apoA-I values the risk increased about 6-fold in a stepwise fashion compared to those with lowest apoB and highest apoA-I values. The highly significant results were similar for men and women and remained after adjusting for age, TC and TG. The figure clearly illustrate that the risk is about the same for those with an increased apoB at highest apoA-I levels, as the risk for those with lowest apoA-I levels but with low apoB values. Thus the figure illustrates the importance of measuring both apoB and apoA-I to get correct information on MI risk level. In the right part of the figure the same results can also be depicted as a straight line (semi-log scale) showing the impact of higher apo-ratio versus increased risk of MI. We also found that apoB was significantly better to predict risk than LDL-C especially for those with low values for LDL-C.

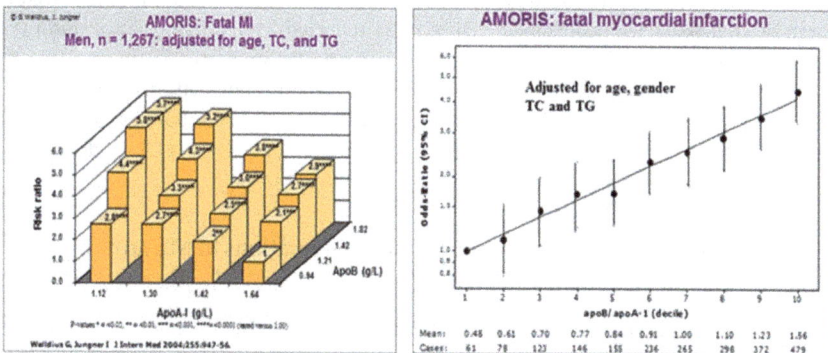

Figure 4. Left; The AMORIS study: Fatal myocardial infarction (Risk ratio) is related to increasing values of apoB and decreasing values of apoA-I. The values are adjusted for age, TC and TG. Similar pattern is seen for men and women (reference 3). Right; The AMORIS study: Fatal myocardial infarction is related to increasing values of the apoB/apoA-I ratio. The values are adjusted for age, gender, TC and TG. (Both figures from reference 3).

With increasing values of the apo-ratio there was a parallel increase in apoB, LDL-C, non-HDL-C and TG **(Figure 5, left)** and a decrease in apoA-I and HDL-C values **(Figure 5, right).** This figure illustrates that an increasing apo-ratio indirectly also indicate the contribution of the other lipids as risk factors. In multivariate analyses the apo-ratio is the strongest of all lipid-related variables and is thus the best summarizing risk variable.

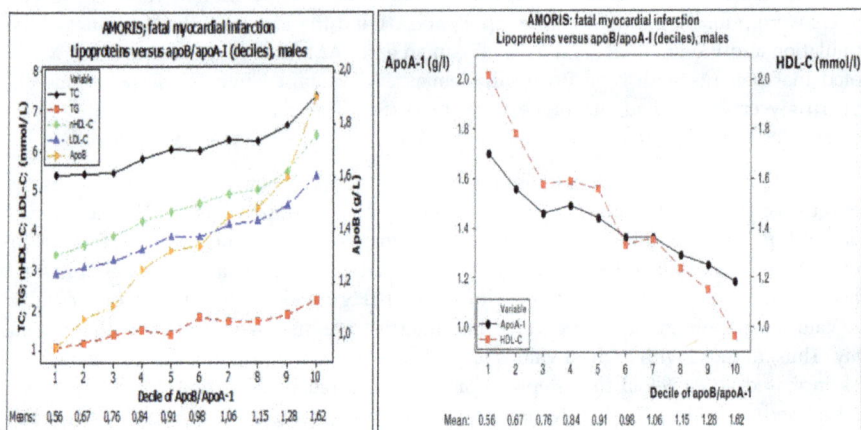

Figure 5. Left; the apo-ratio in deciles (x-axis) versus different atherogenic lipid fractions (y-axis). Right; the apo-ratio in deciles (x-axis) versus values for HDL-C and apoA-I (y-axis). Both figures from the AMORIS study (in reference 3).

In collaboration with Sniderman we have also published data from AMORIS showing that the apo-ratio has a significantly stronger relation with MI than any other lipid-based ratio (3,4,21).

In another AMORIS cohort including 69,029 men and 57,167 women who were followed for a mean of 10.3 years we determined LDL size as reflected by the LDL-C/apoB ratio (87). Because LDL size did not add predictive information to the apo-ratio, it appears that this apo-ratio also captures the risk related to LDL size. These findings add to our previously published results from AMORIS that indicates that the apo-ratio is the best single lipid-related summary index of risk and that TC, TG, non-HDL-C, and LDL-C do not add significant predictive power to the apo-ratio.

7.2. The apo-ratio and inflammatory risk factors – relations to MI, stroke and heart failure in the AMORIS study

The risk relation to age during prolonged follow-up was also studied in an AMORIS population (n = 149,121) free of previous MI at blood sampling. They were followed from 1985 to 2002 with respect to n = 6,794 first cases of MI. The mean value of the apo-ratio for men was 1.0 and for women 0.85 at baseline. In collaboration with Holme we found that the apo-ratio was somewhat stronger for those developing non-fatal than fatal MI (88). The risk was also stronger associated with the apo-ratio in those < 65 years of age than above, but risk remained significantly related to the apo-ratio also in the older population. In multivariate analyses the apo-ratio was a better predictor than TC/HDL-C. Furthermore, the apo-ratio added clinically significant information to TC/HDL-C in men as reflected by a net reclassification improvement (NRI) of 9.4% (P < 0.0001). Furthermore, also in patients developing heart failure, a common complication after MI, the apo-ratio is the best lipid-related variable to classify risk especially in men (89).

Subsequently we have shown that for the inflammation marker haptoglobin (Hp) has strong relations with MI, stroke and heart failure in the AMORIS cohort (90). There were 11,216 men and 4,291 women who had a first MI, 8,463 men and 6,072 women who had a first stroke, and 4,670 and 3,634 who had a first heart failure, respectively. Based on 4,254 MI cases the risk of MI was about 4.5 times higher in the upper joint quartile of the apo-ratio as compared to the lower, whereas this relative risk for Hp was about 4.1. However, the attributable risk for the apo-ratio is higher since more subjects were classified into the top joint quartile of TC and the apo-ratio (12.8%) than that of TC and Hp (8.8%) and into the lower joint quartiles (12.1%) and (6.4%), respectively.

In another AMORIS-based cohort of 65,050 subjects Holme et al. (91) developed an inflammatory score comprising white blood cell count, haptoglobin and in a subgroup also CRP. After 11.8 years follow-up 3,649 MI, 2,663 stroke, 2,690 heart failure, in total 7, 456 MACE, occurred. In multivariate Cox proportional hazards analysis the inflammatory scores added predictive information over and above classical lipids such as TC and TG. Based on the apo-ratio, which was a stronger marker of CVD risk than conventional lipids, the inflammatory score added significant information value measured by net reclassification improvement, especially for those with the higher values for these variables. However, there was no statistically significant biological interaction between lipoproteins and the inflammatory markers. These data indicate that routinely used markers of inflammation in combination with the apo-ratio could be used in daily medical practice to assess CV risk.

We have also published data of lipid- and the apo-ratio from three cultures (Sweden, Iran, US) showing that the apo-ratio is highest in the Swedes (the AMORIS cohort) but similar in the Americans (NHANES) and Iranians (92). By contrast, the TC/HDL-C ratio is highest in the Iranians, intermediate in the Americans and lowest in the Swedes. There were similar associations of the pro-atherogenic and anti-atherogenic lipoproteins between the genders and variation with age in these three different cultures. These data indicate that complete characterization of lipoproteins requires measurement of apoB and apoA-I as well as lipoprotein lipids.

7.3. The apo-ratio in relation to chronic kidney disease and MI risk in the AMORIS study

Some previous studies have shown apoB to be increased and apoA-I to be decreased in patients with renal insufficiency. In the much larger AMORIS study Holzmann et al. (93) performed in 142,394 middle-aged mainly healthy men and women it was shown that the apo-ratio, the TC/HDL-C ratio, and non-HDL-C all are strong predictors of first MI, among both men and women, with or without chronic kidney disease (CKD). Those with the lowest glomerular filtration rate (estimated GFR mL/min/1.73 m², n = 5,838) had the highest apo-ratio. In Receiver Operator Characteristics (ROC) analysis the area under the curve (AUC) for the apo-ratio was 0.77 for men and 0.83 for women without CKD, and 0.65 and 0.74 among men and women with CKD, respectively analyses. These and other data reflect a certain advantage in the prediction of MI for the apo-ratio as compared to conventional lipids. Furthermore, the findings also indicate the presence of severe atherosclerosis both in the kidney and in the coronary arteries.

7.4. The apo-ratio and risk of stroke in the AMORIS study

High LDL-C is a major risk factor for MI. However, LDL-C is rarely increased in those who suffer any type of stroke. A low HDL-C and some abnormalities in either apoB and/or apoA-I have previously been found in patients with ischaemic stroke (94-99). In 2006 Walldius et al. published the first report on risk of stroke based on the AMORIS-population (100). The relationships between different types of fatal stroke and the lipid fractions, apoB, apoA-I and the apo-ratio were examined in 98,722 men and 76,831 women followed for a mean of 10.3 years. High apoB and low apoA-I values were significantly related to risk of stroke. The odds ratio comparing the upper 10th vs. the 1st decile of the apo-ratio for all strokes adjusted for age, gender, TC and TG was 2.07 (95% CI: 1.49–2.88, p < 0.0001). The apo-ratio was linearly related to the risk of stroke although the slope was less than observed for the risk of fatal MI (**Figure 6, left**). Low apoA-I was a common abnormality in all stroke subtypes including subarachnoidal and haemorrhagic strokes. In multivariate analyses the apo-ratio was a significantly stronger risk predictor than TC/HDL-C and LDL-C/HDL-C ratios.

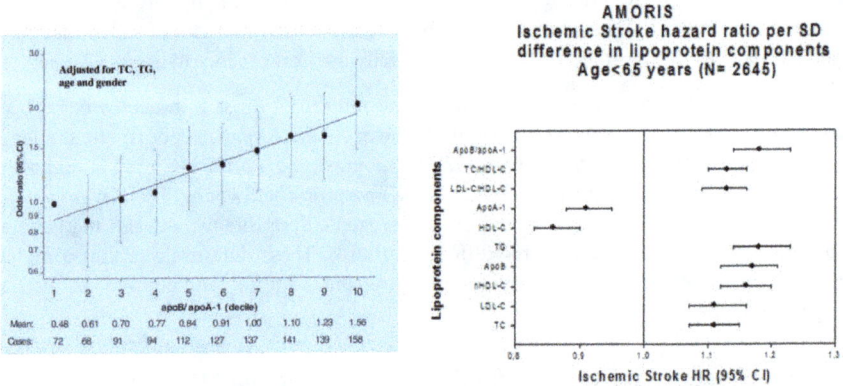

Figure 6. Risk of total stroke (left) (reference 3 and 100) and ischemic stroke (right) (reference 101). Both figures from the AMORIS study.

In a prospective follow-up study (mean observation age 11.8, range 7–17 years) based on the AMORIS population (n = 148,600). Holme et al. focused on risk of fatal and non-fatal ischaemic and haemorrhagic stroke in relation to all lipids and apos (101). Hazard ratio of non-fatal and fatal ischaemic and haemorrhagic stroke for 1 SD difference in lipoprotein components was calculated by gender, adjusted for age, MI, diabetes and hypertension. Ischaemic stroke was more common than haemorrhagic stroke (5:1), but case fatality was higher in haemorrhagic stroke. The apo-ratio, non-HDL-C and TG as well as low HDL-C and a high TC/HDL-C ratio were all predictors of ischemic stroke **(Figure 6, right)** and all cerebrovascular events (n=7,480) with somewhat stronger relations for non-fatal than fatal events. The apo-ratio was significantly stronger than the TC/HDL-C ratio in the patients with ischaemic stroke as reflected by chi-squared information value, adjusted for hypertension, diabetes, AMI, age and gender. The strongest association was for ischaemic stroke in those < 65 years of age and also for those with LDL-C < 3.0 mmol/L. There were no lipid relations to risk of haemorrhagic stroke other than a high apo-ratio related to risk in women.

7.5. Other findings from AMORIS indicating that the apo-ratio predicts CV risk

In addition, the risk of death from aortic aneurysms (n = 241) was significantly related to the apo-ratio (p< 0.0039) (3) adding to the importance of the apo-ratio as a predictor of severe ischaemic complications related to atherosclerosis. In that paper we also noted that, there was no relationship between the apo-ratio and risk of cancer (n = 4,423), motor vehicle accidents (n =100) or dementia (n = 255).

8. The apo-ratio in case-control CV risk studies

8.1. The INTERHEART study and risk of MI

The largest case-control study which has been performed is the INTERHEART study (58) comprising 15,152 patients with a first MI compared to 14,820 subjects from 52 countries world-wide matched for age, gender, ethnicity and continent. The aim of the study was to investigate which of the nine most common risk factors had the strongest relation to risk of MI and also which of the factors was most prevalent (highest Population Attributable Risk). These factors were: lipids primarily measured as the apoB/apoA-I ratio, smoking, diabetes, hypertension, abdominal obesity, psychosocial, fruits and vegetables, exercise, and alcohol. They found that all these risk factors were statistically related to risk.

The strongest **(Figure 7, left, (Table 1, left)** and also the most prevalent risk factor **(Table 1, right)**, was the apo-ratio both in men and women in each of the 52 countries worldwide. The apo-ratio plus smoking variables explained 70% of the entire risk which amounted to 90% for all nine risk factors taken together.

In a subsequent paper (59) they also showed that the apo-ratio had the strongest relation to MI-risk of all other measured lipids **(Figure 7, right, top panel)**. They also showed a significantly stronger relationship to MI risk for the apo-ratio than the TC/HDL-C ratio **(Figure 7, right; bottom panel)**. It was also shown that apoA-I had better diagnostic power than HDL-C over a wider range of low to high values.

Based on the findings and impact of these risk factors on risk of MI the INTERHEART Modifiable Risk Score (IHMRS) was developed based on age, the apo-ratio, smoking –

present, smoking – second hand, diabetes and hypertension with a range of points from 0-32 (102).

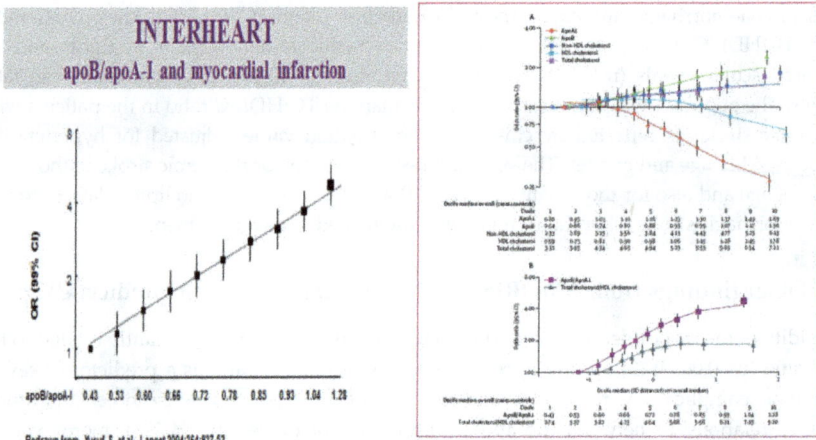

Figure 7. The INTERHEART study. Risk (Odds ratio, y-axis) versus the apoB/apoA-I ratio (left) (reference 58), and single lipids, apolipoproteins and their ratios (right) (reference 59).

INTERHEART Risk of AMI in relation to common risk factors (Yusuf S et al. Lancet 2004;364:937)		INTERHEART Risk of AMI in relation to common risk factors (Yusuf S et al. Lancet 2004;364:937)	
• **Risk factor adj. for all**	**OR (99 % CI)**	• **Risk factor adj. for all**	**PAR (99 % CI)**
• ApoB/apoA-I	3.25 (2.81-3.76)	• ApoB/apoA-I	49.2 (43.8-54.5)
• Smoking	2.87 (2.58-3.19)	• Smoking	35.7 (32.5-39.1)
• Diabetes	2.37 (2.07-2.71)	• Diabetes	9.9 (8.5-11.5)
• Hypertension	1.91 (1.74-2.10)	• Hypertension	17.9 (15.7-20.4)
• Abd.obesity	1.62 (1.45-1.80)	• Abd.obesity	20.1 (15.3-26.0)
• Psychosocial	2.67 (2.21-3.22)	• Psychosocial	32.5 (25.1-40.8)
• Veg.&Fruits	0.70 (0.62-0.79)	• Fruits & Veg.	13.7 (9.9-18.6)
• Exercise	0.86 (0.76-0.97)	• Exercise	12.2 (5.5-25.1)
• Alcohol	0.91 (0.82-1.02)	• Alcohol	6.7 (2.0-20.2)
• **Combined**	**129 (90-185)**	• **Combined**	**90 (88-92)**

Table 1. The INTERHEART study. Risk of myocardial infarction (AMI); Odds ratios for nine conventional risk factors (left) and Population Attributable Risk for nine conventional risk factors (right) (booth tables reprinted from reference 58).

The IHMRS was positively associated with incident MI in a large cohort of people at low risk for CV disease (12% increase in MI risk with a 1-point increase in score). The data were internally validated and the discrimination was tested (ROC c-statistic 0.69, 95% CI: 0.64-0.74) or even higher values up to 0.79 in certain global areas. Results were consistent across ethnic groups and geographic regions. A non-laboratory-based score has also been supplied. The IHMRS demonstrated clinical credibility, evidence of accuracy, and evidence of generality.

In an analysis of 15,780 patients from the INTERHEART study (103) it was shown that HbA1c was a useful diagnostic tool of risk and the levels increased with increasing apo-ratio from 0.75-0.84 for each quintile increase of HbA1c from <5.4 – >6.12% (p < 0.0001). Most of the MI patients had values in the highest HbA1c quintile. The advantage of using the apo-ratio in India (104), Latin America (105), Puerto Rico (106), and Africa (107) based on the INTERHEART study designs has been useful for evaluating CV risk and should be valuable in treating risk in these countries but also elsewhere in the world.

8.2. The INTERSTROKE study

The standardized INTERSTROKE case-control study was performed in 22 countries worldwide (108). Cases were patients with acute first stroke (within 5 days of symptoms onset and 72 hours of hospital admission). Controls had no history of stroke, and were matched with cases for age and sex. In 3,000 cases (n = 2,337, 78%, with ischaemic stroke; n = 663, 22% with intracerebral haemorrhagic stroke) and 3,000 controls, significant risk factors for all stroke were: history of hypertension, current smoking, waist-to-hip ratio, diet risk score, regular physical activity, diabetes mellitus, alcohol intake, psychosocial stress and depression, cardiac causes, and the apo-ratio in falling order. Together, these risk factors accounted for 88.1% of the population attributable   risk for all stroke. Increased concentration of non-HDL-C was not associated with risk of ischaemic stroke, but was associated with reduced risk of intracerebral haemorrhagic stroke, whereas increased concentration of apoB was associated with increased risk of ischaemic stroke, but was not associated with risk of intracerebral haemorrhagic stroke. The apo-ratio was a stronger predictor of ischaemic stroke than was ratio of non-HDL-C/HDL-C.

8.3. Other studies on stroke and atherosclerosis in the carotid arteries

Kostapanos et al. (109) studied 163 patients aged 70 years (88 men) with a first-ever acute ischemic/non-embolic stroke and 166 volunteers (87 men) with no history of CV disease. Compared with subjects with an apo-ratio in the lowest quartile, those within the highest quartile had a 6.3-fold increase in the odds of suffering an ischemic stroke (p<0.001). This association remained significant after controlling for sex, age, smoking status, body mass index, waist circumference, glucose and insulin levels, the presence of hypertension and diabetes mellitus, and lipid profile parameters (adjusted OR = 3.02; 95% CI 5.16-7.83; p = 0.02). The findings support elevated apo-ratio as an independent predictor of ischemic stroke in individuals over age 70.

Park et al. (110) studied 464 statin or fibrate naïve Korean patients with acute ischemic stroke: intracranial (ICAS, n = 236), extracranial (n = 44), and no cerebral atherosclerotic stenosis (n = 184). The ICAS group showed a significantly higher apo-ratio than the other two groups. The apo-ratio of 0.93 was substantially increased in patients with advanced ICAS (3 or more intracranial stenoses), the highest quartile of the apo-ratio was an independent predictor of ICAS (OR, 2.13; 95% CI, 1.05 - 4.33). A dose–response relationship (multivariate analysis) was observed between the presence of advanced ICAS and the apo-ratio quartiles (ORs, 4.03, 4.88, and 7.79, for the fourth quartile versus the first quartile). Patients having more metabolic syndrome components indicating MetS were more likely to have ICAS, advanced ICAS, and a higher apo-ratio (p < 0.001 for all). Thus, a higher apo-ratio is a predictor of ICAS rather than of extracranial atherosclerotic stenosis or no cerebral atherosclerotic stenosis. The apo-ratio might be a biomarker for ICAS in Asian patients with stroke.

8.4. The ISIS-study relating the apo-ratio to risk of MI

This ISIS case–control study was conducted among 3,510 acute MI patients (without prior vascular disease, diabetes, or statin use) in UK hospitals and 9,805 controls (60). Relative risks (age, sex, smoking, and obesity-adjusted) were more strongly related to apoB than to LDL-C and, given apoB, more strongly negatively related to apoA-I than to HDL-C. The apo-ratio was substantially more informative about risk than LDL-C/HDL-C, TC/HDL-C, non-HDL-C, and TC. Relative risks within several subgroups of patients showed no clear heterogeneity of effect with respect to sex, smoking, or BMI. The strongest effects were seen in those aged 30-49 years but even at ages 70-79, a 2SD higher apo-ratio was associated with a highly significant (P < 0.00001) relative risk. Furthermore, the apo-ratio, if untreated, is stable over time. Given the usual value of apoB, the usual value of LDL-C (indicating sdLDL particles) the risk was significantly higher. They concluded that single measurements of apoB and apoA-I are more predictive than single measurements of LDL-C and HDL-C and that the apo-ratio is the single best predictor of all lipid fractions is consistent with previously reviewed results including the AMORIS study (3,44).

9. Other studies showing strong prediction of CV risk by the apo-ratio

In our previous review from 2006 (3) we commented results from several prospective risk studies all showing an important diagnostic improvement of CV risk using apos and the apo-ratio over conventional lipids most commonly also adjusted for other confounders. The Dutch EPIC-Norfolk study (111) published in 2007 was performed in 1,511 apparently healthy controls and in 869 cases who had developed a non-fatal or fatal MI. They showed that in a head to head analysis of TC/HDL-C ratio versus the apo-ratio the Odds ratio for linear trend for quartiles was non-significant for the lipid-ratio but strongly significant for the apo-ratio, p < 0.006. These analyses were adjusted for sex, age, and time of enrollment and was adjusted for diabetes (yes or no), body mass index, smoking status (yes or no), systolic blood pressure, C-reactive protein level, and log-transformed triglyceride level. The apo-ratio added significant predictive value above that of the Framingham risk score since the area under the receiver-operating characteristic curve was 0.594 for Framingham risk score alone vs. 0.613 for Framingham risk score plus the apo-ratio, p < 0.001. Despite the fact that the difference was strongly significantly in favor of the apo-ratio the authors concluded that this was only a small increase. However, the authors pointed out that the apo-ratio is also useful since it can be applied in non-fasting samples.

The German MONICA/Kora Augsburg study (112) showed that in 1,414 men and 1,436 women without prior MI and a median follow up of 13 years the TC/HDL-C ratio predicted

MI risk. In addition, the apo-ratio was significantly related to increased risk of MI adjusted for age, smoking, alcohol, BMI, diabetes and hypertension.

In the American Thrombo study and its follow-up (113) both high apoB and low apoA-I predicted risk of re-infarction. In a follow-up they found that apoB was the strongest risk factor in those who manifested the MetS (114). However, in the German GRIPS (115), the results were negative in that LDL-C in multivariate analysis was found to be a stronger determinant of risk than apoB and the apo-ratio. This is, in fact, one of the very few studies to be found that shows LDL-C to be significantly better than apos in predicting risk. In the South Wales Cearphilly studies (116), although significant prediction was seen for apoB and apoA-I, the addition of apos did not improve prediction MI. In both of these two studies the number of events was below 300.

In the Swedish ULSAM studies (117,118) they showed that the risk of MI increased in parallel with increasing values of the apo-ratio. In those who had values for the ratio of <0.67 the incidence of MI was 9.5%, those who had ratios of 0.67–0.86 had an incidence of 17.7%, those with ratios of 0.87–1.23 had an incidence of 30.7%, and those with apo-ratio values >1.24 had an incidence of 44.8%. These risk values correspond well with those found in the AMORIS study (3,44). A risk prediction score was derived from one half of the population sample from the ULSAM cohort including systolic blood pressure, smoking, family history of MI, serum pro-insulin, and the apo-ratio. The score was highly predictive for future MI in the other half of the population that was not used for generating the score. The ULSAM score performed slightly better than the Framingham and PROCAM scores (evaluated as areas under the receiver operating curves; Framingham, 61%; PROCAM, 63%; ULSAM, 66%; $p < 0.08$). The authors also reported from the 30-year follow up of patients in the ULSAM study that ECG abnormalities were risk markers after the first 20 years of follow up but also that the apo-ratio and blood pressure remained significant risk predictors over three decades (118).

Ingelsson et al. (119) in the US Framingham study found that after a median follow-up of 15.0 years, 291 participants, 198 of whom were men, developed various manifestations of CHD. In multivariate models adjusting for non-lipid risk factors, the apo-ratio predicted CHD (HR per SD increment, 1.39; 95% CI 1.23-1.58 in men and HR, 1.40; 1.16-1.67 in women), but risk ratios were similar for the TC/HDL-C ratio (HR, 1.39; 1.22- 1.58 in men and HR, 1.39; 1.17-1.66 in women) and for LDL-C/HDL-C (HR, 1.35; 1.18-1.54 in men and HR, 1.36; 1.14-1.63 in women). In both genders, models using the apo-ratio were comparable with but not better than that for other lipid ratios. The apo-ratio did not predict CHD risk in a model containing all components of the Framingham risk score including the TC/HDL-C ratio. They concluded that the apo-ratio adds no incremental utility over this lipid ratio. Notably, there were few hard events in this small study, a fact that may restrict the interpretation of the results.

In India Goswami et al. (120) studied 100 patients with MI who were age-matched with 100 healthy control subjects. The exponential value of the regression coefficient beta for the apo-ratio was 11.9, as compared to 4.4 for the LDL-C/HDL-C ratio, 3.5 for the TC/HDL-C ratio and 2.2 for the TC/HDL-C ratio. The findings suggested that the apo-ratio is a better discriminator of CAD risk in the atherosclerosis-prone Indian population, than any of the conventional lipid ratios. They suggested that the apo-ratio should be an alternative to other lipid ratios in the risk assessment in patients with CAD.

In a comparative observational study by Agoston-Coldea et al. (121) on 289 subjects were divided into two groups: 144 subjects with old MI, and 145 subjects without CHD, but with CV risk factors. The multivariate analysis indicated that apoB over 1.7 g/L are closely correlated with MI (p = 0.001) independent of age, smoking, diabetes, hypertension, lipid TC/HDL-C and the LDL-C/HDL-C ratio. The protective effect of apoA-I was also significant (p = 0.004) in multivariate analysis. They concluded that the predictive value of the apo-ratio is superior to that of serum lipid fractions and that the apo-ratio therefore should be introduced in current clinical practice.

In the prospective case-cohort study (PREVEND cohort) (122) 6,948 subjects without previous CHD they studied the risk factors predicting major coronary events. The age- and sex-adjusted HR was 1.37 (95% CI, 1.26-1.48) for the apo-ratio and 1.24 (1.18-1.29) for the TC/HDL-C ratio (both p < 0.001). The risks of the two ratios were only marginally attenuated by additional controlling for traditional risk factors TG, hypertension, diabetes, obesity and smoking), hs-CRP and albuminuria.

In a Korean study by Kim et al (123) they studied the association between plasma lipids, and apolipoproteins and coronary artery disease: a cross-sectional study in a low-risk Korean population in 544 subjects. In the lowest quartile of TC, TG and LDL-C, and the highest quartile of HDL-C, only the apo-ratio was associated with CAD in both men and women. They concluded that the apo-ratio is the only variable that differentiates the patients with CAD from those without and, furthermore, gives additional information to that supplied by traditional lipid risk factors in a low-risk Korean population.

Agoston-Coldea et al. (124) studied 208 patients (100 men and 108 women), with and without previous MI by coronary angiography. They showed that the apo-ratio had a stronger correlation with MI than the TC/HDL-C ratio. Multivariate analysis performed with adjustments for conventional risk factors, showed that the levels of apoB, the apo-ratio and Lp(a), are significant independent CV risk factors. Therefore they recommend that the apo-ratio and Lp(a) should be included in clinical practice.

10. Meta-analysis of studies on CV risk

In 2006 Thomson and Danesh published a meta-analysis based on data from 23 relevant prospective studies in which apoB, apoA-I and the apo-ratio were associated with risk of MI (8). They compared risk in the top versus the bottom tertile of baseline values. The relative risks were; apoB 1.86 (95% CI 1.55-2.22, cases n = 6,320), apoA-I 1.62 (1.43-1.83, cases n = 6,333), and the apo-ratio 1.86 (1.55-2.22, cases n = 3,730). ApoB and the apo-ratio were directly related to risk, whereas apoA-I was protective. In that study no results were given for any lipids.

In 2009 the Emerging Risk Factor Collaboration (ERFC) published an extended meta-analysis in which they included 302,430 men and women without previous vascular disease from 68 long-term prospective studies, mostly in Europe and North America (10). During 2.79 million person-years of follow-up, there were 8,857 nonfatal MI, 3,928 coronary heart disease deaths, 2,534 ischemic strokes, 513 hemorrhagic strokes, and 2,536 unclassified strokes. Half of the studies included less than 100 events, and the largest study (ARIC) included 871 cases. In 22 studies on risk of MI and in 8 studies on risk of ischemic stroke they had also measured apoB, apoA-I and the apo-ratio. In 91,307 individuals with 4,499 MI and in 8 studies with 60,571 individuals and 1,192 cases they could compare how well the TC/HDL-C ratio and apoB, apoA-I and the apo-ratio were related to these CV events. In all of these comparisons non-HDL-C, HDL-C, the non-HDL-C/HDL-C ratio, apoB, apoA-I and the apo-ratio, adjusted for age and sex, were significantly related to risk of both MI and ischemic stroke. When additionally adjusted also for blood pressure, smoking, BMI, hypertension, and other lipid markers the HR was 1.50 (95%CI, 1.38-1.62) for the non-HDL-C/HDL-C ratio and 1.49 (1.39-1.60) for the apo-ratio. Interestingly, adjusting for these confounders changed the HR only marginally. These data show that the lipid- and apo-ratios give similar and significant prediction of risk. Furthermore they also found that apoB had similar risk as non-HDL-C, and apoA-I had similar risk as HDL-C. The ERFC authors concluded that both lipid- and apo-ratios can be used even in the non-fasted state since the apo-ratio and the lipid ratio give similar information. Furthermore, they also discuss that there may be important advantages for using apolipoproteins.

Importantly, the ERFC did not include the three largest studies on risk of MI and stroke related to lipids, apos and the apo-ratio. These are the studies; AMORIS, n = 6,794 first cases of AMI (88), and n = 4,470 first ischemic stroke (101), INTERHEART, cases n = 15, 152 for first MI (59), n = 2,337 for first stroke (108), and the ISIS study, n = 3,510 for first MI (60). These studies were excluded because a complete set of confounding variables were not available (AMORIS), or that two studies were case-control studies (INTERHEART and ISIS). The findings in ERFC are therefore restricted to the results based on only prospective studies with many fewer number of events (total n = 5,691) compared to these much larger studies also covering a world-wide population (AMORIS, INTERHEART, INTERSTROKE and ISIS) (total n = 32,263). So adding all these results to those obtained in the ERFC studies the advantages of the apo-ratio as risk predictor may be even more compelling. Such advantages for clinical use are commented in several sections below and are summarized in the discussion. Results from a recent ERFC publication are also included and discussed in page 39.

11. Relations between the apo-ratio and the metabolic syndrome, glucose - insulin metabolism and diabetes - Risk predictors for CV manifestations

11.1. Metabolic syndrome (MetS) and diabetes

In subjects with the MetS and in patients with diabetes several studies have been performed indicating advantages of using apos, especially apoB, over conventional lipids. In our previous review (3) we summarized these results from Stewart et al. (125), Korean studies (126-128) and studies from India (129) and Canada (130). In these papers the highest values for the apo-ratio were found in those who had most manifestations of the MetS. The apo-ratio was also related to atherosclerosis verified by angiography even if LDL-C values were low. In the Swedish ULSAM study (131) at the 26.8 year follow-up 462 patients had developed MI. The apo-ratio was highest in those who developed a MetS, and their apo-ratio was inversely related to glucose disposal. These findings were independent of LDL-C and smoking. Both the apo-ratio and MetS independently predicted MI.

Sierra-Johnson et al. (132) studied 2,955 adults (mean age 47 years; 1,457 women) without diabetes from the US NHANS III population. The apo-ratio was an independent predictor of insulin resistance after adjustment for age and race, and remained significant after further adjustment for MetS components including TG, HDL-C, traditional and inflammatory risk factors. They recommended that the apo-ratio should be recognized and implemented in future clinical guidelines. In the follow-up paper (133) of a multi-ethnic representative subset of 7,594 US adults (mean age 45 years; 3,881 men, 3,713 women) there were 673 CV deaths of which 432 were from CHD. Both the apo-ratio (HR 2.14, 95% CI, 1.11 – 4.10) and the TC/HDL-C ratio (HR 1.10, 1.04 – 1.16) were related to CHD death. Only apoB (HR 2.01, 1.05 – 3.86) and the apo-ratio (HR 2.09, 1.04 – 4.19) remained significantly associated with CHD death after adjusting for CV risk factors **(Figure 8 left)**. This suggested that the measurement of apolipoproteins has superior clinical utility over traditional risk markers such as the TC/HDL-C ratio in identifying subjects at risk for fatal CV disease. In addition, the combined elevation of glucose and a high apo-ratio increases the risk of MI as documented in the AMORIS study (3) **(Figure 8, right)**.

Zhong et al. (134) found also in China that the apo-ratio increased significantly with number of MetS components. Belfki et al (135) have shown in a Tunisian population that the apo-ratio increased significantly with each of the components as well as with increasing numbers of components of the MetS after adjusting for age and gender. Similarly, the apo-ratio was associated with insulin resistance.

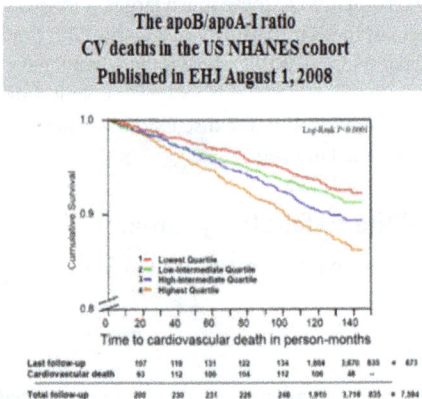

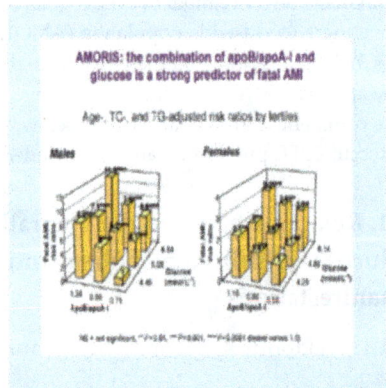

Figure 8. Cumulative survival (y-axis) in relation to quartiles of the apoB/apoA-I ratio in patients with the metabolic syndrome (left) (from NHANES cohort, reference 133). Risk of myocardial infarction in relation to glucose and the apoB/apoA-I ratio (right) (AMORIS study, reference 3).

Based on the findings in subjects with MetS Sniderman and Faraj (136) have argued for including both apoB and apoA-I as stronger risk markers especially compared to LDL-C (often low in MetS), TG and HDL-C. These apos also have strong relations to glucose and insulin homeostasis. Therefore the apo-ratio should be a valid component of the MetS especially since the apo-ratio has so strong predictive value of CV risk. The apo-ratio also summarizes the risk for individuals with MetS into one simple and predictive risk number. In another paper Sniderman et al. (137) have also analyzed pros and cons for using the apo-ratio.

Bruno et al. (138) studied diabetic subjects and they found that apoB and the apo-ratio were associated with CV mortality independently of non-HDL-C. They recommended that apoB and apoA-I should be measured routinely in all people with diabetes, particularly in the elderly.

Bayu et al. (139) studied 224 diabetic patients (85 type 1 and 139 type 2). After adjusting for age, sex, diabetes duration, systolic blood pressure and diabetes medications they found that the apo-ratio was the best predictor of diabetic retinopathy. Traditional lipids improved the ROC area by only 1.8 % whereas the apo-ratio improved the area by 8.2 %.

Enkhma et al. (140) have studied several ethnic groups of European and African Americans and developed a CV risk score which was found to be significantly increased across tertiles of the apo-ratio. They concluded that the apo-ratio differed across ethnicities and was associated with presence of the MetS in both groups. Among African Americans, an elevated apo-ratio independently predicted a greater risk of CAD.

Ounis et al. (141) studied thirty-two obese 13 years old children with 16 subjects who participated in a 8-week training period and 16 subjects serving as a control group. The apo-ratio was positively correlated with TG ($r = 0.46$, $p < 0.01$), blood glucose ($r = 0.48$, $p < 0.01$), waist circumference ($r = 0.34$, $p < 0.01$), systolic ($r = 0.31$, $p < 0.01$) or diastolic ($r = 0.29$, $p < 0.05$) blood pressure and was negatively correlated with HDL-C ($r = 0.51$, $p < 0.01$), Fat max ($r = 0.45$, $p < 0.01$) and VO^2 peak ($r = 0.39$, $p < 0.01$). When adjusted for pubertal stage, the relationships between the apo-ratio and other variables were not significantly altered. The multiple regression analysis showed that the change in total HDL-C is the most significant predictor of the change of the apo-ratio explaining 82% of the variance of its change over the training program.

Gatz et al. (142) studied thirty same-sex twin pairs in which both members were assessed at baseline and one twin subsequently developed dementia, at least 3 years subsequent to the baseline measurement, while the partner remained cognitively intact for at least three additional years. Eighteen of the 30 cases were diagnosed with Alzheimer's disease. Baseline assessments were conducted when twins' average age was 70.6 (SD = 6.8) years. Which twin would develop dementia was predicted by less favorable lipid values defined by higher apoB and higher apo-ratio, poorer grip strength, and − to a lesser extent − higher emotionality on the EAS Temperament Scale. Given the long preclinical period that characterizes Alzheimer's disease, these findings may suggest late life risk factors for dementia. Alternatively, there may be early development of atherosclerosis in critical cerebral arteries based on an elevated apo-ratio over time.

Carnevale-Schianca et al. (143) enrolled 616 patients with normal glucose tolerance (NGT) (273 men and 343 women), and measured insulin resistance, lipid profile, the apo-ratio and the factors compounding the MetS. An unfavorable apo-ratio (> 0.90 for males and > 0.80 for females) was present in 13.9 % of 108 patients with LDL-C < 100 mg/dL. Compared to subjects with lower apo-ratio, they had more elements of MetS and their lipid profile strongly correlated with high CV risk. In NGT individuals with LDL-C < 100 mg/dL, a higher apo-ratio indicated an atherogenic lipid profile, suggesting that LDL-C alone is insufficient to define CV risk. This study demonstrates that the apo-ratio is at least complementary to LDL-C in identifying a more correct CV risk profile of asymptomatic NGT subjects.

Wen et al. (144) measured high sensitive hsCRP, apoB, apoA-I, and the profiles of coronary angiograms, echocardiography and oral glucose tolerance tests (OGTT)s as well as traditional risk factors in 1,757 cardiology patients. The hsCRP or the apo-ratio were significantly correlated with the presence and severity of angiographic profiles, the levels of left ventricular (LV) ejection fraction, LV mass and LV mass index, and the presence of abnormal OGTT. The combination of the apo-ratio and hsCRP had greater correlation with abnormal glucose metabolism than its individual components in patients with normal fasting glucose, and was an independent predictor for coronary artery disease.

12. The apo-ratio and relations to atherosclerosis, vascular functions and inflammation

In many clinical conditions coronary arteriography, carotid ultrasound (CIMT), endothelial function, calcium scoring (CAC) and even more recently Intra Vascular Ultrasound (IVUS) studies of the coronary arteries has been related to lipid- and apo-abnormalities. Coronary and femoral plaques also contain apos (34-36). Many of these studies indicate that apos are more closely related to the amount of atherosclerosis than conventional lipids. Relevant studies are commented below.

In the Uppsala PIVUS study by Andersson et al. (145) the prevalence of carotid plaque was investigated. In 942 free living 70 year old men (n = 469) and women (n = 473) an ultrasound was performed. A plaque was defined by at least 50% increase of the intima-media thickness (IMT). Plaques were slightly more prevalent in men (n = 322) than in women (n = 293). Individuals with plaques had significantly higher the apo-ratio (p = 0.013), LDL-C/HDL-C-ratio (p = 0.04), LDL-C (p = 0.02), higher levels of fasting blood glucose (p = 0.02), Framingham risk score (p < 0.0001), higher levels of systolic blood pressure, (p < 0.0001), and also a higher average of pack-years of cigarette smoking (p = 0.008) after adjustment for gender and statin use. No significant differences were seen for HDL-C, diastolic blood pressure or BMI. The inflammatory markers oxidized LDL, TNF alpha, and leucocyte count as well as insulin resistance (HOMA) were increased.

In another subsample of 70 years old men (n = 124) and women (n = 123) who did not use lipid-lowering drugs from the PIVUS study (146) were investigated whether the amount of visceral (VAT) or subcutaneous adipose tissue (SAT) independently of the other can determine the apo-ratio. VAT and SAT areas were assessed using magnetic resonance imaging. Their adipose tissue areas were related to their levels of apoB, apoA-I and the apo-ratio. ApoA-I levels were independently related to the VAT area (r = −0.33, p < 0.0001) whereas the apoB levels were not (r = 0.102, p = 0.07). The VAT area was independently significantly (r = 0.25, p = 0.001) related to the apo-ratio in the multiple regression analysis whereas the SAT area was not. This observation may indicate that VAT is metabolically active possibly through decreased adiponectin levels. The VAT metabolism seems more related to abnormalities in the apo-ratio which also may be a consequence of abnormal glucose-insulin metabolism as discussed above in other studies on the MetS.

Schmidt and Wikstrand (147) reported that in a multi-variable analysis including all baseline variables only the apo-ratio (p = 0.003) and serum insulin (p = 0.026) were significantly related to IMT composite progression rate indicating that the apo-ratio is an important risk factor for predicting atherosclerotic progression rate during very long-term follow-up in clinically healthy middle-aged men.

Reis et al (148) have studied factors that may influence MetS and development of obesity. They performed weighted Pearson partial correlation coefficients for waist circumference, log-transformed leptin, and insulin vs. metabolic, inflammatory, and thrombogenic CV risk factors among men and women aged 40 years and older, NHANES III. They found that apoB was positively correlated with waist, leptin and insulin both in men and women, whereas apoA-I was significantly and negatively related to these risk markers. These findings may indicate that the apo-ratio can summarize the lipid abnormalities into one number. The results were adjusted for age, ethnicity, smoking, physical activity, alcohol intake and time of fasting.

Junyent et al. (149) assessed carotid intima-media thickness (CIMT) and plaque in relation to classical risk factors and apoA-I and apoB levels in 131 unrelated patients with familial hypercholesterolemia (FCHL), 27 with prior CVD and 190 age- and sex-matched control subjects. By multivariate analysis in a model with all risk factors, inclusive of the MetS, independent associations of CIMT were age, the apo-ratio, systolic blood pressure, fasting glucose, family history of CVD and TC/HDL-C ratio (r² = 0.475, p < 0.001). The strongest determinant of IMT was the apo-ratio (β = 0.422, p < 0.001). The findings support the atherogenicity of the lipid phenotype in FCHL beyond associated risk factors. They also have implications for diagnosis and management of CVD risk in this condition.

Vladimirova-Kitova et al. (150) have found that carriers of a LDL-receptor defective gene have a higher carotid IMT and apo-ratio than non-carriers, whereas no difference between the groups was found with respect to the level of other lipid parameters, ADMA, total homocysteine, cell adhesion molecules, and % flow mediated dilation. Thus the apo-ratio is a predictor of IMT in carriers of this LDL-receptor gene.

Dahlen et al. (151) performed the CARDIPP-1 primary care study a study in 247 patients with type 2 diabetes, aged 55-66 years. They found that there was a significant association between the apo-ratio and CIMT in middle-aged patients with in type 2 diabetes. The association was independent of conventional lipids, hsCRP, glycaemic control and use of statins.

In the study by Rasouli et al. (152) 138 men and 126 women aged 40-70 years, were classified as CAD cases or controls, according to the results of coronary angiography. The severity of CAD was scored on the basis of the number and extent of lesions in coronary arteries. The results indicate that the apo-ratio, apoB and Lp(a) are independent risk factors for CAD and are superior to any of the cholesterol ratios. They suggested using the apo-ratio as the best marker of CAD in clinical practice.

Smith et al. (153) compared the body composition and the apo-ratio in migrant Asian Indians white Caucasians in Canada. Indian men and women had a higher apo-ratio than Caucasians (p = 0.0003)]. Of interest, there were also significant correlations between the apo-ratio and WHR in all groups, except the Indian women.

Both in children and adults obesity either defined by BMI or waist/hip ratio has been found to be directly related to apoB and the apo-ratio, and indirectly to apoA-I levels (154-156).

In the Cardiovascular Risk in Young Finns Study (157) they measured CIMT and brachial endothelial function in 879 subjects. They determined whether apoB and apoA-I measured in childhood and adolescence could predict atherosclerosis in adulthood. In subjects aged 12 to 18 years at baseline, apoB and the apo-ratio were directly (p < 0.001) related and apoA-I was inversely (p = 0.01) related with adulthood IMT. In subjects aged 3 to 18 years at baseline, apoB (p = 0.02) and the apo-ratio (p < 0.001) were inversely related, and apoA-I (p = 0.003) was directly related to adulthood flow mediated dilatation. Adjustment for age, gender, blood pressure, BMI, TG, insulin, CRP and brachial diameter at baseline did not change these relations. The apo-ratio measured in adolescence was stronger than the LDL-C/HDL-C or non-HDL-C/HDL-C ratios (c-values, 0.623 vs. 0.569, p = 0.03) in predicting increased CIMT in adulthood. The authors concluded that apoB and apoA-I measured in children and adolescents reflect an abnormal lipoprotein profile that may predispose to the development of subclinical atherosclerosis later in life. These markers are therefore useful in pediatric lipid risk assessment.

In a cross-sectional and 6-year prospective data from the cardiovascular risk in young Finns study (aged 24 to 39 years) (158) they studied metabolic risk variable MetS and their associations with CIMT. ApoB, CRP, and type II secretory phospholipase A2 enzyme activity were significantly higher and apoA-I lower in subjects with MetS (n = 325) than in subjects without MetS (n=858) indicating that the apo-ratio may summarize the risk into one number. In prospective analysis both MetS and high apoB predicted (p < 0.0001) incident high CIMT. The association between MetS and incident high CIMT was attenuated by about 40% after adjustment with apoB. Adjustments with apoA-I, CRP, or type II secretory phospholipase A2 did not diminish the association. Thus, the atherogenicity of MetS in this population assessed by incident high CIMT is mainly mediated by elevated apoB, but not inflammatory markers.

In the Swedish study Wallenfeldt et al. studied the relationships between abnormalities in lipoprotein concentrations in 338 apparently healthy 58-year-old men with manifestations of the MetS (159). Those who had an apo-ratio > 0.74, irrespective of blood pressure and smoking, had a significant progression (untreated) of the IMT values of the carotids over a

3-year period. Thus CIMT is a non-invasive simple, sensible and useful method to follow dynamic progression of atherosclerosis. Furthermore, the level of the apo-ratio is a strong predictor of these atherosclerotic changes in the arterial wall. Thus, values of the apo-ratio > 0.74 may alert the treating doctor to the need of adequate lipid-lowering therapy.

In a Japanese study (160) sixty-six type 2 diabetic patients with carotid atherosclerosis and 66 age- and sex-matched patients without carotid atherosclerosis were compared. They concluded that the combination of apoB and HOMA-R is a superior marker of carotid atherosclerosis compared with LDL-C alone in patients with type 2 diabetes.

Kim et al. (161) have studied 757 stroke patients undergoing coronary artery bypass grafting. They found that prevalence of asymptomatic carotid stenosis > 50% and > 70% was 26.4 % and 8.6%, respectively. In multivariate analysis, plasma levels of the apo-ratio and homocysteine were independently associated with carotid stenosis. Receiver operating characteristic curve (ROC) analysis indicated area under the curve values of 0.708 (the apo-ratio), 0.678 (Lp(a)), and 0.689 (homocysteine).

Ajeganova et al. (162) have studied patients with rheumatoid arthritis (RA) that commonly are affected by premature atherosclerosis including development of xanthomas. They studied 114 patients, age 50.6 years, 68.4% women, with recent RA (< 12 months after symptoms onset) and they were assessed at 0, 3, 12, 24 and 60 months after RA diagnosis. Plaque detection was positively associated with age and smoking (ever). After adjustment, a longitudinal approach demonstrated an independent positive prediction of CIMT by the apo-ratio (p = 0.030), but negative prediction by apoA-I (p = 0.047). Higher levels of the pro-atherogenic apo-ratio and apoB and low anti-PC (IgM antibodies against phosphorylcholine) were independently associated with bilateral carotid plaque p = 0.002, 0.026 and 0.000, respectively). Both baseline and longitudinal levels of other inflammatory/disease-related factors failed to show significant associations with the study outcomes.

13. Effects of lipid-lowering therapy on change of apoB, apoA-I and the apo-ratio

The mode of actions of statins and their effects on lipids and apos is reviewed in more detail elsewhere (163-165). The most commonly used drugs today are the statins that can reduce apoB synthesis and increase apoA-I synthesis and turnover. In clinical practice simvastatin and pravastatin are the most commonly used statins since they are now available as generics. They can reduce apoB up to about 20% and increase apoA-I by about 2-5% and a bit more for simvastatin. The most effective apoB-reducing statins are atorvastatin and rosuvastatin which lower the apoB-values by about 40-45% and 45-50%, respectively. Best increase in apoA-I concentrations is obtained by rosuvastatin which can increase the value by about 10-15% depending on baseline values, the lower the higher is the increase (163-165). Commonly for all statins there is a strong dose-response relationship, except for atorvastatin where higher doses commonly result in lowering of HDL-C and apoA-I values. The strongest lowering effects of the apo-ratio is obtained by rosuvastatin which lowers this ratio by about 50 %, followed by atorvastatin about 40-45 %, and simvastatin and pravastatin up to 30 %.

14. Prediction of outcome in statin trials using LDL-C or the apo-ratio

LDL-C has been the primary focus in lipid-lowering trials for more than two decades. A vast number of studies, both in primary and secondary prevention, have shown that there is a close relationship between LDL-C and CV event rates, the lower the LDL-C, the lower is the risk (163-165). In several of these trials also apoB, apoA-I and the apo-ratio have been measured. When explaining the relationship of each lipid fraction and each apo-fraction to CV event reduction virtually all lipids as well as apoB and apoA-I and the apo-ratio are significantly related to outcome. However, LDL-C is much weaker predictor than apoB and any lipid ratio. The best relationship with CV risk reduction is the apo-ratio. Examples from several trials are presented below.

In the AFCAPS/TexCAPS study (52,53), lovastatin 20-40 mg/d or placebo were given to 3,304 patients with rather normal LDL-C but low HDL-C values. ApoB decreased by 18.9 % and apoA-I increased by 7.2 %. At 5 years, there was a 37 % decrease in the relative risk for having a first acute coronary event in the lovastatin versus placebo group. In a head to head analysis it was found that apoB was better than LDL-C, p < 0.01, apoA-I was better than HDL-C, p < 0.01, and the apo-ratio was better than the TC/HDL-C ratio, p <0.01 in explaining the event reduction (Figure 9, left). In this study it made no difference to which treatment group the patients were assigned, conventional diet – placebo or the lovastatin group. The apo-ratio value on treatment was the only lipid-related marker that was significantly related to outcome (Figure 9, right).

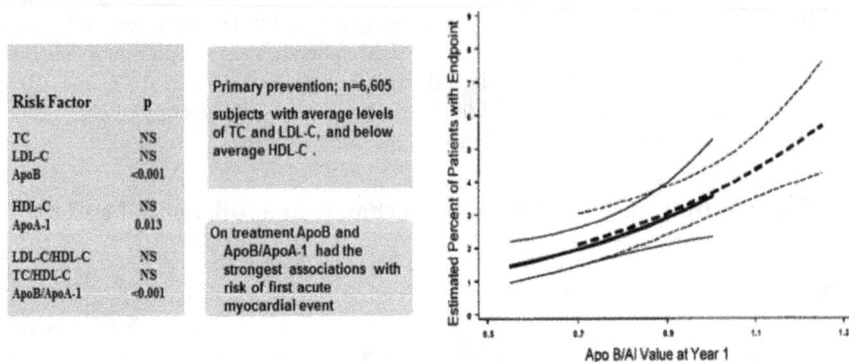

Risk Factor	p
TC	NS
LDL-C	NS
ApoB	<0.001
HDL-C	NS
ApoA-1	0.013
LDL-C/HDL-C	NS
TC/HDL-C	NS
ApoB/ApoA-1	<0.001

Primary prevention; n=6,605 subjects with average levels of TC and LDL-C, and below average HDL-C .

On treatment ApoB and ApoB/ApoA-1 had the strongest associations with risk of first acute myocardial event

Figure 9. The AFCAPS/TexCAPS study. The apoB/apoA-I ratio was the best predictor of outcome expressed in head to head analyses versus lipids (left) (reference 52). Risk remaining during treatment (lovastatin versus placebo) in relation to obtained values for the apoB/apoA-I ratio (right) (reference 53).

In the LIPID trial pravastatin reduced CHD mortality by 24 % and total mortality by 22 % (3,4,54). The TC/HDL-C and the apo-ratios on treatment were considerably better in explaining outcome than either LDL-C or HDL-C. The values of the apo-ratio had strongest relations to event reduction.

To which target LDL-C values should lipid-lowering aim? In the ACCESS study (166) therapy reduced LDL-C levels to 'normal – target levels'. However, such therapy only reached apoB levels to about the 50th percentile of a population. This means that the patients were not optimally treated by using LDL-C at recommended guideline levels. These results may illustrate that apoB would be a better target if proper target levels have been proposed.

To reach targets in guidelines has been further investigated in a more recent paper by Vodnala et al. (167). They applied the ATP III guidelines, including Framingham Risk Scores to determine whether patients met non-HDL-C goals upon referral. In order to reach targets for non-HDL-C among patients (n = 5,692) most high- and many intermediate-risk patients goals would require more aggressive treatment to reach either the TC/HDL-C = 3.5 or the apo-ratio = 0.50 goals. Thus, a more intense therapy using better target goals, i.e. apoB or the apo-ratio, than the conventional LDL-C or non-HDL-C would most likely add clinical value and better treatment effects.

Van den Bogaard et al. (168) studied 9,247 patients (mean age 61 years, 81% males), participating in the Treatment to New Targets (TNT) trial in which the effects of 80 mg versus 10 mg atorvastatin was compared. The association between lipoprotein components and the risk of cerebrovascular events after the first year into the trial was investigated. All lipoprotein components, except LDL-C, showed a significant gradient for incidence of cerebrovascular events with increasing quartiles of the lipoprotein component. If the lipoprotein components were treated as continuous variables, the adjusted HR for cerebrovascular events for 1 SD difference in 1-year lipoprotein components were for LDL-C 1.13 (95%CI, 1.02–1.25), for HDL-C 0.86 (0.76–0.97), for apoB 1.17 (1.04–1.28), for apoA-I 0.83 (0.74–0.94), for TC / HDL-C 1.22 (1.10–1.34) and for the apo-ratio 1.24 (1.12–1.37). The apo-ratio was superior to TC/ HDL-C, because adding the apo-ratio to TC/HDL-C improved prediction, whereas adding TC/HDL-C to the apo-ratio did not. These findings are consistent with the AMORIS study linking the apo-ratio to risk of stroke (99,100), and are also similar to results from the combined data of TNT and IDEAL showing that TC/HDL-C and the apo-ratios are more closely associated with CVD than any of the individual lipoprotein parameters. They concluded that in coronary heart disease patients receiving intensive lipid-lowering treatment, the on-treatment apo-ratio provides the strongest association with incidence of cerebrovascular events followed by TC / HDL-C. They also stated that as current European and US guidelines only acknowledge LDL-C as a therapeutic target and HDL-C and triglycerides as risk markers it will be up to future guideline committees to implement these new parameters as risk predictors and to define new treatment targets based on these apolipoproteins.

Kastelein et al. (169) showed in a post hoc analysis that combined data from 2 prospective, randomized clinical trials in which 10 001 TNT and 8,888 ("Incremental Decrease in End Points through Aggressive Lipid Lowering"- IDEAL) patients with established coronary heart disease were assigned to atorvastatin 10 mg/d or atorvastatin 80 mg/d. In models with LDL-C, non-HDL-C and apoB were positively associated with cardiovascular outcome, whereas a positive relationship with LDL-C was lost. In a model that contained non-HDL-C and apoB, neither was significant owing to collinearity. Inclusion of measurements of apoA-I further strengthened the relationships. The TC/HDL-C and the apo-ratio in particular were each more closely associated with outcome than any of the individual pro-atherogenic lipoprotein parameters **(Table 2)**. In a pair-wise COX model comparison of the two ratios the TC/HDL-C was non-significant but the apo-ratio was significant, p<0.001. However, the authors mainly conclude that these data support the use of non-HDL-C or apoB as novel treatment targets for statin therapy, but do not believe that the apo-ratio is yet a valid risk variable because of uncertainness of the impact of risk of HDL-C and apoA-I. Furthermore, they state that in the absence of interventions that have been proven to consistently reduce CVD risk through raising plasma levels of HDL-C or apoA-I, it seems premature to consider the ratio variables as clinically useful. These conclusions merit further comments in the discussion. However, clearly the apo-ratio comes out as the best CVD predictor as manifested by their data when all head-to-head comparisons are performed between various lipids and apos.

LDL cholesterol	0.95	0.87–1.05	0.33	Apolipoprotein B	0.94	0.85–1.04	0.26
Apolipoprotein B	1.24	1.13–1.36	<0.001	Apolipoprotein B/A-I	1.30	1.20–1.39	0.001
Non-HDL cholesterol†	1.00	0.93–1.07	0.94	Total/HDL cholesterol	1.00	0.92–1.10	0.91
Apolipoprotein B/A-I	1.24	1.17–1.32	0.001	Apolipoprotein B/A-I	1.24	1.13–1.36	0.001

Table 2. TNT-IDEAL pooled data. Head to head comparisons between various lipids, apolipoproteins and ratios (redrawn from reference 169).

Holme et al. (170) studied the ability of apolipoproteins to predict new-onset of congestive heart failure (HF) in statin-treated patients with coronary heart disease (CHD) in the IDEAL study based on 8,326 patients of whom 185 subjects had a HF event. Variables related to LDL-C carried less predictive information than those related to HDL-C, and apoA-I which was the single variable most strongly associated with HF. LDL-C was less predictive than both non-HDL-C and apoB. The apo-ratio was most strongly related to HF after adjustment for potential confounders, among which diabetes had a stronger correlation with HF than did hypertension. The apo-ratio was 2.2 times stronger associated than that of diabetes. Calculation of the net reclassification improvement (NRI) index revealed that about 3.7 % of the patients had to be reclassified into more correct categories of risk once the apo-ratio was added to the adjustment factors. The reduction in risk by intensive lipid-lowering treatment as compared to usual-dose simvastatin was well predicted by the difference in apo-ratio on-treatment levels mostly through the reductions in apoB. Thus, both apoB, apoA-I and the apo-ratio had additional clinical value above lipids in predicting risk of HF.

Holme et al. (171) also looked into the ability of apoB, apoA-I or the apo-ratio to predict new coronary heart disease (CHD) events in patients with CHD on statin treatment in the IDEAL trial comparing the effects of atorvastatin 80 mg/d to that of simvastatin 20-40 mg/d to prevent CHD subsequent major coronary events (MACE). Variables related to LDL-C carried more predictive information than those related to HDL-C, but LDL-C was less predictive than both non-HDL-C and apoB. Of all lipoprotein variables, the apo-ratio was the best predictor of MACE during statin treatment. The apo-ratio carried as much information as apoB, apoA-I, LDL-C, and HDL-C together. However, for estimating differences in relative risk reduction between the treatment groups, apoB and non-HDL-C were the strongest predictors. They recommended that measurements of apoB and apoA-I should be more widely available in clinical praxis.

Results from the recently published ASTEROID Trial (172) showed that in patients with acute coronary syndromes treated with rosuvastatin 40 mg daily for 2 years a significant (p < 0.001) regression was found of the atherosclerotic burden in the coronary arteries (intravascular ultrasound). In these patients LDL-C was reduced from 3.35 mmol/L (130 mg/dl) to 1.55 mmol/L (60 mg/dl), p < 0.001 and the apo-ratio was reduced from high 0.95 to low 0.49, p < 0.001. These results indicate that the risk related to the apo-ratio risk was reduced from the eighth risk decile to the first decile, i.e. to normality.

Nicholls et al. (173) presented data based on 4 studies in which IVUS was used in 1,455 coronary patients. They were given lipid-lowering with either atorvastatin, simvastatin, pravastatin and rosuvastatin (strongest lipid-lowering). A highly significant regression of coronary atheroma volume over a two year period was recorded. They stated that "Reducing the ratio of apoB to apoA-I was the strongest lipid predictor of changes in atheroma burden in patients treated with a statin". Thus, even small, but clinically important changes in atheroma volume, can be identified by IVUS techniques and also by closely related changes in the strongest marker of lipoprotein metabolism, i.e. the apo-ratio.

Tani et al. in Japan performed a 6-month prospective study of 64 patients with coronary artery disease treated with pravastatin (174). The plaque volume, assessed by IVUS, decreased by 12.6% (p < 0.0001). A significant decrease of 6.4 % and 14.6 % was found in the serum level of apoB and the apo-ratio (p < 0.0001 and p <0.0001, respectively, vs baseline), and apoA-I increased by 14.0 % (p < 0.0001). A stepwise regression analysis revealed that the change in the apo-ratio was an independent predictor of the change in coronary plaque volume (p < 0.0023). They concluded that a decrease in the apo-ratio is a simple predictor for coronary atherosclerotic regression: the lower the apo-ratio, the lower the risk of coronary atherosclerosis.

Taskinen et al. studied diabetic patients treated with fenofibrate (the FIELD study,175). Lipid ratios and the apo-ratio performed significantly better than any single lipid or apolipoprotein in predicting CVD risk during treatment. In the placebo group, the variables best predicting CVD events were non-HDL-C/HDL-C, TC/HDL-C (HR 1.21, p < 0.001 for both), the apo-ratio (HR 1.20, p < 0.001), LDL-C/HDL-C (HR 1.17, p < 0.001), HDL-C (HR 0.84, p < 0.001) and apoA-I (HR 0.85, p < 0.001). In the fenofibrate group, the first four predictors were very similar (the apo-ratio was fourth), followed by non-HDL-C and apoB.

In the JUPITER primary prevention trial (176) rosuvastatin 20 mg versus placebo was given to patients with initial LDL-C levels < 3.4 mmol/L and hsCRP > 2 mg/dL. Already after a medium time of treatment of 1.9 years, the trial was stopped for safety reasons since the actively treated patients benefitted by a highly significant risk reduction in MACE by 50 %. It should be pointed out that several thousand patients, those first recruited into the trial, participated for more than three to four years in the trial. LDL-C was reduced to 1.4 mmol/L and the apo-ratio was reduced from 0.95 to 0.49, p < 0.001. This indicates that "normal values" for the apo-ratio should be in the order of < 0.50 in order to obtain as low future risk as possible.

In a recent publication from JUPITER the authors reported that LDL-C, non-HDL-C, apoB and lipid-ratios as well as the apo-ratio had about similar predictive value of remaining risk during treatment with rosuvastatin (177). However, in subgroup analyses they reported that apoA-I had a greater capacity to define remaining risk than HDL-C. Furthermore, they also found that any lipid-related ratios had a greater predictive value than single values of LDL-C, non-HDL-C or apoB. In addition, if LDL-C values reached < 100 mg/dL or < 70 mg/dL, or if non-HDL-C targets were reached < 130 mg/dL or < 100 mg/dL, the only lipid-related variable or ratio that still was associated with remaining significant risk was the apo-ratio. These data, although the number of events is small in the sub-cohorts, indicate that the apo-ratio is a realistic and a valid predictor of risk and may be better than conventional lipids. However, the authors indicated that differences were small and that LDL-C and non-HDL-C were still sufficiently good as targets for treatment despite the fact that the results were in favor of the apo-ratio.

15. Treating CV risk patients to new targets using apolipoproteins

The apo-ratio, as shown in this paper, has commonly been shown to predict CV risk equally well or, in fact, more commonly even significantly better than conventional lipids in both prospective and treatment studies. So, which cut levels and targets of the apo-ratio should be recommended in the clinic to indicate CV risk before and after treatment? Since there is an almost linear increase (semi-log scale) in risk with increasing values of the apo-ratio from both AMORIS and the INTERHEART studies **(Figure 10)** it is clear that at values of the apo-ratio > 0.90 (values should be given in two decimals in order not to lose important information) there is a considerable increase in risk, whereas values from 0.70 to about 0.90 are indicative of a moderate risk. Values for men < 0.70 and for fertile females < 0.60 can be more normal especially if no other risk factors are present. The "ideal-biologically normal values" are rather < 0.50 as also documented in lipid-lowering trials in which CV events have been successfully reduced (176,177). So the target values during therapy must focus on these levels, the lower the apo-ratio the better is the therapy.

Lipids and apos are commonly correlated as also manifested in the AMORIS study (3 and others). In order to simplify for the physicians to learn what a value of LDL-C corresponds to regarding apoB **(Figure 11, left)**, a table has been compiled based on data from AMORIS also for the relationship between LDL-C and the apo-ratio **(Figure 11, right)**. A value of the apo-ratio of 0.80 roughly corresponds to a value for LDL-C of 3.0 mmol/L, and an apo-value of about 0.50 corresponds to LDL-C value 1.6 mmol/L for men and about 0.1 units lower for females. Notably, there is a large deviation from this correlation line. Those having a higher apoB or a higher apo-ratio at all levels of LDL-C (above the line) in general have a much higher CV risk than those below the line. Further details and relations between apolipoproteins, lipids and their relations to CV risk, and cut- and target levels of apoB and apoA-I have been reviewed (3). Since the target level for LDL-C according to many guidelines is set at LDL-C < 1.6 mmol/L, a target and normal value of the apo-ratio < 0.50 seems to be a realistic number.

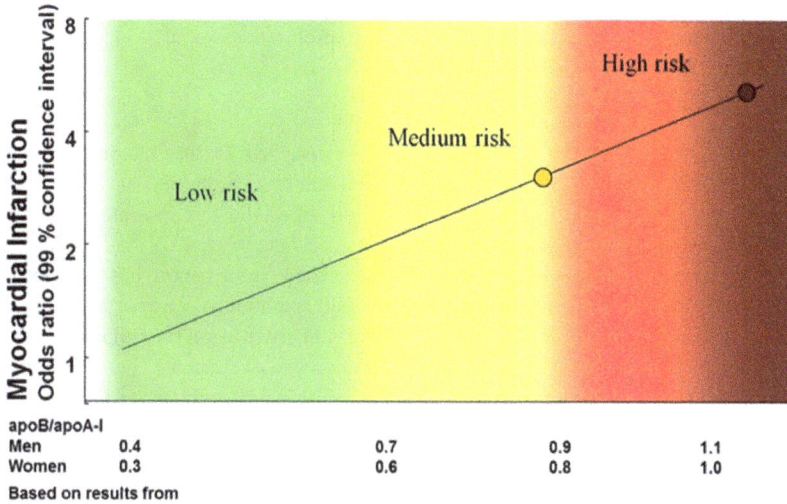

Figure 10. This line of risk of myocardial infarction is based on the findings in the AMORIS (reference 3) and the INTERHEART (reference 58) studies. Tentative cut-values are indicated in green (low risk), yellow (medium risk), and red (high risk). Values for a particular patient can be indicated by the dots on the line. During lipid-lowering treatment it is easy to monitor how a patient moves upwards or downwards in the risk line for the apo-ratio.

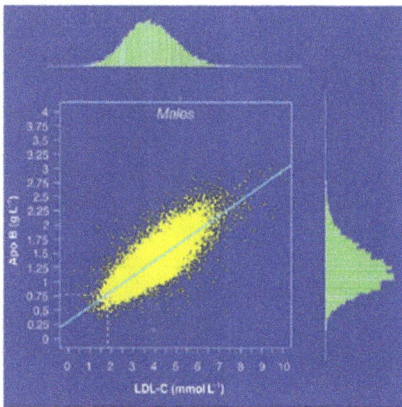

AMORIS; LDL-C correlates to the apoB/apoA-I ratio

LDL-C (mmol/L)	apoB/apoA-I men	apoB/apoA-I women
1.6	0.510	0.414
2.0	0.595	0.491
2.6	0.722	0.606
3.0	0.806	0.684
3.6	0.933	0.799
4.0	1.017	0.877

Figure 11. Data from the AMORIS study. Relations between LDL-C and apoB (left), and LDL-C and the apoB/apoA-I levels (right). Various cut-levels of LDL-C correspond to apoB and apoB/apoA-I values (both figures from reference 3).

How much can effective lipid-lowering therapy reduce apoB and the apo-ratio and how much can apoA-I be increased? Physical exercise and diet, if effective and longstanding, can reduce apoB by 5-10 % at the most and the apo-ratio by about 5% and increase apoA-I by about 5%. For more information see reference 3 and data on effects of statins in section 13 above.

16. Discussion

Today, LDL-C, non-HDL-C and lipid ratios are prioritized in international guidelines although apoB has also been mentioned in a few guidelines (9,18,38,178-180). In this review evidence is given indicating that apoB, apoA-I, and especially the apo-ratio, are at least equally good, or often even better than conventional lipids to predict CV risk prospectively and during lipid-lowering treatment. Much of this new information has not yet been included in any previously published meta-analyses. It is therefore of importance to review these data obtained from countries in the whole world in order to get the full information of what apolipoproteins can deliver for CV risk prediction and evaluation. These findings and advantages are summarized below and also briefly in **Table 3 (I) and (II).**

The biological relevance of using apoB and apoA-I as markers of CV risk is convincing since these proteins are carriers of lipids in the circulation and deliver C to peripheral tissues including the arterial wall (mediated by apoB). ApoA-I can remove C from the subendothelial space for breakdown and removal through the bile and further GI excretion. ApoA-I has also my other protective actions as summarized above and can thereby modify or inhibit inflammation and atherogenesis provoked by oxidation and modification of LDL-C with apoB (3,64). Thus, both apoB and apoA-I are biologically and patho-physiologically strongly active in normal biology and in plaque formation.

There are also methodological advantages of using these apos (3,23,28,65). Direct measurements of apoB and apoA-I by internationally standardized methods are available, analysis can be performed even if taken from patients in the non-fasted state, apos can be trustfully analyzed on frozen samples, the errors of the methods are not dependent on TG levels, and the methodological errors are usually low. Furthermore, costs for the direct analysis can be low as is the case in many countries. Importantly, the apo-ratio reflects the whole lipoprotein spectrum of virtually all phenotypes (type III patients need additional definition of risk) into one number. No sort like mg/dL or mmol/L has to be given for the ratio, which may otherwise be difficult to convert to understandable numbers in different countries. In summary, just one number of the apoB/apoA-I ratio indicates the risk level, similar numbers in all parts of the world. The higher the number, the higher is the risk (3) as also supported by a vast number of studies summarized in this paper. The cardiovascular risk line related to increasing value of the apo-ratio seems to be very similar world-wide **(Figure 10).**

ApoB, which indicates the number of potentially atherogenic particles, mainly sdLDL particles, has in a majority of publications been shown to be a better predictor of CV risk than LDL-C (9,11,13,14,44,59-62,88,89 and others) but in several instances apoB and non-HDL-C seem to indicate similar CV risk (10). One explanation why apoB may be better than non-HDL-C in risk prediction may be due to the fact that larger VLDL- and IDL-C-containing particles may have less potential to penetrate into the arterial walls than

Biological relevance
ApoB and apoA-I are carriers of lipids into and out of the arterial wall. Major pathophysiological mechanims are dependent on these proteins and how they can be modified (apoB) and be protective due to defensive actions.

Methodological advantages
Methods for apoB and apoA-I internationally standardized. Methodological errors of apoB and apoA-I are generally <5%. Fasting is not needed. High TG does not interfere. Frozen samples can be analyzed. The apo-ratio; no sort like mmol/L or mg/dL is needed. One number indicates the "cholesterol balance", easy to remember and act upon. Identifies sdLDL particle numbers. The apo-ratio reflects the risk associated with an imbalance between atherogenic and anti-atherogenic lipoproteins.

Relations to CV diseases
Strong predictors of myocardial infarction, stroke, heart failure, and also related to risk of renal failure and aortic aneurysms

Table 3 (I)

Relations to CV risk factors
Strong associations with abdominal obesity, metabolic syndrome and both diabetes

Relations to CV risk in univariate and multivariate analyses
Risk relationships for individual apos ant the apo-ratio commonly remained after adjustment for multiple conventional risk factors.

Relations to lipids and lipoproteins as predictors of CV risk
Lipids and lipid-based ratios are rarely significantly better than apos or the apo-ratio. However, apos and the apo-ratio are at least as good as lipids and lipid ratios, but commonly significantly better than lipids to predict CV risk

Relations to atherosclerosis
Strong associations with atherosclerosis in carotid arteries (IMT), coronary atherosclerosis (angiography and IVUS), femoral plaques and impaired endothelial function. Predicts progression and regression of carotid and coronary atherosclerosis

Relations to lipid-lowering treatment
Predicts outcome in statin trials equally well or commonly better than conventional lipids and lipoproteins

Table 3 (II)

Table 3. (I) and (II). Summary of findings supporting the use of apoB, apoA-I and the apo-ratio.

sdLDL particles. In fact, in a number of large publications including meta-analyses apoB has been shown to be a stronger predictor than the next best predictor non-HDL-C (9,11,13,14,61,62). LDL-C is only the third best predictor of future CV risk according to major analyses (9,62) and so also during statin treatment (169-171). However, in another meta-analysis they found non-HDL-C to be better than LDL-C and apoB during statin treatment (64). In a majority of these studies data have been adjusted for age and gender as well as other confounding risk factors like blood pressure, smoking, obesity and commonly also diabetes and other lipids.

Direct comparative data of HDL-C versus apoA-I is more sparse and is still much debated due to the complexity of HDL metabolism. ApoA-I has often similar predictive value as HDL-C as presented in the ERFC meta-analysis (10). However, especially in the large INTERHEART study, apoA-I over the whole range of HDL-values was a better predictor of risk than HDL-C (59). Similarly, in the AFCAPS/TexCAPS statin study apoA-I was a stronger determinant of risk than HDL-C (52,53).

What about the lipid-ratios versus the apo-ratio, which has strongest relations to CV risk? Importantly, all ratios and especially the apo-ratio predict prospective risk better than any single lipid variable (3,4,7,22,44,52,53,56,59,60,80,88,89,100,101,169-171 and others). Similarly, during statin treatment ratios also beat single lipoproteins in predicting risk (168-171,174). That should be obvious since ratios has a greater potential to find subjects at risk in whom the anti-atherogenic capacity of HDL-C or apoA-I are deranged. These data are also obtained when controlling for confounders i.e. conventional risk factors. Thus, the apo-ratio may have a better potential to identify subjects with different phenotypes than a single lipoprotein fraction. These strong findings in favor of any ratio, especially the apo-ratio, are strangely enough, not considered important in any international guideline despite the fact that ratios virtually in all studies in which ratios have been used outperform the results obtained by single lipoprotein fractions. Why this unscientific approach by guidelines committees?

Results from meta-analyses are generally well trusted but can also be questioned regarding selection criteria for including studies, acceptance of analytical and diagnostic methods used in each of the studies, primary and secondary variables used as major endpoints as well as the general conclusions drawn from the analyses. The results from the first ERFC meta-analysis have been taken to indicate that the apo-ratio and the TC/HDL-C ratio are equally good predictors of risk (10) and that apoB is equally good as risk predictor as non-HDL-C and apoA-I is equally predictive as HDL-C. The authors also open for future use of apolipoproteins especially in evaluation of risk of MI. In these risk conditions they found that apoB and apoA-I could be more useful in men than women, and in subjects with high TC, in those with low HDL-C, in individuals with hypertension, and in those with intermediate CV risk (Framingham risk score) apos can also be useful. They also found that the apo-ratio was a better predictor of CHD than stroke. However, in a recent publications in JAMA they conclude that the TC/HDL-C ratio had stronger predictive power than the apo-ratio when these ratios were added to conventional risk factors.

Two major critical views against these JAMA (see footnotes a/ and b/ below) papers may be raised that unfavorably affect the trust of using apos as risk predictors. Many early studies on apos included in these meta-analyses had large methodological errors (not internationally standardized) which may affect the conclusions on the credibility to use apos as risk predictors. This is unfair to the modern apo-technology which has much lower methodological errors.

Furthermore, in the ERFC studies they pooled non-fatal MI, all CHD fatalities, peripheral vascular diseases, and even all strokes, especially haemorrhagic stroke and unidentified stroke in very old people, into the primary variable "cardiovascular events". Such pooling of events considerably dilute the potential of adequate information yielded by an appropriately measured apos and the apo-ratio. This is especially the case for patients with risk of MI and in those suffering ischemic strokes in which positive diagnostic values have been obtained for apos as summarized in previously commented studies.

The authors also discuss some potential problems with introduction of apos such as need for education, lack of availability of apo-methods in the most laboratories, standardization problems as well as additional costs for such methods. All these aspects and possible problems must obviously be considered when new diagnostic tests shall be introduced for clinical use in risk evaluation. Yes, education is mandatory and may take time, but such problems must not over shade the importance of innovation of analytical tools. Costs can be significantly reduced if apo-tests become standard analyses. In fact, many biochemists already now favor these analyses over conventional lipids as documented previously in this paper.

Another criticism of the ERFC-studies is related to which studies were excluded (lack of confounding variables or case-control studies) from the meta-analyses in ERFC. Thus, major studies like AMORIS (44), INTERHEART (58) and ISIS (60) were not included in the ERFC meta-analysis. Neither were their positive results commented in the discussion on risk of MI despite the fact that these three studies have six times as many well defined events than those in the ERFC studies. In all these large studies apoB, apoA-I and especially the apo-ratio, due to their large number of events, were each significantly stronger predictors than conventional lipids. In ERFC there were many studies, but few of these studies showed significant differences between lipids and apos due to few well defined hard events. Neither did ERFC point out that the apo-ratio also seems to be the best variable to describe the remaining CV risk after statin treatment. This has been shown especially in the statin trials like AFCAPS/TexCAPS (52,53), IDEAL (169-171), TNT (168-170), CARDS (181), and JUPITER (176) as well as in studies on regression of atherosclerosis during lipid-lowering therapy (172-174) as pointed out above. In most of these studies the data were also adjusted for age, gender, conventional lipids and lipoproteins as well as other major risk factors like blood pressure, smoking, obesity and diabetes. Grundy simply concludes in the JAMA editorial

---

a/ The Emerging Risk Factors Collaboration. Lipid-Related Markers and Cardiovascular Disease Prediction. JAMA. 2012;307(23):2499-2506.
b/ Scott M. Grundy. Editorial: Use of Emerging Lipoprotein Risk Factors in Assessment of Cardiovascular Risk. 2012;307(23):2540-2541.

that conventional risk factors plus LDL-C, and possibly one more risk factor, is enough as tools for prediction of risk – a very conservative approach which is so (too!) common in US!

How much do confounders/risk factors impact on the results from all these CV risk studies? Importantly, in ERFC (10) and also in the majority of studies cited above, the impact of adjusting for major confounders was very small and only changed the risk (HR, RR or OR) to a minor degree. In fact, the apolipoproteins added value, measured as net reclassification index (NRI) in several large studies (88,91,170). This indicates that apolipoproteins, especially the apo-ratio, could change the numbers of individuals either to a higher or a lower CV risk compared to conventional lipids. Newly developed risk algorithms based on the apo-ratio have also been developed showing at least equal predictive or even better values than conventional risk algorithms (102,117,118). Thus, apoB, apoA-I and the apo-ratio can already now be used in clinical settings.

In the present review the apo-ratio has been shown to be closely related to many different types of CV events in prospective studies. These common diseases are myocardial infarction, stroke, especially ischemic stroke, heart failure, renal failure, aortic aneurysms, development of diabetes, including retinopathy (Table 3 I and II). However, in the meta-analyses published so far only CV events have been chosen as endpoints and other manifestation of CV risk related to atherosclerosis have been excluded.

Is the apo-ratio useful in predicting various metabolic and inflammatory conditions commonly underlying atherosclerosis and its future consequences? In fact, the apo-ratio has also been found to be a valuable summarizing index of lipid-abnormalities and their complications in a large number of studies of the MetS and/or diabetes (125-139,143,144). In addition, the apo-ratio values are also increased in patients with hypertension, obesity, in pubertal children and in those with heredity for CV diseases (130,141). The apo-ratio is also more closely than lipids related to atherosclerosis in a large number of studies in which different techniques like coronary angiography, arterial wall thickness obtained by ultrasound techniques in the carotid arteries (CIMT values) or even in the coronary arteries by intravascular ultrasound (IVUS) and arterial abnormalities such as the endothelial dysfunction have been used (145-162). Thus, in all these disease or risk situations the apo-ratio may identify those at an increased risk even better than what is currently performed by using LDL-C or the recently recommended non-HDL-C.

The newest research data on the apo-ratio have not yet been reviewed by international guideline committees. Thus, so far, in the newest guidelines developed over the last few years non-HDL-C and apoB are mentioned, and accepted for clinical use, whereas the apo-ratio is still waiting for acceptance (9,35,178,179).

In conclusion; with all the new knowledge presented in this paper about the strong relations between apoB, apoA-I, and the apo-ratio, and CV risk as well as other disease manifestations, it is proposed, as many researchers have already done, that these strong risk predictors/factors/markers are included in new guidelines. In many disease conditions and manifestations of atherosclerosis apolipoproteins are at least equally informative, and often better than LDL-C, non-HDL-C and lipid ratios in predicting risk. It is realized that there will be pedagogical hurdles, but it should be possible to educate physicians, patients and health providers to understand that these apolipoproteins are markers of normal and abnormal cholesterol metabolism. The apo-ratio simply reflects the "balance between the bad cholesterols and the good cholesterols" technically measured by apolipoproteins. The apo-ratio is a valid cardiovascular risk index (CRI) that reflects the level of CV risk for virtually all patients with different lipid phenotypes, the higher the value of the apo-ratio, the higher is the risk. Finally, targeting lower values (about 0.50) of the apo-ratio during therapy may more correctly identify who is at risk or not at risk, and how high is the risk? Does the risk depend on the atherogenic apoB, or the anti-atherogenic apoA-I or rather on the most informative value i.e. the apo-ratio which summarizes the level of risk in a simple way? Since physicians usually only manage to effectively evaluate and trust one laboratory marker, the apo-ratio is such a valid marker. By simply plotting the value for a given patient on the risk line you can easily follow improvement during therapy and also motivate the patient to improve values to normal levels **(Figure 10)**. New guidelines should at least contain equally objective information (cut-values and target values) on how to use apoB, apoA-I, and the apo-ratio as on lipids so that physicians can choose whichever diagnostic marker of risk they prefer. Gradually this new apolipoprotein-based risk classification with a focus on the apoB/apoA-I ratio may, or rather should be, introduced in clinical practice.

Author details

Göran Walldius

Department of Epidemiology, Institute of Environmental Medicine (IMM),
Karolinska Institutet, Stockholm, Sweden

Acknowledgement

The close collaboration since 1985 with Ingmar Jungner, the former head of CALAB laboratories in Stockholm, Sweden and the founder of the AMORIS database, is gratefully acknowledged. Many of the publications from the AMORIS study presented in this paper were also developed together with Ingar Holme, Are Aastveit, both from Oslo, Norway, Werner Kolar and Eugen Steiner, CALAB, Stockholm, Sweden, Allan Sniderman, Montreal, Canada and Niklas Hammar, Stockholm, Sweden. Their scientific knowledge and friendship is much appreciated. Much of the AMORIS work has been supported by grants from Gunnar and Ingmar Jungner Foundation for Laboratory Medicine, Stockholm, Sweden, Bure Hälsa & Sjukvård Ltd, Stockholm; and AstraZeneca, Mölndal, Sweden.

Abbreviations

apos	apolipoproteins
apoB	apolipoprotein B
apoA-I	apolipoprotein A-I
apo-ratio	apoB/apoA-I ratio
C	cholesterol

CV   cardiovascular
VLDL   Very Low Density Lipoprotein
IDL   Intermediate Density Lipoprotein
LDL   Low Density Lipoprotein
sdLDL   small dense LDL
TC   total cholesterol
HDL   high density lipoprotein
hsCRP   high sensitivity CRP (C-reactive protein)
IVUS   intravascular ultrasound
CIMT   carotid intima media thickness
CRI   Cardiovascular Risk Index
HR   Hazards Ratio
NRI   Net Reclassification Index
MACE   Major Coronary Events
RR   Relative Risk

17. References

[1] Elovson J, Chatterton JE, Bell GT, Schumaker VN, Reuben MA, Puppione D L, Reeve Jr
     JR, Young NL. Plasma very low density lipoproteins contain a single molecule of
     apolipoprotein B. J Lipid Res 1988; 29:1461-1473.

[2] Walldius G, Jungner I. Apolipoprotein B and apolipoprotein A-I: risk indicators of
     coronary heart disease and targets for lipid-modifying therapy. J Intern Med 2004;
     255(2): 188-205.

[3] Walldius G, Jungner I. The apoB/apoA-I ratio: a strong, new risk factor for
     cardiovascular disease and a target for lipid-lowering therapy – a review of the
     evidence. J Intern Med 2006; 259: 493-519.

[4] Sniderman AD, Furberg CD, Keech A, Roeters van Lennep JE, Frohlich J, Jungner I,
     Walldius G. Apolipoproteins versus lipids as indices of coronary risk and as targets for
     statin therapy treatment. Lancet 2003; 361: 777-780.

[5] Walldius G, Jungner I. Is there a better marker of cardiovascular risk than LDL
     cholesterol? Apolipoprotein B and A-I – new risk factors and targets for therapy. Nutr
     Metab Cardiovasc Dis 2007; 17: 565-571.

[6] Walldius G, Jungner I. ApoB, apoA-I, and the apoB/apoA-I ratio as predictors, markers,
     and factors for cardiovascular risk. The Fats of Life, LVDD/AACC. www.aacc.org. 2007;
     XXI(1): 11-21.

[7] Rader DJ, Hoeg JM, Brewer HB Jr. Quantitation of plasma apolipoproteins in the
     primary and secondary prevention of coronary artery disease. Ann Intern Med 1994;
     120: 1012-1025.

[8] Thompson A, Danesh J. Associations between apolipoprotein B, apolipoprotein AI, the
     apolipoprotein B/AI ratio and coronary heart disease: a literature-based meta-analysis
     of prospective studies. J Intern Med. 2006; 259: 481-492.

[9] Contois JH, McConnell JP, Sethi AA, Csako G, Devaraj S, et al. Apolipoprotein B and cardiovascular disease risk: position statement from the AACC Lipoproteins and Vascular Diseases Division Working Group on Best Practices. Clin Chem 2009; 55(3): 407-419.

[10] Di Angelantonio E, Sarwar N, Perry P, Kaptoge S, Ray KK, et al. The Emerging Risk Factors Collaboration. Major lipids, apolipoproteins, and risk of vascular disease. JAMA 2009; 302(18): 1993-2000. Note also the added supplement.

[11] Contois JH, Warnick GR, Sniderman AD. Review Article; Reliability of low-density lipoprotein cholesterol, non-high-density lipoprotein cholesterol, and apolipoprotein B measurement. J Clin Lipidol 2011; 5: 264-272.

[12] Olofsson SO, Wiklund O, Borén B. Apolipoproteins A-I and B: biosynthesis, role in the development of atherosclerosis and targets for intervention against cardiovascular disease. Vascular Health and Risk Management 2007; 3(4): 491-502.

[13] Sniderman A, McQueen M, Contois J, Williams K,. Furberg CD. Why is non-high-density lipoprotein cholesterol a better marker of the risk of vascular disease than low-density lipoprotein cholesterol? Journal of Clinical Lipidology 2010; 4: 152-155.

[14] Sniderman A, Williams K, de Graaf J. Non-HDL C equals apolipoprotein B: except when it does not! Curr Opin Lipidol 2010; 21: 518-524.

[15] Rader DJ, Davidson M, Caplan RJ, Pears JS. Lipid and apolipoprotein ratios: association with coronary artery disease and effects of rosuvastatin compared with atorvastatin, pravastatin, and simvastatin. Am J Cardiol 2003; 91(Suppl.): 20C-24C.

[16] Friedewald WT, Levy RI, Fredrickson DS. Estimation of the concentration of low-density lipoprotein cholesterol in plasma, without use of the preparative ultracentrifuge. Clin Chem 1972; 18: 499-502.

[17] Stein EA, Sniderman AD, Laskarzewski P. Assessment of Reaching Goal in Patients with Combined Hyperlipidemia: Low-Density Lipoprotein Cholesterol, Non-High-Density Lipoprotein Cholesterol, or Apolipoprotein B. Am J Cardiol 2005; 96(suppl): 36K–43K.

[18] Myers GL, Christenson RHM, Cushman M, Ballantyne CM, Cooper GR, et al. National Academy of Clinical Biochemistry Laboratory Medicine Practice Guidelines: Emerging Biomarkers for Primary Prevention of Cardiovascular Disease. Clin Chem 2009; 55: 378-384.

[19] Warnick GR. Editors'Notes. Fats of Life, LVDD/ AACC, www.aacc.org. 2010; XXIV(3):2-5, 33-37.

[20] Sniderman AD, Walldius G, Jungner I. The four horsemen of cholesterol versus ApoB: Good care versus better care. Fats of Life. LVDD/ AACC, www.aacc.org. 2003; XVII(4): 9-24.

[21] Sniderman AD, Jungner I, Holme I, Aastveit A, Walldius G. Errors that result from using the TC/HDL-C ratio rather than the apoB/apoA-I ratio to identify the lipoprotein-related risk of vascular disease. J Intern Med 2006; 259: 455-461.

[22] Walldius G, Jungner I, Aastveit AH, Holme I, Furberg CD, Sniderman AD. The apoB/apoA-I ratio is better than the cholesterol ratios to estimate the balance between plasma pro-atherogenic and anti-atherogenic lipoproteins and to predict coronary risk. Clin Chem Lab Med 2004; 42: 1355-1363.

[23] Scharnagl H, Nauck M, Wieland H, März W. The Friedewald formula underestimates LDL cholesterol at low concentrations. Clin Chem Lab Med 2001; 39: 426-431.

[24] Otvos JD. Why cholesterol measurements may be misleading about lipoprotein levels and cardiovascular disease risk – clinical implications of lipoprotein quantification using NMR spectroscopy. J Lab Med 2002; 26: 544-550.

[25] van Deventer HE, Miller WG, Myers GL, Sakurabayashi I, Bachmann LM, Caudill SP, Dziekonski A, Edwards S, Kimberly MM, Korzun WJ, Leary ET, Nakajima K, Nakamura M, Shamburek RD, Vetrovec GW, Warnick GR, Remaley AT. Non–HDL Cholesterol Shows Improved Accuracy for Cardiovascular Risk Score Classification Compared to Direct or Calculated LDL Cholesterol in a Dyslipidemic Population. Clinical Chemistry 2011; 57(3): 490-501.

[26] Marcovina SM, Albers JJ, Henderson LO, Hannon WH. International Federation of Clinical Chemistry standardization project for measurements of apolipoproteins A-I and B: III. Comparability of apolipoprotein A-I values by use of International Reference Material. Clin Chem 1993; 39: 773-781.

[27] Liu J, Sempos CT, Donahue RP, Dorn J, Trevisan M, Grundy SM. Non–High-Density Lipoprotein and Very-Low-Density Lipoprotein Cholesterol and Their Risk Predictive Values in Coronary Heart Disease. Am J Cardiol 2006; 98: 1363-1368.

[28] Marcovina SM, Albers JJ, Kennedy H, Mei JV, Henderson LO, Hannon WH. International Federation of Clinical Chemistry standardization project for measurements of apolipoproteins A-I and B: IV. Comparability of apolipoprotein B values by use of International Reference Material. Clin Chem 1994; 40: 586-592.

[29] Marcovina S, Packard CJ. Measurement and meaning of apolipoprotein AI and apolipoprotein B plasma levels. J Intern Med 2006; 259: 437-446.

[30] Zambon A, Brown BG, Deeb SS, Brunzell JD. Genetics of apolipoprotein B and apolipoprotein AI and premature coronary artery disease. J Intern Med 2006; 259: 473-480.

[31] Gatz M, Reynolds CA, Finkel D, Pedersen NL, Walters E. Dementia in Swedish Twins: Predicting Incident Cases. Behav Genet 2010; 40:768-75, DOI 10.1007/s10519-010-9407-4.

[32] Vladimirova-Kitova LG, Deneva-Koicheva TI. Increased Intima-Media Thickness in Carriers of the LDL-Receptor Defective Gene versus Noncarriers with Newly Detected Asymptomatic Severe Hypercholesterolemia. Echocardiography 2011; 28: 223-234.

[33] Middelberg RPS, Spector TD, Swaminathan R, Snieder H. Genetic and Environmental Influences on Lipids, Lipoproteins, and Apolipoproteins Effects of Menopause. Arterioscler Thromb Vasc Biol. 2002; 22: 1142-1147.

[34] Schmidt C, Fagerberg, B Wikstrand J, Hulthe J. ApoB/apoA–I ratio is related to femoral artery plaques and is predictive for future cardiovascular events in healthy men. Atherosclerosis 2006; 189: 178-185.

[35] Hoff HF. Apolipoprotein Localization in Human Cranial Arteries, Coronary Arteries, and the Aorta. Stroke 1976; 7(4): 390-393.

[36] Fogelstrand P, Borén J. Retention of atherogenic lipoproteins in the artery wall and its role in atherogenesis; Review. Nutrition, Metabolism & Cardiovascular Diseases 2012; 22: 1-7.

[37] Contois JH, McNamara JR, Lammi-Keefe CJ, Wilson PWF, Massov T, Schaeffer EJ. Reference intervals for plasma apolipoprotein B determined with a standardized commercial immunoturbidimetric assay: results from the Framingham Offspring Study. Clinical Chemistry 1996; 42: 515-523.

[38] Grundy SM, Cleeman JI, Merz CNB et al. Implications of recent clinical trials for the National Cholesterol Education Program Adult Treatment Panel III Guidelines. J Am Coll Cardiol 2004; 44: 720-732.

[39] Avogaro P, Bon GB, Cazzolato G, Quinci GB, Sanson A, et al. Variations in apolipoproteins B and AI during the course of myocardial infarction. Eur J Clin Invest 1978; 8: 121-129.

[40] Avogaro P, Cazzolato G, Bon GB, Berlussi F, Quinci GB. Values of apoA-I and apoB in humans according to age and sex. Clin Chim Acta 1979; 95: 511-515.

[41] Sniderman A, Shapiro S, Marpole S, Skinner B, Teng B, et al. Association of coronary atherosclerosis with hyperapobetalipoproteinemia (increased protein but normal cholesterol levels in human plasma low density (beta) lipoproteins. Proc Natl Acad Sci, USA 1980; 77: 604-608.

[42] Lamarche B, Moorjani S, Lupien PJ et al. Apolipoprotein A-1 and B levels and the risk of ischemic heart disease during a five- year follow-up of men in the Que'bec Cardiovascular Study. Circulation 1996; 94: 273-278.

[43] Sharrett AR, Ballantyne CM, Coady SA et al. Coronary heart disease prediction from lipoprotein cholesterol levels, triglycerides, lipoprotein(a), apolipoproteins A-I and B, and HDL density subfractions: The Atherosclerosis Risk in Communities (ARIC) Study. Circulation 2001; 104: 1108-1113.

[44] Walldius G, Jungner I, Holme I, Aastveit AH, Kolar W, Steiner E. High apolipoprotein B, low apolipoprotein A-I, and improvement in the prediction of fatal myocardial infarction (AMORIS study): a prospective study. Lancet 2001; 358: 2026-2033.

[45] Moss AJ, Goldstein RE, Marder VJ, Sparks CE, Oakes D, Greenberg H, et al. Thrombogenic factors and recurrent coronary events. Circulation 1999; 99: 2517-2522.

[46] Corsetti JP, Zareba W, Moss AJ, Sparks CE. Apolipoprotein B determines risk for recurrent coronary events in postinfarction patients with metabolic syndrome. Atherosclerosis 2004; 177: 367-373.

[47] Talmud PJ, Hawe E, Miller GJ, Humphries SE. Nonfasting apolipoprotein B and triglyceride levels as a useful predictor of coronary heart disease risk in middle-aged UK men. Arterioscler Thromb Vasc Biol 2002; 22: 1918-1923.

[48] Shai I, Rimm EB, Hankinson SE et al. Multivariate assessment of lipid parameters as predictors of coronary heart disease among postmenopausal women. Potential implications for clinical guidelines. Circulation 2004; 110: 2824-2830.

[49] Pischon T, Girman CJ, Sacks FM, Rifai N, Stampfer MJ, Rimm EB. Non-high-density lipoprotein cholesterol and apolipoprotein B in the prediction of coronary heart disease in men. Circulation 2005; 112: 3375-3383.

[50] Benn M, Nordestgaard BG, Boje Jensen G, Tybjærg-Hansen A. Improving Prediction of Ischemic Cardiovascular Disease in the General Population Using Apolipoprotein B. Arterioscler Thromb Vasc Biol. 2007; 27: 661-670.

[51] Pedersen TR, Olsson AG, Faergeman O, et al. Lipoprotein changes and reduction in the incidence of major coronary heart disease events in the Scandinavian Simvastatin Survival Study (4S). Circulation 1998; 97: 1453-1460.

[52] Gotto AM, Whitney E, Stein EA et al. Relation between baseline and on-treatment lipid parameters and first acute major coronary events in the Air Force/Texas Coronary Atherosclerosis Prevention Study (AFCAPS/TexCAPS). Circulation 2000; 101: 477-484.

[53] Gotto, AM Jr. Establishing the benefit of statins in low-to-moderate—risk primary prevention: The Air Force/Texas Coronary. Atherosclerosis Prevention Study (AFCAPS/TexCAPS). Atherosclerosis Supplements 2007; 8: 3-8.

[54] Simes RJ, Marschner IC, Hunt D et al. Relationship between lipid levels and clinical outcomes in the long-term intervention with pravastatin in ischemic disease (LIPID) trial. To what extent is the reduction in coronary events with pravastatin explained by on-study lipid levels? Circulation 2002; 105: 1162-1169.

[55] van Lennep JE, Westerveld HT, van Lennep HWO, Zwinderman AH, Erkelens DW, van der Wall EE. Apolipoprotein concentrations during treatment and recurrent coronary artery disease events. Arterioscler Thromb Vasc Biol 2000; 20: 2408-2413.

[56] Holme I, Aastveit AH, Hammar N, Jungner I, Walldius G. Relationships between lipoprotein components and risk of ischaemic and haemorrhagic stroke in the Apolipoprotein MOrtality RISk study (AMORIS). J Intern Med 2009; 265: 275-287.

[57] Sniderman AD, Jungner I, Holme I, Aastveit A, Walldius G. Errors that result from using the TC/HDL-C ratio rather than the apoB/apoA-I ratio to identify the lipoprotein-related risk of vascular disease. J Intern Med 2006; 259: 455-461.

[58] Yusuf S, Hawken S, Ôunpuu S, Dans T, Avezum A, et al., on behalf of the INTERHEART Study Investigators. Effect of potentially modifiable risk factors associated with myocardial infarction in 52 countries (the INTERHEART study): case-control study. Lancet 2004; 364: 937-952.

[59] McQueen MJ, Hawken S, Wang X, Ounpuu S, Sniderman A, et al. Lipids, lipoproteins, and apolipoproteins as risk markers of myocardial infarction in 52 countries (the INTERHEART study): a case-control study. Lancet 2008; 372: 224-233.

[60] Parish S, Peto R, Palmer A, Clarke R, Lewington S, et al. The joint effects of apolipoprotein B, apolipoprotein A1, LDL cholesterol, and HDL cholesterol on risk: 3510 cases of acute myocardial infarction and 9805 controls. Eur Heart J 2009; 30(17): 2137-2146. First published online June 11, 2009,doi:10.1093/eurheartj/ ehp221.

[61] Barter PJ, Ballantyne CM, Carmena R, Castro Cabezas M, Chapman MJ, Couture P, De Graaf J, Durrington PN, Faergeman O, Frohlich J, Furberg CD, Gagne C, Haffner SM, Humphries SE, Jungner I, Krauss RM, Kwiterovich P. Marcovina S, Packard CJ, Pearson TA, Reddy KS, Rosenson R, Sarrafzadegan N, Sniderman AD, Stalenhoef AF, Stein E, Talmud PJ, Tonkin AM, Walldius G, Williams KMS. ApoB versus cholesterol in estimating cardiovascular risk and in guiding therapy: report of the thirty-person/ten-country panel. J Intern Med 2006; 259: 247-258.

[62] Sniderman AD, Williams K, Contois JH, Monroe HM, McQueen MJ, de Graaf J, Furberg CD. A Meta-analysis of LDL-C, non-HDL-C, and apoB as markers of cardiovascular risk. Circ Cardiovasc Qual Outcomes. 2011; 4: 337-345.

[63] Sniderman AD, St-Pierre A, Cantin B, Dagenais GR, Depre's J-P, Lamarche B. Concordance/discordance between plasma apolipoprotein B levels and the cholesterol indexes of atherosclerotic risk. Am J Cardiol 2003; 91: 1173-1177.

[64] Boekholdt SM, Arsenault BJ, Mora S, Pedersen TR, LaRosa JC, Nestel PJ, Simes RJ, Durrington P, Hitman GA, Welch KMA, DeMicco DA, Zwinderman AH, Clearfield MB, Downs JR, Tonkin AM, Colhoun HM, Gotto Jr AM, Ridker PM, Kastelein JJP. Association of LDL Cholesterol, Non–HDL Cholesterol, and Apolipoprotein B Levels With Risk of Cardiovascular Events Among Patients Treated With Statins. A Meta-analysis. JAMA 2012; 307(12): 130213-130209.

[65] Barter PJ, Rye K-A. The rationale for using apoA-I as a clinical marker of cardiovascular risk. J Intern Med 2006; 259: 447-454.

[66] Walldius G, Jungner I. Apolipoprotein A-I versus HDL cholesterol in the prediction of risk for myocardial infarction and stroke Curr Opin Cardiol 2007; 22: 359-367.

[67] Asztalos BF, Tani M, Schaefer EJ. Metabolic and functional relevance of HDL subspecies. Curr Opin Lipidol 2011; 22: 176-185.

[68] Larsson A, Carlsson L , Axelsson J. Low diurnal variability of apolipoprotein A1, apolipoprotein B and apolipoprotein B/apolipoprotein A1 ratio during normal sleep and after an acute shift of sleep. Clinical Biochemistry 2008; 41: 859-862.

[69] Elliott DA, Shannon Weickert C, Garner B. Apolipoproteins in the brain: implications for neurological and psychiatric disorders. Clin Lipidol. 2010; 51(4): 555-573. doi:10.2217/CLP.10.37

[70] Kavoa AE et al. Qualitative characteristics of HDL in young patients of an acute myocardial infarction. Atherosclerosis 2012; 220: 257-264.

[71] Sung K-C, Rhee E-J , Kim H, Park JB, Kim Y-K, Rosenson RS. Prevalence of low LDL-cholesterol levels and elevated high-sensitivity C-reactive protein levels in apparently healthy Korean adults. Nutrition, Metabolism & Cardiovascular Diseases. Available on line July 23, 2011. 2011; doi:10.1016/ j.numecd.2011.03.006.

[72] Liangpunsakul S, Rong QI, M.S, Crabb DW, Witzmann F. Relationship Between Alcohol Drinking and Aspartate Aminotransferase: Alanine Aminotransferase (AST:ALT) Ratio, Mean Corpuscular Volume (MCV), Gamma-Glutamyl Transpeptidase (GGT), and Apolipoprotein A1and B in the U.S. Population. J. Stud. Alcohol Drugs, 2010; 71: 249-252.

[73] Schlitt A, Blankenberg S, Bickel C et al. Prognostic value of lipoproteins and their relation to inflammatory markers among patients with coronary artery disease. Int J Cardiol 2005; 102: 477-485.

[74] Held C, Hjemdahl P, Rehnqvist N , Björkander I, Forslund L, Brodin U , Berglund L, Angelin B. Cardiovascular prognosis in relation to apolipoproteins and other lipid parameters in patients with stable angina pectoris treated with verapamil or metoprolol. Results from the Angina Prognosis Study in Stockholm (APSIS) Atherosclerosis 1979; 135: 109-118.

[75] Sharp DS, Burchfield CM, Rodriguez BL, Sharrett AR, Sorlie PD, Marcovina SM. Apolipoprotein A-I predicts coronary heart disease only at low concentrations of high-density lipoprotein cholesterol: an epidemiological study of Japanese-Americans. Int J Clin Lab Res 2000; 30: 39-48.

[76] Luc G, Bard J-M, Ferrie`res J et al. Value of HDL cholesterol, apolipoprotein A-I, lipoprotein A-I, and lipoprotein A-I/A-II in prediction of coronary heart disease: the PRIME Study.Prospective Epidemiological Study of Myocardial Infarction. Arterioscler Thromb Vasc Biol 2002; 22: 1155-1161.

[77] Patel JV, Abraheem A, Creamer J, Gunning M, Hughes EA, Lip GYH. Apolipoproteins in the discrimination of atherosclerotic burden and cardiac function in patients with stable coronary artery disease. European Journal of Heart Failure 2010; 12: 254-259. doi:10.1093/eurjhf/hfp202.

[78] Wedel H, McMurray JJV, Lindberg M, Wikstrand J, Cleland JGF, Cornel JH, Dunselman P, Hjalmarson Å, Kjekshus J, Komajda M, Kuusi T, Vanhaecke J, Waagstein F on behalf of the CORONA Study Group. Predictors of fatal and non-fatal outcomes in the Controlled Rosuvastatin Multinational Trial in Heart Failure (CORONA): incremental value of apolipoprotein A-1, high-sensitivity C-reactive peptide and N-terminal pro B-type natriuretic peptide. European J Heart Fail 2009; 11: 281-291 doi:10.1093/eurjhf/hfn046.

[79] Chien K-L, Sung F-C, Hsu H-C, Su T-C, Lin R-S, Lee Y-T. Apolipoprotein A-I and B and Stroke Events in a Community-Based Cohort in Taiwan. Report of the Chin-Shan Community Cardiovascular Study. Stroke. 2002; 33: 39-44.

[80] Van der Steeg WA, Boekholdt SM, Stein EA, El-Harchaoui K, Stroes ESG, Sandhu MS, Wareham NJ, J. Jukema JW, Luben R, Zwinderman AH, Kastelein JJP, Khaw K-T. Role of the Apolipoprotein B–Apolipoprotein A-I Ratio in Cardiovascular Risk Assessment: A Case–Control Analysis in EPIC-Norfolk. Ann Intern Med. 2007; 146: 640-648.

[81] Walldius G, Jungner I. Apolipoprotein A-I versus HDL cholesterol in the prediction of risk for myocardial infarction and stroke. Curr Opin Cardiol 2007; 22: 359-367.

[82] Jungner I, Marcovina SM, Walldius G, Holme I, Kolar W, Steiner E. Apolipoprotein B and A-I values in 147 576 Swedish males and females, standardized according to the World Health Organization – International Federation of Clinical Chemistry First International Reference Materials. Clin Chem 1998; 44: 1641-1649.

[83] Walldius G, Jungner I, Kolar W, Steiner E. Apolipoprotein B and total serum cholesterol levels in 41 000 males and females. Clin Chem 1990; 36(6): 952-956.

[84] Jungner I, Walldius G, Holme I, Kolar W, Steiner E. Apolipoprotein B and A-1 in relation to serum cholesterol and triglycerides in 43,000 Swedish males and females. Int J Clin Lab Res 1992; 21: 247-255.

[85] Walldius G, Jungner I, Kolar W, Holme I, Steiner E. High cholesterol and triglyceride values in Swedish males and females: increased risk of fatal myocardial infarction: First report from the AMORIS (Apolipoprotein related MOrtality RISk) Study. Blood Press 1992; I (suppl 4): 35-42.

[86] Walldius G, Aastveit A, Jungner I. Hypercholesterolemia and hypertriglyceridemia— greatest cardiac risk in subjects with high apoB/apoA-I levels. Original Research Article. International Congress Series, 2004; 1262: 203-206.

[87] Jungner I, Sniderman AD, Furberg C, Aastveit AH, Holme I, Walldius G. Does Low-Density Lipoprotein Size Add to Atherogenic Particle Number in Predicting the Risk of Fatal Myocardial Infarction? Am J Cardiol 2006; 97: 943-946.

[88] Holme I, Aastveit AH, Jungner I, Walldius G. Relationships between lipoprotein components and risk of myocardial infarction: age, gender and short versus longer follow-up periods in the Apolipoprotein MOrtality RISk study (AMORIS). J Intern Med 2008; 264: 30-38.

[89] Holme I, Aastveit AH, Hammar N, Jungner I, Walldius G. Lipoprotein components and risk of congestive heart failure in 84 470 men and women in the Apolipoprotein MOrtality RISk study (AMORIS). Eur J Heart Fail 2009; 11: 1036-1042. Published Oct 3, 2009. Doi:10.1093/eurjhf/hfp129

[90] Holme I, Aastveit AH, Hammar N, Jungner I, Walldius G. Haptoglobin and risk of myocardial infarctiion, stroke and congestive heart failure in 342,125 men and women in the Apolipoprotein MOrtality RISk study (AMORIS). Ann Med 2009; 41: 522-532.

[91] Holme I, Aastveit AH, Hammar N, Jungner I, Walldius G. Inflammatory markers, lipoprotein components and risk of major cardiovascular events in 65,005 men and women in the Apolipoprotein MOrtality RISk study (AMORIS). 2010: 213(1): 299-305. Atherosclerosis 2010; doi:10.1016/j.atherosclerosis.2010.08.049.

[92] Solhpour A, Parkhideh S, Sarrafzadegan N, Asgary S, Williams K, Jungner I, Aastveit A, Walldius G, Sniderman A. Levels of lipids and apolipoproteins in three cultures. Atherosclerosis 2009; 207: 200-207.

[93] Holtzmann MJ, Jungner I, Walldius G, Ivert T, Nordqvist T, Östergren J, Hammar N. Dyslipidemia is a Strong Predictor of Myocardial infarction in Subjects with Chronic Kidney Disease. Ann Med 2010; Early Online, 1–8. Published in Annals of Medicine, 2012; 44: 262–270.

[94] Bhatia M, Howard SC, Clark TG et al. Apolipoproteins as predictors of ischaemic stroke in patients with a previous transient ischaemic attack. Cerebrovasc Dis 2006; 21: 323-328.

[95] Qureshi AI, Giles WH, Croft JB, Guterman LR, Hopkins LN. Apolipoproteins A-1 and apoB and the likelihood of non-fatal stroke and myocardial infarction – data from The Third National Health and Nutrition Examination Survey. Med Sci Monit 2002; 8: CR311-316.

[96] Chien KL, Sung FC, Hsu HC, Su TC, Lin RS, Lee YT. Apolipoprotein A-I and B and stroke events in a community-based cohort in Taiwan: report of the Chin-Shan Community Cardiovascular Study. Stroke 2002; 33: 39-44.

[97] Sharobeem KM, Patel JV, Ritch AES, Lip GYH, Gill PS, Hughes EA. Elevated lipoprotein (a) and apolipoprotein B to AI ratio in South Asian patients with ischaemic stroke. Int J Clin Pract 2007; 61: 1824-1828.

[98] Koren-Morag N, Goldbourt U, Graff E, Tanne D. Apolipoproteins B and AI and the risk of ischemic cerebrovascular events in patients with pre-existing atherothrombotic disease. J Neurol Sci 2008; 270: 82-87.

[99] Hankey GJ. Potential new risk factors for ischemic stroke. What is their potential? Stroke 2006; 37: 2181-2188.

[100] Walldius G, Aastveit AH, Jungner I. Stroke mortality and the apoB/apoA-I ratio: results of the AMORIS prospective study. J Intern Med 2006; 259: 259-266.

[101] Holme I, Aastveit AH, Hammar N, Jungner I, Walldius G. Relationships between lipoprotein components and risk of ischaemic and haemorrhagic stroke in the

Apolipoprotein MOrtality RISk study (AMORIS). J Intern Med. 2009; 265: 275-287. doi: 10.1111/j.1365-2796.2008.02016.xJ.

[102] McGorrian C, Yusuf S, Islam S, Jung H, Rangarajan S, Avezum A, Prabhakaran D, Almahmeed W,Rumboldt Z, Budaj A, Dans AL, Gerstein HC, Teo K, Anand SS on behalf of the INTERHEART Investigators Estimating modifiable coronary heart disease risk in multiple regions of the world: the INTERHEART Modifiable Risk Score. European Heart Journal Advance Access published December 22, 2010. European Heart Journal, doi:10.1093/eurheartj/ehq448. Published in European Heart Journal 2011; 32: 581-590.

[103] Gerstein HC, Islam S, Anand S, Almahmeed A, Damasceno A, Dans A, Lang CC, Luna MA, McQueen M, Rangarajan S, Rosengren A, Wang X, Yusuf S. Dysglycaemia and the risk of acute myocardial infarction in multiple ethnic groups: an analysis of 15,780 patients from the INTERHEART study. Diabetologia 2010; 53: 2509-2517, DOI 10.1007/s00125-010-1871-0.

[104] Joshi P, Islam S , Pais P, Reddy S, Dorairaj P, Kazmi K, Pandey MR, Haque S, Mendis S, Rangarajan S, Yusuf S. Risk Factors for Early Myocardial Infarction in South Asians Compared With Individuals in Other Countries. JAMA 2007; 297: 286-294.

[105] Lanas F, Avezum A, Bautista LE, Diaz R, Luna M, Islam S, Yusuf S for the INTERHEART Investigators in Latin America. Risk Factors for Acute Myocardial Infarction in Latin America. The INTERHEART Latin American Study. Circulation 2007; 115: 1067-1074.

[106] Kabagambe EK, Baylin A, Campos H. Nonfatal Acute Myocardial Infarction in Costa Rica. Modifiable Risk Factors, Population-Attributable Risks, and Adherence to Dietary Guidelines. Circulation. 2007; 115: 1075-1081.

[107] Steyn K, Sliwa K, Hawken S, Commerford P, Onen C, Damasceno A, Ounpuu S, Yusuf S for the INTERHEART Investigators in Africa. Risk Factors Associated With Myocardial Infarction in Africa The INTERHEART Africa Study. Circulation 2005; 112: 3554-3561.

[108] O'Donnell MJ, Xavier D, Liu L, Zhang H, Chin SL, Rao-Melacini P, Rangarajan S, qul Islam S, Pais P, McQueen MJ, Mondo C, Damasceno A, Lopez-Jaramillo P, Hankey GJ, Dans AL, Yusoff K, Truelsen T, Diener HC, Sacco RL, Ryglewicz D, Czlonkowska A, Weimar C, Wang X, Yusuf S on behalf of the INTERSTROKE investigators. Risk factors for ischaemic and intracerebral haemorrhagic stroke in 22 countries (the INTERSTROKE study): a case-control study. Published Online June 18, 2010 DOI:10.1016/S0140- 6736(10)60834-3. Published in Lancet 2010; 376: 112–23.

[109] Kostapanos MS, Christogiannis LG, Bika E, Bairaktari ET, Goudevenos JA, Elisaf MS, Milionis HJ. Apolipoprotein B-to-A1 Ratio as a Predictor of Acute Ischemic Nonembolic Stroke in Elderly Subjects. J Stroke Cerebrovasc Dis. 2010; 19: 497-502.

[110] Park J-H, Hong K-S, Lee E-J, Lee J, Kim D-E. High Levels of Apolipoprotein B/AI Ratio Are Associated With Intracranial Atherosclerotic Stenosis. Stroke 2011; 42: 3040-3046.

[111] Boekholdt SM, van der Steeg WA, Stein EA et al. The ratio of apolipoproteins B to A-I and the risk of future coronary artery disease in apparently healthy men and women; the EPIC-Norfolk prospective population study. Ann Intern Med. 2007; 146: 640-648.

[112] Meisinger C, Loewel H, Mraz W, Koenig W. Prognostic value of apolipoprotein B and A-I in the prediction of myocardial infarction in middle-aged men and women: results from the MONICA/KORA Augsburg cohort study. Eur Heart J 2005; 26: 271-278.

[113] Moss AJ, Goldstein RE, Marder VJ et al. Thrombogenic factors and recurrent coronary events. Circulation 1999; 99: 2517-2522.

[114] Corsetti JP, Zareba W, Moss AJ, Sparks CE. Apolipoprotein B determines risk for recurrent coronary events in postinfarction patients with metabolic syndrome. Atherosclerosis 2004; 177: 367-373.

[115] Cremer P, Nagel D, Mann H et al. Ten-year follow-up results from the Goettingen Risk, Incidence and Prevalence Study (GRIPS): I. Risk factors for myocardial infarction in a cohort of 5790 men. Atherosclerosis 1997; 129: 221-230.

[116] Sweetnam PM, Bolton CH, Downs LG et al. Apolipoproteins A-I, A-II and B, lipoprotein(a) and the risk of ischaemic heart disease: the Caerphilly Study. Eur J Clin Invest 2000; 30: 947-956.

[117] Dunder K, Lind L, Zethelius B, Berglund L, Lithell H. Evaluation of a scoring scheme, including proinsulin and the apolipoprotein B/apolipo-protein AI ratio, for the risk of acute coronary events in middle-aged men: Uppsala Longitudinal Study of Adult Men (ULSAM). Am Heart J 2004; 148: 596-601.

[118] Ström-Möller C, Zethelius B, Sundström J, Lind L. Impact of follow-up time and re-measurement of the electrocardiogram and conventional cardiovascular risk factors on their predictive value for myocardial infarction. J Intern Med 2006; 260: 22-30.

[119] Ingelsson E, Schaefer EJ, Contois JH, McNamara JR, Sullivan L,Keyes MJ, Pencina MJ, Schoonmaker C, Wilson PWF, D'Agostino RB, Vasan RS. Clinical Utility of Different Lipid Measures for Prediction of Coronary Heart Disease in Men and Women. JAMA 2007; 298(7): 776-785.

[120] Goswami B, Rajappa M, Mallika V, Kumar S, Shukla DK. Apo-B/apo-AI ratio: a better discriminator of coronary artery disease risk than other conventional lipid ratios in Indian patients with acute myocardial infarction. Acta Cardiol 2008; 63(6): 749-755.

[121] Agoston-Coldea L. Apolipoproteins A-I and B-markers in coronary risk evaluation. Rom J Intern Med 2007; 45(3): 251-258.

[122] Kappelle P.J.W.H., Gansevoort R.T., Hillege J.L., Wolffenbuttel B.H.R., Dullaart R.P.F. on behalf of the PREVEND study group. Apolipoprotein B/A-I and total cholesterol/high-density lipoprotein cholesterol ratios both predict cardiovascular events in the general population independently of non- lipid risk factors, albuminuria and C-reactive protein. J Intern Med 2011; 269(2): 232-242.

[123] Kim H-K, Chang S-A, Choi E-K, Kim Y-K, Kim H-S, Sohn D-W, Oh B-H, Lee M-M, Park Y-B, Choi Y-S. Association between plasma lipids, and apolipoproteins and coronary artery disease: a cross-sectional study in a low-risk Korean population. International Journal of Cardiology 2005; 101: 435-440.

[124] Agoston-Coldea L, Mocan T, Gatfossé M, Dumitrascu DL. The correlation of apolipoprotein B, apolipoprotein B/apolipoprotein A-I ratio and lipoprotein(a) with myocardial infarction. Cent Eur J Med 2008; 3(4): 422-429. DOI: 10.2478/s11536-008-0057-3.

[125] Stewart MW, Humphriss DB, Mitcheson J, Webster J, Walker M, Laker MF. Lipoprotein composition and serum apolipoproteins in normoglycaemic first-degree relatives of noninsulin dependent diabetic patients. Atherosclerosis 1998; 139: 115-121.

[126] Kim BJ, Hwang ST, Sung KC et al. Comparison of the relationships between serum apolipoprotein B and serum lipid distributions. Clin Chem 2005; 51: 2257-2263.

[127] Sung K.-C., Rhee E.-J., Kim H., Park, J.-B., Kim, Y.-K., Woo, S. Wilson A.M. An elevated apolipoprotein B/AI ratio is independently associated with microalbuminuria in male subjects with impaired fasting glucose. 2011; 21(8): 610-616.

[128] Kim H-K, Chang S-A, Choi E-K et al. Association between plasma lipids, and apolipoproteins and coronary artery disease: a cross-sectional study in a low-risk Korean population. Int J Cardiol 2005; 101: 435-440.

[129] Snehalatha C, Ramachandran A, Sivasankari S et al. Is increased apolipoprotein B-A major factor enhancing the risk of coronary artery disease in type 2 diabetes? J Assoc Physicians India (JAPI) 2002; 50: 1036-1038.

[130] Solymoss BC, Bourassa MG, Campeau L et al. Effect of increasing metabolic syndrome score on atherosclerotic risk profile and coronary artery disease angiographic severity. Am J Cardiol 2004; 93: 159-164.

[131] Lind L, Vessby B, Sundström J. The apolipoprotein B/AI ratio and the metabolic syndrome independently predict risk for myocardial infarction in middle-aged men. ArteriosclerThromb Vasc Biol 2006; 26: 406-410.

[132] Sierra-Johnson J, Romero-Corral A, Somers VK., Lopez-Jimenez F, Wall¬dius G, et al. ApoB/apoA-I ratio: an independent predictor of insulin resistance in US non-diabetic subjects. Eur Heart J 2007; 28: 2637-2643. doi:10.1093/eurheartj/ehm360.

[133] Sierra-Johnson J, Fisher RM, Romero-Corral A, Somers VK, Lopez-Jimenez F, Öhrvik J, Walldius G, Hellenius ML, Hamsten A. Concentration of apolipoprotein B is comparable with the apolipoprotein B/apolipoprotein A-I ratio and better than routine clinical lipid measurements in predicting coronary heart disease mortality: findings from a multi-ethnic US population. Eur Heart J. 2009; 30: 710-717.

[134] Zhong L, Li Q, Jiang Y, Cheng D, Liu Z, Wang B, Luo R, Cheng Q, Qing H. The ApoB/ApoA1 Ratio is Associated with Metabolic Syndrome and its Components in a Chinese Population. Inflammation 2010; 33(6): 353-358. DOI:10.1007/s10753-010-9193-4.

[135] Belfki H , Ben Ali S , Bougatef S, Ben Ahmed D, Haddad N , Jmal A, Abdennebi M , Ben Romdhane H. The Apolipoprotein B/Apolipoprotein A 1 ratio in relation to metabolic syndrome and its components in a sample of the Tunisian population. Experimental and Molecular Pathology 2011; 91: 622-625.

[136] Sniderman AD, Faraj M. Apolipoprotein B, apolipoprotein A-I, insulin resistance and the metabolic syndrome. Curr Opin Lipidol 2007; 18: 633-637.

[137] Sniderman AD, Kiss R. The strength and limitations of the apoB/apoA-I to predict the risk of vascular disease: a Hegelian analysis. Curr atheroscl rep. 2007; 9(4): 261-265.

[138] Bruno G, Merletti F, Biggeri A, Bargero G, Prina-Cerai S, Pagano G, Cavallo-Perin P. Diabetologia 2006; 49: 937-944.

[139] Bayu M, Sasongo TY, Wong STY, Nguyen TT, Kawaskai R, Jenkins A, Shaw J, Wang JJ. Serum Apolipoprotein AI and B are stronger biomarkers of diabetic retinopathy than traditional lipid. Diabetes Care 2011; 34: 474-479.

[140] Enkhma B, Anuurad E, Zhanga Z, Pearson TA, Berglund L. Usefulness of Apolipoprotein B/Apolipoprotein A-I Ratio to Predict Coronary Artery Disease Independent of the Metabolic Syndrome in African Americans. Am J Cardiol 2010; 106: 1264-1269.

[141] Ounis OB, Elloumi M, Makni E, Zouhal H, Amri M, Tabka Z, Lac G. Exercise improves the ApoB/ApoA-I ratio, a marker of the metabolic syndrome in obese children. Acta Pædiatrica 2010; 99: 1679-1685.

[142] Gatz M, Reynolds CA, Finkel D, Pedersen NL, Walters E. Dementia in Swedish Twins: Predicting Incident Cases. Behav Genet 2010; 40: 768-775, DOI 10.1007/s10519-010-9407-4.

[143] Carnevale Schianca GP, Pedrazzoli R, Onolfo S, Colli E, Cornetti E, Bergamasco L, Fra GP, Bartoli E. ApoB/apoA-I ratio is better than LDL-C in detecting cardiovascular risk. Nut Metabol Cardiovasc Dis 2011; 21: 406-411.

[144] Wen ZZ, Geng DF, Luo JG, Wang JF. Combined use of high-sensitivity C-reactive protein and apolipoproteins B/apolipoprotein A-1 ratio prior to elective coronary angiography and oral glucose tolerance tests Clinical Biochemistry Clinical Biochemistry 2011; 44: 1284-1291.

[145] Andersson J, Sundström J, Kurland L, Gustavsson T, Hulthe J, et al. The Carotid Artery Plaque Size and Echogenicity are Related to Different Cardiovascular Risk Factors in the Elderly. Lipids 2009; 44: 397-403.

[146] Barbier CE, Lind L, Ahlström H, Larsson A, Johansson L. Apolipoprotein B/A-I ratio related to visceral but not to subcutaneous adipose tissue in elderly Swedes Atherosclerosis 2010; 211: 656-659.

[147] Schmidt C, Wikstrand J. High apoB/apoA-I ratio is associated with increased progression rate of carotid artery intima-media thickness in clinically healthy 58-year-old men: Experiences from very long-term follow-up in the AIR study. Atherosclerosis 2009; 205: 284-289.

[148] Reis JP, Macera CA, Wingard DL, Araneta MRG, Lindsay SP, Marshall SJ. The relation of leptin and insulin with obesity-related cardiovascular risk factors in US adults. Atherosclerosis 2008; 200: 150-160.

[149] Junyent M , Zamb'on D , Gilabert R , Cof'an M , N'u~nez I , Rosa E. Carotid atherosclerosis in familial combined hyperlipidemia associated with the APOB/APOA-I ratio. Atherosclerosis 2008; 197: 740-746.

[150] Vladimirova-Kitova LG, Deneva-Koicheva TI. Increased Intima-Media Thickness in Carriers of the LDL-Receptor Defective Gene versus Noncarriers with Newly Detected Asymptomatic Severe Hypercholesterolemia. Echocardiography 2011; 28: 223-234.

[151] Dahlén E. M., Länne T, Engvall J, Lindström T, Grodzinsky E, Nyström FH, Östgren CJ. Carotid intima-media thickness and apolipoprotein B/apolipoprotein A-I ratio in middle-aged patients with Type 2 diabetes 247 patients with Type 2 diabetes, aged 55–66 years, in the Cardiovascular Riskfactors in Patients with Diabetes—a Prospective study in Primary care (CARDIPP-1) study. Diabet Med 2009; 26: 384-390.

[152] Rasouli M, Kiasari AM, Mokhberi V. The ratio of apoB/apoAI, apoB and lipoprotein(a) are the best predictors of stable coronary artery disease. Clin Chem Lab Med 2006; 44(8): 1015-1021.

[153] Smith J, Cianflone K, Al-Amri M, Sniderman A. Body composition and the apoB/apoA-I ratio in migrant Asian Indians and white Caucasians in Canada. Clin Sci 2006; 111: 201-207. doi:10.1042/CS20060045 201.

[154] Maffeis C, Pietrobelli A, Grezzani A, Provera S, Tato` L. Waist circumference and cardiovascular risk factors in prepubertal children. Obes Res 2001; 9: 179-87.

[155] Gardner CD, Tribble DL, Young DR, Ahn D, Fortmann SP. Associations of HDL, HDL2, and HDL3 cholesterol and apolipoprotein A-I and B with lifestyle factors in healthy men and women: The Stanford Five City Project. Prev Med 2000; 30: 346-356.

[156] Okosun IS, Prewitt TE, Liao Y, Cooper RS. Association of waist circumference with ApoB to ApoA-I ratio in black and white Americans. Int J Obes Relat Metab Disord 1999; 23: 498-504.

[157] Juonala M, Viikari JSA, Kähönen M, Solakivi T, Helenius H, Jula A, Marniemi J, Taittonen L, Laitinen T, Nikkari T, Raitakari OT. Childhood Levels of Serum Apolipoproteins B and A-I Predict Carotid Intima-Media Thickness and Brachial Endothelial Function in Adulthood The Cardiovascular Risk in Young Finns Study. J Am Coll Cardiol 2008; 52: 293-299.

[158] Mattsson N, Magnussen CG, Rönnemaa T, Mallat Z, Benessiano J, Jula A, Taittonen L, Kähönrn M, Juonala M, Viikari JSA,Raitakari OT. Metabolic Syndrome and Carotid Intima-Media Thickness in Young Adults: Roles of Apolipoprotein B, Apolipoprotein A-I, C-Reactive Protein, and Secretory Phospholipase A2: The Cardiovascular Risk in Young Finns Study. Arterioscler Thromb Vasc Biol. 2010; 30: 1861-1866.

[159] Wallenfeldt K, Bokemark L, Wikstrand J, Hulthe J, Fagerberg B. Apolipoprotein B/apolipoprotein A-I in relation to the metabolic syndrome and change in carotid artery intima-media thickness during 3 years in middle-aged men. Stroke 2004; 35: 2248-2252.

[160] Matsumoto K, Fujita N, Nakamura K, Senoo T, Tominaga T, Ueki Y. Apolipoprotein B and insulin resistance are good markers of carotid atherosclerosis in patients with type 2 diabetes mellitus. Diabet Res Clin Pract 2008; 82: 93-97.

[161] Kim SJ, Song P, Park JH, Lee YT, Kim WS, Park YG, Bang OY, Chung C-S, Lee KH, Kim G-M. Biomarkers of Asymptomatic Carotid Stenosis in Patients Undergoing Coronary Artery Bypass Grafting. Stroke 2011; 42: 734-739, DOI: 10.1161/STROKEAHA.110.595546.

[162] Ajeganova S, Ehrnfelt C, Alizadeh R, Rohani M, Jogestrand T, Hafström I, Frostegård J. Longitudinal levels of apolipoproteins and antibodies against phosphorylcholine are independently associated with carotid artery atherosclerosis 5 years after rheumatoid arthritis onset—a prospective cohort study. Rheumatology 2011; 50: 1785-1793. Advance Access published July 9, 2011.

[163] Koha KK, Sakumab I, Quonc MJ. Review; Differential metabolic effects of distinct statins. Atherosclerosis 2011; 215: 1-218.

[164] Nicholls SJ, Brandrup-Wognsen G, Palmer M, Barter PJ. Meta-analysis of Comparative Efficacy of Increasing Dose of Atorvastatin Versus Rosuvastatin Versus Simvastatin on Lowering Levels of Atherogenic Lipids (from VOYAGER). Am J Cardiol 2010; 105: 69-76.

[165] Faergeman O, Hill L, Windler E, Wiklund O, Asmar R, Duffield E, Sosef F on behalf of the ECLIPSE study investigators. Efficacy and Tolerability of Rosuvastatin and

Atorvastatin when Force-Titrated in Patients with Primary Hypercholesterolemia. Results from the ECLIPSE Study. Cardiology 2008; 111: 219-228. DOI: 10.1159/000127442.

[166] Ballantyne CM, Andrews TC, Hsia JA, Kramer JH, Shear C for the ACCESS Study Group. Correlation of non-high-density lipoprotein cholesterol with apolipoprotein B: effect of 5 hydroxymethylglutaryl coenzyme A reductase inhibitors on non-high-density lipoprotein cholesterol levels. Am J Cardiol 2001; 88: 265-269.

[167] Vodnala R, Bard RL, Krishnan SM, Jackson EA, Rubenfire M, Brook RD. Potential effects on clinical management of treatment algorithms on the basis of apolipoprotein-B/A-1 and total/high-density lipoprotein-cholesterol ratios. J Clin Lipidol 2011; 5: 159-165. DOI: 10.1111/j.1365-2362.2010.02387.x

[168] Van den Bogaard B, Van den Born B-JH, Fayyad R, Waters DD, DeMicco DA, et al., on behalf of the Treating to New Targets investigators. On-treatment lipoprotein components and risk of cerebrovascular events in the Treating to New Targets study. Eur J Clin Invest 2010 Sep 27, doi:10.1111/j.1366-2362.2010.02387.x On line.www TNT. Published Eur J Clin Invest 2011; 41(2): 134-142.

[169] Kastelein JJP, van der Steeg WA, Holme I, Gaffney M, Cater NB, Barter P, Deedwania P, Olsson AG, Boekholdt M, Demicco DA, Szarek M, LaRosa JC, Pedersen TR, Grundy SM, for the TNT and IDEAL Study Groups. Lipids, Apolipoproteins, and Their Ratios in Relation to Cardiovascular Events With Statin Treatment. Circulation 2008; 117: 3002-3009.

[170] Holme I, Strandberg TE, Faergeman O, Kastelein JJP, Olsson AG, Tikkanen MJ, Lytken Larsen M, Lindahl C, Pedersen TR, on behalf of the Incremental Decrease in End Points Through Aggressive Lipid Lowering Study Group. Congestive heart failure is associated with lipoprotein components in statin-treated patients with coronary heart disease. Insights from the Incremental Decrease in End points Through Aggressive Lipid Lowering Trial (IDEAL). Atherosclerosis 2009; 205: 522-527.

[171] Holme I, Cater NB, Faergeman O, Kastelein JJP, Olsson AG, et al., on behalf of the Incremental decrease in endpoints through aggressive lipid-lowering study group. Lipoprotein predictors of cardiovascular events in statin-treated patients with coronary heart disease. Insights from the Incremental Decrease in End-points through Aggressive Lipid-lowering Trial (IDEAL). Ann Med 2008; 40: 456-464.

[172] Nissen, SE, Nicholls SJ, Sipahi et al. Effect of very highintensity statin therapy on regression of coronary atherosclerosis. The ASTEROID trial. JAMA 2006; 295: 1556-1565. March 13, 2006; Epub ahead of print.

[173] Nicholls SJ, Tuzcu EM, Sipahi I, Grasso AW, Schoenhagen P, et al. Statins, High-Density Lipoprotein Cholesterol, and Regression of Coronary Atherosclerosis. JAMA 2007; 297: 499-508.

[174] Tani S, Nagao K, Anazawa T, Kawamata H, Furuya S, Takahashi H, Iidaa K, Matsumoto M, Washioa T, Kumabe N, Hirayama A. Relation of Change in Apolipoprotein B/Apolipoprotein A-I Ratio to Coronary Plaque Regression After Pravastatin Treatment in Patients With Coronary Artery Disease. Am J Cardiol 2010; 105: 144-148.

[175] Taskinen MR, Barter PJ, Ehnholm C, Sullivan DR, Mann K, Simes J, Best JD, Hamwood S, Keech AC on behalf of the FIELD study investigators. Ability of traditional lipid ratios and apolipoprotein ratios to predict cardiovascular risk in people with type 2 diabetes. Diabetologia 2010; 53: 1846-1855, DOI 10.1007/s00125-010-1806-9.

[176] Ridker PM, Danielson E, Fonseca FAH, et al, on behalf of the JUPITER Trial Study Group. Reduction in C-reactive protein and LDL cholesterol and cardiovascular event rates after initiation of rosuvastatin: a prospective study of the JUPITER trial. Lancet 2009; published online March 29, 2009. DOI:10.1016/S0140-6736(09)60447-5. Published in Lancet 2009; 373: 1175–82

[177] Mora S, Glynn RJ, Boekholdt M,. Nordestgaard BG, Kastelein JJP, Ridker PM.On-Treatment Non–High-Density Lipoprotein Cholesterol, Apolipoprotein B, Triglycerides, and Lipid Ratios in Relation to Residual Vascular Risk After Treatment With Potent Statin Therapy JUPITER (Justification for the Use of Statins in Prevention: An Intervention Trial Evaluating Rosuvastatin) J Am Coll Cardiol, 2012; 59: 1521-1528, doi:10.1016/j.jacc.2011.12.035

[178] Reiner Z, De Backer G, Graham I, Taskinen M-R, Wiklund O, Agewall S, Alegria E, Chapman MJ, Durrington P, Erdine S, Halcox J, Hobbs R, Kjekshus J, Perrone Filardi P, Riccardi G, Storey RF, Wood D.Review: ESC/EAS Guidelines for the management of dyslipidaemias The Task Force for the management of dyslipidaemias of the European Society of Cardiology (ESC) and the European Atherosclerosis Society (EAS). Atherosclerosis 2011; 217S: S1-S44.

[179] Genest J, Frohlich J, Fodor G, McPherson R for the Working Group on Hypercholesterolemia and other Dyslipidemias. Recommendations for the management of dyslipidemia and the prevention of cardiovascular disease: summary of the 2003 update. Can Med Assoc J 2003; 169: 921-924.

[180] Connelly PW, Poapst M, Davignon J et al. Reference values of plasma apolipoproteins A-I and B, and association with nonlipid risk factors in the populations of two Canadian provinces: Quebec and Saskatchewan. Can J Cardiol 1999; 15: 409-418.

[181] Charlton-Menys V, Betteridge DJ, Colhoun H, Fuller J, France M, Hitman GA, Livingstone SJ, Neil HAW, Newman CB, Szarek M, DeMicco DA, Durrington PN. Apolipoproteins, cardiovascular risk and statin response in type 2 diabetes: the Collaborative Atorvastatin Diabetes Study (CARDS). Diabetologia 2009; 52: 218-225. DOI 10.1007/s00125-008-1176-8.

Chapter 6

Approaches to Access Biological Data Sources

Assia Rharbi, Khadija Amine, Zohra Bakkoury,
Afaf Mikou, Anass Kettani and Abdelkader Betari

Additional information is available at the end of the chapter

1. Introduction

In recent years, technological revolutions in genomics and proteomics have revolutionized the work of researchers in molecular biology. Through various techniques of data generation, they have at their hand in the web a very large amount of information contained in public and heterogeneous data sources. Each source has content organized around a particular data type like sequences in Uniprot (for proteins) and Genbank (for gene and mRNA), protein structure in PDB (Protein Data Bank) and publications in biomedical Medline. Their content is heterogeneous in the sense that a similar data can be represented differently in two data sources (eg different names for the same gene). More data sources have a variety in terms of structure, and there are sources of structured data, such as relational databases or semi-structured sources like XML and unstructured sources such as databases composed of flat files. That is to say that a biologist who wishes to obtain information from these sources have to question these one by one, then copy and analyze the data collected, and manage redundancy, complementarities of the information and inconsistencies. Today, one of the greatest challenges of bioinformatics is to enable biologists to effectively access multiple data sources, each with a different pattern. Various approaches have been adopted to unify access to various data sources given a query. Several systems have been produced from data warehouses, a federation of databases or mediators.

In this work, we are interested in mediation systems. Such systems offer to the user a uniform and centralized view of distributed data, this view may also reflect a more abstract, condensed, qualitative data and therefore more meaningful to the user. These mediation systems are also very useful in the presence of heterogeneous data, because they seem to use a homogeneous system.

We aim to assist biologists in their research through the development of a generic tool for the integration of heterogeneous genomic data distributed over the web, and we are placed in a very particular context that is the study of cardiovascular disease and especially familial hypercholesterolemia. This is a disorder of high LDL ("bad") cholesterol that is passed down through families, which means it is inherited. This disease is caused by a genetic mutation of certain lipoproteins. Indeed, these lipoproteins (called LDL) carry the 2/3 of cholesterol circulating in the blood; they deliver cholesterol to tissues by a system of recognition between Apo lipoprotein Band a receiver: the LDL receptor (lock and key system) that allows the entry of LDL and their cholesterol content in cells. When the LDL receptor (LDL-R) is weak (about one mutation), LDL accumulates in the blood and artery walls causing familial hypercholesterolemia (HF). So knowing these different mutations by biologists, can greatly facilitate the molecular screening of the disease and therefore to find the proper treatment. However, to answer such a query: "What are the mutations that cause familial hypercholesterolemia (HF)?" The biologist hasto make a fastidious search in disparate and heterogeneous databases which requires a considerable investment time.

This chapter is structured as following:

- First we present the background of the project
- Second we focus on the problem of heterogeneity of data sources and biological characteristics of these sources.
- Third we present the state of the art of data integration, problems and constraints of this integration and the various existing approaches to solve this problem.
- And fourth we expose studied scenario, the realization and perspectives.

2. General context

Since the completion of the human genome sequencing in April 2003, we observe the accumulation of an outsize amounts of genomic and proteomic data on the web often syntactically and semantically heterogeneous and difficult to capitalize.

Information about genes provides access to their corresponding proteins. In addition, all diseases are associated with alterations in the structure or function of such proteins. A good knowledge of protein structure provides insight into their function.

Bioinformatics has become an important tool to explore genomic data by relying heavily on computer systems. It suggests methods and software's for biological data storage and processing. Actually, it is acquiring and organizing data, developing software for the analysis, comparison and modeling of these data and analysis results produced by bioinformatics software to infer new biological knowledge, in collaboration with biologists.

This work contributes to facilitate to biologists searching among heterogeneous and distributed data in public and / or private data sources on the web. In particular, it helps them to analyze proteins, by building a platform for integrating biological data. This will provide a tracking system to target special proteins involved in a disease known as familial hypercholesterolemia and thus, to better understand the biological activity of these macromolecules.

Familial hypercholesterolemia disease results from mutations in the LDLR gene. The LDLR gene provides instructions for making a protein called a low-density lipoprotein receptor.

This type of receptor binds to particles called low-density lipoproteins (LDLs), commonly known as bad cholesterol. By removing low-density lipoproteins from the bloodstream, these receptors play a critical role in regulating cholesterol levels. When the LDL receptor (LDL-R) is deficient, LDL accumulated in arteries induces the familial hypercholesterolemia (HF) pathology. So, in biology knowing these different mutations can greatly facilitate the molecular screening of the disease and thus find appropriate treatment.

3. Biological data sources

Number of data sources and tools available to biologists on the web has grown dramatically in recent years. This huge number of available data along with heterogeneous information generated wide variety of access interfaces, and also a profound heterogeneity.

3.1. Genomic databases

There are two types of databanks, those that correspond to a set of heterogeneous data so-called "databases" and those more homogeneous established around a specific theme.

Also, to avoid confusion we will distinguish between semantic databases, general [2] and specialized [3]databases.

For specific requirements related to the activity of a group, or to bibliographic compilations, many specific databases were created in laboratories. In some cases, these databases have been developed continuously; others have not been updated and disappeared as they represented a specific need. Still others are unknown or poorly known and are waiting for further investigation.

All these specialized databases of interest may vary considerably from one base to another according to their size. In most of the case, these bases correspond to a combination compared of generalist databases such as: Swiss-Prot, GenBank. DDBJ (DNA Data Bank of Japan), EMBL (European Molecular Biology Laboratory) which are used very often. It is important to know, that according to the field of activity or the genomics research, the surveyed banks are not necessarily the same. The genomic libraries contain various information that may include:

- Characteristics of proteins or genes such as localization of the gene in the cell: LocusLink5, the 3D structure of protein: Protein Data Bank (GDP) and Molecular Modeling Database (MMDB) or its biological function. More specifically, some databases contain information about a specific family of protein such as "Enzyme8" which include exclusively enzyme type proteins.
- Some phenotypes (specific genes, morphological feature, clinical syndrome ...) or more specifically some genetic diseases: Online Mendelian Inheritance in Man (OMIM);
- Specific species or families of species: FlyBase, Reptilia, Saccharomyces Genome Database (SGD), Mouse Genome Database (MGD);
- The medical literature (banks abstracts): Medline, PubMed.

Table1 and Table 2 give two examples of genomic databases along with protein database.

Designation	Location	Roles	Comments	Web sites and references
Nucleic bases				
EMBL	European Bioinformatics Institute (EBI) Europe	More than 1 million records (January 1998) for more than 15,500 species. The predominant species: Homo sapien,Caenorhabditis elegans, Saccharomyces cerevvisae ...	Information's search tools: SRS, System Retrieval System and via a web interface on EBI, through BLAST et FASTA software	Accessible via the web site: http://www.ebi.ac.uk/ebi_docs/embl_db/ebi/topembl.html
Specific genomic resources				
SGD Saccharomyces Genome Database	Works of Cherry and al, 1998.	Online resources on molecular biology and S.cerevisiae genetic	Numerous research help functions on line	Accessible via the web site: http://genome-www.stanford.edu/Saccharomyces/

Table 1. Genomic databases

Designation	Location	Roles	Comments	Web sites and references
Primary databases				
PIR Protein Information Resource	National Biomedical Research Foundation	Sequences collecting to detect evolutionary relationship between proteins	The current structure includes 4 compartments : PIR1, PIR2, PIR3 and PIR4	Accessible via the web site http://nbrfa.georgetown.edu/pir/
Composite databases				
NRDB Non-redundant Database	National Center Biotechnology Information USA	Composed of GenPept (derived from GenBank), PDB sequences, SWISS-PROT, SPupdate, PIR, and GenPeptupdate (update of GenPept). NRDB is the default database of BLAST and NCBI service	Accessible via the web site: http://www.ncbi.nlm.nih.gov/Web/NRDB/	NRDB Non-redundant Database

Table 2. Protein databases

3.2. Characteristics of biological data sources

The diversity of information distributed sources and their heterogeneity are the one of the main problem that the web users have to face. This heterogeneity may result from the size or structure of the sources (structured sources: relational databases, partially structured sources: XML documents, or unstructured: texts), the access mode and query, or semantic heterogeneity: between concept maps, and implicit or explicit underlying ontology's.

Biological sources have a large heterogeneity at different levels:

- Syntactic: because of the different formats for describing the content sources usually ASN.1 (formal notation for describing data transmitted via exchange protocols), (eg Enter), but also more standard formats such as XML (eg GenBank).
- Semantic which covers several aspects. First, it concerns the focus. However, each base focuses on a type of biological object (eg, the focus of Swiss-Prot is the protein, the focus of GenBank is gene, and the PDB is the 3D protein structure).
- Then, according to the base, the same information is not represented with the same level of detail: some bases are generalists (eg Swiss-Prot in general on proteins) while others are more specialized (eg SGD (Saccharomyces Genome Database) on yeast proteins).
- The final aspect of semantic heterogeneity is related to the diversity of nomenclature modes. Different vocabularies are used to annotate the sequences and the reliance on such annotations is seldom complete. Moreover, within a same database, there are a several names for each single entity (protein, gene). The name of an entity may depend on the disease to which it is linked or to its inventor.
- Source query language: another form of heterogeneity comes from query languages. Languages are often simple forms (combinations of words to search in a text), in the case of portals or simple databases. But one can also find structured languages such as SQL or OQL.
- Protocols for collecting data that are different such as CGI / HTTP or FTP. Access to web sources is limited to the entry forms and their underlying programs
- The tools offered by the Web: there are many tools for text searching and sequence comparison algorithms such as BLAST (Basic Local Alignment Search Tool), FASTA15 or LASSAP16.

4. State of the art of approaches to integration

A data integration system remedies to the problems associated with the expansion of public data sources by giving the possibility to have a unified view of them. Such a system is the interface between user and data sources simplifying requests to perform (a request to query all sources covered by the system). The user is not obliged to know where the data are and how they are structured.

4.1. Current integration approaches

There are two major approaches for integration of information: (1) the data warehouse (DW) or materialized approach and (2) virtual approach (mediator based).In DW approach, huge amount of historic data is stored in the DW. In the virtual approach, on the other hand, the data is not materialized, but rather is globally manipulated using views. Each of these approaches is suitable in some kinds of applications.

4.1.1. Data warehouse

DW is a powerful tool for decision support and querying the data because it explicitly stores information from heterogeneous sources locally. However, some external data, such as new product announcements from opponents and currency exchange rates, may be needed to support the accuracy of the business decisions. We should not neglect the importance of such data to avoid the problems of incompleteness, inexact, or sometimes wrong results. Warehousing huge and frequently changed information is a big challenge for the following reasons.

Firstly, since the data in the DW is loaded in snapshots and the DW is a huge information repository. Secondly, as the data sources change frequently, the maintenance becomes a complicated and costly issue

Here are two examples of using data warehouses:

- Genomics Unified Schema, GUS [4] is a system for creating a data warehouse focused on molecular biology;
- Gene Expression Data Warehouse, GEDAW [5] is a warehouse dedicated to the analysis of the transcriptome of human liver.

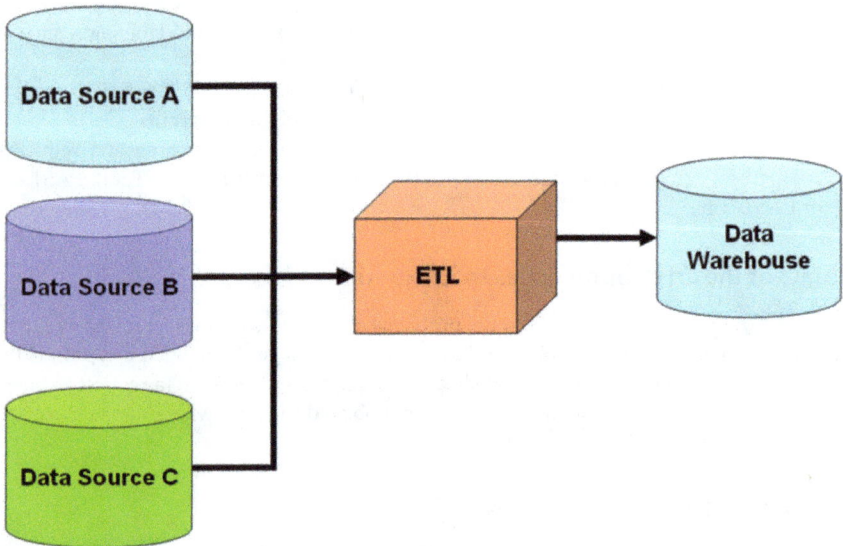

Figure 1. Simple schematic for a data warehouse

4.1.2. The virtual approach (mediator based)

In this approach, the actual data resides in the sources, and queries against the integrated 'virtual' view will be decomposed into sub queries and posed to the sources. This approach is preferred over the materialized approach DW when the information sources change very often. On the other hand, the DW approach may be desired when a quick query answer is required and the information sources change rarely.

The most important step in the construction of a mediator is the creation of the global schema. The mapping consists on the relations between the global schema and local sources. Specification of this mapping, depending on the method, determines the difficulty of query reformulation and the facility of adding or removing sources within the system. Two methods are commonly used to determine the global schema

- GAV (Global As View) approach: In this approach, each concept of the global schema is mapped to a query over data sources. In other words, when the user presents his/her query over the integrated schema, the data corresponds to a concept in the integrated schema, which can actually be answered from the data sources through a specific query. The query processing in GAV is easy, since it just unfold each concept in the integrated schema in the user query with the associated query over the sources, but this approach does not help much when the sources change or grow very often, since these factors affect the mappings and require restricting the integrated schema.
- LAV (Local As View) approach: LAV approach defines the mapping in the other way around; each concept in the data sources is defined in terms of a query over the integrated schema. This makes query processing more difficult, since in this case, the system does not know explicitly how to reformulate the concepts in the integrated view expressed in the user query in terms of the data sources. On the other hand, changes or incremental growth in the sources will not lead to reconstruction of the integrated schema, and need only to modify the mappings

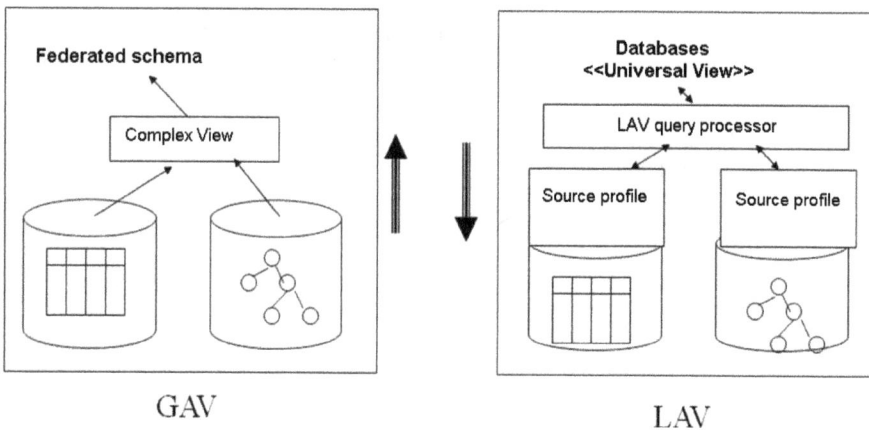

Figure 2. GAV vs. LAV

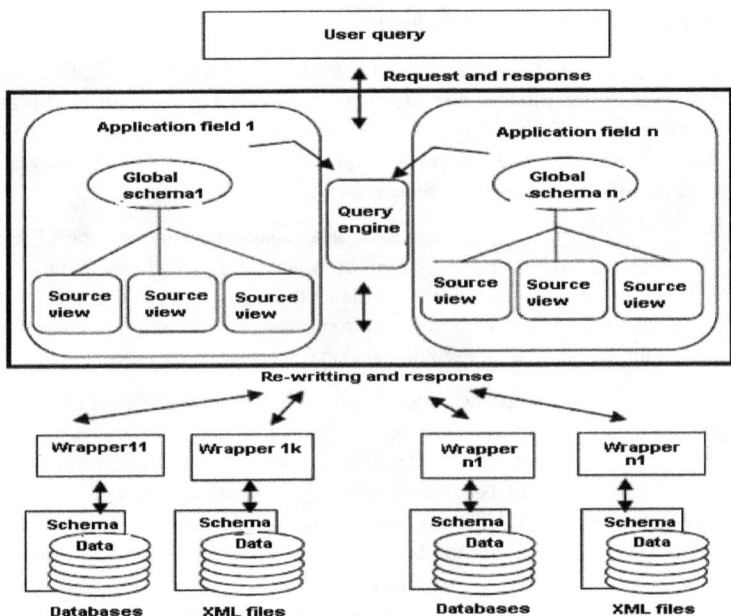

Figure 3. Architecture of a mediator

In fact, these two approaches are not opposite, but complementary; depending on the problem to be solved. To integrate a few sources, most of which are stable, better to use the GAV method. By cons, as part of a large-scale integration, the LAV method is preferable as a material change at a local source with little or no impact on the global schema.

Two examples of systems integration based mediator:

• *Tambis (Transparent Access To Multiple Bioinformatics Information Sources) [6]* is an integration system coupled to an ontology that allows for better interoperability between sources;

K2/BioKleisli [7] is a system based on CPL (Collection Programming Language) is a query language for high-level querying multiple sources.

4.1.3. The multi agents approach

This approach was used in GID-IGC *(Integrated Genomic Database - Genome Information System) project*. The proposed architecture uses a network of agents communicating each with other via CORBA and KQML. All have a specific function, such as *EIA* (External Interface Agent) that manages the user interface, or *SCA* (Dial Selector

Agent) witch decompose the global query into sub-queries for local data sources. This approach is very modular and easily extensible.

4.1.4. Navigating between sources

This approach is based on what users usually do when searching for information on the web, which involves a search page to page by clicking the mouse. In practice, queries generated for this type of tool are converted into path expressions. The data banks are then integrated based on their cross-references. These expressions can answer the query of the user according to different levels of satisfaction.

A reference is a link between two data sources (Figure 4), a bridge between the information relating on the same object or the same concept. It can be done through an identifier of an external source or a URL (Unified Resource Locator). If the link can be browsed in both directions it is a cross-reference ("cross-reference").

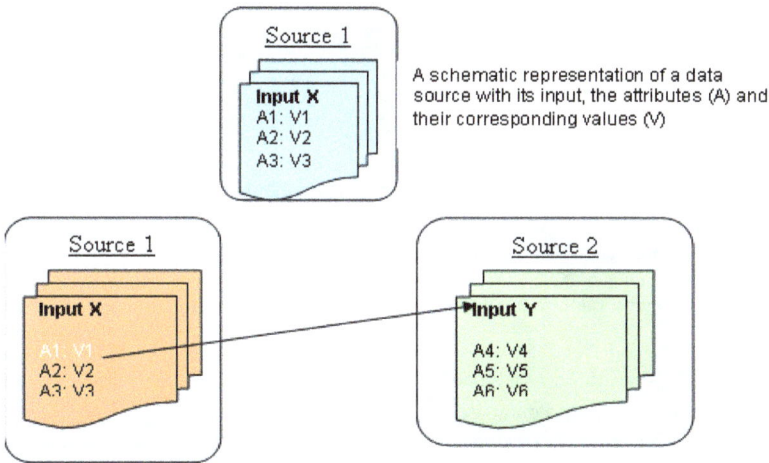

Figure 4. Navigating between sources

4.2. Adopted approach

In this work, we are interested in mediation systems. Such systems offer a uniform and centralized view of distributed data. This view may also reflect a more abstract, condensed, qualitative data and therefore more meaningful to the user. These mediation systems are also very useful in the presence of heterogeneous data, because they seem to use a homogeneous system.

In this architecture, each component provides a set of features, which, together will help to satisfy the user request at the end.

- The mediator is a software module that directly receives the user's request. It has to locate the necessary information to answer the query, resolve schematic and semantic conflicts, query different sources and integrate the partial results in a consistent and coherent response. This is the most complex component but only one instance of it is necessary (unlike multiple adapters). It provides access to multiple data sources as if it was a single one and offers this consultation through multiple languages and ontologies.

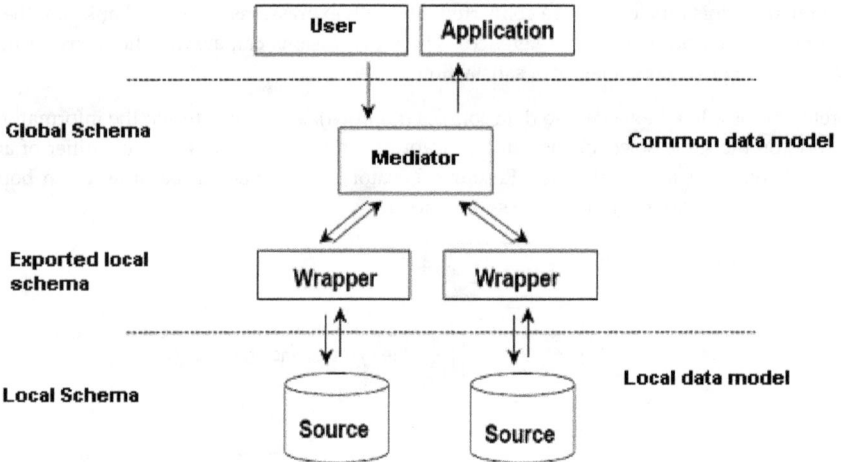

Figure 5. Adopted approach

This is a crucial component that allows a local system to distribute its information to a community of users.

- The adapter allows the presentation of data in the mediation's syntactic format. So it's an interface for querying a database using a standardized language (pivot language).
- Data source: Represent the sources and banks of biological information. A data source can be described by its:
    - Location: Reference, communication protocol, access technique (JDBC, ODBC API), support (DBMS, web pages)
    - Type of data it manages: structured (relational, object), semi-structured (XML, OEM), unstructured (image, multimedia)
    - Ability to query: SQL, OQL, search
    - Results Format: XML, HTML, relationships, texts

5. Studied scenario

After exploring the different integration systems that solve the problem of heterogeneous biological data sources, this section describes the scenario we have chosen to work on.

5.1. Biologist's need

The objective of our work is to develop an integration system for biological data with an application on familial hypercholesterolemia. Such system should facilitate access to multiple data sources available on the Web, in a transparent and uniform way by giving biologists a single virtual source that summarizes the sites of interest to the application.

In order to satisfy this biologist's need we studied their current way to work. We first focused on existing tools, data sources they use and their functional specifications.

Tools: they use, mainly:

- CHARMM (Chemistry at Harvard Macromolecular Mechanics) [8]: This program offers a wide choice for the production and analysis of molecular simulations. It simulates the standard energy minimization of a given structure and the production of a molecular dynamics trajectory.
- VMD (Visual Molecular Dynamic) [9]: This software is used to visualize the molecules available on the web.

Data sources: The focus was mainly on the following sources:

- PDB (Protein Data Bank) [10]: It's a worldwide collection of data on three-dimensional structure of biological macromolecules: Proteins and nucleic acids. The PDB is the primary source of structural biological data. It allows access to 3D structures of pharmaceutical interest proteins.
- PubMed [11]: it's a free search engine giving access to the MEDLINE bibliographic database, gathering citations and abstracts of biomedical research.

5.2. Adopted scenario

The adopted scenario consists on building a local database "homemade" that would include unorganized data already available in the biologist's laboratory and for our particular case data related to LDL receptor mutations. This goes along with the research we're going to do on the Web using the mediation system:

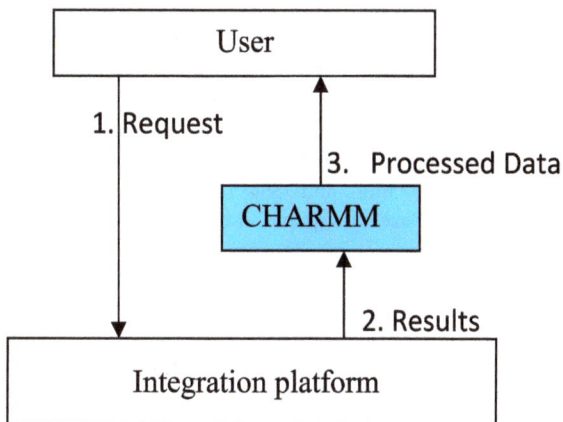

Figure 6. The adopted scenario

The integration platform is the mediation system. It will query data sources related to cardiovascular diseases especially familial hypercholesterolemia namely: PDB and PubMed. The results of the query will be processed by the tool CHARMM before being presented to the user [12, 13].

5.3. Selected data sources

PubMed [8]: Is the leading bibliographic data search engine of all fields of biology and especially medicine. It was developed by the National Center for Biotechnology Information (NCBI), and is hosted by the National Library of Medicine U.S. National Institutes of Health. PubMed is a free search engine giving access to the MEDLINE bibliographic database, gathering citations and abstracts of biomedical research.

The MEDLINE database in April 2007 had more than 15 million citations from 1950 published in 5000 biomedical journals (journals in biology and medicine) distinct. It is the database of reference for biomedical sciences. As with other indexes, including a citation in PubMed has no content. In addition to MEDLINE, PubMed also provides access to:

- OldMedline for articles before 1966
- Citations of all articles, even "irrelevant" (that is to say, covering topics such as plate tectonics or astrophysics) from certain MEDLINE journals, primarily those published in major newspapers of general science or biochemical (such as Science and Nature, for example).
- Citations being listed before indexing in MEDLINE or McSH, or passage or status "off topic"
- Older citations selected for MEDLINE journal from which they arise (when they are supplied electronically by the publisher)
- Articles submitted to PubMed Central for free

Most citations include a link to the full article when it is available (eg PubMed Central). PubMed is a search engine that allows users to search in the MEDLINE database; this information is also available from private organizations such as Ovid and Silverplatter, among others. PubMed is free since the mid-1990s. For optimal use of PubMed, it is necessary to have an understanding of his core, MEDLINE, and especially the MeSH vocabulary used for indexing articles in MEDLINE.

We can also find in PubMed information about the log, which can search by title, subject, short title, NLM ID, ISO abbreviation, and ISSN (International Standard Serial Number) written and electronic. The database "newspaper" includes all newspapers Enter Base.

The major interest of these bibliographic databases is that:

- Their bodies are used to identify recent publications in scientific journals.
- They help to establish bibliographies (lists of relevant articles) on a subject or author.
- They are portals to access full text documents available on the Internet.
- The bibliographic databases used to find references to documents, select, print or export them to other software. They may also propose to order documents or provide access to full text.

PDB (Protein Data Bank): The databank on proteins of Research Collaborator for Structural Bioinformatics, more commonly known as Protein Data Bank or PDB is a worldwide collection of data on the three-dimensional (or 3D structure) of biological macromolecules : protein essentially, and nucleic.

Founded in 1971 by Brookhaven National Laboratory, the Protein Data Bank was transferred in 1998 to the Research Collaborator for Structural Bioinformatics (RCSB), which consists of Rutgers University, the University of Wisconsin at Madison, National Institute of Standards and Technology (NIST) and the "San Diego Supercomputer Centre." The PDB originally contained (in 1971) 7 structures. The number of structures deposited has grown since the 1980s. Indeed, at that time, the crystallographic techniques have improved, the structures determined by NMR have been added, and the scientific community has changed its view on data sharing.

The PDB containedon 28-04-2008, 50480 structures. The data are from the original pdb format, and in recent years are also mmCif format, specifically developed for structural data from the PDB. From 2000 to 3000 structures are added each year. The bank contains files for each molecular model. These files describe the exact location of each atom of the macromolecule studied, that is to say, the Cartesian coordinates of the atom in a three-dimensional coordinate.

Each model is referenced in the bank by a unique identifier to 4 characters, the first is always a numeric character, the next three being alphanumeric characters. This identifier is called **"pdb code"**.

Several formats exist for PDB files:

The PDB format: it is the original format. The guide of this format has been revised several times; the current version is version 2.2[14], which has existed since 1996. Originally pdb format was dictated by the width and the use of punch cards for computers. Consequently, each line contains exactly 80 characters.

Pdb file format is a text file where each column has its meaning: Each parameter is positioned so immutable. Thus, the first 6 columns, that is to say the first 6 characters for a given line, determine the scope of the file. Found for example in the fields " TITLE_ "(That is to say, the title of the macromolecule of interest)," KEYWDS "(The keywords of the entry)," EXPDTA "Which provides information on the experimental method used," SEQRES "(The sequence of the protein under study)," ATOM_ "Or" HETATM "Fields containing all information related to a particular atom.

Pdb format limitations: Format in 80 columns pdb files is relatively restrictive. The maximum number of atoms in a pdb file is 99999, since there are only 5 columns allocated for the numbers of atoms. Similarly the number of residues per chain is at most 9999: There are only 4 columns allowed for this figure. The number of channels is limited to 62: A single column is available, and possible values are one of the 26 letters of the alphabet in upper or lower case, or one of the digits 0 through 9. As this format has been defined, these limitations did not seem restrictive, but they have been taken several times during the deposition of extremely large structures, such as viruses, ribosome, and multienzyme complexes.

MmCIF format: The growing interest in the development of database and electronic publications in the late 1980s has created the need for a more structured, standardized, open-ended and high quality data from the PDB. In 1990, the International Union of Crystallography IUCr extended to macromolecules data representation used to describe crystal structures of molecules of low molecular weight. This representation is called CIF, for Crystallographic Information File. The dictionary mmCIF (macromolecular Crystallographic Information File) published in 1996, was then developed.

In MmCIF format, each field of each section of a pdb file is represented by a description of a characteristic of an object, which includes both the name of the characteristic (eg _struct.entry_id), and the content of the description (pdb code: 1cbn). Which we can call "name-value". It is easy to convert, without loss of information, an mmCIF file format pdb, since all information is directly analyzed. It is not possible, however, to completely automate the conversion of a pdb file format mmCIF, since many mmCIF descriptors are either absent from the PDB file, either in this field " REMARK "Who can not always be analyzed. The contents of fields " REMARK " is indeed separated according to different mmCIF dictionary entries, in order to preserve the completeness of the information contained in such Materials and Methods section (crystal characteristics, refinement method ...) or in the description of the biologically active molecule or other molecules (substrate, inhibitor, ...)

The mmCIF dictionary contains over 1700 entries, which are much safer not all used in a single PDB file. All field names are preceded by the character "underscore"(_), In order to differentiate the values themselves. Each name corresponds to an mmCIF dictionary entry, where the characteristics of the object are exactly defined.

Pdbml format: This format is pdbml adaptation to XML data format bps and contains the entries described in the dictionary "PDB Exchange Dictionary". This dictionary contains the same entries as the mmCIF dictionary, in order to take into account all data managed and distributed by the PDB. This format can store much more information on models than pdb format.

Data retrieval: The files describing molecular models can be downloaded from the website of the PDB and visualized using various software such as Rasmol [15], Jmol [16], chime [17] or an extension VRML [18] (plugin) a browser. The website of the PDB also contains resources for teaching, on structural genomics and other useful software.

5.4. The global schema

By studying and exploring the previous sources and by combining data from genome sources, we have identified all data that define the dictionary related to familial hypercholesterolemia disease (Table 3). From this data dictionary and business rules (as defined and established by experts in the field of biology), we extracted the major biological entities useful for our study. These entities are not independent and form a semantic graph with nodes reflecting relationships between these entities.

Property	Description
code_biblio	Library code
auteur_biblio	Author of publication
date_biblio	Date of publication
volume_biblio	Volume of structure
langue_biblio	Language
contribution_biblio	Contribution
journal_biblio	Newspaper
revue_scint_biblio	Journal
livre_biblio	Book
Cd_proceeding_biblio	CD procedure
nom_recepteur	Name of receiver
nom_mutation	Name of mutation
classe_mutation	Class of mutation
nom_proteine	Name of the protein
longueur_proteine	Length of the protein
type_proteine	Protein type
structure_summary	Summary of the structure
structure_title	Under the structure
nom_molecule	Name of the molecule
author_name	Name of author
date_depot	Date Filed
date_release	Date of publication
derniere_release	Last update
Resolution	Resolution
Compound	Compound
Classification	Classification
molecule_chain_type	Channel Type
experimental_methode	Experimental method of resolution (RX, NMR)

Table 3. Data Dictionaries

Health Aspects of Lipoproteins

Bibliographie

code_biblio	varchar(100) <pk>
auteur_biblio	varchar(100)
date_biblio	varchar(100)
volume_biblio	varchar(100)
langue_biblio	varchar(100)
contribution_biblio	varchar(100)
journal_biblio	varchar(100)
revue_scient_biblio	varchar(100)
livre_biblio	varchar(100)
cd_proceeding_biblio	varchar(100)

dispose

nom_proteine	varchar(100) <pk,fk1>
titre_structure	varchar(100) <pk,fk2>

FK_DISPOSE_DISPOSE2_STRUCTUR

FK_DISPOSE_DISPOSE_PROTEINE

StructureProtéines

titre_structure	varchar(100) <pk>
resume_structure	varchar(100)
nom_molecule	varchar(100)
nom_auteur	varchar(100)
date_depot	varchar(100)
date_release	varchar(100)
derniere_release	varchar(100)
experimental_methode	varchar(100)
molecule_chaine_type	varchar(100)
classification	varchar(100)
compound	varchar(100)
resolution	varchar(100)

Protéines

nom_proteine	varchar(100) <pk>
longeur_proteine	varchar(100)
type_proteine	varchar(100)
sequence_proteine	varchar(100)

FK_DECRITS__DECRITS_D_BIBLIOGR

décrits-dans

code_maladie	varchar(255) <pk,fk1>
code_biblio	varchar(100) <pk,fk2>

FK_ARNM_TRADUIT_PROTEINE

FK_DECRITS__DECRITS_D_HYPERCHO

ARNm

nom_ARNm	varchar(100) <pk>
nom_proteine	varchar(100) <fk1>
nom_gene	varchar(100) <fk2>
longeur_ARNm	varchar(100)
type_ARNm	varchar(100)
sequence_ARNm	varchar(100)

Hypercholestérolémie

code_maladie	varchar(255) <pk>

FK_MUTATION_CAUSER_HYPERCHO

FK_ARNM_TRANSCRIT_GENES

Mutations

nom_mutation	varchar(255) <pk>
code_maladie	varchar(255) <fk1>
nom_recepteur	varchar(100) <fk2>
classe_mutation	varchar(100)

Gênes

nom_gene	varchar(100) <pk>
longeur_gene	varchar(100)
type_gene	varchar(100)
sequence_gene	varchar(255)

FK_MUTATION_CONCERNE_RECEPTEU

FK_RECEPTEU_EST_GENES

Récepteurs

nom_recepteur	varchar(100) <pk>
nom_gene	varchar(100) <fk>
type_recepteur	varchar(100)

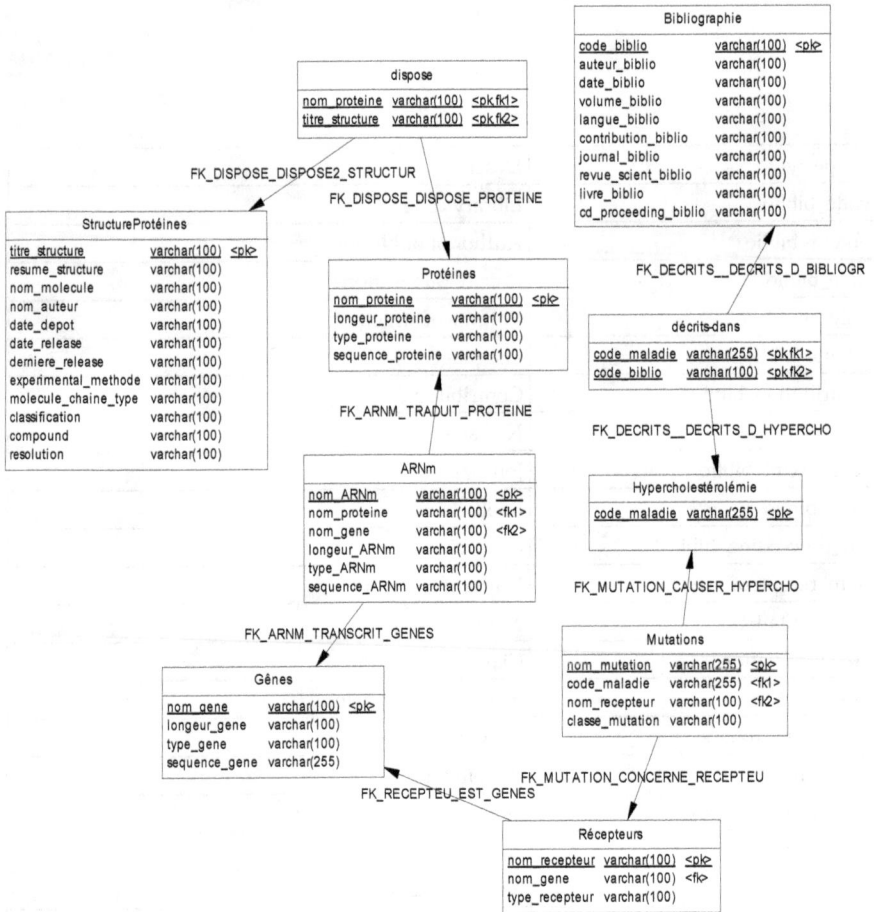

Figure 7. Global Schema

From the global schema, it is possible to make our request and submit it to SQL mediator for treatment. For example, the query that givesthe associated protein mutated gene and publications on familial hypercholesterolemia is expressed as follows in SQL:

```
Select nom_proteine, journal_biblio, auteur_biblio, date_biblio, langue_biblio
From mRNA, gene g, b bibliography
Where a.nom_gene = g.nom_gene
And g.nom_gene in (select nom_gene from recepteurs r, mutation m
            Where r.nom_recepteur = m.nom_recepteur)
```

For its execution, the query is first submitted to the mediator which is responsible for locating sources and queries them through the wrapper or the associated adapter.It should be noted that the only access point to our sources for interrogation is a web form that, once processed through a wrapper, gives us the local sources that we describe below.

5.5. Analysis of the query

In the global query, 6 attributes are involved: Nom_proteine, nom_gene, journal_biblio, auteur_biblio, date_biblio, langue_biblio shown in the following table along with the sources (PubMed (S1), and PDB (S2).

Attributes	Lists of sources
journal_biblio, auteur_biblio, date_biblio,langue_biblio	S1
nom_proteine	S2
nom_gene	S2, S1

Table 4. Identification of sources

From these sources, we can extract a local schema, for example:

S1_L (journal_biblio, auteur_biblio, date_biblio, langue_biblio, nom_gene),

S2_L (nom_proteine, nom_gene)

From a programming point of view, S1_L and S2_L represent wrapper sources.

Next section describes the realization in which we develop wrappers, submit queries to the mediator, and combine the final result to be presented to the user.

6. Realization

We define four steps in the realization:

- The development of wrapper
- Definition of "global schema"
- Correspondence between local tables and global schema
- Results analysis

6.1. Step 1: Development of wrappers

A wrapper is a program that envelops the execution of another program in the way that the environment can be more suitable. The mediator requests the various databases via wrappers that will extract information from websites of interest. It is necessary to create a wrapper for each specific database.

The sources that we identified in section 4 will be integrated through wrappers. There are different types of wrappers depending on the type of pages they incorporate. These can be either text files or XML files (Extensible Mark-up Language). It is necessary to know the structure of these files and know where the information is located (after any tag, for example). Developing wrappers is linked to functional specifications of the sources presented earlier.

Wrappers allow therefore the extraction of data to be represented in tables. Indeed, we declare the objects and their attributes for each site based on data provided. From all this information, local schemas (relational) for each of these databases are established.

Various programs were written in java. Even if a wrapper has been created for each database, they all have the same main structure. To fill out the fields of tables, the wrapper accesses the Web site to integrate the page and look for keywords behind which is the value to extract. Wrappers are of two types depending on the format of the sources   : Either text wrappers or XML wrappers (Figure 8).

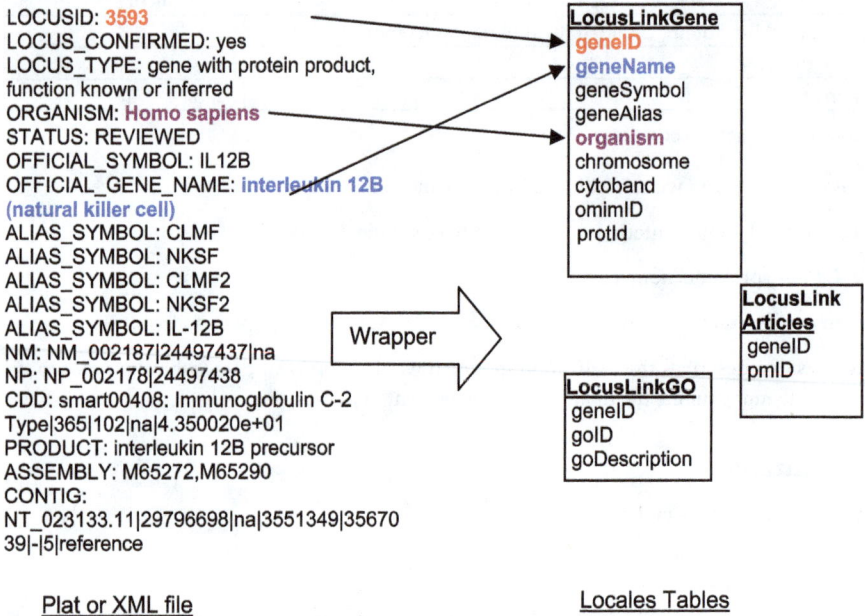

LOCUSID: 3593
LOCUS_CONFIRMED: yes
LOCUS_TYPE: gene with protein product,
function known or inferred
ORGANISM: Homo sapiens
STATUS: REVIEWED
OFFICIAL_SYMBOL: IL12B
OFFICIAL_GENE_NAME: interleukin 12B
(natural killer cell)
ALIAS_SYMBOL: CLMF
ALIAS_SYMBOL: NKSF
ALIAS_SYMBOL: CLMF2
ALIAS_SYMBOL: NKSF2
ALIAS_SYMBOL: IL-12B
NM: NM_002187|24497437|na
NP: NP_002178|24497438
CDD: smart00408: Immunoglobulin C-2
Type|365|102|na|4.350020e+01
PRODUCT: interleukin 12B precursor
ASSEMBLY: M65272,M65290
CONTIG:
NT_023133.11|29796698|na|3551349|35670
39|-|5|reference

LocusLinkGene
geneID
geneName
geneSymbol
geneAlias
organism
chromosome
cytoband
omimID
protId

LocusLink Articles
geneID
pmID

LocusLinkGO
geneID
goID
goDescription

Wrapper

Plat or XML file   Locales Tables

Figure 8. Presentation of a wrapper

Finally, a program that generates and initializes (gives the starting values for all wrappers) is created to coordinate everything. This program is also written in Java and integrates all the wrappers and their relationships. We thus obtain a set of local tables (Provisional) performed by the wrappers.

6.2. Step 2: Definition of "global schema"

It is therefore necessary to build the global schema that will be the only interface for user. Indeed, the user does not know absolutely how the data are integrated. The global schema is a set of relational tables that are defined using local tables (for information). This schema was introduced in the previous section.

6.3. Step 3: Matches between local tables and global schema

As the various local tables have been filled by the wrappers and the global schema has been established, we should now define the correspondence rules between them in order to implement global schema with the extracted information from local schemas. The problem is that several sources may correspond to a business table (we must then join conditions on these tables) or otherwise a source may have several tables trades. For this, we use the Medience server tool.

Medience Server (Figure 5.2) is a complete environment that treats all matching problems (different formats, different representation of business information, dispersal of information described in a single business table). It is a "virtual database", because it does not store information but analyse the user needs. This tool will serve as a mediator that is to say that it will be the unique interface for the user as it will both integrate databases, present data and also offers possibility to loop and see only some information tables of interest to the user. The use of this tool goes in three steps:

1. The first step consists on recording data sources and creating associated source tables using source files provided by wrappers. The global schema is also implemented by local tables. We define the attributes of all tables.
2. In the second part we define the correspondence rules from source tables to the global schema tables. The supply of each table in the global database is defined from the records of source tables. It is thus possible to standardize results coming from various sources by the definition of a standard type.
3. The final step is the verification of all components and installation of all matching rules to make it operational (Figure 9).

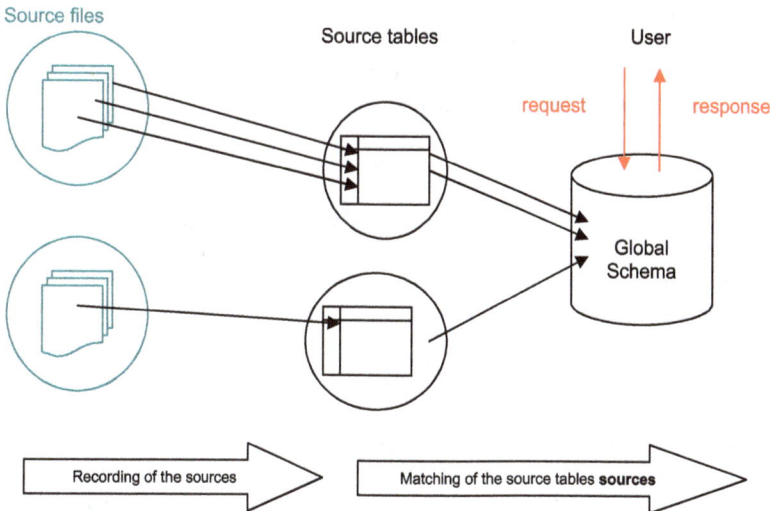

Figure 9. Architecture of Medience

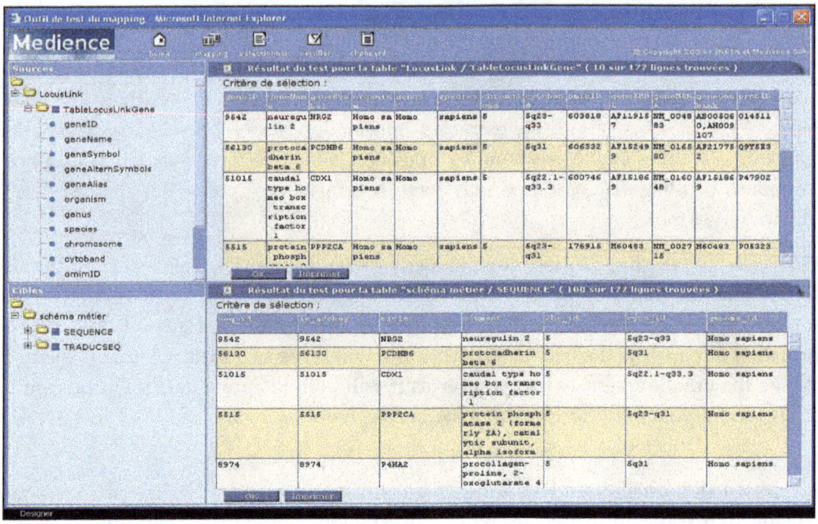

Figure 10. An example of Medience interface [19]

6.4. Step 4: Data analysis

It is therefore possible through a platform like Medience to integrate data sources (BD, Excel files, and text files) and view the results in a tabular form. Now, we can process to the analysis of the results. For this, the definition of demand in terms of mining must be decided: How can we use the data provided. Medience offers the possibility to ask tables on the global schema in SQL way. It also offers the ability to define views on these tables and keep a small part that is particularly interesting. It is possible to use our tool to answer the question like what is the protein associated with the mutated gene responsible for familial hypercholesterolemia and related publications.

7. Conclusion

The objective of this work was to develop a system for integrating biological data with an application on familial hypercholesterolemia disease. Such a system should facilitate access to multiple data sources available on the Web, in a transparent and uniform way, giving biologists a single virtual source that summarizes all relevant data sources for the application.

This chapter describes the solution adopted to achieve such a system, where the main elements have been identified, and a computer deployment scenario developed. Among different existing integration approaches, we adopted the mediator approach to integrate data sources. In this approach the most important step is the construction of the global schema as the mediator has to process queries at runtime in order to integrate data sources. We first studied the biologists' needs by exploring different scenarios and we identified with their help various data sources involved.

A study of these sources was necessary in order to build our global schema. From the diagram established, we formulated our SQL query as we built various adapters associated with different sources and at the end we have submitted this request to the mediator for treatment.

As prospects, we have to implement and test this solution and combine the final result of the mediator and that of the tool CHARMM before presenting to the user.

We are currently expanding the platform by integrating other proteins involved in cardiovascular diseases which are the main cause of mortality in the world. In particular, we are investigating a protein called paraoxonase-1 (PON1) which plays an important role in the cardiovascular diseases prevention.

PON1 is an HDL associated enzyme synthesized in the liver and distributed in the blood. It catalyzes the hydrolysis of modified lipids in both HDL (known as good cholesterol) and LDL (known as bad cholesterol) particles and protects them from oxidative modifications, and subsequently reducing the risk of atherosclerosis.

Further bioinformatics analysis including molecular simulations are performed on the PON1 enzyme to better understand the structure activity relationship and also to explore the mutated proteins (genetic polymorphism associated with heart disease) responsible for the weak activity revealed trough the clinical study in both diabetic and coronary patients from Morocco.

Author details

Assia Rharbi and Zohra Bakkoury
Equipe : AMIPS Ecole Mohammadia des Ingénieurs,
Université Mohammed V, Agdal, Rabat – Morocco

Afaf Mikou
Laboratoire GAIA, Spectroscopie Faculté des Sciences Ain Chock –
Université Hassan II, Casablanca – Morocco

Khadija Amine
Laboratoire GAIA, Spectroscopie Faculté des Sciences Ain Chock -
Université Hassan II, Casablanca – Morocco
Laboratoire de Recherche sur les Lipoprotéine et l'Athérosclérose, Unité de Recherche Associée au
CNRS-URAC 34-, Faculté des Sciences Ben Msik-Casablanca, Université Hassan II Mohammedia,
Morocco

Anass Kettani
Laboratoire de Recherche sur les Lipoprotéine et l'Athérosclérose, Unité de Recherche Associée au
CNRS-URAC 34-, Faculté des Sciences Ben Msik-Casablanca, Université Hassan II Mohammedia,
Morocco

Abdelkader Betari
ENSA Oujda, Université Mohammed Premier Oujda, Morocco

8. References

[1] M. El Messal, K. Aït Chihab, R. Chater, JC. Vallvé, F. Bennis, A. Hafidi, J. Ribalta, M. Varret, M. Loutfi, JP. Rabès, A. Kettani, C. Boileau, L. Masana, A. Adlouni. Familial Hypercholesterolemia in Morocco: first report of mutations in the LDL receptor gene. J Hum Genet.48 (4):199-203, 2003

[2] http://www.dsi.univ-paris5.fr/bio2/autof2/cha2_1.htm : Bases de données biologiques / Banques généralistes

[3] http://www.dsi.univ-paris5.fr/bio2/autof2/cha2_2.htm : Bases de données biologiques / Banques spécialisées

[4] www.gusdb.org : Genomics Unified Schema (GUS)

[5] Emilie Guérin, Gwenaëlle Marquet, Anita Burgun, Olivier Loréal et Fouzia Moussouni, GEDAW : un environnement intégré pour l'analyse du Transcriptome, JOBIM 2005

[6] http://www.cs.man.ac.uk/~stevensr/tambis/ : TAMBIS

[7] Thomas Hernandez, Subbarao Kambhampati, Integration of Biological Sources: Current Systems and Challenges Ahead, SIGMOD september 2004

[8] http://www.charmm.org/: CHARMM (Chemistry at HARvard Macromolecular Mechanics)

[9] http://www.ks.uiuc.edu/Research/vmd/: VMD (Visual Molecular Dyamic)

[10] www.rcsb.org/pdb/

[11] http://www.ncbi.nlm.nih.gov/sites/entrez?db=PubMed&itool=toolbar

[12] Assia Rharbi, Zohra Bakkoury, Afaf Mikou, Anass Kettani, Abdelkader Betari, and Omar Boucelma, Intégration des données génomiques pour la maladie d'hypercholestérolémie familiale, Journées Scientifiques en Bio-informatique (JSB'2007)

[13] Assia Rharbi, Zohra Bakkoury, Afaf Mikou, Anass Kettani, Abdelkader Betari, and Omar Boucelma, Intégration des données appliquée au domaine biologique, Cinquième Conférence sur les Systèmes Intelligents : Théories et Application (SITA'08)

[14] http://www.rcsb.org/pdb/file_formats/pdb/pdbguide2.2/guide2.2_frame.html

[15] http://rasmol.org/

[16] http://jmol.sourceforge.net/

[17] http://fr.wikipedia.org/w/index.php?title=Chime&action=edit&redlink=1

[18] http://www.w3.org/MarkUp/VRML/

[19] F.-M. Colonna, Thèse : "Intégration de données hétérogènes et distribuées sur le Web et applications à la biologie", Université Paul Cézanne (Aix-Marseille III), Décembre 2008

Hyper- and Dyslipoproteinemias

Chapter 7

Lipoproteins and Cardiovascular Diseases

Adebowale Saba and Olayinka Oridupa

Additional information is available at the end of the chapter

1. Introduction

1.1. What are lipids?

Lipids consists of a broad group of naturally occurring molecules that include fats, waxes, sterols including cholesterol, fat-soluble vitamins (such as vitamins A, D, E, and K), monoglycerides, diglycerides, triglycerides, phospholipids, and others. Lipids were previously known as sources of energy storage and the building blocks for cell membrane. Lipids are now known to play several key roles in intracellular signalling, membrane trafficking, hormonal regulation, blood clotting (Muller-Roeber and Pical, 2002; Vance and Vance, 2002; Fahy *et al.*, 2009). All lipids may be defined as hydrophobic or amphiphilic small molecules. The amphiphilic nature of some lipids allows them to form structures such as vesicles, liposomes, or membranes in an aqueous environment. Biological lipids originate entirely or in part from two distinct types of biochemical subunits, which are ketoacyl and isoprene groups (Fahy *et al.*, 2009).

Lipids typically do not travel alone in the blood. Instead, it binds to a protein that transports it to its destination in the body. The complex formed by the binding of lipid to protein i.e. lipoprotein, makes lipids water soluble, which enables its transportation in blood. The lipoprotein particle is composed of an outer shell of phospholipids, which renders the particle soluble in water; a core of fats called lipid, including cholesterol and a surface apoprotein (apolipoprotein). Ideally, the lipoprotein aggregates should be described in terms of the different protein components (apolipoprotein) because this determines the overall structures and metabolism of the lipoprotein, and the interactions with receptor molecules in liver and peripheral tissues. The apolipoprotein molecule enables tissues to recognize and take up the lipoprotein particle. However, lipoproteins are classified based on their characteristic density on ultracentrifugation, which has been used to segregate the different lipoprotein classes. Lipoproteins are broadly classified as high density lipoprotein (HDL), low density lipoprotein (LDL), intermediate density lipoprotein (IDL), very low density lipoprotein (VLDL) and chylomicrons (CM). Each of these particles perform different functions and can be detrimental (VLDL, IDL, LDL) or beneficial (HDL) to the cardiovascular system.

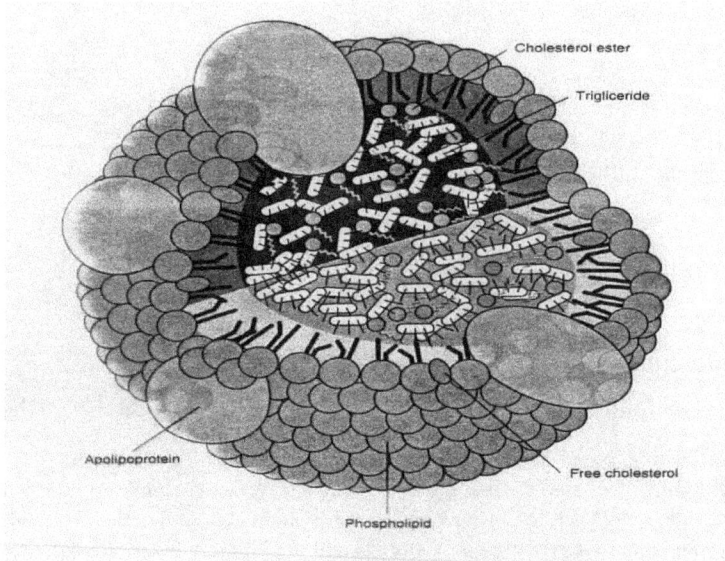

Figure 1. Structure of Lipoprotein available from http://www.campbell.edu

	CM	VLDL	IDL	HDL
Density (g/ml)	< 0.94	0.94 – 1.006	1.006 – 1.063	1.063 -1.210
Diameter (Á)	6000 - 2000	600	250	70-120
Total lipid (wt %) *	99	91	80	44
Triacylglycerols	85	55	10	6
Cholesterol esters	3	18	50	40
Cholesterol	2	7	11	7
Phospholipids	8	20	29	46

*Most of the remaining materials comprise the various apolipoproteins

Table 1. Physical properties and lipid compositions of lipoprotein classes

1.2. Role of cholesterol in membrane dynamics

It is relevant to establish the important of cholesterol in the body to be able to relate the various metabolic events associated with cholesterol and its homeostasis. Mammalian cell membranes contain varying proportions of cholesterol depending on organelle and cell type. These levels are tightly controlled by lipid transfer, through both vesicular and protein-bound pathways. With its rigid sterol backbone, cholesterol preferentially locates among saturated membrane lipids that have straight, elongated hydrocarbon chains rather than among kinked, unsaturated species. The presence of cholesterol in the membrane increases lateral ordering of lipids, reducing permeability and fluidity and potentially restricting diffusion of membrane proteins. Its distribution is not uniform within a membrane: regions of high cholesterol and corresponding low fluidity are termed lipid rafts. These areas act as platforms for the assembly of signalling complexes within the membrane and have been implicated in the development of numerous disease processes, notably arteriosclerosis and cancer (Di Vizio *et al.*, 2008; Ikonen, 2008).

2. Cardiovascular disease and risk factors

Elevated plasma levels of low density lipoprotein (LDL) and low levels of high density lipoprotein (HDL) poses a major risk of development of cardiovascular diseases (Grundy *et al.*, 1999). A dietary intake of saturated fat and a sedentary lifestyle has been associated with about 31% of coronary heart disease and 11% of stroke in humans. According to the Framingham Heart Study and other studies (Wilson *et al.*, 1998), the major and independent risk factors for coronary heart disease (CHD) are cigarette smoking of any amount, elevated blood pressure, elevated serum total cholesterol and low-density lipoprotein cholesterol (LDL-C), low serum high-density lipoprotein cholesterol (HDL-C), diabetes mellitus, and advancing age. More recently, a review by Patrick and Uzick (2001) documented new risk factors for CHD which included levels of circulating homocysteine, fibrinogen, C-reactive protein (CRP), endogenous tissue plasminogen-activator, plasminogen-activator inhibitor type I, lipoprotein(a), factor VII and certain infections such as *Chlamydia pneumonia*. These studies showed that the total risk of an individual is the summation of all major risk factors.

Other factors contributing to the total risk for CHD are categorized as conditional risk factors and predisposing risk factors. The conditional risk factors are associated with increased risk for CHD, although their causative, independent, and quantitative contributions to CHD have not been well documented. The predisposing risk factors are those that worsen the independent risk factors. Two of these risk factors; obesity and physical inactivity, are designated major risk factors by the American Heart Association (AHA) (Fletcher *et al.*, 1996; Eckel, 1997). The adverse effects of obesity are worsened when it is expressed as abdominal obesity, an indicator of insulin resistance. These risk factors apply before clinical manifestation of coronary atherosclerotic diseases. The clinical significance of these risk assessment is to identify high-risk patients who require attention, motivate patients to adhere to risk-reduction therapies and modify the intensity of risk reduction effort required in potential patients (Grundy *et al.*, 1999).

2.1. Lipoproteins, cholesterol and atherosclerosis

Cholesterol is a building block of the outer layer of cell membranes. Cholesterol is a waxy steroid of fat that is produced in the liver or intestines. It is used to produce hormones and cell membranes and is transported in the blood plasma of all mammals (Leah, 2009). As an essential structural component of mammalian cell membranes, it is required to establish proper membrane permeability and fluidity. In addition, cholesterol is an important component for the manufacture of bile acids, steroid hormones, and vitamin D. Cholesterol is the principal sterol synthesized by animals; however, small quantities can be synthesized in other eukaryotes such as plants and fungi. It is almost completely absent among prokaryotes including bacteria (Pearson *et al.*, 2003).

Owing to its limited solubility in water, cholesterol is transported in blood in lipoproteins. The lipoprotein outer layer is formed of amphiphilic cholesterol and phospholipid molecules, studded with proteins, surrounding a hydrophobic core of triglycerides and cholesterol esters. Lipoproteins are specifically targeted to cells by distinct apolipoproteins on their surface that bind to specific receptors. Low density lipoprotein (LDL) contains the highest level of cholesterol. LDL receptors in peripheral tissues bind LDL, triggering its endocytosis, lysosomal targeting and hydrolysis. When cells have abundant cholesterol, LDL receptor synthesis is inhibited by the sterol regulatory element binding proteins (SREBP) pathway (Wang *et al.*, 1993; Yokoyama *et al.*, 1993; Brown and Goldstein, 2009).

The biosynthesis of cholesterol is intensely regulated in the body with negative feedback of plasma cholesterol levels. The molecular basis of this regulation was set out by Michael Brown and Joseph Goldstein, earning them the Nobel Prize in Physiology and Medicine in 1985 (Leah, 2009). A key irreversible step of cholesterol synthesis is catalyzed by HMG-CoA reductase. Transcription of the HMG gene is controlled by SREBPs, transcription factors that bind sterol regulatory elements. SREBPs are only able to enter the nucleus when cholesterol levels fall. At other times they are tied up in a complex that includes Scap (SREBP-cleavage activating protein), an escort protein with a cholesterol-binding motif that senses cellular cholesterol levels. The SREBP pathway is now implicated in multiple regulatory aspects of lipid formation and metabolism (Brown and Goldstein, 2009).

2.2. Atherogenicity of lipoprotein sub-fractions

The first stages of cholesterol build up in the blood vessels (atherosclerosis) occur when LDL particles circulating in the blood penetrate through the inner lining of blood vessels and become trapped in the artery wall. The normal function of LDL is to deliver cholesterol to cells, where it is used in membranes or for the synthesis of steroid hormones. Cells take up cholesterol by receptor-mediated endocytosis. LDL binds to a specific LDL receptor and is internalized in an endocytic vesicle. Receptors are recycled to the cell surface, while hydrolysis in an endolysosome releases cholesterol for use in the cell. The liver removes LDL and other lipoproteins from the circulation by receptor-mediated endocytosis.

Deregulation of cholesterol levels results in the existence of more LDL in the blood than can be taken up by LDL receptors. Excess LDL is oxidized and taken up by macrophages, forming foam cells that can become trapped in the walls of blood vessels along with cells of inflammation (Zioncheck *et al.*, 1991; Young and McEneny, 2001). Fatty streaks, consisting of subendothelial collection of foam cells are initially formed in blood vessels. Small, dense LDL particles are more atherogenic than large, buoyant LDL particles, and oxidation of LDL also increases its atherogenicity. In addition, LDL belongs to the group of lipoproteins that contain apolipoprotein (apo) B-100. Some of the particles in this highly heterogeneous group contain other apolipoproteins, such as apo C-II, apo C-III, and apo E. Furthermore, some particles are larger and rich in triglycerides (large VLDL), whereas others are smaller and rich in cholesteryl esters (small VLDL, IDL). It is now known that remnant lipoproteins containing apo C-III are highly atherogenic and may be more specific measures of coronary heart disease (CHD) risk assessment than plasma triglycerides (Carmena *et al.*, 2004).

The end result is the formation of an atherosclerotic plaque which occludes the endothelial lumen and impedes blood flow, leading to myocardial infarction, the major cause of heart attacks and strokes. Although LDL levels correlate with heart attack risk, high density lipoprotein (HDL) has an inverse ratio of risk because HDL particles transport cholesterol to the liver for excretion. Modern cholesterol tests distinguish the LDL/HDL ratio as well as the overall level (Barter *et al.*, 2007). Other sub-fractions of lipoproteins such as chylomicrons, IDL and VLDL may enter the endothelial spaces due to their sizes, thus contribute substantially to development of atherosclerotic plaques. They may also increase prothrombotic factors, triggering cardiovascular diseases (Brunzell *et al.*, 2008).

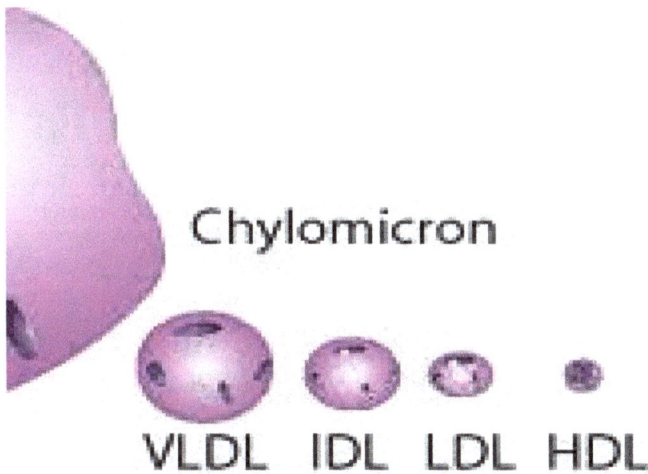

Figure 2. Sizes of Lipoproteins available at http://www.sigmaaldrich.com/european-export.html

Triglyceride-rich lipoproteins comprise a great variety of nascent and metabolically modified lipoprotein particles differing in size, density, and lipid and apolipoprotein composition. Studies have shown an inverse relationship between the size of lipoproteins and their ability to cross the endothelial barrier to enter the arterial intima. Chylomicrons and large VLDLs are probably not capable of entering the arterial wall. On the other hand, small VLDL and IDL can enter the arterial intima. Therefore, certain triglyceride rich lipoproteins are atherogenic, whereas others are not. A large body of evidence suggests that small VLDLs and IDLs are independently associated with atherosclerosis (Carmena *et al.*, 2004).

2.3. Apolipoproteins in lipoproteins

Apolipoproteins are the carrier proteins for lipoproteins and they consist of a single polypeptide chain often with relatively little tertiary structure. They are required to solubilise the non-polar lipids in the circulation and in some instances to recognise specific receptors. They are classified as Apo A1, A2, A4, A5, B48, B100, C1, C2, C3, D, E, H, J, L, M and Apo (a). Most apolipoproteins are synthesised by the liver and intestine.

Apolipoprotein	Molecular weight	Lipoprotein	Function
Apo AI	29,100	HDL	Lecithin: cholesterol acyltransferase (LCAT) activation. Main structural protein. Binds ABCA1 on macrophages
Apo AII	17,400	HDL	Enhances hepatic lipase activity
Apo AIII	46,000	CM	
Apo AIV	46,000	HDL, CM	Inhibits food intake in CNS
Apo AV	39,000	HDL	Enhances triacylglycerol uptake
Apo B48	241,000	CM	Derived from Apo B100 gene by RNA editing, lacks the LDL receptor binding site
Apo B100	512,000	LDL, IDL, VLDL	Binds to LDL receptor
Apo CI	7,600	VLDL, CM	Activates LCAT
Apo CII	8,900	VLDL, CM	Activates lipoprotein lipase
Apo CIII	8,750	VLDL, CM	Inhibits lipoprotein lipase
Apo D	33,000	HDL	Closely associated with LCAT, progesterone binding
Apo E	34,000	HDL	At least 3 forms. Binds to LDL receptor
Apo(a)	300,000 – 800,000	LDL, Lp(a)	Linked by disulfide bonds to apo B100 and similar to plasminogen, associated with premature coronary artery disease and stroke
Apo H	50,000	Chylomicrons	Involved with triacylglycerol metabolism
Apo M		HDL	Transports sphingosine-1-phosphate

Table 2. Classes of apolipoproteins, their molecular weight and functions

2.3.1. Apolipoprotein A

Apo A are subdivided into apo AI, AII, AIII, AIV and AV. Apo AI and AII are the major apolipoproteins in the HDL particle. HDL is primarily saddled with the responsibility of removing excess cholesterol from peripheral tissues and delivering it to the liver for excretion in bile as bile acids in a process known as reverse cholesterol transport. Apo AI is synthesised mainly by the liver and also by the intestine. The protein consists of 243 amino acids arranged as eight α-helical segments of 22 amino acids which have 11 –mer repeats, no disulfide bonds or glycosylations. The helices are believed to be amphipathic in nature with both hydrophobic and hydrophilic faces. This nature enhances its interaction with the lipid and aqueous phases. Apo AII on the other hand is found as a twin chain of 77 amino acids each, linked by disulfide bonds. It enhances the activity of hepatic lipase, thus increasing lipoprotein metabolism. Apo III is found in chylomicrons.

2.3.2. Apolipoprotein E

Three isoforms of this apolipoprotein exist and they are all synthesised mainly by the liver and also by several tissues such as arterial wall, brain and adipose tissue. They are important for homeostasis of lipid and lipoproteins in blood circulation as well as their metabolism in these tissues. Apo E is required for the clearance of VLDL remnant (IDL) from circulation in the liver. Other suggestions on its involvement with immune response and inflammation have also been put forward.

2.3.3. Apolipoprotein B

Two types of apolipoprotein B are synthesised from the intestine and liver as apo B100 which has the full length of 4536 amino acid residues and a truncated form with 48% of the full length known as apo B48. These proteins which are synthesised on ribosomes, an organelle located on the surface of rough endoplasmic reticulum are translocated through the reticular membrane into the lumen of endoplasmic reticulum. Assembly of VLDL occurs here by accretion of lipids to the core of apo B particle. This occurs in three distinct stages as the apo B grows bigger, forming the pre-VLDL to VLDL 2 which eventually grows to become the triacylglycerol-rich VLDL 1 or chylomicrons, the most energy dense substances in the body. VLDL 2 is a triacylglycerol-poor version of VLDL and its assembly occurs in golgi bodies. This is transported to basolateral membrane of the intestinal cells where final assembly of VLDL 1 or chylomicrons occur and these are secreted into the lamina propria of the intestinal cells by reverse exocytosis.

Apo B100 and B48 are large and water-insoluble, and are the only non-exchangeable apolipoproteins. They are major components of chylomicrons and VLDL, and usually remain with their lipid aggregates throughout their passage in plasma and several metabolic changes occurring during their circulation in plasma. Chylomicrons are usually transported through the intestinal lymphatic system and flow into blood circulation via the left subclavian vein. As apo B carries the VLDL or chylomicrons through the blood stream, their triacylglycerol content are removed by peripheral tissues via enzymatic activity of lipoprotein lipase, located in the endothelial wall. This makes free fatty acids available for energy production in muscle and some are stored in adipose tissue. Apo B48 remains with the remnant of the lipoprotein particle, along with dietary cholesterol and apo E. the lipoprotein remnant is eventually cleared from circulation in the liver by an apo E dependent receptor-mediated reaction.

ApoB-100 is one of the largest monomeric proteins known and it is the major apolipoprotein component common to the atherogenic lipoproteins [VLDL, LDL, IDL and Lp(a)] (Boerwinkle et al., 1992; Carmena et al., 2004). Apo B100 differs from apo B48 by the presence of LDL receptor site on the apo B100 molecule. Apo B-100 is encoded for by the Apo B gene and mutations in this gene cause familial hypercholesterolemia, an autosomal hereditary metabolic disorder. The level of apo B-100 is a stronger predictor of risk than LDL in humans. Apo B-100 is speculated to mediate delivery of the cholesterol content of lipoproteins to cells via an unknown mechanism. It is well established that Apo B-100 is associated with atherogenic lipoproteins, thus the number of Apo B-100 can be used to determine the risk of atherosclerosis and CHD in individuals. Apo B-100/apo A-I ratio are strongly and positively related to increased risk of fatal myocardial infarction. Apolipoprotein A-I (apo A-I) is the major apolipoprotein in the HDL structure. The Apo B-100/apo A-I ratio is especially valuable in patients with normal or low LDL concentrations, a frequent observation in type 2 diabetes mellitus which may present with hypertriglyceridemia and hyper-apo B concentrations (Carmena et al., 2004).

2.3.4. Apolipoprotein C

Apolipoprotein C is subdivided into three and each has its own distinct function. Apo CI is involved with activation of Lecithin: cholesterol acyltransferase (LCAT) along with apo AI. This enzyme converts free cholesterol into cholesterol ester, which enhances the incorporation of cholesterol into the lipid core of a lipoprotein particle, particularly in assembly of HDL particle. The enzyme is mainly bound to HDL and LDL in plasma. Apo CII and CIII have antagonistic activity to each other which are required in regulation of lipoprotein lipase activity. Apo CII is required for activation of lipoprotein lipase, while CIII inhibits lipoprotein lipase activity. Apo CII is believed to open a lid-like region of the enzyme which allows the active site to hydrolyse the fatty acid ester bonds of triacylglycerols. In addition to inhibition of lipoprotein lipase, apo CIII also inhibits the binding of lipoproteins to receptors at the cell surface, thereby decreasing hydrolysis of triacylglycerols. High levels of apo CIII have been associated with elevated serum levels of triacylglycerols (hypertriglyceridemia).

2.3.5. Transfer of apolipoproteins in lipoprotein homeostasis

Lipids enter blood circulation bound to apolipoproteins as chylomicrons or VLDL which as secreted into the blood stream from the intestines. Chylomicrons or VLDL consist mainly of apo B100 and B48, but also consist of some apo AI, along with other apolipoprotein which will be discussed. These lipoproteins carry triacylglycerol-rich cholesterols to the peripheral tissues to provide sources of energy and for storage, while HDL carries excess cholesterol from peripheral tissue to the liver for excretion in bile acids. Immediately chylomicrons enter blood circulation, an exchange of apolipoproteins occurs between chylomicrons and HDL. The apo AI content of chylomicrons is exchanged for the apo C and E content of HDL. Apo C content is required for activation and inhibition of lipoprotein lipase which hydrolyses the triacylglycerol content of chylomicrons and VLDL, while apo E is needed for the receptor mediated clearance from circulation. Circulation of VLDL and chylomicrons in the blood stream exposes the particles to enzymatic release of triacylglycerols from the lipoprotein core and excess cholesterol is removed from cells. The triacylglycerol- poor and cholesterol rich LDL remnant produced is potentially toxic to the body and needed to be safely cleared from blood circulation. The main concern for this lipoprotein is its toxic effect on the cardiovascular system. The liver scavenges and disposes chylomicrons remnant more effectively than the LDL particles, a mechanism put in place by the body to get rid of the more atherogenic particle of the two; chylomicrons remnant. LDL particles are mostly removed by other mechanisms involving HDL.

2.3.6. Apolipoprotein(a)

Apolipoprotein(a) itself is a large glycoprotein that exhibits size heterogeneity among individuals with isoforms that range between 180 – 700kDa in size. Apo(a) genotypes were determined using a newly developed pulsed-field gel electrophoresis method which distinguished 19 different genotypes at the apo(a) locus. The apo(a) gene itself was found to account for virtually all the genetic variability in plasma Lp(a) levels (Boerwinkle *et al.*, 1992). The apo(a) cDNA contains multiple tandem copies of a sequence that encodes a cysteine-rich protein motif called a kringle. The particular repeated kringle in apo(a) is designated kringle 4 because it closely resembles the fourth kringle in plasminogen, with the protease domain of apo(a) containing 88% amino-acid identity to plasminogen. McLean *et al.* (1992) proposed that the apo(a) isoforms are of different sizes because of variations in the numbers of kringle 4-encoding repeats in the apo(a) gene. The molecular mass of apo(a) protein varies from 187 kDa for an apo(a) that contains 12 kringle 4 domains, to 662 kDa for an apo(a) that contains 50 kringle 4 domains (Carmena *et al.*, 2004).

2.4. Lipoprotein A [Lp(a)]

Lipoprotein [Lp(a)] is a variant of LDL with an additional apolipoprotein in the structure. Lp(a) is essentially an LDL particle with a large glycoprotein, apolipoprotein (a) [apo(a)] attached to it (McLean *et al.*, 1987; Loscalzo *et al.*, 1990; Boerwinkle *et al.*, 1992; Palabrica *et al.*, 1995). Lp(a) resembles low density lipoprotein (LDL) in lipid composition, but it is distinguished by the presence of apo(a) which is bound by a disulfide linkage to apolipoprotein B-100, a ligand in the LDL molecule by which LDL binds to its receptor. Lp(a) levels has been demonstrated to have a clear association with development of atherosclerosis and other cardiovascular diseases (Zenker *et al.*, 1986; Danesh *et al.*, 2000; Berglund and Anuurad, 2008; Danik *et al.*, 2008). The postulated atherogenicity of Lp(a) is probably due to the presence of apo(a) component of the Lp(a) molecule. A study showed that the removal of apo(a) from Lp(a) particles result in a lipoprotein with greatly enhanced affinity for the LDL receptor (Armstrong *et al.*, 1985).

2.4.1. Synthesis of lipoprotein(a)

Apo(a) is expressed by liver cells (hepatocytes), and the assembly of apo(a) and LDL particles seems to take place at the outer hepatocyte surface. The half-life of Lp(a) in the circulation is about 3 to 4 days (Rader *et al.*, 1993) and this particle varies in blood concentration from one individual to another from <0.2 - >200 mg/dL. Ethnicity is a factor, with those of Asian and African origin averaging the highest concentrations. Within ethnic groups, individual elevation of Lp(a) is directly associated with increased risk of cardiovascular diseases (Sandholzer *et al.*, 1991; Chien *et al.*, 2008). Lp(a) is usually unaffected by factors like age, blood pressure, and total cholesterol.

2.4.2. Similarity between lipoprotein(a) and plasminogen

The structure of Lp(a) is similar to plasminogen, a naturally occurring glycoprotein that participates in dissolving of clots that form in the bloodstream, and tissue plasminogen activator (tPA). Lp(a) competes with plasminogen for its binding site, leading to reduced fibrinolysis (Loscalzo *et al.*, 1990; Palabrica *et al.*, 1995). Also because Lp(a) stimulates secretion of PAI-1 it leads to thrombogenesis. In addition, because of LDL cholesterol content, Lp(a) contributes to atherosclerosis (Schreiner *et al.*, 1993; Sotiriou *et al.*, 2006) and ultimately a cardiovascular risk factor (Berglund and Ramakrishnan, 2004).

2.4.3. Correlation between apolipoprotein size and Lp(a) concentration

There is a general inverse correlation between the size of the apo(a) isoform and the Lp(a) plasma concentration (Bowden *et al.*, 1994; Kraft *et al.*, 1996) which is caused by a variable rate of degradation before the apo(a) protein has matured for Lp(a) assembly (White *et al.*, 1994). The plasma concentration of Lp(a) is unaffected by many physiological, pharmacological, and environmental factors that affect the levels of other plasma lipoproteins (Albers *et al.*, 1977). A genetic determination of plasma Lp(a) levels was strongly suggested due to this lack of environmental and physiological influences. Consistent with this formulation, early genetic studies suggested that the presence of Lp(a) in plasma was inherited as a single autosomal dominant trait (Berg and Mohr, 1963; Iselius *et al.*, 1981), with an estimated heritability level ranging from 0.75 to 0.98 (Boerwinkle *et al.*, 1992).

Plasma Lp(a) concentrations vary 1000-fold between individuals and represent a continuous quantitative genetic trait with a skewed distribution in Caucasian populations (Utermann, 1989). A study was conducted by Lackner *et al.* (1991) in which the apo(a) gene of members of 12 Caucasian families were segregated. It was found that within a given family, sibling pairs with identical apo(a) genotypes tended to have very similar plasma Lp(a) levels (Lackner *et al.*, 1991). However, individuals with the same apo(a) genotypes who were members of different families often had significantly different plasma concentrations of

Lp(a). Taken together, these observations suggest that the apo(a) gene is the major determinant of plasma Lp(a) levels and that cis-acting DNA sequences at or near the apo(a) locus, other than the number of kringle 4 repeats, contribute importantly to plasma Lp(a) concentrations (Boerwinkle *et al.*, 1992). Variation in the hypervariable apo(a) gene on chromosome 6q2.6-q2.7 and interaction of apo(a) alleles with defective LDL-receptor genes explain a large fraction of the variability of plasma Lp(a) concentrations (Utermann, 1989).

Furthermore, the size of the apo(a) glycoprotein varies in individuals and this size is inversely correlated with the plasma level of Lp(a). The reason for the inverse correlation between the size of the apo(a) gene and level of plasma Lp(a) is not known, but a variation of length within the kringle 4-encoding region of the apo(a) gene may account for a greater proportion of the inter-individual variation in plasma Lp(a) concentrations. Also, the number of kringle 4 repeats in the gene may not have a direct effect on plasma Lp(a) concentration (Boerwinkle *et al.*, 1992; Brunner *et al.*, 1996). A study conducted on apo(a) gene of mamorset monkeys showed a plasma Lp(a) concentration of a very wide range of over a 100-fold, but only one apo(a) isoform (Guo *et al.*, 1991). This may be explained by the differences in the composition of kringle 4 sequence of apo(a) genes in which individuals may have same sizes of apo(a) alleles but different plasma Lp(a) concentrations. The frequency of recombination activity in this locus may be responsible for the variation in their kringle 4 composition and number which may have marked effect on synthesis and /or degradation of Lp(a) (Boerwinkle *et al.*, 1992).

2.5. Role of oxidation in atherogenesis

Oxidative stress, especially LDL oxidation has been suggested for almost three decades as the most probable aetiology of atherosclerosis (Steinbrecher *et al.*, 1984). Markers of LDL oxidation in plasma, particularly circulating oxidized LDL and auto-antibodies against oxidized LDL, could be used to assess the development of atherosclerosis in patients (Carmena *et al.*, 2004). Circulating oxidized LDL is additive to the global risk score based on age, sex, total and HDL cholesterol, diabetes mellitus, hypertension, and smoking as a useful marker for identifying persons at risk for CAD (Holvoet *et al.*, 2001; Toshima *et al.*, 2000).

A study has associated circulating oxidized LDL with both subclinical atherosclerosis (clinically silent ultrasound assessed atherosclerotic changes in the carotid and femoral arteries) and inflammatory variables (C-reactive protein and the inflammatory cytokines interleukin-6 and tumor necrosis factor-α). This conclusion supports the concept that oxidatively modified LDL may play a major role in development of atherosclerosis (Hulthe and Fagerberg, 2002). It has been proposed that, because of the antigenic properties of oxidized LDL, the anti-oxidized LDL antibody titer could represent a useful index of in vivo LDL oxidation. Autoantibodies against oxidized LDL have been reported to be associated with atherosclerosis, but existing reports are still conflicting. Some studies have reported a positive relationship between autoantibodies against oxidized LDL and CHD (Sherer *et al.*, 2001) whereas another did not (Leinonen *et al.*, 1998). There is a strong cross-reactivity between autoantibodies against oxidized LDL and anticardiolipin antibodies, which have been positively associated with CHD (Erkkila *et al.*, 2000).

2.6. Relationship between insulin resistance, diabetes, and small, dense LDL

Cardiovascular heart disease risk is usually significantly increased when elevated levels of small, dense LDL accompanied by hypertriglyceridemia, reduced HDL-cholesterol levels, abdominal obesity, and insulin resistance. Results from the Que'bec Cardiovascular Study have indicated that persons displaying elevated plasma concentrations of insulin and apo B together with small, dense LDL particles showed a remarkable increase in CHD risk (Lamarche *et al.*, 1999).

Sensitivity to insulin in diabetic and non-diabetic individuals was assessed using nuclear magnetic resonance (NMR) spectroscopy. Insulin resistance had profound effects on lipoprotein size and an increase in serum triglycerides. The lipid profile revealed a 2- to 3-fold increase in concentrations of large VLDL particles with no change in medium or small VLDL, increase in overall LDL particle concentration with more small LDL particle size and reduced large LDL. In type 2 diabetes, these alterations could be attributed primarily to the underlying insulin resistance. These changes in the NMR lipoprotein subclass profile predictably increased the risk of cardiovascular disease but were not fully apparent in the conventional lipid profile (Garvey *et al.*, 2003). The Diabetes Atherosclerosis Intervention Study (DAIS) (Vakkilainen *et al.*, 2003) showed that lipid-modifying treatment decreased the angiographic progression of coronary atherosclerosis in subjects with type 2 diabetes. This effect was related in part to the correction of lipoprotein abnormalities. Compared with placebo, fenofibrate treatment significantly increased LDL particle size and HDL cholesterol and decreased plasma total cholesterol, LDL cholesterol, and triglyceride concentrations. The final LDL particle size was inversely correlated with the increase in percentage diameter stenosis (Vakkilainen *et al.*, 2003).

The Pittsburgh Epidemiology of Diabetes Complications Study on whether NMR lipoprotein spectroscopy improves the prediction of coronary artery disease (CAD) in patients with childhood-onset type 1 diabetes, independently of conventional lipid and other risk factors showed that both lipid mass and particle concentrations (NMR spectroscopy) of all VLDL subclasses, small LDL, medium LDL, and medium HDL were increased in CAD cases compared with controls, whereas large HDL was decreased. Mean LDL and HDL particle sizes were also less in CAD cases (Soedamah-Muthu *et al.*, 2003; Carmena *et al.*, 2004).

2.7. Dyslipoproteinaemia

Dyslipoproteinaemia is a term broadly used for derangement in lipid and lipoprotein metabolism, which may either be hyperlipoproteinaemia or hypolipoproteinaemia. Dyslipoproteinaemias are generally classified as familial (primary) or acquired (secondary). This chapter discusses hyperlipoproteinaemia with its close relevance to development of cardiovascular diseases. Primary hyperlipoproteinaemias are of genetic origin and may be due to a mutation in a receptor protein which presents as inborn errors of lipid metabolism, and includes common hypercholesterolemia, combined familiar hyperlipidemia, familiar hypercholesterolemia, familiar hypertriglyceridemia, VLDL remnants hyperlipidemia and primary chylomicronaemia (Garmendia, 2003). Primary hyperlipoproteinaemia was first classified by Fredrickson and Lees (1965) and this classification was adopted by World Health Organization (WHO). They divided primary hyperlipoproteinaemias into four types and details are show in Table 3 below.

The secondary hyperlipoproteinemias also mimic primary types and may present with similar symptoms. Secondary dyslipoproteinaemias are usually due to other underlying causes that lead to alterations in plasma lipid and lipoprotein metabolism, including hypothyroidism, diabetes mellitus, nephrotic syndrome, chronic biliary obstruction, renal insufficiency. Some drugs modify lipid metabolism and these include alcohol, beta-adrenergic blockers, diuretics, progestagens, corticosteroids (Garmendia, 2003). Treatment of the underlying cause or discontinuation of offending drug may resolve the dyslipoproteinaemia. Lipid and lipoprotein abnormalities are common observations and are regarded as modifiable risk factors for development of cardiovascular diseases.

Hyperlipo-proteinaemia	Sub-type	Classification	Defect	Lipoprotein increased	Treatment
Type I	a, c	Familial hyperchylo-micronemia	↓ Lipoprotein lipase (LpL)	Chylomicrons	Diet control
	b	apoprotein CII deficiency	Altered Apo CII		
Type II	a	Familial hypercholesterolemia	LDL receptor deficiency	LDL	Bile acid sequestrants, statins,niacin
	b	Familial combined hyperlipidemia	↓ LDL receptor or and ↑Apo B	LDL and VLDL	Statins, niacin,fibrate
Type III		Familial dysbetalipo-proteinaemia	Apo E2 synthesis	IDL	Fibrate, statins
Type IV		Familial hypertriglyce-ridaemia	↑VLDL and ↓LpL	VLDL	Fibrate, niacin, statin
Type V			↑VLDL and ↓LpL	VLDL and chylomicrons	Niacin, fibrate

Adapted from Fredrickson classification of hyperdyslipoproteinaemia (Fredrickson and Lees, 1965).

Table 3.

2.8. Theories of atherogenesis

Arteries are blood vessels that carry oxygenated blood from the heart to all tissues of the body. The arterial wall is composed of three layers, namely the intima (inner lining), media and adventitia. A single layer of endothelial cells line the inner surface of the intima, forming a barrier to blood cells and plasma flowing within the blood vessel. Atherosclerosis is characterized by lesions in the intima of arteries, seen as raised fibrous plagues ranging in colour from pearly gray to yellowish gray. The cellular components of the plaque include a cell similar to the adjacent endothelial cell, macrophages, fibrinogen from which fibrin is formed and white blood cells intersparsed between dense connective tissue which consist majorly of collagen fibers. The cells within and around the plaque are usually lipid ladened. Atherosclerosis poses a high risk not just because it can close up an artery, slowing down or entirely restricting blood flow, but may also lead to thrombus formation. A thrombus is a complex aggregation of platelets, red and white blood cells in a fibrin network. Several theories have emanated, suggesting the actual pathogenesis of atherosclerosis.

Schoenhagen (2006) documented the different theories that have been postulated in the course of history. In 1851, a scientist, Rokitansky suggested the encrustation theory or thrombogenesis in which it is said that atherosclerosis began in the intima of arteries with the deposition of thrombus. This is followed by the organisation of the thrombus through infiltration of fibroblast, secondarily followed by deposition of lipid. The German pathologist, Rudolf Virchow postulated the insudation or inflammation theory in 1856, a different initiation of atherosclerosis. It was suggested that infiltration of fatty substances from the blood stream into the arterial wall leads to deposition of cholesterol which acts as an irritant, causing inflammation and the proliferation of cells. The cholesterol deposits act as irritant in the arterial intima, initiating inflammatory process as macrophages are incriminated as key role players in the phases of the disease. This theory was further supported by the work of N.N. Anitschkow in 1933 where he discovered that a disease resembling human atherosclerosis could be reproduced in rabbits with high serum cholesterol or LDL levels. He thus stated this occurrence may be as a result of defects in metabolism of lipids and lipids Schoenhagen, 2006).

The flow theory relates the circulation of blood in vessels to its effect on arterial wall. It stated that lesions occurred more often at curved, branching, or bifurcated sites, generally at regions of perturbed blood flow. Other hypotheses that arose from the flow theory include the stagnation point hypothesis by Fox and Hugh (1966), high wall shear stress hypothesis proposed by Fry (1968), low wall shear stress hypothesis by Caro et al. (1969), diminished lateral pressure hypothesis by Texon (1980) and the convection-diffusion hypothesis.

All the theories above were established based on three methods. Atherosclerotic plaques from autopsy findings from individuals of both sexes, various ages and race with different diseases which included hyperlipidaemia, diabetes and hypertension were considered. Epidemiological studies of factors which promote or prevent development of atherosclerosis, and finally experimental pathology which established the sequence of lesion development or regression were considered. None of these theories entirely explains the pathogenesis of atherosclerosis, but each has explained an aspect of this process.

2.9. Prevention and treatment of cardiovascular diseases

The contributing factors to development of cardiovascular diseases are numerous as mentioned in the risk factors above. Lowering of plasma cholesterol levels is usually the first line of intervention for prevention and treatment of cardiovascular diseases. Dramatic successes have been recorded with cholesterol-lowering therapy which may suggest that maintenance of low cholesterol levels is sufficient to prevent development of atherosclerosis or reversing an established disease condition (Brunzell *et al.*, 2008). Different approaches have been used for prevention and treatment of this condition, some are enumerated below.

2.9.1. Role of High density lipoprotein-cholesterol (HDL-C)

High density lipoprotein-cholesterol (HDL-C) is the smallest of the lipoprotein sub-fractions. It is however, the most complicated and diverse of the lipoproteins. It is the major lipoprotein which transports excess cholesterol from the plasma to the liver for excretion or utilization in the liver and other hormone producing regions of the body. Excess cholesterol is eliminated from the body via the liver, which secretes cholesterol in bile or converts it to bile salts (Toth, 2005; Tall, 2008). Also, its anti-inflammatory property protects LDL from oxidation and limits the concentrations of oxidized components, which may pose as atherogenic treats. HDL-C have been associated with reduced risk of cardiovascular events (Duffy and Rader, 2009; Khera *et al.*, 2011). HDL-C plays a key role in the reverse transportation of cholesterol by accepting cholesterol from lipid-laden macrophages (Lehrke *et al.*, 2007). In the study conducted by Khera *et al.* (2011), the ability of HDL to promote cholesterol efflux from macrophages was strongly and inversely associated with both subclinical atherosclerosis and obstructive coronary artery disease. It was also discovered that the associations persisted after adjustment for traditional cardiovascular risk factors, including the levels of HDL cholesterol and apolipoprotein A-I. HDL has several protein constituents which are exchangeable with other lipoproteins, and it acquires different apolipoproteins in the process of maturation such as apo AII, AIV, AV, CI, CII, CIII and E which results in generation of diverse HDL particles with various metabolic functions. In addition to being carrier proteins for the HDL particle, these proteins have protective roles which they play against cardiovascular diseases, such as by acting as anti-inflammatory regulators to limit the activity of pro-inflammatory cytokines.

Nascent HDL is synthesised and secreted by the liver and small intestine. It travels in the circulation where it gathers cholesterol to form mature HDL, which then returns the cholesterol to the liver via various pathways. Apolipoprotein A-I (ApoA-I) is the major protein component of high density lipoprotein (HDL) in plasma. The protein is encoded for by APOAI gene (Breslow *et al.*, 1982, Arinami *et al.*, 1990). Defects in this gene have been associated with HDL deficiencies (HUGO Gene Nomenclature Committee, 2011). This protein increases the efflux of cholesterol from tissue to liver where it is excreted. A few individuals were reported to produce a HDL ApoA-I protein variant called ApoA-I Milano, an abnormal and apparently more efficient apolipoprotein. It has low measured HDL-C levels yet very low rates of cardiovascular events even with high blood cholesterol values

(Franceschini *et al.*, 1981). Apo CI, CII and apo E also accumulate in the nascent HDL particle, which serves as a store for these apolipoproteins in circulation.

Phospholipids are transferred from macrophages by a specific transporter molecule known as ATP-binding cassette transport protein A1 (ABCA-1) into the core of the lipoprotein, and cholesterols are extracted from the cells by a transporter protein derived from macrophages in the sub-endothelial spaces of tissues; ABCG-1 transporter. The eventual maturation of the HDL particle is dependent on the lecithin: cholesterol acyltransferase (LCAT), an enzyme activated by apoAI, which catalyses the formation of cholesterol esters from cholesterol. Mobilization of free cholesterol and phospholipids from IDL and LDL continues until a matured, spherical HDL particle is formed. Endocytosis of the matured HDL into hepatocytes occurs and the cholesterol and cholesterol esters are transported via a facilitated transfer to distinct pools within the cell. The modified HDL particles are secreted back into circulation where they can further acquire cholesterol before they re-circulate to the liver. The complete reverse cholesterol transport occurs with the addition of apo E to the HDL particle which facilitates their uptake and catabolism.

2.9.2. Lipoprotein Lipase as an anti-atherogenic agent

Activation of lipoprotein lipase (LpL) activity has been reported to have anti-atherogenic activity. Lipoprotein lipase (LpL) is a rate-limiting enzyme found on the surface of endothelial cells. It is polypeptide with 839 amino acids and an extracellular domain which binds to apo B100 and apo E. LpL catalyses the hydrolysis of the triacylglycerol (TAG) component of circulating chylomicrons and very low density lipoproteins (VLDL). The enzyme digests the TAG to fatty acids and monoglycerides. This provides non-esterified fatty acids and 2-monoacylglycerol which can be utilised immediately by cells for energy production or synthesis of other lipids. Unutilized fatty acids may be bound to circulating albumin and released slowly to meet future cellular requirements. Glycerol produced from LpL activity is transported back to the liver and kidneys, where it is converted to dihydroxyacetone phosphate in the alternative glycolytic pathway. The fatty acids from LpL activity in the muscle may diffuse into cells to be oxidized to two-carbon units or used to re-synthesis TAG which are stored in adipose cells (Clee *et al.*, 2000; Tsutsumi, 2003). Significant LpL activity occurs in muscle, adipose tissue and lactating mammary glands. Accumulation of VLDL remnants (IDL with apo B100 and apo E are converted to LDL with further loss of triacylglycerols. Both carrier proteins are necessary for recognition of IDL and LDL by the LDL receptors in the liver, after which they are taken up into hepatocytes by endocytosis and catabolized.

Research carried out over the past two decades have not only established a central role for LpL in the overall lipid metabolism and transport but have also identified additional, non-catalytic functions of the enzyme. Furthermore, abnormalities in LpL function have been found to be associated with a number of pathophysiological conditions, including atherosclerosis, chylomicronaemia, obesity, Alzheimer's disease, and dyslipidaemia associated with diabetes, insulin resistance, and infection (Mead *et al.*, 2002).

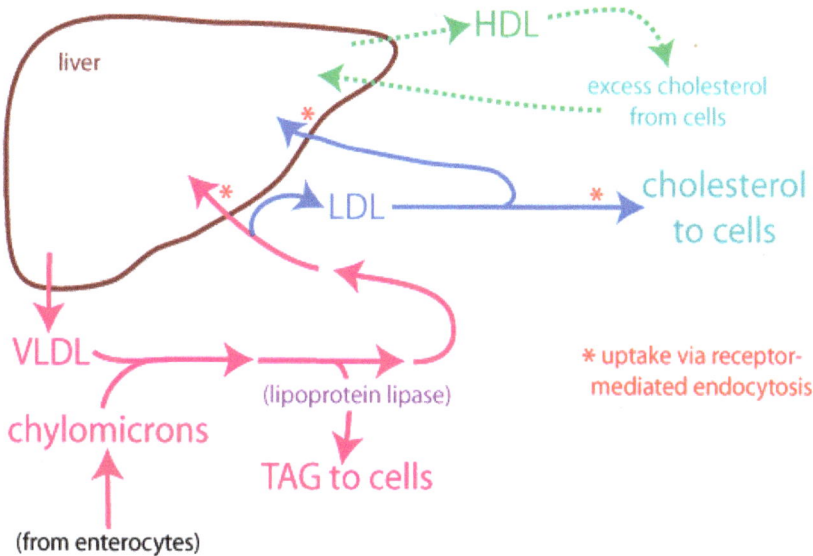

Figure 3. Summary of the fate of Lipoprotein sub-fractions (Adapted from
http://courses.washington.edu/conj/bess/cholesterol/liver.html)

LpL encodes lipoprotein lipase, which is expressed in heart, muscle, and adipose tissue. LpL
functions as a homodimer, and has the dual functions of triglyceride hydrolysis and
ligand/bridging factor for receptor-mediated lipoprotein uptake. Through catalysis, VLDL is
converted to IDL and then to LDL. Severe mutations that cause LpL deficiency result in type
I hyperlipoproteinemia, while less extreme mutations in LpL are linked to many disorders
of lipoprotein metabolism.

LpL isozymes are regulated differently depending on the tissue. For example, insulin is
known to activate LpL in adipocytes and its placement in the capillary endothelium. By
contrast, insulin has been shown to decrease expression of muscle LpL (Kiens *et al.*, 1989).
The form that is in adipocytes is activated by insulin, whereas that in muscle and
myocardium is activated by glucagon and epinepherine. This helps to explain why during
fasting, LpL activity increases in muscle tissue and decreases in adipose tissue. After
feasting, the opposite occurs (Braun and Severson, 1992; Mead *et al.*, 2002).

The concentration of LpL displayed on endothelial cell surface cannot be regulated by
endothelial cells, as they neither synthesize nor degrade LpL. Instead, this regulation occurs
by managing the flux of LpL arriving at the lipolytic site and being released into circulation
attached to lipoproteins (Braun and Severson, 1992; Goldberg, 1996). The typical
concentration of LpL in plasma is in the nanomolar range. Lipoprotein lipase deficiency
leads to hypertriglyceridemia (elevated levels of triglycerides in the bloodstream) and

decreased high density lipoprotein activity (Clee *et al.*, 2000; Tsutsumi, 2003; Okubo *et al.*, 2007). Diets high in refined carbohydrates have been shown to cause tissue-specific overexpression of LpL. This has been implicated in tissue-specific insulin resistance and consequent development of type 2 diabetes mellitus & obesity.

2.9.3. Influence of Hormones on plasma LDL-C and Lp(a) levels

Several studies have reported conflicting reports on the effect of hormonal replacement therapy on plasma LDL-C and Lp(a) levels (Taskinen *et al.*, 1996; Shlipak *et al.*, 2000; Vigna *et al.*, 2002). In a cohort study conducted by Danik *et al.* (2008), the effect of hormone replacement therapy (HT) on Lp(a) and cardiovascular risk was investigated. It was reported that the relationship of high Lp(a) levels with increased cardiovascular disease is modified by hormonal therapy. These data suggest that the predictive utility of Lp(a) is markedly attenuated among women taking HT and may inform clinicians' interpretation of Lp(a) values in such patients. It was noteworthy that the effect of hormonal therapy was observed only in women with high LDL cholesterol levels, in agreement with previous studies suggesting an interaction between Lp(a) and LDL cholesterol (Berglung and Anuurad, 2008).

3. Summary

Elevated serum LDL-C and low levels of HDL-C are known as major and independent risk factors for CHD. Small, dense lipoprotein sub-fractions have been reported to have atherogenic potentials, with particular reference to Lipoprotein(a) [Lp(a)], a variant of low density lipoprotein (LDL). Other atherogenic sub-fractions of lipoproteins are VLDL, LDL, and IDL. These sub-fractions are characterized by the presence of apolipoprotein B-100, with an additional apolipoprotein known as apo(a) in the Lp(a) structure. Apo(a) is structurally and functionally similar to plasminogen and it accounts for virtually all the genetic variability in plasma Lp(a) levels. Variation of length within the kringle 4-encoding region of the apo(a) gene may account for a greater proportion of the inter-individual variation in plasma Lp(a) concentrations, with a strong genetic involvement inherited as a single autosomal dominant trait.

In the course of the pathogenesis of atherosclerosis, oxidized LDL is taken up by macrophages and into endothelial cells. This leads to formation of atherosclerotic plaques which precedes development of CHD. Oxidized LDL is antigenic and titres of auto-antibodies against oxidized LDL in plasma can be used as indicator of a positive association with CHD. A positive association has also been established between plasma level of LDL, specifically oxidized LDL and other risk factors contributing to development of CHD. Such risk factors were identified as hypertriglyceridemia, reduced HDL-C levels, abdominal obesity and insulin resistance. Treatment of CHD can be achieved by lowering of plasma cholesterol levels which has been achieved by cholesterol-lowering therapy, suggesting that maintenance of low cholesterol levels may sufficiently prevent or reverse an established atherosclerosis. Increasing plasma levels of HDL-C has been reported to also be of benefit.

HDL-C has anti-inflammatory activity which may prevent oxidation of LDL and it plays a key role in reverse transportation of cholesterol from lipid-laden macrophages.

The enzyme Lipoprotein lipase (LpL) may be useful in chemotherapy or prophylaxis of CHD. LpL has anti-atherogenic activity by its dual functions of triglyceride hydrolysis and ligand/brigding factor for receptor-mediated lipoprotein uptake. Activation of the enzyme is dependent on the tissue, resulting in variability of its activity. Hormonal replacement therapy may also be of benefit to patients with CHD and related diseases, but reported on current findings are conflicting.

Author details

Adebowale Saba and Olayinka Oridupa
University of Ibadan, Nigeria

4. References

Albers, J. J.; Adolphson J. L. & Hazzard W. R. (1977). Radio-immunoassay of human plasma Lp(a) lipoprotein. *J. Lipid Res.* Vol 18, pp 331-338. ISSN 0022-2275

Arinami, T.; Hirano, T.; Kobayashi, K.; Yamanouchi, Y. & Hamaguchi, H. (1990). Assignment of the apolipoprotein A-I gene to 11q23 based on RFLP in a case with a partial deletion of chromosome 11, del(11)(q23.3----qter). *Hum. Genet.* Vol. 85, No. 1, pp 39–40. PMID 1972696

Armstrong, V.W., Walli, A.K. & Seidel, D. 1985. Isolation, characterization, and uptake in human fibroblasts of an apo (a)-free lipoprotein obtained on reduction of lipoprotein (a). *J Lipid Res*, Vol 26, pp 1314-1323. ISSN 0022-2275

Barter, P., Gotto, A.M., LaRosa, J.C., Maroni, J., Szarek, M., Grundy, M.S.S.M., Kastelein, J.J.P., Vera Bittner, V. & Fruchart J. (2007). HDL Cholesterol, Very Low Levels of LDL Cholesterol, and Cardiovascular Events. *N. Engl. J. Med.* Vol 357, pp 1301-1310 (2007). /doi:10.1056/NEJMoa064278

Berg, K. & Mohr, J. (1963). Genetics of the Lp system. *Acta Genet.* Vol. 13, pp 349-360. doi:10.1159/000151817

Berglund, L. & Ramakrishnan, R. (2004). Lipoprotein(a): an elusive cardiovascular risk factor. *Arterioscler. Thromb. Vasc. Biol.* Vol. 24, No.12, pp 2219–26. doi:10.1161/ 01.ATV.0000144010.55563.63. PMID 15345512

Berglund, L. & Anuurad, E. (2008). Role of lipoprotein(a) in cardiovascular diseases: Current and future perspectives. *J Am Coll Cardiol*, Vol 52, pp 132-134, doi:10.1016/j.jacc.2008.04.008.

Boerwinkle, E., Leffert, C.C., Lin, J., Lackner, C., Chiesa, G. & Hobbs, H.H. (1992). Apolipoprotein (a) gene accounts fro greater than 90% of the variation in plasm lipoprotein (a) concentrations. *J Clin Invest* Vol 90, pp 52-60. 0021-9738/92/07/0052/09

Bowden, J.F., Pritchard, P.H., Hill, J.S. & Frohlich, J.J. (1994). Lp(a) concentration and apo(a) isoform size. Relation to the presence of coronary artery disease in familial hypercholesterolemia. *Arterioscler Thromb,* Vol 14, pp 1561. doi: 10.1161/01.ATV.14.10.1561

Braun, J.E. & Severson, D.L. (1992). Regulation of the synthesis, processing and translocation of lipoprotein lipase. *Biochem J* Vol. 287, No.2, (October, 1992) pp 337–47. PMC1133170

Breslow, J.L., Ross, D., McPherson, J., Williams, H., Kurnit, D., Nussbaum, A.L., Karathanasis, S.K. & Zannis, V.I. (1982). Isolation and characterization of cDNA clones for human apolipoprotein A-I. *Proc. Natl. Acad. Sci. U.S.A.* Vol. 79, No. 22, pp 6861–5. doi:10.1073/pnas.79.22.6861. PMC 347233. PMID 6294659. http://www.pubmedcentral.nih.gov/articlerender.fcgi?tool=pmcentrez&artid=347233.

Brown, M.S. & Goldstein, J.L. (2009). Cholesterol feedback: from Schenheimer's bottle to Scap's MELADL. *J. Lipid Res.* (April 2009), Vol. 50, S15-S27. doi:10.1194/jlr.R800054-JLR200

Brunner, C., Lobentanz, E.M., Pethö-Schramm, A., Ernst, A., Kang, C., Dieplinger, H., Müller, H.J. & Utermann, G. (1996). The number of identical kringle IV repeats in apolipoprotein(a) affects its processing and secretion by HepG2 cells. *J. Biol. Chem.* Vol. 271, No. 50, pp 32403–10. doi:10.1074/jbc.271.50.32403. PMID 8943305.

Brunzell, J.D., Davidson, M., Furberg, C.D., Goldberg, R.B., Howard, B.V., Stein, J.H. & Witztum, J.L. (2008). Lipoprotein management in patients with cardiometabolic risk: Consensus conference report from the American Diabetes Association and the American College of Cardiology Foundation. *J Am Coll Cardiol* Vol. 51, pp 1512-1524. doi:10.1016/j.jacc.2008.02.034. Available from http://content.onlinejacc.org/cgi/content/full/51/15/1512

Carmena, R., Duriez, P. & Fruchart, J.C. (2004). Atherogenic lipoprotein particles in atherosclerosis. *Circulation* Vol. 109, pp 2-7. doi: 10.1161/01.CIR.0000131511.50734.44. Available from http://circ.ahajournals.org/content/109/23_suppl_1/III-2

Chien, K.L., Hsu, H.C., Su, T.C., Sung, F.C., Chen, M.F. & Lee, Y.T. (2008). Lipoprotein(a) and Cardiovascular Disease in Ethnic Chinese: The Chin-Shan Community Cardiovascular Cohort Study. *Clinical Chemistry* Vol. 54, pp 285-291, 2008. doi:10.1373/clinchem.2007.090969

Clee, S.M., Bissada, N., Miao, F., Miao, L., Marais, A.D., Henderson, H.E., Steures, P., McManus, J., McManus, MCManus, B., LeBoeuf, R.C., Kastelein, J.J.P. & Hayden, M.R. (2000). Plasma and vessel wall lipoprotein lipase have different roles in atherosclerosis. *J Lipid Res* Vol. 41, pp 521-531. Available from http://www.jlr.org/content/41/4/521.abstract

Danesh, J., Collins, R. & Peto, R.. (2000). Lipoprotein(a) and Coronary Heart Disease: Meta-Analysis of Prospective Studies. *Circulation.* Vol. 102, pp 1082-1085 doi: 10.1161/01.CIR.102.10.1082

Danik, S., Rifai, N., Buring, J.E. & Ridker P.M. (2008). Lipoprotein(a), Hormone Replacement Therapy, and Risk of Future Cardiovascular Events. *J Am Coll Cardiol,* Vol. 52, pp 124-131, doi:10.1016/j.jacc.2008.04.009. ISSN 0735-1097/08.

Di Vizio, D., Solomon, K.R. & Freeman, M.R. (2008). Cholesterol and cholesterol-rich membranes in prostate cancer: an update. *Tumori* Vol 5, pp 633-639. Available from

http://www.tumorionline.it/allegati/00386_2008_05/fulltext/1%20-
%20Di%20Vizio%20%28633-639%29.pdf

Duffy, D. & Rader, D.J. (2009). Update on strategies to increase HDL quantity and function. Nat Rev Cardiol; Vol. 6, pp 455-463. Available from www.ncbi.nlm.nih.gov

Eckel, R.H. (1997). Obesity and heart disease: a statement for healthcare professionals from the Nutrition Committee, American Heart Association. *Circulation*.Vol. 96, pp 3248 – 3250. doi: 10.1161/01.CIR.96.9.3248. Available from http://circ.ahajournals.org/content/96/9/3248.full

Fahy, E., Subramaniam, S., Murphy, R., Nishijima, M., Raetz, C., Shimizu, T., Spener, F., Van Meer, G., Wakelam, M. & Dennis, E.A. (2009). Update of the LIPID MAPS comprehensive classification system for lipids. *Journal of Lipid Research* Vol. 50, S9–S14. doi:10.1194/jlr.R800095-JLR200. PMID 19098281

Fredrickson, D.S., Lees, R.S. (1965). A system for phenotyping hyperlipoproteinaemia. *Circulation* Vol 31, No. 3, pp 321-327. Doi: 10.1161/01.CIR.31.3.321. PMID 14262568

Fletcher, G.F., Balady, G., Blair, S.N., Blumenthal, J., Caspersen, C., Chaitman, B., Epstein, S., Froelicher, E.S.S., Froelicher, V.F., Pina, I.L. & Pollock, M.L. (1996). Statement on exercise: benefits and recommendations for physical activity programs for all Americans: a statement for health professionals by the Committee on Exercise and Cardiac Rehabilitation of the Councilon Clinical Cardiology, American Heart Association. *Circulation*, Vol. 94, pp 857– 862. doi: 10.1161/01.CIR.94.4.857. Available from http://circ.ahajournals.org/content/94/4/857.long

Franceschini, G., Sirtori, M., Gianfranceschi, G. & Sirtori, C.R. (1981). Relation between the HDL apoproteins and AI isoproteins in subjects with the AIMilano abnormality. *Metab. Clin. Exp.* Vol 30, No. 5, pp 502–9. doi:10.1016/0026-0495(81)90188-8. PMID 6785551.

Garmendia, F. (2003). Advances in the knowledge and treatment of dyslipoproteinnemias. *An. Fac. Med.* Vol 64, No. 2, pp 101-106.

Garvey, W.T., Kwon, S., Zheng, D., Shaughnessy, S., Wallace, P., Hutto, A., Pugh, K., Jenkins, A.J., Klein, R.L. & Liao Y. (2003). Effects of insulin resistance and type 2 diabetes on lipoprotein subclass particle size and concentration determined by nuclear magnetic resonance. *Diabetes.* Vol. 52, pp 453–462. Available from http://diabetes.diabetesjournals.org/content/52/2/453.full.pdf

Goldberg, I.J. (1996). Lipoprotein lipase and lipolysis: central roles in lipoprotein metabolism and atherogenesis. J *Lipid Res* Vol. 37, No. 4, pp 693–707. Available from http://www.ncbi.nlm.nih.gov/pubmed/8732771

Guo, H.C., Michel, J.B., Blouquit, Y. & Chapman, M. J. (1991). Lipoprotein(a) and apolipoprotein(a) in a new world monkey, the common marmoset (callithrix jacchus): Association of variable plasma lipoprotein(a) levels with a single apolipoprotein(a) isoform. *Arterioscler. Thromb.* Vol. 11, pp 1030-1041. doi:10.1161/01.ATV.11.4.1030 Available from http://atvb.ahajournals.org/content/11/4/1030

Grundy, S.M., Paternak, R., Greenland, P., Smith, S. & Fuster, V. (1999). Assessment of cardiovascular risk by use of multiple-risk-factor assessment equations. *Circulation* Vol. 100, pp 1481-1492. doi: 10.1161/01.CIR.100.13.1481. Available from http://circ.ahajournals.org/content/100/13/1481

Holvoet, P., Mertens, A., Verhamme, P., Bogaerts, K., Beyens, G., Verhaeghe, R., Collen, D., Muls, E. & Van de Werf, F. (2001). Circulating oxidized LDL is a useful marker for identifying patients with coronary artery disease. *Arterioscler Thromb Vasc Biol,.* Vol. 21, pp 844–848. doi: 10.1161/01.ATV.21.5.844. Available from http://atvb.ahajournals.org/content/21/5/844.full

HUGO Gene Nomenclature Committee (HGNC), (2011). APOA1 apolipoprotein A-I (*Homo sapiens*). Gene ID: 335, protein coding. http://www.ncbi.nlm.nih.gov/sites/entrez?Db=gene&Cmd=ShowDetailView&TermToSe arch=335 Accessed on 9 November, 2011.

Hulthe, J. & Fagerberg, B. (2002). Circulating oxidized LDL is associated with subclinical atherosclerosis development and inflammatory cytokines (AIR Study). *Arterioscler Thromb Vasc Biol,* Vol. 22, pp 1162–1167. doi: 10.1161/01.ATV.0000021150.63480.CD. Available from http://atvb.ahajournals.org/content/22/7/1162.abstract?ijkey=80aaca3f2daac9fbc15623e16 18c678b0a8a2db5&keytype2=tf_ipsecsha

Ikonen, E. (2008). Cellular cholesterol trafficking and compartmentalization. *Nature Rev. Mol. Cell Biol.* Vol. 9, pp 125-138. doi:10.1038/nrm2336

Iselius, L., Dahlen, G. H., De Faire, U. & Lundman, T. (1981). Complex segregation analysis of the Lp(a)/pre-.l3-lipoprotein trait. *Clin. Genet.*Vol. 20, pp 147-151. DOI: 10.1111/j.1399-0004.1981.tb01820.x. available from http://onlinelibrary.wiley.com/doi/10.1111/j.1399-0004.1981.tb01820.x/abstract

Kiens, B., Lithell, H., Mikines, K.J. & Richter, E.A. (1989). Effects of insulin and exercise on muscle lipoprotein lipase activity in man and its relation to insulin action. *J. Clin. Invest.* Vol. 84, No. 4, pp 1124–9. 0021-9738/89/10/1124/06. Available from http://www.ncbi.nlm.nih.gov/pmc/articles/PMC329768/pdf/jcinvest00485-0080.pdf

Khera, A.V., Cuchel, M., de la Llera-Moya, M., Rodrigues, A., Burke, M.F., Jafri, K., French, B.C., Phillips, J.A., Mucksavage, M.L., Wilensky, R.L., Mohler, E.R., Rothblat, G.H. & Rader, D.J. (2011). Cholesterol Efflux Capacity, High-Density Lipoprotein Function, and Atherosclerosis. *N Engl J Med*, Vol. 364, pp127-135. PMID 21226578. Available from www.ncbi.nlm.nih.gov/pubmed/21226578

Kraft, H.G., Lingenhel, A., Köchl, S., Hoppichler, F., Kronenberg, F., Abe, A., Mühlberger, V., Schönitzer, D. & Utermann, G. (1996). Apolipoprotein(a) kringle IV repeat number predicts risk for coronary heart disease. *Arterioscler Thromb Vasc Biol*, Vol. 16, pp 713. PMID 8640397. Available from http://www.ncbi.nlm.nih.gov/pubmed/8640397

Lackner, C., E. Boerwinkle, Leffert, Rahmig, T. & Hobbs, H. H. (1991). Molecular basis of sapolipoprotein(a) isoform size heterogeneity as revealed by pulsed-field gel electrophoresis. *J. Clin. Invest.*, Vol. 87, pp 2077-2086. PMID 1645755. Available from http://www.ncbi.nlm.nih.gov/pubmed/1645755

Lamarche, B., Lemieux, I. & Despres, J.P. (1999). The small, dense LDL phenotype and the risk of coronary heart disease: epidemiology, pathophysiology and therapeutic aspects. *Diabetes Metab.* Vol. 25, pp 199–211. PMID 10499189. Available from http://www.ncbi.nlm.nih.gov/pubmed/10499189?dopt=Abstract

Leah, E. (2009). Cholesterol. Lipidomics Gateway. doi:10.1038/lipidmaps.2009.3. Available from http://www.lipidmaps.org/update/2009/090501/full/lipidmaps.2009.3.html.

Lehrke, M., Millington, S.C., Lefterova, M., Cumaranatunge, R.G., Szapary, P., Wilensky, R., Rader, D.J., Lazar, M.A. & Reilly, M.P. (2007). CXCL16 is a marker of inflammation, atherosclerosis, and acute coronary syndromes in humans. *J Am Coll Cardiol*, Vol. 49, pp 442-449. doi:10.1016/j.jacc.2006.09.034. Available from http://content.onlinejacc.org/cgi/content/full/49/4/442

Leinonen, J.S., Rantalaiho, V., Laippala, P., Wirta O, Pasternack A, Alho H, Jaakkola O, Yla-Herttuala S, Koivula T & Lehtimaki T. (1998). The level of autoantibodies against oxidized LDL is not associated with the presence of coronary heart disease or diabetic kidney disease in patients with non-insulindependent diabetes mellitus. *Free Radic Res.*, Vol. 29, pp 137–141. PMID 9790515. Available from http://www.ncbi.nlm.nih.gov/pubmed/9790515?dopt=Abstract

Loscalzo, J., Weinfeld, M., Fless, G.M. & Scanu, A.M. (1990). Lipoprotein(a), fibrin binding, and plasminogen activation. *Arteriosclerosis*, Vol. 10, (March/April, 1990), pp 240-245. Available from atvb.ahajournals.org/content/10/2/240.full.pdf

McLean, J.W., Tomlinson, J.E., Kuang, W.J., Eaton, D.L., Chen, E.Y., Fless, G.M., Scanu, A.M. & Lawn, R.M. (1987). cDNA sequence of human apolipoprotein(a) is homologous to plasminogen. *Nature* Vol. 300, No. 12, (November, 1987) pp 132-137.

Mead, J.R., Irvine, S.A. & Ramji, D.P. (2002). Lipoprotein lipase: structure, function, regulation, and role in disease. *J Mol Med*, Vol. 80, No. 12, pp 753–69. PMID 12483461. Available from http://www.ncbi.nlm.nih.gov/pubmed/12483461

Muller-Roeber, B. & Pical, C. (2002). Inositol Phospholipid Metabolism in Arabidopsis. Characterized and Putative Isoforms of Inositol Phospholipid Kinase and Phosphoinositide-Specific Phospholipase C. *Plant Physiol.*, Vol. 130, No. 1, pp 22-46 doi: 10.1104/pp.004770

Okubo, M., Horinishi, A., Saito, M., Ebara, T., Endo, Y., Kaku, K., Murase, T. & Eto, M.A. (2007). A novel complex deletion-insertion mutation mediated by Alu repetitive elements leads to lipoprotein lipase deficiency. *Mol. Genet. Metab.*, Vol. 92, No. 3, pp 229–33. PMID 17706445. Available from http://www.ncbi.nlm.nih.gov/pubmed/17706445

Palabrica, T.M., Liu, A.C., Aronovitz, M.J., Furie, B., Lawn, R.M. & Furie, B.C. (1995). Antifibrinolytic activity of apolipoprotein(a) in vivo: human apolipoprotein(a) transgenic mice are resistant to tissue plasminogen activator-mediated thrombolysis. *Nat Med*, Vol. 1, pp 256. Available from http://www.nature.com/naturemedicine

Patrick, L. & Uzick, M. (2001). Cardiovascular disease: C-reactive protein abd the inflammatory disease paradigm: HMG-CoA reductase inhibitors, alpha-tocopherol, red yeast rice and olive oil polyphenols. A review of the literature. *Altern Med Rev* Vol. 6, No. 3, pp 248-271. Available from http://www.thorne.com/altmedrev/.fulltext/6/3/248.pdf

Pearson, A., Budin, M. & Brocks, J.J. (2003). Phylogenetic and biochemical evidence for sterol synthesis in the bacterium Gemmata obscuriglobus. *Proc. Natl. Acad. Sci. U.S.A.* Vol. 100, No. 26, pp 15352–7. doi:10.1073/pnas.2536559100. PMC 307571. PMID 14660793.

Rader, D.J., Cain, W., Zech, L.A., Usher, D. & Brewer, H.B. (1993). Variation in lipoprotein(a) concentrations among individuals with the same apolipoprotein (a) isoform is determined by the rate of lipoprotein(a) production. *J. Clin. Invest.* Vol. 91, No. 2, pp 443–7. doi:10.1172/JCI116221. PMC 287951. PMID 8432853

Sandholzer, C., Hallman, D.M., Saha, N., Sigurdsson, G., Lackner, C., Császár, A., Boerwinkle, E. & Utermann, G. (1991). Effects of the apolipoprotein(a) size polymorphism on the lipoprotein(a) concentration in 7 ethnic groups. *Hum. Genet.* Vol. 86, No. 6, pp 607–14. doi:10.1007/BF00201550. PMID 2026424.

Schreiner, P.J., Morrisett, J.D., Sharrett, A.R., Patsch, W., Tyroler, H.A., Wu, K. & Heiss, G. (1993). Lipoprotein(a) as a risk factor for preclinical atherosclerosis. *Arterioscler. Thromb.*, Vol. 13, No. 6, pp 826–33. doi:10.1161/01.ATV.13.6.826. PMID 8499402

Schoenhagen, M. (2006). Current developments in atherosclerosis research. Misra Schoenhagen (editor). Published by Nova Science Inc., New York. ISBN 1-59454-493-X. Available from http://www.novapublishers.com

Sherer, Y., Tenenbaum, A., Praprotnik, S., Shemesh J, Blank M, Fisman EZ, Harats D, George J, Levy Y, Peter JB, Motro M, Shoenfield Y. (2001). Coronary artery disease but not coronary calcification is associated with elevated levels of cardiolipin, beta-2-glycoprotein-I, and oxidized LDL antibodies. *Cardiology*, Vol. 95, pp 20–24. PMID 11385187. Available from http://www.ncbi.nlm.nih.gov/pubmed/11385187?dopt=Abstract

Shlipak, M.G., Simon, J.A., Vittinghoff, E., Lin, F., Barrett-Connor, E., Knopp, R.H., Levy, R.I. & Hulley, S.B. (2000). Estrogen and progestin, lipoprotein(a), and the risk of recurrent coronary heart disease events after menopause. *JAMA*, Vol. 283, pp 1845-1852. URL: http://jama.ama-assn.org/cgi/content/abstract/283/14/1845

Soedamah-Muthu, S.S., Chang, Y.F., Otvos, J., Evans, R.W. & Orchard, T.J. (2003). Lipoprotein subclass measurements by nuclear magnetic resonance spectroscopy improve the prediction of coronary artery disease in type 1 diabetes: a prospective report from the Pittsburgh Epidemiology of Diabetes Complications Study. *Diabetologia*, Vol. 46, pp 674–682. PMID 12743701. Available from http://www.ncbi.nlm.nih.gov/pubmed/12743701?dopt=Abstract

Sotiriou, S.N., Orlova, V.V., Al-Fakhri, N., Ihanus, E., Economopoulou, M., Isermann, B., Bdeir, K., Nawroth, P.P., Preissner, K.T., Gahmberg, C.G., Koschinsky, M.L. & Chavakis, T. (2006). Lipoprotein(a) in atherosclerotic plaques recruits inflammatory cells through interaction with Mac-1 integrin". *FASEB J.* Vol. 20, No 3, pp 559–61. PMID 16403785. Available from http://www.ncbi.nlm.nih.gov/pubmed/16403785

Steinbrecher, U.P., Parthasarathy, S., Leake, D.S., Witztum, J.L. & Steinberg, D. (1984). Modification of lowdensity lipoprotein by endothelial cells involves lipid peroxidation and degradation of low density lipoprotein phospholipids. *Proc Natl Acad Sci U S A.*, Vol. 81, pp 3883–3887. Available from http://www.pnas.org/content/81/12/3883.abstract?ijkey=fea45b830edf328270f68e98a3cc0 3d2bb008eff&keytype2=tf_ipsecsha

Tall, A.R. (2008). Cholesterol efflux pathways and other potential mechanisms involved in the athero-protective effect of high density lipoproteins. *J Intern Med*, Vol. 263, No. 3, pp 256-273. ISSN 0954-6820

Taskinen, M.R., Puolakka, J., Pyorala, T., Luotola, H., Bjorn, M., Kaariainen, J., Lahdenpera, S. & Ehnholm, C. (1996). Hormone replacement therapy lowers plasma Lp(a) concentrations. Comparison of cyclic transdermal and continuous estrogen-progestin regimens. *Arterioscler Thromb Vasc Biol*, Vol. 16, pp 1215-1221. doi: 10.1161/ 01.ATV.16.10.1215. Available from http://atvb.ahajournals.org/content/16/10/1215.full

Toshima, S., Hasegawa, A., Kurabayashi, M., Itabe, H., Takano, T., Sugano, J., Shimamura, K., Kimura, J., Michishita, I., Suzuki, T. & Nagai, R. (2000). Circulating oxidized low density lipoprotein levels: a biochemical risk marker for coronary heart disease. *Arterioscler Thromb Vasc Biol.*, Vol. 20, pp 2243–2247. doi: 10.1161/01.ATV.20.10.2243. Available from http://atvb.ahajournals.org/content/20/10/2243.abstract?ijkey=4a5cafa55c6ce067f39ca5eb 803d7801adb5f339&keytype2=tf_ipsecsha

Toth, P. (2005). The good cholesterol High-Density Lipoprotein. *Circulation*, Vol. 111, No. 5, pp e89-e91. Available from http://circ.ahajournals.org/cgi/content/full/111/5/e89.

Tsutsumi, K. (2003). Lipoprotein lipase and atherosclerosis. *Curr Vasc Pharm*, Vol. 1, pp 11-17. ISSN 1570-1611/03

Utermann, G. (1989). The mysteries of lipoprotein(a). *Science*, Vol. 246, No. 4932, pp 904-10. Available from www.sciencemag.org

Vakkilainen, J., Steiner, G., Ansquer, J.C., Aubin, F., Rattier, S., Foucher, C., Hamsten, A &, Taskinen, M. (2003). Relationships between low-density lipoprotein particle size, plasma lipoproteins, and progression of coronary artery disease: the Diabetes Atherosclerosis Intervention Study (DAIS). *Circulation.*, Vol. 107, pp 1733–1737. doi: 10.1161/01.CIR.0000057982.50167.6E. Available from http://circ.ahajournals.org/content/107/13/1733.abstract?ijkey=401352270488f9ca9f3caafa f43179b7a352181c&keytype2=tf_ipsecsha

Vance, D.E. & Vance, J.E. (2002). Biochemistry of lipids, lipoproteins and membranes. Elsevier, Amsterdam. ISBN 978-0-444-53219-0.

Vigna, G.B., Donega, P., Zanca, R., Barban, A., Passaro, A., Pansini, F., Bonaccorsi, G., Mollica, G. & Fellin, R. (2002). Simvastatin, transdermal patch, and oral estrogen-progestogen preparation in early-postmenopausal hypercholesterolemic women: a randomized, placebo-controlled clinical trial *Metabolism*, Vol. 51, pp 1463-1470.

Wang, X., Briggs, M. R., Hua, X., Yokoyama, C., Goldstein, J. L. & Brown. M. S. (1993). Nuclear protein that binds sterol regulatory element of LDL receptor promoter: II. Purification and characterization. *J. Biol. Chem.* Vol. 268, pp 14497–14504.

White, A.L., Rainwater, D.L., Hixson, J.E., Estlack, L.E. & Lanford, R.E. (1994). Intracellular processing of apo(a) in primary baboon hepatocytes. *Chem. Phys. Lipids*, Vol. 67-68, pp 123–33. doi:10.1016/0009-3084(94)90131-7. PMID 8187206.

Wilson, P.W., D'Agostino, R.B., Levy, D., Belanger, A.M., Silbershatz, H. & Kannel, W.B. (1998). Prediction of coronary heart disease using risk factor categories. *Circulation.*, Vol. 97, pp 1837–1847. PMID 9603539. Available from http://www.ncbi.nlm.nih.gov/pubmed/9603539

Young, I.S. & McEneny, J. (2001). Lipoprotein oxidation and atherosclerosis. Biochem Soc Trans 29 (2): 358-362. PMID 11356183. Available from http://www.ncbi.nlm.nih.gov/pubmed/11356183

Yokoyama, C., Wang, X., Briggs, M. R., Admon, A. , Wu, J., Hua, X., Goldstein, J. L. & Brown, M. S. (1993). SREBP-1, a basic helix-loop-helix leucine zipper protein that controls transcription of the LDL receptor gene. *Cell*. Vol. 75, pp 187–197. PMID 8402897. Available from http://www.ncbi.nlm.nih.gov/pubmed/8402897

Zenker, G., Koltringer, P., Bone, G., Niederkorn, K., Pfeiffer, K. & Jurgens, G. (1986). Lipoprotein(a) as a strong indicator for cerebrovascular disease. *Stroke*, Vol. 17, pp 942-945. doi: 10.1161/01.STR.17.5.942. Available from http://stroke.ahajournals.org/content/17/5/942.

Zioncheck, T.F., Powell, L.M., Rice, G.C., Eaton, D.L. & Lawn, R.M. (1991). Interaction of recombinant apolipoprotein(a) and lipoprotein(a) with macrophages. *J Clin Invest*, Vol. 87, pp 767. 0021-9738/91/03/0767/05

Chapter 8

Lipoproteins Impact
Increasing Cardiovascular Mortality

Jelena Umbrasiene, Ruta-Marija Babarskiene and Jone Vencloviene

Additional information is available at the end of the chapter

1. Introduction

Cardiovascular diseases – the main result of the generalized atherosclerosis are the leading cause of global mortality all over the world [1,2]. The number of atherosclerotic diseases - an ischemic stroke, coronary heart disease and peripheral artery disease increases every year [1]. Possibly, due to increase in the population age, better health care and improved survival the prevalence of heart diseases is still so high [3]. The cardiovascular mortality in the most developed countries also is very high [2,3]. About half of all deaths occures due to cardiovascular diseases, it's an over 4,35 million deaths each year in the 53 member states of the World Health Organization European Region and more than 1,9 million deaths each year in the European Union [2]. Moreover there is a 35 billion euros damage due to working people production loss regarding to cardiovascular morbidity and mortality [2]. The cardiovascular mortality is still a problem not only in the European Union, but in the other developed countries as well. Atherosclerotic coronary artery disease was the most common cause of death in the United States in 2004. Men were more often affected, than women by a ratio of 4:1 and after age of 70 by ratio 1:1 [4]. In 2000 about 37 % of death in Canada were due to cardiovascular diseases [3]. They are still the main cause of mortality in Lithuania, as in the older Western European countries as well [5,6]. At the last decade, cardiovascular morbidity and mortality in Lithuania has not declined (Figure 1) [7,8].

In 2008 in Lithuania standartized cardiovascular mortality rate was 520,1 per 100 000 population (Figure 2) [8]. Although in the last years cardiovascular mortality has a tendency to decrease, it's still very high [7]. Lithuanian mortality from coronary artery disease rate in 2008 was 321,29 per 100 000 population (Figure 3) [7,8]. By the statistic data from the Lithuanian Institute of Hygiene, in 2011 56,3% of the people have died from cardiovascular disease in Lithuania. In 2011, 20944 men and 20093 women have died, 47,7% and 62,7% due to coronary artery disease respectively [9].

The main cardiovascular disease - coronary heart disease - highly associated with an increased cardiovascular mortality, hospitalisation and patients disability, significantly raising the cost of medical care [6]. In 2009 it was 4283,39 per 100 000 population hospital discharge for cardiovascular diseases and 1311,8 for coronary artery disease in Lithuania (Figure 4,5) [8].

In 2000 in Canada 7,3 billion dollars (17%) of total direct health care costs and 12,3 billion (14,5%) dollars of total indirect health care costs for all disease categories were attributed to cardiovascular diseases [3]. In the European Union, the economic cost of cardiovascular diseases in direct and indirect healthcare goes to 192 billion euros annually [1]. A total annual cost for person is vary from 50 euros in Malta to 600 euros in Germany, and 372 euros in average [2].

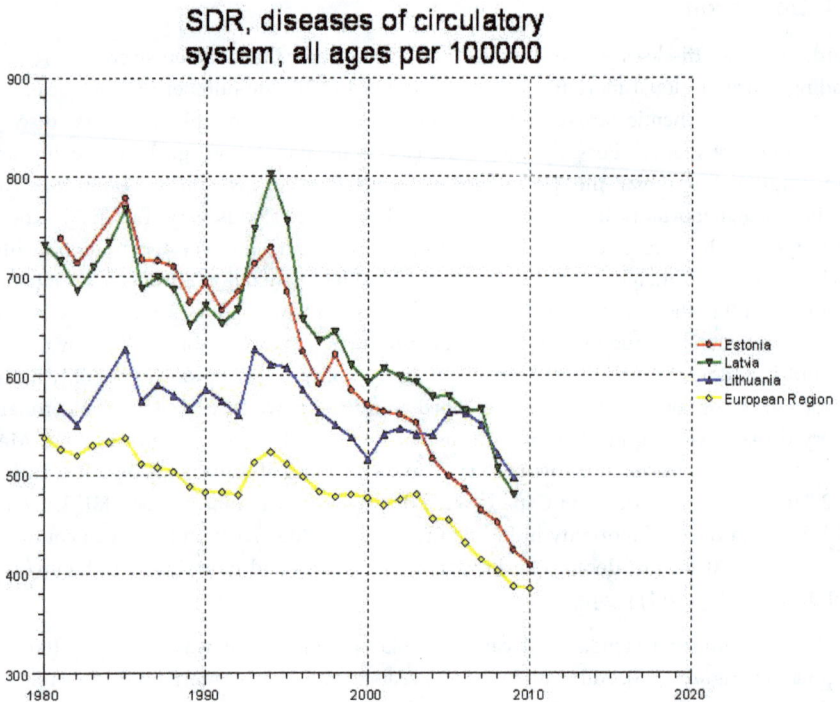

SDR – standartized death rate

Figure 1. Age standartized cardiovascular mortality rate for Baltic States and all European Region dynamic.

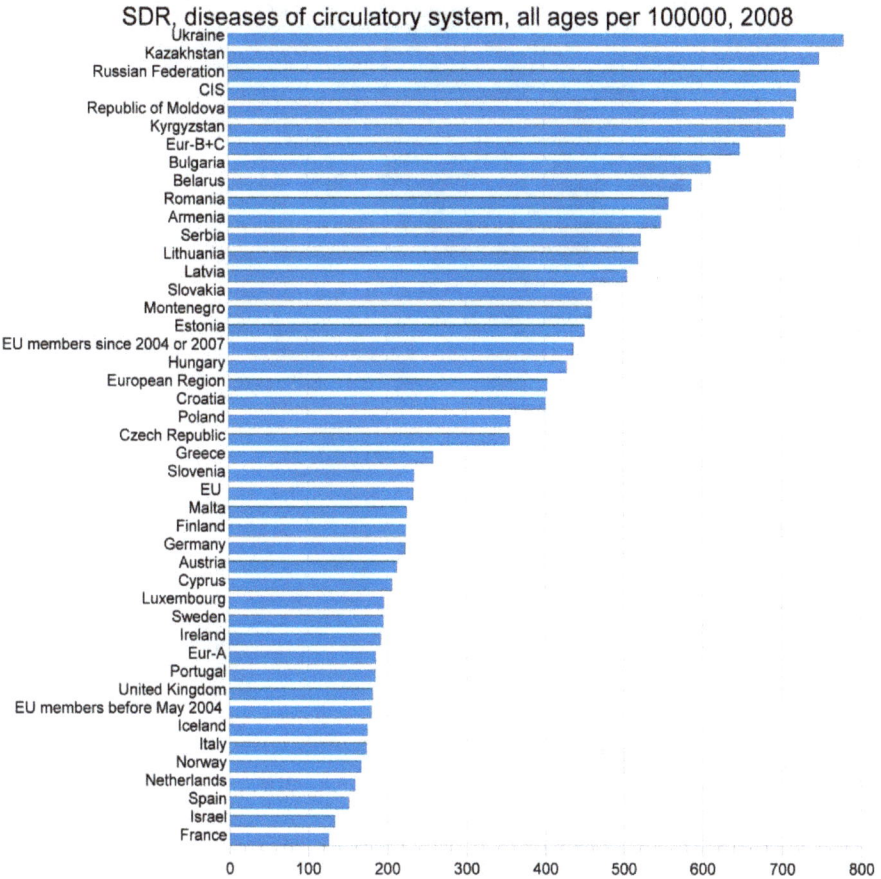

SDR, diseases of circulatory system, all ages per 100000, 2008

Ukraine
Kazakhstan
Russian Federation
CIS
Republic of Moldova
Kyrgyzstan
Eur-B+C
Bulgaria
Belarus
Romania
Armenia
Serbia
Lithuania
Latvia
Slovakia
Montenegro
Estonia
EU members since 2004 or 2007
Hungary
European Region
Croatia
Poland
Czech Republic
Greece
Slovenia
EU
Malta
Finland
Germany
Austria
Cyprus
Luxembourg
Sweden
Ireland
Eur-A
Portugal
United Kingdom
EU members before May 2004
Iceland
Italy
Norway
Netherlands
Spain
Israel
France

0   100   200   300   400   500   600   700   800

SDR – standardized death rate

Figure 2. Age standardized cardiovascular mortality rate per 100 000 population, 2008.

SDR, ischaemic heart disease, all ages per 100000, 2008

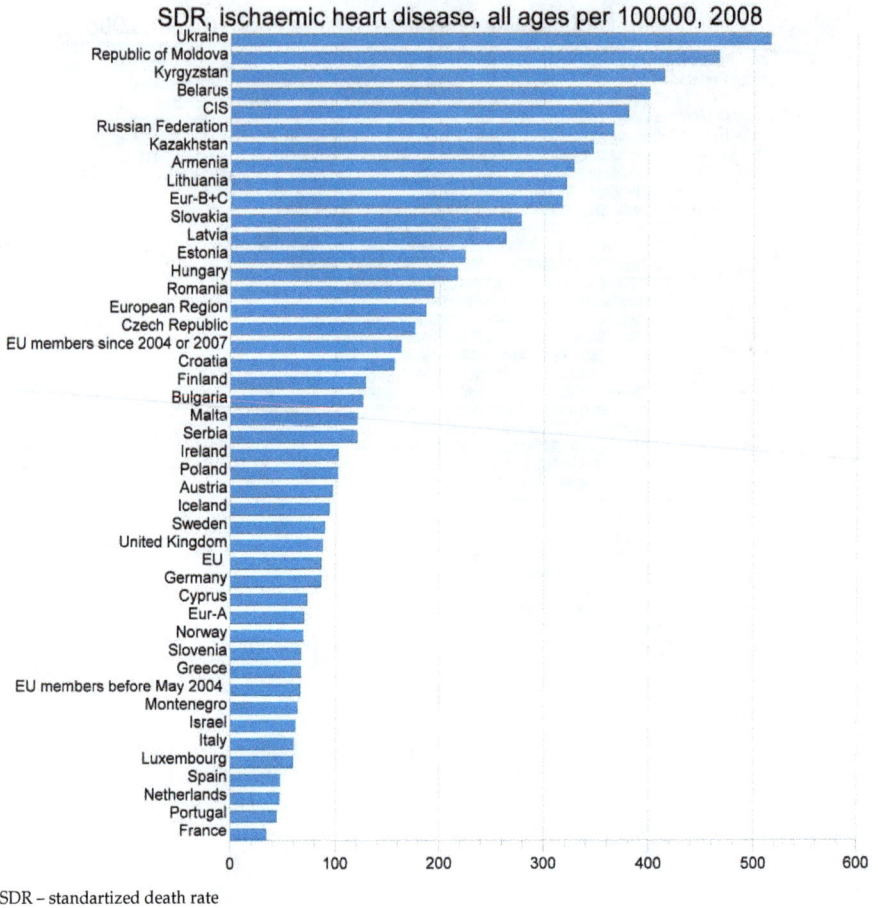

SDR – standartized death rate

Figure 3. Age standartized mortality rate for coronary artery disease per 100 000 population, 2008.

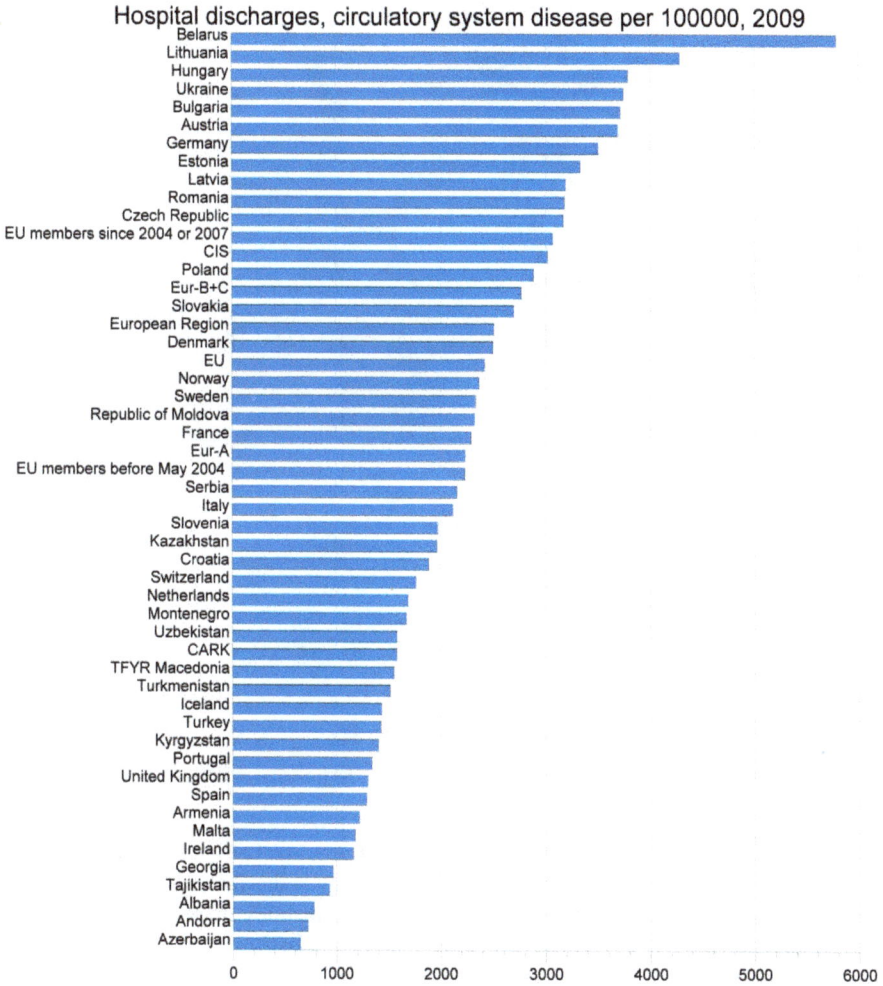

Figure 4. Hospital discharges for the patients with cardiovascular diseases in 2009, per 100 000 population.

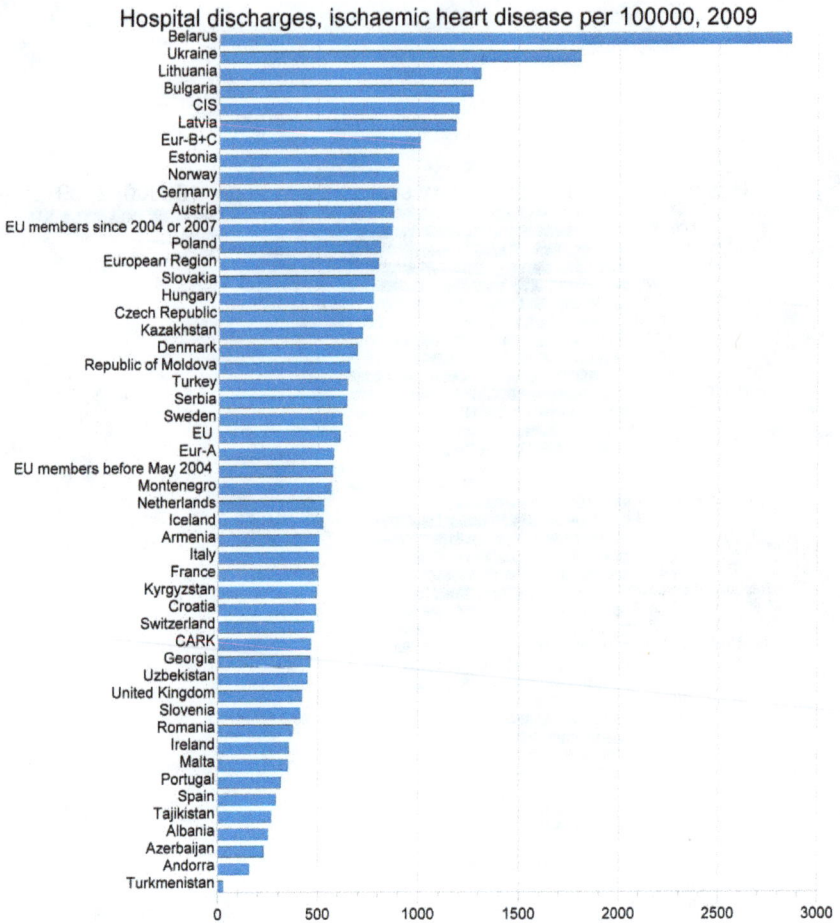

Figure 5. Hospital discharges for the patients with coronary artery disease in 2009, per 100 000 population.

Epidemiological studies have evaluated a number of important risk factors for coronary artery disease, such as positive family history, particulary in the age less than 40 for men, and 50 for women, age, male gender, blood lipids abnormalities, diabetes, hypertension, loss of physical activity, smoking and others, not so substantial (high sensitivity C-reactive protein, hyperfibrinogenemia etc.) [4,6,10-12]. Reducing one or more of these risk factors reduces the risk of major cardiac event accordingly [4]. There are a lot of evidence that lipoprotein disorder is the main pathogenesis of atherosclerosis. This relationship that was estimated century ago by Anitschkow is still important today [3,13]. Variuos epidemiological studies demonstrated a strong association between dyslipoproteinemia and coronary heart disease. There is a strong relation between serum cholesterol concentration level and the coronary heart disease risk [14]. The Multiple Risk Factor Intervention Trial (MRFIT) in USA with 356222 men with different cardiovascular risk factors and 6-year follow-up period have shown that elevated total cholesterol blood concentration significantly increases cardiovascular risk [14,15]. In 2008, the authors published the report about the continuous follow-up for 25 years. The main finding was that total cholesterol is continuous and strong independent predictor for cardiovascular mortality. Estimated increased cardiovascular mortality risk at every total cholesterol level from 160 mg/dl (about 4,14 mmol/l) and higher [15]. Abnormal lipids metabolism or excessive intake of cholesterol especially with a genetic predisposition, initiates the atherosclerosis. A lot of clinical studies established total cholesterol and low density lipoprotein cholesterol are associated with a great risk of coronary heart disease. The reduce of total cholesterol by 10% decreasing the risk of ischemic heart disease by 25% within 5 years [16]. Low density lipoprotein cholesterol reduction not only decreases cardiovascular events, but reduce total mortality as well [3,17]. Furthermore, large randomized controlled clinical trials established the low density lipoprotein cholesterol lowering benefits [10]. It is proved, the reduce of low density lipoprotein cholesterol by 1 mmol/l, decreasing the risk of acute cardiac events by 20%, cardiovascular mortality by 22% [1,11,16-18]. Treatment of lipoproteins disorder also decrease the development of new lesion, regenerates endothelial function and signally reduce cardiovascular events in treated patients [4]. However, the data based on the National Health And Nutrition Examination Survey (NHANES) study from 2005-2008 have estimated that 71 million adults (33,5%) in the USA had elevated low density lipoprotein cholesterol level, but only 34 million (48,1%) were treated and 23 million (33,2%) had reached target low density lipoprotein value. Though, comparing this data to the data from NHANES study in 1999-2002, the number of people with elevated low density lipoprotein level treated with lipids-lowering medications increased from 28,4% to 48,1% between 1999-2002 and 2005-2008 periods. The prevalence of controled low density lipoprotein increased from 14,6% to 33,2% [17]. Although, statins significantly reduce low density lipoprotein cholesterol and coronary heart disease risk, substantial residual cardiovascular risk remains, even with very aggressive low density lipoprotein cholesterol values reduction [11,19,20]. However, atherosclerosis pathogenesis is multiple. It depends not only on low density lipoprotein cholesterol level, but also on genetic, environmental factors, infections, lifestyle factors and other diseases or condition [10-12]. More than a hundred different risk factors for atherosclerosis are estimated today. Although it is known many risk factors for coronary heart disease, the most of them are modifiable. Such as smoking cessation, treatment of dyslipidaemia, lowering of blood pressure can prevent the progression of atherosclerosis and major cardiovascular events [4]. One of the most important mechanisms of the atherosclerosis pathogenesis is Endothelial dysfunction [21]. In the early stages of atherosclerosis endothelian-dependent vasorelaxation disturbes due to oxidative stress and reduced nitric oxide bioavailability. Monocytes and T-lymphocytes adhesion occures. These inflammatory cells penetrate the cell wall, as well as lipid accumulation in the walls of blood vessels takes place. The inflammation and lipids accumulation make a plaque unstable, so it may occlude the vessel. Endothelial dysfunction is observed not only in the initial stage, but also in all other stages of atherosclerosis as well [21-23]. However, the main risk factors still are male gender and older age (more common in women in menopause), heredity, hypertension, diabetes, smoking, stress, obesity, lack of physical activity, elevated low density lipoprotein cholesterol and total cholesterol and decrease high density lipoprotein cholesterol levels [6,10-12]. Numerous epidemiological studies have found reduced high density lipoprotein cholesterol as an independent risk factor for cardiovascular disease [24]. The Framingham study evaluated 43-44% increasing coronary events in patients with high density lipoprotein cholesterol < 40 mg/dL (1,03 mmol/l) [25]. Patients whose high density lipoprotein cholesterol less than 0,9 mmol/l (35 mg/dL) have 8 times higher risk of cardiovascular disease, versus those, whose high density lipoprotein cholesterol more than 1,68 mmol/l (65 mg/dL) [26]. Studies demonstrates that declined high density lipoprotein cholesterol levels are relatively common in general population. 16-18% of men and 3-6% of women have a high density lipoprotein cholesterol level less than 0,9 mmol/l (35 mg/dL) [20]. Moreover, the reduced high density lipoprotein cholesterol level is a component of the metabolic syndrome – the great predictor of high cardiovascular risk. Experimental studies have found high density lipoprotein cholesterol as a potential antiatherogenic by following characteristics. Estimated high density lipoprotein cholesterol facilitates reverse cholesterol transport and delivers cholesterol from the smooth muscles into hepatic cholesterol uptake. So, harmfull atherogenic cholesterol parts, such as low density lipoprotein cholesterol, are catabolized and neutralized [27-29]. High density lipoprotein cholesterol acts as an antioxidant, reducing vascular oxidative stress and has anti-inflammatory properties, reducing vascular inflammation due to atherosclerosis. There are evidence high density lipoprotein cholesterol has a vasoprotective effect, facilitates blood vessel relaxation, play an important role in the inhibition of white blood cells chemotaxis and adhesion. Also it is known about an anti-apoptotic effect of high density lipoprotein cholesterol on endothelial cells. High density lipoprotein cholesterol enhances the proliferation and migration of Endothelial cells and endothelial progenitor cells and thereby promotes the restoration of the endothelium's integrity. Finally, it has an antiplatelet/profibrinolitic effect, in this way reducing platelet aggregation and inactivating coagulation cascade [20,27-29]. Despite the evidence that reduced high density lipoprotein cholesterol is associated with an increased cardiovascular morbidity and mortality, the major guidelines in cardiology still do not recommend to initiate the treatment of dyslipidemia on high density lipoprotein cholesterol.

So, dyslipoproteinemia is a major risk factor for atherosclerosis and coronary artery disease. Its' proper recognition and management can significantly reduce cardiovascular and total mortality rates [12]. Follow the American Heart Association and the National Heart, Lung and Blood Institute and the Adult Treatment Panel III guidelines it is recommended to start treat from the low density lipoprotein cholesterol. Recent clinical studies provide supporting evidence for low density lipoprotein cholesterol target values of less than 2,5 mmol/l (< 100 mg/dl) for the prevention of coronary artery disease for the high cardiovascular risk patients and less thant 1,8 mmol/l (< 70 mg/dl) for the very high cardiovascular risk patients [1]. Studies demonstrate the significant decrease of atherosclerosis with aggressive reduction of low density lipoprotein cholesterol level in patients with coronary artery disease [3]. Only achieved target low density lipoprotein cholesterol value it is recommended to take care of high density lipoprotein cholesterol. Studies evaluated, that high density lipoprotein cholesterol level more than 60 mg/dl (about 1,5 mmol/l) significantly reduce cardiovascular risk and can be named as „inverse risk factor" [21]. The target high density lipoprotein cholesterol is over 1,03 mmol/l (40 mg/dL) for men and more than 1,29 mmol/l (50 mg/dL) for women [20].

2. Lipoproteins disorder as a risk factor for cardiovascular mortality

Today there are more than one lipoproteins disorder classification. The Frederickson, Lees and Levy's one was based on the lipoprotein fraction after separation by electrophoresis. This classification recognized chylomicrons, very low density cholesterol and low density cholesterol. However, the main limitation of this classification, that it does not include high density lipoprotein cholesterol. That's why the World Health Organisation, the European Atherosclerosis Society and the National Cholesterol Education Program have classified lipoproteins disorder on the basis of the absolute plasma level of lipids (total cholesterol and trygliceride) and lipoprotein cholesterol level (low density lipoprotein cholesterol and high density lipoprotein cholesterol) [12,30]. This classification sustained on biochemical characteristics of lipoproteins and lipids. The plasma lipids do not circulate freely in plasma. They are bound to proteins and transported as macromolecular complexes called lipoproteins [24]. In these complexes lipids are surrounded by a stabilizing coat of phospholipid. There are five principal types of lipoprotein particles in the blood: very low density lipoproteins, intermediate density, low density, high density lipoproteins and chylomicrons. They are structurally different by electrophoretic mobility and density after separation in the ultracentrifuge and by the function [14,24]. The lipoprotein density depends on amount of fats contained within it [31].

Chylomicrons are the largest lipoproteins and synthesized in the small intestine from dietary fat and cholesterol [14,24,31]. They contain triglyceride from the intestine and a small amount of cholesterol. The main task of chylomicrons to transport the digestion products of dietary fat to the liver and peripheral tissue, where they are needed as a source of energy. In the circulation triglycerides are removed from chylomicrons via the action of lipoprotein lipase. If present in large amounts, such as after a fatty meal, chylomicrons cause the plasma to appear milky. Very low density lipoproteins are synthesized in the liver continuously and consists of triglyceride and cholesterol. Like chylomicrons they function primarily to distribute triglycerides to target sites such as adipose tissue and skeletal muscle where they are used for storage and energy [31]. It is the main body source of energy in prolonged fasting [14]. Like chylomicrons, they are removed due to lipoprotein lipase action. With removal of triglycerides and protein, very low density lipoproteins are converted to low density lipoproteins. High plasma levels of very low density lipoprotein cholesterol are to be found in familial hypertriglyceridaemia, diabetes mellitus, in people with a depressed thyroid function and in people with a high alcohol intake [31]. Intermediate density lipoproteins – one of the source of low density lipoproteins production. Last-mentioned are the main particles of lipids. They can deposit lipids into the arterial wall and initiate atherosclerosis. Low density lipoprotein cholesterol are cholesterol-rich particles. About 70% of plasma cholesterol find in this form. Low density lipoprotein cholesterol have a main role in transporting the cholesterol manufactured in the liver to the tissues, where it is used. When low density lipoprotein cholesterol binds to low density lipoprotein cholesterol receptors on the cell surface, low density lipoprotein cholesterol is taken into the cell and broken down into free cholesterol and amino acids. Disorders involving a defect in or lack of low density lipoprotein cholesterol receptors are usually characterised by high plasma cholesterol levels. In the case of the inherited familial hypercholesterolemia the cholesterol excess cannot be cleared efficiently from the blood and therefore accumulates, caused coronary heart disease. And the last particles – high density lipoproteins are produced in liver and intestine. They are composed of 50% protein, with phospholipid and cholesterol as the remainder [31]. They transport lipids away from the periphery. The transfer of pro-atherogenic particles, such as very low density lipoproteins to the liver for the reverse cholesterol transport is one of the most important role of high density lipoprotein cholesterol. In this process harmful pro-atherogenic particles are transporting from the periphery to the liver for the reverse cholesterol transport and neutralizing [14,21]. It is well known low density lipoprotein cholesterol is one of the major factor for the development of atheroma. Atherosclerotic plaque consist of accumulated intracellular and extracellular lipids, smooth muscle cells, connective tissue, and glycosaminglycans.

There are two main hypotheses to explain the pathogenesis of atherosclerosis: the lipid hypothesis and the chronic endothelial injury hypothesis. Both of them are interrelated. The endothelial dysfunction is an initial stage of atherosclerosis, occures due to oxidative stress and sub-endothelial accumulation of lipids. Low density lipoprotein cholesterol undergo oxidation and become local cytotoxic. Macrophages migrate into the sub-endothelial space, take up lipids and become "foam" cells. The earliest detectable lesion of atherosclerosis is the fatty strip. This strip consists of foam cells full of lipids. As the process progress, the smooth muscle cells also migrate into the lesion. At this stage, the lesion may be hemodynamically insignificant. But endothelial dysfunction exists and it's ability to limit the entry of lipoproteins into the vessel is impaired [4]. So, the elevation of plasma low density lipoprotein cholesterol level results in penetration of low density lipoprotein cholesterol into the vessel wall, lipids accumulation in macrophages and smooth muscle cells. Endothelial injury produces loss of endothelium, adhesion of platelets to subendothelium, aggregation of platelets, chemotaxis of monocytes and T-cell lymphocytes, and release of growth factors that induce migration of smooth muscle cells from media to intima, where synthesize connective tissue and proteoglycans and forms a fibrous plaque. Low density lipoprotein cholesterol is cytotoxic and may cause endothelial injury and stimulate smooth muscle growth. Touched endothelial cell are functionally impaired and increase the uptake of low density lipoprotein cholesterol from plasma [24]. Growing atherosclerotic plaque may cause a severe stenosis that can progress to total arterial occlusion. Eventually the plaque may become calcified. Some plaques, reached in lipids and inflammatory cells, as macrophages, covered with a thin fibrous cap may undergo spontaneous rupture, resulting in cascade of events, stimulates thrombosis and ends in acute ischaemic event [4,24].

Hypercholesterolemia occures either from overproduction or defective clearance of very low density lipoprotein cholesterol or from increased conversion of very low density lipoprotein cholesterol to low density lipoprotein cholesterol. Overproduction of very low density lipoprotein cholesterol by liver may be caused by obesity, diabetes, alcohol consumption, nephrotic syndrome or genetic disorders. Each of this conditions can result in increased low density lipoprotein cholesterol and trygliceride levels. When dietary cholesterol reaches the liver, elevates intracellular cholesterol level. Due to this, low density lipoprotein cholesterol receptor synthesis is suppressed. This suppression occurs at the level of transcription of the low density lipoprotein cholesterol gene. A reduced number of receptors results in higher levels of plasma low density lipoprotein cholesterol and total cholesterol [24]. Today, it appears, that high density lipoprotein cholesterol assist in the mobilization of low density lipoprotein cholesterol [4]. Cardiovascular risk increases progressively with elevated of low density lipoprotein cholesterol and with a decrease in high density lipoprotein cholesterol level. Studies demonstrated that each decrease in high density lipoprotein cholesterol level by 1 mg/dl (0,0259 mmol/l) elevating cardiovascular risk by 2-3%. [20]. And contrarily, each increase in high density lipoprotein cholesterol level by 1 mg/dl (about 0,02 mol/l) lowering cardiovascular mortality by 6%, independently of low density lipoprotein cholesterol level [20,32]. Similarly, The Treating to New Target study evaluated high density lipoprotein cholesterol as a more significantly predictive for cardiovascular events comparing with low density lipoprotein cholesterol [21,33]. High density lipoprotein cholesterol is protective through multiple mechanisms. There are some new points in the high density lipoprotein cholesterol role and effects on atherosclerosis. Recently published studies showed the antooxidative role of high density lipoprotein cholesterol. Due to this effect the reduction of vascular oxidative stress is occured. It is thought, this can contribute to the atheroprotective effects. Supposedly, high density lipoprotein cholesterol decreases inflammatory process, stops the proliferation and migration of endothelial cells and has anti-apoptotic effects on them. All of this contributes to the anti-atherosclerotic effect [21]. High density lipoprotein cholesterol also affects the platelets function and haemostatic cascade [14].

The prevalence of hypercholesterolemia differ in the world. In 1996 in Taiwan 41,5% men and 19,6% of women had abnormal rates of plasma cholesterol [34]. In 1995 in Holand 19,2% of men and 12,4% of women have total cholesterol more than 6,5 mmol/l [35]. In 1999-2000 in Europe the European Action on Secondary Prevention through Intervention to Reduce Events (EUROASPIRE) study have been performed. The prevalence of high total cholesterol have been declined in Europe 1995-2000 from 86,2 till 58,8% In 2000 58,8% of the population have total cholesterol more than 5,0 mmol/l [36]. In 2006-2007 The European Society of Cardiology carried out the EUROASPIRE III survey in 76 medical centers in 22 European countries (Belgium, Bulgaria, Cyprus, Croatia, The Czech Republic, Finland, France, Germany, Greece, Hungary, Ireland, Italy, Latvia, Lithuania, Poland, the Netherlands, Romania, Russian Federation, Slovenia, Spain, Turkey and UK). The total 13935 participants with established coronary artery disease were reviewed and 8966 were interviewed 6 months after acute coronary event. 76,5% of all the patients had elevated total cholesterol and low density lipoprotein cholesterol. For 51% the total cholesterol more than 4,5 mmol/l was estimated and only for one half of them a total cholesterol goal (<4,5 mmol/l) was reached with lipid-lowering medications, despite of the rather high rate of the statins prescription (80,7%). So, this study showed that one of the most important risk factor for cardiovascular mortality - dyslipoproteinemia control was inadequate and most of the patients did not achieve the targets defined in the guidelines [37]. By the data from MONICA project with 39 population from the 21 countries, in 2003 the main total cholesterol in Kaunas, Lithuania was near 6 mmol/l for men and 6,5 mmol/l for women. Females from Lithuania were at the top of all countries, whereas the men were about an average. 15 countries evaluated higher levels of total cholesterol for men [38]. One of the meta-analysis showed that the highest protective effect can be get treating high risk patients with very high total cholesterol level [39]. Regarding to the existing evidence on lipoproteins disorder, treatment and also due to increased cardiovascular mortality in Lithuania we started the clinical data. At this study we have evaluated independent risk factors for one year cardiovascular mortality for the patients with acute and chronic coronary syndromes. Lipoprotein disorder was one of the most important risk factor for one year cardiovascular mortality.

3. Methods

A total of 3268 patients with coronary heart disease who were selected for this study. The data was collected by a standardized questionnaire. A total of 1865 (728 women and 1137 men) with acute and chronic coronary heart disease, male and female, aged from 20 years till more than 80 years were reexamined after one year. Risk factors for coronary heart disease were evaluated. Lipoprotein disorder was definable as low density lipoprotein cholesterol level in twelve-hour fasting venous blood samples more than 3 mmol/l, total cholesterol level – more than 5,2 mmol/l, high density lipoprotein cholesterol level less than 1,2 mmol/l for women and less than 1,0 mmol/l for men. Due to medical history and data on admission patients were attributed to chronic or acute coronary syndrome. The myocardial infarction and unstable angina were attributed to acute coronary syndromes. Stable angina – to chronic coronary syndrome. Myocardial infarction was diagnosed according to the World Health Organisation guidelines: angina pain and equivalent, ischemic signs on ECG (Q wave, ST and T changes) and an increase in troponin I more than 0,05 mg/l. Unstable angina diagnosis confirmed with the angine syndrome, ischemic changes on the ECG without increasing enzymes in the blood and with angiography assessment of the coronary artery. Stable angina determined using a standard clinic, ECG, exercise test and angiography.

4. Statistical analysis

The statistical analysis was performed using SPSS (Statistical Package for Social Science) version 13 and Microsoft Office Excel 2003 statistical programs. Descriptive statistics was used for the quantitative data analysis. Categorical data have been summarized as frequencies and percentages, and for comparisons, chi-square test have been used. Univariate and multivariate logistic regression analysis was used for the risk assessment. One year mortality risk was evaluated by isolated and standardized odds ratios with 95% confidence interval (CI).

5. Results

The data from 1865 patients with chronic and acute coronary syndromes was analysed. For more than a half of the patients an acute coronary syndrome was diagnosed. The participants were mostly men (61%). 54,7% of the patients had a reduced level of high density lipoprotein cholesterol (less than 1,0 mmol/l for male, and less than 1,2 mmol/l for female), for about 32% an increased total cholesterol and low density lipoprotein cholesterol levels for each were evaluated. About 20,5% of the patients with decreased high density lipoprotein cholesterol level have elevated low density lipoprotein cholesterol level together. Nearly 90% of women with diagnosed dyslipidemia had a reduced high density lipoprotein cholesterol, whereas total cholesterol and low density lipoprotein cholesterol levels were elevated in about one-third of the females. The proportion in these atherogenic lipids in men was about 30% for everyone. 7,6% of the patients had died within one year. The one year cardiovascular mortality was similar for men and women, also for the patients with acute or chronic coronary syndrome. Nearly 22% of died patients had an increased levels of total cholesterol and low density lipoprotein cholesterol. For more than 67% of them the decrease of high density lipoprotein cholesterol was evaluated. The majority of the patients (50,5%) with acute coronary syndrome and more than 80% with stable angina had a reduced high density lipoprotein cholesterol (Table 1).

	High LDL[1] n(%)	Low HDL[2] n(%)	High TC[3], n(%)	Total, n(%)	one-year CV mortality, n(%)
Total	588 (31,5)	1021 (54,7)	594 (31,8)	1865 (100)	
Medical history					
Acute coronary syndrome	371(35,3)	531 (50,5)	359 (34,2)	1050(56,3)	90(8,6)
Chronic coronary syndrome	217(36,5)	490(82,5)	235(39,6)	815(43,6)	52(8,7)
Gender					
Female	241(33,1)	654(89,8)	264(36,2)	728(39)	55(7,5)
Male	347(30,5)	367(32,2)	330(29)	1137(61)	87(7,6)
Age groups					
< 70 years	419(35)	585(48,8)	433(36,2)	1197(64,2)	70(49,3)
70-80 years	144(25,6)	368(65,6)	136(24,2)	561(30,1)	56(39,4)
> 80 years	25(24,7)	63(62,4)	24(23,8)	101(5,4)	15(10,6)
One year CV mortality	31(21,8)	96(67,6)	31(21,8)	142(7,6)	

LDL- low density lipoprotein cholesterol, HDL – high density lipoprotein cholesterol, CV – cardiovascular; [1]>3,0 mmol/l, [2]< 1,0 for males; < 1,2 mmol/l for females, [3]> 5,2 mmol/l

Table 1. Patients baseline characteristics.

For the 34-40% of the patients elevated low density lipoprotein cholesterol and total cholesterol were diagnosed (Figure 6,7). Patients distribution due to age shows Figure 8.

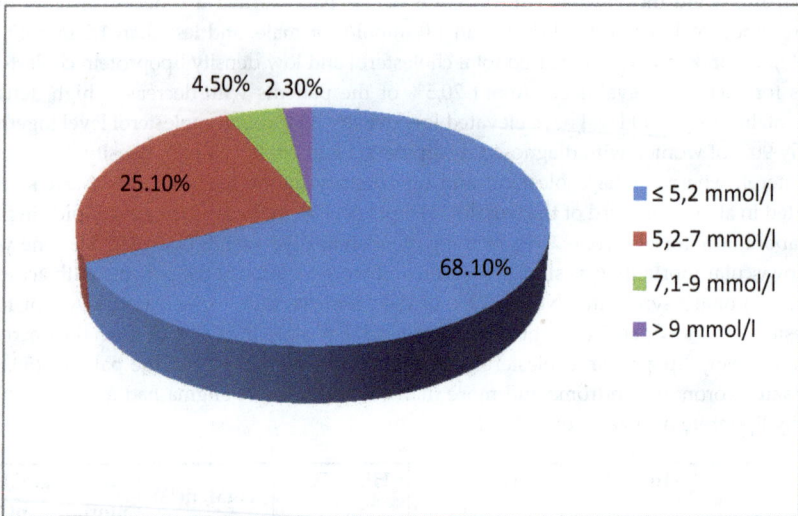

Figure 6. Part of the patients with different total cholesterol level.

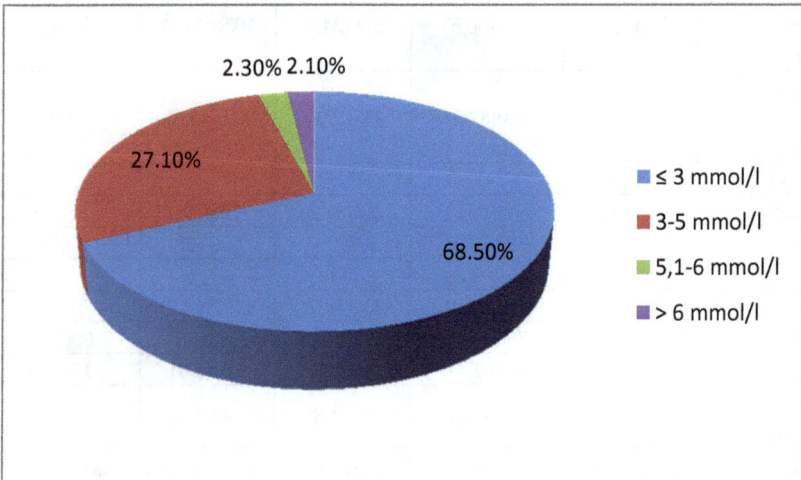

Figure 7. Part of the patients with different low density lipoprotein cholesterol level.

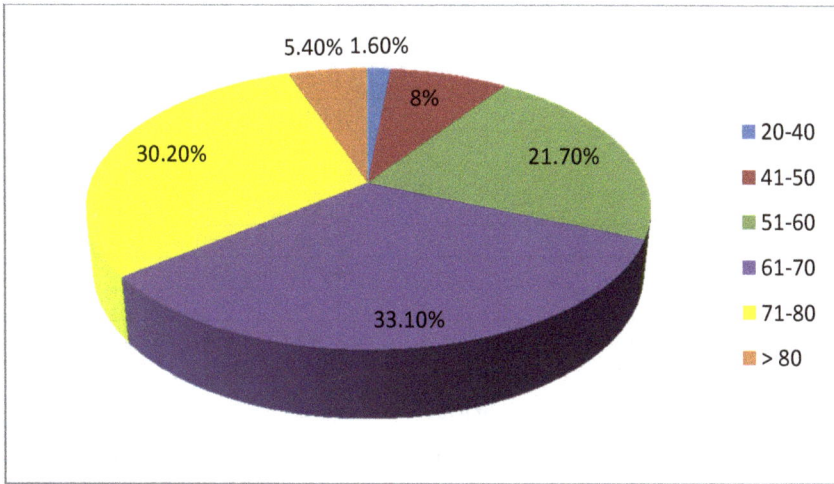

Figure 8. Patients distribution due to the age.

About 35-36% of the patients younger than 70 years, high total cholesterol and low density lipoprotein cholesterol levels were evaluated, the high density lipoprotein cholesterol have been decreased in nearly 49%. For the seniors (more than 70 years), the elevated total cholesterol and low density lipoprotein cholesterol were not so common (24-25% of the patients), but the reduced high density lipoprotein cholesterol was present more frequently (62-65%) (Table 1). Our data evaluated 1,8 times greater independent one year cardiovascular mortality risk for the patients with decreased high density lipoprotein cholesterol level (1,800, 95%CI 1,251-2,591, p=0,002). However, the assessment of the increased general values of the total cholesterol (more than 5,2 mmol/l) and low density lipoprotein cholesterol (more than 3,0 mmol/l) reduced mortality risk (0,575, 95%CI 0,382-0,868, p=0,008 and 0,585, 95%CI 0,388-0,882, p=0,01 respectively) (Table 2).

Risk factor	OR (95%CI)	p value
TC	0,575(0,382-0,868)	0,008
LDL	0,585 (0,388-0,882)	0,01
HDL	1,800 (1,251-2,591)	0,002

TC – total cholesterol, LDL- low density lipoprotein cholesterol, HDL – high density lipoprotein cholesterol

Table 2. Independent cardiovascular one year mortality rate.

Though, one year cardiovascular death risk elevates with the increase of these parameters (total cholesterol more than 9,0 mmol/l, low density lipoprotein cholesterol more than 6,0 mmol/l), although not significant (1,742, 95%CI 0,718-4,224, p=0,22 for total cholesterol more than 9,0 mmol/l and 1,167, 95%CI 0,408-3,339, p=0,773 for low density lipoprotein cholesterol more than 6,0 mmol/l). It is believed, the absence of the statistical significans due to small sampe size (Table 3).

Risk factor	OR (95%CI)	p value
TC		
5,2-7 mmol/l	0,465(0,286-0,759)	0,002
7,1-9 mmol/l	0,670(0,266-1,689)	0,396
> 9 mmol/l	1,742(0,718-4,224)	0,22
LDL		
3-5 mmol/l	0,524(0,333-0,825)	0,005
5,1-6 mmol/l	0,788(0,240-2,588)	0,694
> 6 mmol/l	1,167(0,408-3,339)	0,773
HDL		
< 1,0 for males; < 1,2 mmol/l for females	1,800(1,251-2,591)	0,002

TC – total cholesterol, LDL- low density lipoprotein cholesterol, HDL – high density lipoprotein cholesterol

Table 3. Independent cardiovascular one year mortality rate depending on lipoproteins level.

For the patients with acute coronary syndrome one year cardiovascular mortality rate insignificantly increases with the total cholesterol more than 9,0 mmol/l (2,578, 95%CI 0,931-7,136, p=0,068) and low density lipoprotein cholesterol more than 5 mmol/l (1,030, 95%CI 0,305-3,481, p=0,963 for the level of 5-6 mmol/l, and 2,023, 95%CI 0,668-6,130, p= 0,213 for the level more than 6,0 mmol/l). Simillary, high density lipoprotein cholesterol less than 1,0 mmol/l for men and less than 1,2 mmol/l for women increases one year cardiovascular mortality 1,4 times insignificantly (1,444, 95%CI 0,932-2,239, p=0,1) (Table 4).

Risk factor	OR (95%CI)	p value
TC		
>5,2 mmol/l	0,765(0,476-1,230)	0,269
5,2-7 mmol/l	0,633(0,364-1,102)	0,106
7,1-9 mmol/l	0,768(0,269-2,195)	0,623
> 9 mmol/l	2,578(0,931-7,136)	0,068
LDL		
> 3 mmol/l	0,724(0,451-1,164)	0,183
3-5 mmol/l	0,613(0,360-1,041)	0,07
5,1-6 mmol/l	1,030(0,305-3,481)	0,963
> 6 mmol/l	2,023(0,668-6,130)	0,213
HDL		
< 1,0 for males; < 1,2 mmol/l for females	1,444(0,932-2,239)	0,1

TC – total cholesterol, LDL- low density lipoprotein cholesterol, HDL – high density lipoprotein cholesterol

Table 4. Cardiovascular one year mortality rate for the patients with acute coronary syndrome depending on lipoproteins level.

For the patients with chronic coronary artery disease, only reduced high density lipoprotein cholesterol increased mortality risk, and this was great and significant (3,378, 95%CI 1,623-7,028, p= 0,001). It surprised the increase of the total cholesterol and low density lipoprotein cholesterol reduced one year cardiovascular mortality (almost in all groups significantly, p<0,05) (Table 5).

Risk factor	OR (95%CI)	p value
TC		
>5,2 mmol/l	0,247(0,097-0,628)	0,003
5,2-7 mmol/l	0,183(0,056-0,595)	0,005
7,1-9 mmol/l	0,420(0,056-3,160)	0,005
> 9 mmol/l	0,667(0,087-5,124)	0,697
LDL		
> 3 mmol/l	0,276(0,108-0,705)	0,007
3-5 mmol/l	0,320(0,125-0,818)	0,017
5,1-6 mmol/l	-	-
> 6 mmol/l	-	-
HDL		
< 1,0 for males; < 1,2 mmol/l for females	3,378(1,623-7,028)	0,001

TC – total cholesterol, LDL- low density lipoprotein cholesterol, HDL – high density lipoprotein cholesterol

Table 5. Cardiovascular one year mortality rate for the patients with chronic coronary syndrome depending on lipoproteins level.

Both in men and women with reduced high density lipoprotein cholesterol elevated mortality risk was evaluated (2,044, 95%CI 0,622-6,716, p= 0,239 for women, and 2,303, 95%CI 1,483-3,577, p< 0,001 for men respectively). For females high total cholesterol (more than 9,0 mmol/l) and low density lipoprotein cholesterol (more than 6,0 mmol/l) insignificantly increased one-year cardiovascular mortality risk. For men, relevant total cholesterol level was more than 7,0 mmol/l and low density lipoprotein cholesterol more than 5,0 mmol/l, insignificantly (Table 6,7).

Risk factor	OR (95%CI)	p value
TC		
>5,2 mmol/l	0,236(0,105-0,530)	0
5,2-7 mmol/l	0,123(0,038-0,4)	0
7,1-9 mmol/l	0,263(0,035-1,963)	0,193
> 9 mmol/l	2,0(0,550-7,239)	0,292
LDL		
> 3 mmol/l	0,372(0,179-0,773)	0,008
3-5 mmol/l	0,283(0,119-0,674)	0,004
5,1-6 mmol/l	0,533(0,07-4,082)	0,544
> 6 mmol/l	1,743(0,375-8,106)	0,479
HDL		
< 1,0 for males; < 1,2 mmol/l for females	2,044(0,622-6,716)	0,239

TC – total cholesterol, LDL- low density lipoprotein cholesterol, HDL – high density lipoprotein cholesterol

Table 6. Cardiovascular one year mortality rate for females depending on lipoproteins level.

Risk factor	OR (95%CI)	p value
TC		
>5,2 mmol/l	0,926(0,568-1,510)	0
5,2-7 mmol/l	0,844(0,484-1,470)	0,548
7,1-9 mmol/l	1,050(0,366-3,017)	0,928
> 9 mmol/l	1,540(0,450-5,272)	0,491
LDL		
> 3 mmol/l	0,755(0,458-1,246)	0,008
3-5 mmol/l	0,722(0,421-1,239)	0,238
5,1-6 mmol/l	1,014(0,233-1,239)	0,985
> 6 mmol/l	0,892(0,207-3,852)	0,879
HDL		
< 1,0 for males; < 1,2 mmol/l for females	2,303(1,483-3,577)	0

TC – total cholesterol, LDL- low density lipoprotein cholesterol, HDL – high density lipoprotein cholesterol

Table 7. Cardiovascular one year mortality rate for males depending on lipoproteins level.

It was noticed, that a decrease in high density lipoprotein cholesterol – is an important and reliable cardiovascular mortality risk factor in middle-aged patients (40-60 years). For the 41-50 years patients the mortality risk increases nearly 5 times when high density lipoprotein cholesterol level declines less than 1,0 mmol/l for men, and less than 1,2 mmol/l for women (4,985, 95%CI 1,230-20,196, p<0,05). In the 51-60 year group the risk of death increases 2,5 times with a similar levels of high density lipoprotein cholesterol significantly (2,572, 95%CI 1,094-6,106, p< 0,05) and with a total cholesterol more than 5,2 mmol/l insignificantly (1,073, 95%CI 0,462-2,495, p=0,87). A similar trend for the high density lipoprotein cholesterol was evaluated for the elderly patients, without significance (due to small sample size) (Table 8,9,10).

Years	OR (95%CI)	p value
20-40		
41-50	0,591(0,147-2,383)	0,46
51-60	0,610(0,238-1,564)	0,303
61-70	0,581(0,258-1,312)	0,191
71-80	0,771(0,395-1,504)	0,445
>80	0,421(0,088-2,012)	0,279

Table 8. Cardiovascular one year mortality rate depending on age groups for the patients with increased low density lipoprotein cholesterol level.

Years	OR (95%CI)	p value
20-40	1,545(0,087-27,358)	0,767
41-50	4,985(1,230-20,196)	0,024
51-60	2,572(1,094-6,106)	0,032
61-70	1,155(0,568-2,347)	0,691
71-80	1,227(0,675-2,232)	0,502
>80	2,745(0,721-10,445)	0,139

Table 9. Cardiovascular one year mortality rate depending on age groups for the patients with decreased high density lipoprotein cholesterol level.

Years	OR (95% CI)	p value
20-40	2,111(0,118-37,722)	0,611
41-50	0,265(0,054-1,293)	0,1
51-60	1,073(0,462-2,495)	0,87
61-70	0,485(0,207-1,137)	0,096
71-80	0,654(0,321-1,334)	0,243
>80	0,448(0,094-2,142)	0,314

Table 10. Cardiovascular one year mortality rate depending on age groups for the patients with increased total cholesterol level.

6. Discussion

In the last decade, lack of evidence on low density lipoprotein cholesterol and high density lipoprotein cholesterol in the pathogenesis of coronary heart disease have appeared. Mostly long-term outcomes were evaluated by the previous studies on lipoprotein disorder. We decided to estimate impact of the dyslipoproteinemia to the one year survival. It is proved by another studies, that patients with very low high density lipoprotein cholesterol have much higher risk of severe cardiovascular event or cardiovascular death comparing with patients with normal high density lipoprotein cholesterol level. Lower high density lipoprotein cholesterol values are associated with a higher great cardiovascular events risk and a greater burden of atherosclerosis, even among the patients with reduced low density lipoprotein cholesterol level [33,40,41]. In another side, very low low density lipoprotein cholesterol level is a significant prognostic factor, improved survival for the patient with acute coronary syndrome and may be a target for the treatment. In this study first and foremost we found that reduced high density lipoprotein cholesterol are highly prevalent in a large cohort of the patients with coronary artery disease and tend to be associated with a significantly higher cardiovascular mortality risk. More than a half of the patients in our study had decreased high density lipoprotein cholesterol, and therefore the higher cardiovascular events and mortality risk, especially for the patients with stable angina. These data are similar to another studies [40]. Results from another studies showed that the prevalence of the elevated low density lipoprotein cholesterol increases with age [17]. By data from our study it is not only the problem for the eldery patients. The prevalence of the impaired low density lipoprotein cholesterol by the gender was similar both for men and women and it was high for the patients with established coronary artery disease, taking notice that elevated low density lipoprotein cholesterol can be managed and controlled successfully with lifestyle changes, medications or a combination both of them. We have found that decreased high density lipoprotein cholesterol level is a significant independent risk factor for cardiovascular one year mortality. Interestingly, in another similar studies reduced high density lipoprotein cholesterol more often were found in young men. In our study 90% of females with coronary artery disease had a decreased high density lipoprotein cholesterol level. Also, insufficient high density lipoprotein cholesterol level more often have been found in eldery people. Although high density lipoprotein cholesterol less than 1,3 mmol/l for women has been widely considered as a cardiovascular risk factor, in the present study we selected a cutoff point of less than 1,2 mmol/l as a lowest high density lipoprotein cholesterol value that allowed us to identify those females at risk of cardiovascular one year mortality. It have been evaluated that about 20% of participants of our study had reduced high density lipoprotein cholesterol with elevated low density lipoprotein cholesterol level together. So, it is let to suspect, that one year cardiovascular mortality risk for them have to be much higher. There are a lot of evidence, that decreased high density lipoprotein cholesterol significantly increases cardiovascular mortality risk in stable patients. Also, there are some studies, showed that reduced high density lipoprotein cholesterol is associated with a higher risk of adverse outcomes [40]. Some reports on lipoproteins did not evaluated cardiovascular mortality due to acute or chronic ischaemic syndrome. Comparing acute coronary syndrome and chronic coronary artery disease patients we have been evaluated the more important role of total cholesterol and low density lipoprotein cholesterol on cardiovascular one year mortality for acute patients, though not significant. In contrast, high density lipoprotein cholesterol was strong independent risk factor both for acute (not significant) and chronic patients. Suprisingly, total cholesterol more than 5,2 mmol/l and low density lipoprotein cholesterol more than 3,0 mmol/l reduced one year mortality risk both for acute and chronic patients significantly. Additionally, the previous studies showed the increased mortality rate due to elevated low density lipoprotein cholsterol, have not comprehensively evaluated the impact of different low density lipoprotein cholesterol and high density lipoprotein cholesterol lipoproteins levels on cardiovascular mortality. Lehto and al. evaluated, that among 35-64 years females with acute myocardial infarction total cholesterol more that 8 mmol/l significantly increases reccurence cardiovascular disease risk [42]. It was a reason to search an impact of different levels of low density lipoprotein cholesterol and total cholesterol on cardiovascular mortality risk for men and women. Our hypothesis was confirmed, as it became clear, that one year cardiovascular mortality risk sharply rises when signally increased total cholesterol more than 9 mmol/l and low density lipoprotein cholesterol more than 6 mmol/l, especially in women. The future major research need to evaluate a different lipoprotein and total cholesterol levels impact in cardiovascular mortality, not only in short term, but in long-term outcomes as well. It seems, the highest levels of lipids, that could be attributed to hereditary dyslipoproteinemia, may be very important predicting cardiovascular mortality rates and reducing a cardiovascular death risk. As it was find earlier, high density lipoprotein cholesterol more predictive for middle-aged men. Similarly, our study evaluated the more important role of decreased high density lipoprotein cholesterol, especially for 51-60 years men with the chronic coronary artery disease for one year cardiovascular mortality.

7. Conclusion

Lipoproteins disorder is the main factor for development of the atherosclerosis and predicts cardiovascular mortality. The most important findings from our data concerns the inverse relationship between the high density lipoprotein cholesterol and cardiovascular mortality rates. This association is characterized by a high degree of generality and strength.

Author details

Jelena Umbrasiene*
Medical Academy, Lithuanian University of Health Sciences, Kaunas, Lithuania,
Kaunas Region Cardiology Society, Kaunas, Lithuania,

Ruta-Marija Babarskiene
Kaunas Region Cardiology Society, Kaunas, Lithuania
Department of Cardiology, Medical Academy, Lithuanian University of Health Sciences, Kaunas, Lithuania

Jone Vencloviene
Institute of Cardiology, Medical Academy, Lithuanian University of Health Sciences, Kaunas, Lithuania,
Vytautas Magnus University, Kaunas, Lithuania

Acknowledgement

We would like to show our gratitude to the company Berlin-Chemie Menarini Group, whose financial encouragement enabled us to publish this paper. The Company had no interest in this research.

8. References

[1] Reiner Z, Catapano AL, Backer GD, Graham I, Taskinen MR, Wiklund O et al. ESC/EAS Guidelines for the management of dyslipidaemias. The Task Force for the management of dyslipidaemias of the European Society of Cardiology (ESC) and the European Atherosclerosis Society (EAS). European Heart Journal 2011;32,1769–1818.
[2] European Heart Health Charter. Available: http://www.heartcharter.eu

---

* Corresponding Author

[3] Fodor JG, Frohlich JJ, Genest JJG, McPherson PR. Recommendations for the management and treatment of dyslipidemia. Report of the Working group on Hypercholesterolemia and Other Dyslipidemias. CMAJ 2000;162(10):1441-7.

[4] Tierney LM, McPhee SJ, Papdakis MA. Current medical diagnosis and treatment. 43 edition. Lange medical books/McGraw-Hill. 2004, p. 326-327.

[5] OECD iLibrary. Health at a glance: Europe 2010. Available: http://ec.europa.eu/health/reports/docs/health_glance_en.pdf

[6] Statins for the prevention of cardiovascular events. National Institute for Health and Clinical Excellence (NICE). January 2006. Available: http://www.nice.org.uk/nicemedia/pdf/TA094guidance.pdf

[7] National Health Council Annual Report 2010. Available: http://www3.lrs.lt/pls/inter/w5_show?p_r=697&p_k=1

[8] European health for all database (HFA-DB). Reviewed 2012-04-09. Available: http://data.euro.who.int/hfadb/

[9] Lithuanian Institute of Hygiene database. Reviewed 2012-04-09. Available: http://www.hi.lt/news/391.html

[10] Grundy SM, Cleeman JI, Merz CN, Brewer HB, Clark LT, Hunninghake DB et al. Implications of recent clinical trials for the National Cholesterol Education Program Adult Treatment Panel III guidelines. Circulation. 2004;110(2):227-239.

[11] Cannon CP, Braunwald E, McCabe CH, Rader DJ, Rouleau JL, Belder R et al. Intensive versus moderate lipid lowering with statins after acute coronary syndromes. N Engl J Med. 2004;350(15):1495-1504.

[12] Braunwald E, Douglas P. Braunwald Heart Disease. A textbook of cardiovascular medicine. The 6th edition. 2005 p.921-939.

[13] Anitschkow N. Über die Veranderungen der Kaninchenaorta bei experimenteller Cholesterinsteatose. Beitr Pathol Anat 1913;56:379-404.

[14] Kumar P, Clark M. Clinical medicine. Fifth edition. Saunders, 2002 p.1104-1107.

[15] Stamler J, Neaton JD. The Multiple Risk Factor Intervention Trial (MRFIT)—importance then and now (reprinted). JAMA 2008 (300);11:1343-1345.

[16] European guidelines on cardiovascular disease prevention in clinical practice: executive summary. Eur Heart J 2007;28:2375-2414.

[17] Kuklina EV, Shaw KM, Hong Y. Vital Signs: Prevalence, Treatment, and Control of High Levels of Low-Density Lipoprotein Cholesterol — United States, 1999–2002 and 2005–2008. Morbidity & Mortality Weekly Report. 2011;60(4):109-114.

[18] Cholesterol Treatment Trialists' (CTT) Collaboration. Efficacy and safety of more intensive lowereing of LDL-cholesterol: a meta-analysis of data from 170,000 participants in 26 randomised trials. Lancet 2010; 376:1670–81.

[19] Ballantyne C, ed. Clinical lipidology. A Companion to Braunwald's Heart Disease. Philadelphia,PA. Saunders 2009, p.56 - 129.

[20] LaRosa JC, Grundy SM, Waters DD, et al. Inten-sive lipid lowering with atorvastatin in patients with stable coronary disease. N Engl J Med. 2005;352 (14):1425-1435.

[21] Singh IM, Shishehbor MH, Ansell BJ. High-density lipoprotein as a therapeutic target. JAMA 2007;298(7):786-798.

[22] Werner N., Bohm M. Importance of high density lipoprotein cholesterol in atherosclerotic disease. www.escardio.org. E-journal. Vol.6, N.18. Available: http://www.escardio.org/communities/councils/ccp/e-journal/volume6/Pages/vol6n18.aspx

[23] Libby P. Inflammation in atherosclerosis. Nature. 2002;420:868-874.

[24] Schachinger V, Britten MB, Zeiher AM. Prognostic impact of coronary vasodilator dysfunction on adverse long-term outcome of coronary heart disease. Circulation. 2000;101:1899-1906.

[25] Beers MH, Berkow R. The Merck manual of diagnosis and therapy. The seventeenth edition. 1999, p. 201-203.

[26] Castelli WP, Garrison RJ, Wilson PW, Abbott RD,Kalousdian S, KannelWB. Incidence of coronary heart disease and lipoprotein cholesterol levels: the Framing-ham Study. JAMA. 1986;256(20):2835-2838.

[27] Sharrett AR, Ballantyne CM, Coady SA, et al. Coro-nary heart disease prediction from lipoprotein cho-lesterol levels, triglycerides, lipoprotein(a), apolipo-proteins A-I and B, and HDL density subfractions: the Atherosclerosis Risk in Communities (ARIC) Study. Circulation. 2001;104(10):1108-1113

[28] Thompson MM, Reed SC, Cockerill GW. Therapeutic approaches to raising plasma HDL-cholesterol levels. Nat Clin Pract Cardiovasc Med. 2004;1:84-89.

[29] Sumi M, Sata M, Miura S, Rye KA, Toya N, Kanaoka Y, Yanaga K, Ohki T, Saku K, Nagai R. Reconstituted high-density lipoprotein stimulates differentiation of endothelial progenitor cells and enhances ischemia-induced angiogenesis. Arterioscler Thromb Vasc Biol. 2007;27:813-818.

[30] Noor R, Shuaib U, Wang CX, Todd K, Ghani U, Schwindt B, Shuaib A. High-density lipoprotein cholesterol regulates endothelial progenitor cells by increasing eNOS and preventing apoptosis. Atherosclerosis. 2007;192:92-99.

[31] Joint British Recommendations on prevention of coronary heart disease in practice: summary. BMJ 2000;320:705-708.

[32] Rosengren A, Eriksson H, Larsson B, Svärdsudd K, Tibblin G, Welin L et al. Secular changes in cardiovascular risk factors over 30 years in Swedish men aged 50: the study of men born in 1913, 1923, 1933 and 1943. J Int Med 2000;247:111-118.

[33] Barter P, Gotto AM, LaRosa JC, Maroni J, Szarek M, Grundy SM, Kastelein JJ, Bittner V, Fruchart JC. HDL cholesterol, very low levels of LDL cholesterol, and cardiovascular events. N Engl J Med. 2007;357:1301-1310.

[34] Wu DM, Chu NF, Sung PK, Sung PK, Lee MS, Tsai JT, et al. Prevalence and clustering of cardiovascular risk factors among healthy adults in a Chinese population: the MJ Health Screening Center Study in Taiwan. Int J Obes 2001;25:1189-1195.

[35] Bakx JC, Van den Hoogen HJM, Deurenberg P, van Doremalen J, van den Bosch WJ. Changes in serum cholesterol levels over 18 years in a cohort of men and women: the Nijmegen cohort study. Prev Med 2000;30:138-145.

[36] Clinical reality of coronary prevention guidelines: a comparison of EUROASPIRE I and II in nine countries. EUROASPIRE I and II Group. European Action on Secondary Prevention by Intervention to Reduce Events. Lancet 2001;35:995-1001.

[37] Kotseva K, Wood D, De Backer G, De Bacquer D, Pyörälä K, Keil U et al. EUROASPIRE III: a survey on the lifestyle, risk factors and use of cardioprotective drug therapies in coronary patients from 22 European countries. Eur J Cardiovasc Prev Rehabil 2009;16(2):121-37.

[38] MONICA monograph: World's largest study of heart disease, stroke, risk factors, and population trends 1979-2002. Tundstall-Pendoe H edition. WHO, Geneva, 2003.

[39] Davey-Smith G, Song F, Sheldon TA. Cholesterol lowering and mortality: the importance of considering initial level of risk. Br Med J 1993;306:1367-1373.

[40] Roe MT,Ou FS, Alexander KP, Newby LK, Foody JM, Gibler WB, Boden WE et al. Patterns and prognostic implications of low high-density lipoprotein levels in patients with non-ST-segment elevation Acute coronary syndromes. Eur Heart J 2008;29(20):2480-2488.

[41] deGoma EM, Leeper NJ, Heidenreich PA. Clinical significance of high-density lipoprotein cholesterol in patients with low low-density lipoprotein cholesterol. J Am Coll Cardiol 2008;51:49-55.

[42] Lehto S, Palomaki P, Miettinen H et al. Serum cholesterol and high density lipoprotein cholesterol distribution in patients with acute myocardial infarction and in the general population of Kuopio province, eastern Finland. J Intern Med 1993;233(2):179-185.

Chapter 9

Linking the Pathobiology of Hypercholesterolemia with the Neutrophil Mechanotransduction

Xiaoyan Zhang and Hainsworth Y. Shin

Additional information is available at the end of the chapter

1. Introduction

This chapter focuses on the potential contributions of the blood-borne neutrophils to hypercholesterolemia-related pathophysiology (e.g., thrombus formation, embolism, heart attack, stroke, etc.). Neutrophils are immersed in the cholesterol-abundant plasma of blood and play critical roles in the acute inflammatory response of the body to infection or tissue damage. Because of their high degree of sensitivity to inflammatory agonists and their arsenal of potent microbicidal and tissue degradative agents, a number of redundant cellular mechanisms exist to control or "turn-off" the inflammatory processes by these cells under physiological (non-pathological) conditions. Failure of these mechanisms leads to sustained levels of cell activity that contribute to a chronic inflammatory phenotype with the continuous release of proteases and cytokines as well as the potential to elicit non-specific damage to host tissues. Alternatively, chronic neutrophil activity may impair tissue perfusion via its effects on the rheological flow behavior of blood particularly in terms of the ability of leukocytes to transit the microcirculation[1]. Such potential damage mechanisms are thought to govern an increasing number of human pathological scenarios (e.g., Alzheimer's, diabetes, vascular disease) that correlate with a chronic inflammatory state. In this regard, chronic inflammation has gained recognition in the scientific community and even in the mainstream national media (e.g., Time[2] and Newsweek[3] magazines) as a common denominator for human diseases. The question is whether the dysregulation of neutrophil activity is a significant component of this potential disease mechanism.

We address this issue from a mechanobiological perspective by presenting evidence that supports a role of impaired neutrophil mechanotransduction of hemodynamics-derived fluid flow in the pathogenesis of hypercholesterolemia-linked diseases. For this purpose, we will first discuss the link between chronic inflammation and hypercholesterolemia and then highlight the neutrophil involvement in the pathophysiology of related cardiovascular diseases, e.g. atherosclerosis and microvascular dysfunction. To further exemplify this link, we will explain the potential mechanism(s) by which cholesterol in blood may impact the biochemical regulation of neutrophil activity at the cellular level. Finally, we will introduce our own evidence as well as those of others pointing to dysregulated neutrophil mechanotransduction as an important component of hypercholesterolemic pathologies.

2. Hypercholesterolemia and chronic inflammation

Hypercholesterolemia is the dominant risk factor for atherosclerosis and its downstream complications (e.g., heart attack, stroke, etc.). Over the past two decades, a wealth of insight has pointed to the development of atherosclerotic lesions in the large vessels (e.g., aorta, carotid, femoral artery, etc.) as occurring at the interface between hypercholesterolemia and inflammation (see reviews[4, 5]). According to the current paradigm, at atheroma-prone sites, inflamed endothelial cells (due to damage or dysregulation) initiate the invasion of blood leukocytes (predominantly, monocytes) and smooth muscle cells (SMCs) into the subendothelial (e.g., the intimal) layer of the vascular wall contributing to atherosclerotic tissue remodeling, thrombosis, and finally embolus formation. The main lipid species that appear to dominate this inflammatory process are modified low-density lipoprotein (LDL) particles, particularly oxidized LDL (oxLDL), which act as potent proatherogenic and proinflammatory factors responsible for not only loading monocyte-derived macrophages with cholesterol but also directly stimulating leukocytes and other vascular wall cells (for a more complete explanation, see reviews[5, 6]).

Hypercholesterolemia also induces chronic inflammation in the microcirculation[7]. Phenotypic changes in the microvasculature are observed long before the appearance of fatty streak lesions in the large arteries of animals placed on high fat (HFD), i.e. proatherogenic, diet[8, 9]. The inflammatory phenotype of the microvessels in hypercholesterolemic animals results in increased basal levels of rolling, adherent, and emigrating leukocytes in the postcapillary venules, predominantly neutrophils, as well as enhanced production of reactive oxygen species (ROS). Hypercholesterolemia also exaggerates microvascular responses to a range of proinflammatory stimuli. For example, the postcapillary venules of LDL receptor deficient (LDLr-/-) mice, a murine model of modest hypercholesterolemia (with 3-fold higher levels of plasma cholesterol compared to their wild-type (WT) counterparts), exhibit enhanced leukocyte adhesion and albumin leakage in response to experimental ischemia/reperfusion injury as compared to those of WT mice[10]. Interestingly, similar phenotypic changes can be observed in the microvasculature of normocholesterolemic animals administered oxLDL[11, 12], suggesting that oxLDL participates in hypercholesterolemia-related microvascular dysfunction.

Although the mechanisms responsible for the induction of inflammation by hypercholesterolemia in both microvessels and larger arteries remain unclear, it appears that they both begin with endothelial dysfunction characterized by reduced vasodilation, a proinflammatory state, and enhanced permeability to macromolecules (e.g., lipids). However, the original triggers for this endothelial dysfunction are still controversial. In this regard, oxLDL, based on its potent proinflammatory effects, has been considered as a candidate that initiates the inflammatory responses. In fact, production and release of ROS and myeloperoxidase (MPO)[13], which play critical roles in the oxidation of LDL-cholesterol conjugates and are tightly controlled under the physiological non-inflamed conditions, increase in response to hypercholesterolemia. The cellular basis for the causality between oxLDL formation and chronic inflammation, however, remains elusive.

Interestingly, the preferential formation of atherosclerotic lesions at bifurcations, severe curvatures, and stenoses in the arterial circulation strongly suggests that the hemodynamic flow environment is an important determinant in atherogenesis. Fluid flow-derived frictional (i.e., shear) stresses imposed on the surfaces of endothelium lining the vascular wall have been shown to serve an atheroprotective function when blood flow is laminar (i.e., smooth and ordered; for a more comprehensive discussion, see review[14]). For example, laminar fluid flow stimulates endothelial production of nitric oxide (NO), with vasodilatory and anti-inflammatory actions[15]. In contrast, oscillatory shear stresses enhance production and release of ROS[16]. In addition, disturbed flows lead to the upregulation of adhesion molecules on the endothelial surface (e.g., intercellular adhesion molecule-1 or ICAM-1) responsible for recruiting leukocytes to the vascular wall[17]. In effect, generation of complex distributions of fluid shear stresses on the vascular wall, such as at sites of bifurcations and branch points, appears to shift the endothelial phenotype from atheroprotective to proatherogenic.

However, complex flow fields are not sufficient for the onset/progression of hypercholesterolemia-related atherosclerosis since, for example, we are born with bifurcations and curvatures but do not develop atherosclerosis from birth. It is, therefore, clear that cardiovascular pathobiology due to hypercholesterolemia occurs at the intersection between vascular cell biology and the surrounding fluid stress environment. In this regard, it may be the sensitivity (i.e., responsiveness) of vascular cells to fluid shear stress that is altered in the face of hypercholesterolemia leading to a proinflammatory and a proatherogenic phenotype. Moreover, the endothelial cells are not the only cells in the vasculature. Neutrophils also exist in the cholesterol-enriched, fluid flow environment of the circulation and are critical for initiating acute inflammation. Recently, a growing body of evidence supporting the involvement of neutrophils has emerged.

3. The neutrophil involvement in hypercholesterolemia-related vascular dysfunction

Neutrophils make up the majority of the nucleated leukocytes in human blood with the remaining being monocytes and lymphocytes. As the principal gatekeepers of the acute inflammatory response of the body's immune system, neutrophils are extremely sensitive to inflammatory stimuli allowing them to rapidly (i.e., on the order of milliseconds) transition from an inactivated to an activated state. Upon activation, the upregulation of cell-cell adhesion molecules (e.g., selectins, integrins) enables neutrophils to roll and adhere onto the endothelium prior to their transmigration (via diapedesis) across the vascular wall and to the target tissues (i.e., sites of infection and tissue injury) where they release a potent array of biochemicals including proinflammatory mediators, ROS, and proteases to fight infection and orchestrate tissue repair. During this process, neutrophils also undergo changes in their physical attributes such as their size, deformability, and adhesiveness. It is these features that cause neutrophils to be major players in the pathobiology of hypercholesterolemia-related cardiovascular diseases for both the macro- and micro- circulations.

3.1. Potential roles of neutrophils in atherosclerosis

While it has long been appreciated that monocytes and their descendants, along with T lymphocytes, mast cells, and platelets, contribute to the development and destabilization of atherosclerotic lesions, only recently has the neutrophil been seriously considered as a contributing factor for disease onset and/or progression. Direct evidence comes from the identification of neutrophils in different locations of atherosclerotic lesions present in hypercholesterolemic mice and humans using antibodies to neutrophil-specific antigens including Ly6G, CD177, and CD66b[18-21]. Neutrophils, in fact, have been reported to accumulate in atheroprone arteries preceding plaque formation in hypercholesterolemic murine models of atherosclerosis[18, 20]. Further evidence of a neutrophilic component in the early stages of atherosclerosis is the positive correlation between the number of circulating neutrophils and lesion sizes[18]. Experimental data also point to neutrophil infiltration into the highly inflamed areas of atherosclerotic arteries during late disease stages[19] with contribution to lesion destabilization and thromboembolus formation[21, 22].

One way chronically activated neutrophils may enter atheroprone regions in the macrocirculation are at sites of disturbed flow and recirculation[23] where their enhanced residence times promote capture at the vascular wall[24]. Alternatively, activated neutrophils may disrupt vascular (i.e., adventitial or medial) wall perfusion in the vasa vasorum (or microcirculation) of large vessels (e.g., aorta) leading to vessel tissue injury followed by atherogenesis (from within the vessel wall to the luminal surface)[25-27]. In these ways, neutrophils may initiate or exacerbate atherosclerosis at different stages via their capability to release vast amounts of ROS and proteins stored in their cytosolic granules[28]. For example, while MPO released by activated neutrophils can reduce the bioavailability of NO[29, 30] and contribute to the onset of endothelial dysfunction, a number of granule proteins, such as LL-37, azurocidin, cathepsin G, and α-defensins, exert direct chemotactic activity for monocytes[31, 32]. Moreover, neutrophil secretory products, e.g., α-defensins, also promote macrophage maturation and activation, contributing to the uptake of oxLDL and the formation of foam cells[28]. Finally, neutrophil-derived proteolytic enzymes, particularly matrix metalloproteinase (MMP) -2 and -9[28], play critical roles in plaque destabilization and eventual rupture.

3.2. Effects of neutrophils on microvascular dysfunction

Similar to situations in large arteries, activated neutrophils promote microvascular dysfunction through the sustained release of proinflammatory, cytotoxic, and degradative agents. However, while leukocytes have no significant effect on the macrovascular flow properties of blood (which is dominated by the substantially greater numbers of erythrocytes), these cells, particularly the neutrophils, influence blood flow in the microcirculation (Figure 1). This is because vessel diameters in the microcirculation are in the range of 6 – 100 μm, which are comparable in size to the diameters of leukocytes.

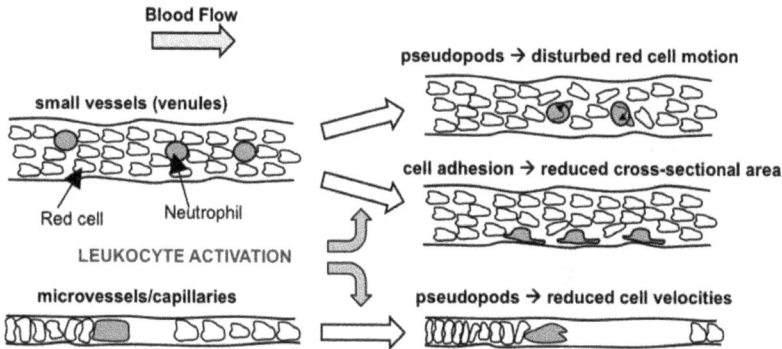

Figure 1. Potential rheological effects of leukocyte activation on the blood flow in the microcirculation. Sustained activation, e.g., due to proinflammatory stimuli, hinders leukocyte passage through the small vessels either by promoting pseudopod projections or through enabling cell adhesion to the vascular wall. Ultimately, these may elevate peripheral resistance and contribute to microvascular dysfunctions (adapted from Shin, H.Y., et al., 2011[79]).

The relatively comparable size scales of the neutrophil and vessel diameters are important to note since the quiescent neutrophil under physiological (i.e., non-inflamed) conditions is capable of efficient transit through the microvessels due to their inherently round, deformable, and non-adhesive state. On the other hand, cell activation physically hinders the passage of neutrophils through the small vessels of the microcirculation[1]. Pseudopods projected by activated neutrophils, while enabling cells to attach to other cells (e.g., endothelial cells, other blood cells) or phagocytose particles, also contribute to a reduction in cell deformability due to their enriched content of F-actin, and increases in geometric size and irregularity[1, 33], all of which serve to increase leukocyte transit time or enhance leukocyte retention in the microvasculature[34, 35]. In turn, activated leukocytes disrupt the motion of erythrocytes, leading to increases in the apparent viscosity of blood and microvascular resistance[1, 34, 36, 37]. Moreover, activated neutrophils are hyperadhesive and exhibit extensive interactions with other leukocytes or platelets, e.g. during hypercholesterolemia[7], which may also enhance the apparent viscosity of blood. Finally, once neutrophils adhere to endothelium, they further increase flow resistance by reducing microvessel diameters (resistance $\propto 1/[diameter]^4$)[34, 37]. The state of neutrophil activation is, thus, a critical determinant of tissue blood flow and perfusion.

In summary, as a result of their arsenal of noxious agents and their effects on microvascular blood flow, it is evident that tight regulation of neutrophil activity is an essential requirement for a healthy circulation. A failure to either prevent or "turn-off" cell activity, e.g., due to hypercholesterolemia, leads to sustained neutrophil activation which has potential impacts not only in terms of the initiation and progression of atherosclerosis in large arteries but also as it relates to microvascular blood flow and downstream tissue perfusion.

4. The influence of cholesterol on neutrophil activity

One way hypercholesterolemia may influence the activation state of neutrophils is to modify the lipid composition of biological membranes. Cholesterol is an essential component of mammalian cell membranes. Approximately 90% of the free (i.e. unesterified) cholesterol in cells resides in the plasma membrane[38]. These sterol molecules not only maintain the integrity of cell membranes, but also play an important role in regulating membrane properties (e.g., microviscosity) and functions (e.g., via their influence on membrane-bound signaling components). In addition to de novo biosynthesis, mammalian cells can take up cholesterol from the extracellular milieu. Exposure to elevated cholesterol levels both in vivo and in vitro enhances cholesterol abundance within the plasma membrane of neutrophils and other blood cells[39-42]. These findings, in conjunction with the wealth of evidence demonstrating the influence of the extracellular cholesterol levels on neutrophil activity, point to membrane cholesterol enrichment as a potential link between hypercholesterolemia and chronic neutrophil activity. To better understand this link, we next describe the possible cholesterol uptake pathways, the influence of cholesterol on physicochemical properties of the cell membrane, and lastly, the influence of cholesterol on neutrophil activity.

4.1. Transport of extracellular cholesterol into the plasma membrane

Due to its insolubility in aqueous media, cholesterol must be transported complexed to carrier molecules, i.e. within the hydrophobic cores of lipoproteins[43]. Lipoproteins (e.g., LDL) in the blood plasma are positioned in close proximity to the circulating blood cells. Conditions that elevate cholesterol-enriched lipoprotein levels may thus favor cholesterol transport into the membranes of these blood cells[39, 42]. In the laboratory, cyclodextrin derivatives (e.g., methyl-β-cyclodextrin or MβCD), synthetic cholesterol carrier molecules, are commonly used to alter membrane cholesterol abundance. Such treatments elicit acute changes in membrane cholesterol levels and downstream cell activity indicating the existence of mechanism(s) that permit rapid transport of cholesterol into nearby cell membranes. Cholesterol uptake may occur by either receptor dependent/independent endocytosis followed by rapid membrane mobilization[44] or direct exchange between the hydrophobic environments of carrier molecules and the lipid bilayer(Figure 2).

4.1.1. Receptor-dependent endocytosis

For a variety of cell types (including leukocytes)[45], LDL-cholesterol is taken up in vivo mainly through LDL receptor (LDLr)-mediated endocytosis. LDLr expression is subject to feedback regulation and, thus, is unlikely to contribute significantly to the overaccumulation of cellular cholesterol[43]. Cellular uptake of cholesterol can also occur via endocytosis mediated by other receptors[43]. Potentially, these pathways can load cholesterol continuously into the cell leading to cholesterol elevations in the plasma membrane[43]. For example, a family of scavenger receptors have been identified on monocytes, macrophages, and SMCs[6] that, by binding to modified LDL (e.g., oxidized and acetylated LDL) with high affinity, account for the majority of cholesterol uptake by these cells[6]. To our knowledge, such scavenger receptors have not yet been identified for neutrophils.

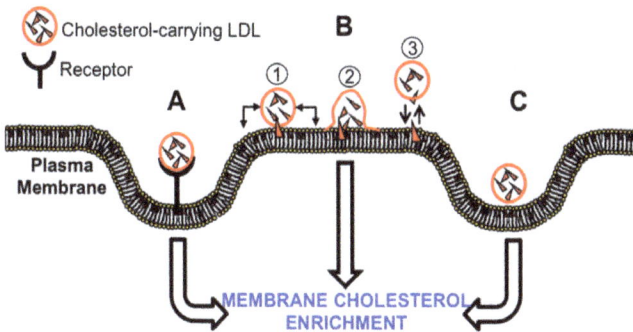

Figure 2. Schematic representation of three possible modes of cholesterol uptake. A: Receptor-mediated endocytosis; B: Direct surface exchange of cholesterol between extracellular carrier molecules and plasma membrane which may occur due to the formation of a transient collision complex without (①) or with (②) membrane fusion or resulting from diffusion across the aqueous phase (③); C: Receptor-independent endocytosis.

4.1.2. Receptor-independent endocytosis

The entire LDL particle can be internalized as a result of fluid or bulk endocytosis without receptor-mediated LDL binding to the cell surface[46]. It is taken up at a rate strictly proportional to its concentration in the extracellular milieu[43]. Alternatively, some LDL, e.g., cationized LDL, can also be taken up by the cell through a non-specific low affinity adsorptive endocytotic process. In this case, endocytosis occurs after cationized LDL binds to the negatively charged membrane surface[47]. For both of these modes of endocytosis, cholesterol transport is not influenced by intracellular cholesterol levels and thus may lead to progressive cholesterol uptake[43].

4.1.3. Cholesterol surface exchange

Cholesterol may also directly enter or exit the plasma membrane[48]. In this case, free cholesterol is exchanged between the hydrophobic cores of the plasma membrane and extracellular carriers (e.g., lipoproteins). The direction of net flux of cholesterol is governed by its concentration gradient between the lipid bilayer and the carrier molecules. Two mechanisms for this surface transfer have been proposed: 1) formation of transient collision complexes with/without membrane fusion and 2) direct diffusion across the aqueous phase. In principle, these transport modes follow similar kinetics with transfer rates depending on the concentrations and structures of both donor (e.g., extracellular LDL) and acceptor (e.g., cells) particles (for more details, see review[48]). This level of complexity contributes, in part, to the diverse half-times ranging from seconds to hours measured for the uptake of cholesterol by human erythrocytes[49]. Finally, this pathway is not under feedback control.

4.2. The influence of cholesterol on the plasma membrane dynamics

In the lipid bilayer, cholesterol orients with its polar hydroxyl group encountering the aqueous phase and the hydrophobic steroid ring parallel to and buried in the hydrocarbon chains of the phospholipids[50]. This unique orientation allows cholesterol to interact with membrane phospholipids and sphingolipids and thus influence their physicochemistry. Along these lines, cholesterol influences the physical and biological properties of the lipid bilayer and, in doing so, impacts the functions of membrane signaling molecules.

4.2.1. Effects of cholesterol on the physical properties of cellular membranes

A key function of cholesterol is to modulate the fluidity (i.e., the inverse of microviscosity) of the lipid bilayer. The close inter-positioning of sterols (i.e., cholesterol) between neighboring membrane phospholipids imposes a degree of immobility on the carbon atoms nearest the membrane surface, while increasing the freedom of motion deep within the hydrophobic core of the membrane[51]. In this regard, membrane fluidity reflects the temperature-dependent influence of cholesterol on the gel to liquid-crystal (i.e., solid-like to fluid-like) phase transition of the lipid bilayer[51]. Under physiological conditions (i.e., 37ºC), biological membranes adopt a liquid crystalline state whereby increases and decreases in cholesterol content reduce and enhance membrane fluidity, respectively[42, 52, 53].

Operationally, membrane fluidity refers to the ensemble of physical properties that govern the motion of the phospholipid molecules in a membrane, including segmental, rotational, lateral, and translational motions[54]. In this fashion, lipid bilayer fluidity can physically influence the dynamics of membrane-associated molecules including proteins that drive downstream cell functions. Two mechanisms have been proposed to explain how membrane fluidity alters the activities and functions of membrane proteins. One mechanism occurs through effects on protein mobility, particularly lateral diffusion which impacts collisional encounters[54]. This effect appears to be of physiological significance particularly for diffusion-controlled processes that are mediated by large membrane proteins. For example, modulation of membrane fluidity influences Ca++-dependent cAMP signaling through changes in protein mobility that governs the coupling between hormone receptors and the adenylate cyclase catalytic unit[55, 56]. Alternatively, membrane fluidity can also influence the structural flexibility of membrane proteins and, thus, their ability to adopt an optimal conformation for activity[57]. Membrane fluidity has, in fact, been reported to impact the conformation-dependent activation of G-protein coupled receptors (GPCRs) on endothelial cells by affecting changes in the protein tertiary structure[58].

4.2.2. Role of cholesterol in lipid raft structure and function

Lipid rafts are nano-scale microdomains that are abundant in the plasma membrane. Structurally, lipid rafts consist of dynamic assemblies of cholesterol and sphingolipids in the outer leaflet of the phospholipid bilayer[59]. The preponderance of saturated hydrocarbon chains in raft sphingolipids renders lipid rafts with a distinct liquid-ordered (i.e., solid-like) phase that is dispersed in the liquid-disordered matrix of the lipid bilayer[60]. One important property of lipid rafts is that they include or exclude proteins to variable extents depending on the raft affinity of proteins[59]. Once individual rafts cluster, they spatially facilitate interactions between raft proteins and expose them to a new membrane environment that is enriched in accessory enzymes and/or second-messenger molecules. In doing so, lipid rafts serve to efficiently initiate and/or amplify signaling cascades. As such, lipid rafts act as signaling platforms that orchestrate outside-in and inside-out signal transduction. Interestingly, these cholesterol-rich microdomains have also been implicated as mechanotransduction centers such as caveolae, a subtype of lipid rafts that reportedly play a role in endothelial mechanotransduction of shear stress and pressure[61-63].

Cholesterol is required to support the formation of lipid rafts and maintain their functionality. It condenses the packing of sphingolipids in the exoplasmic leaflet by occupying the spaces between their saturated hydrocarbon chains near the hydrophilic polar head groups. In this way, cholesterol content and organization influence the stability of lipid rafts with an impact on their capacity to interact with target proteins. Removal or depletion of cholesterol from the plasma membrane using MβCD has been widely used to disrupt rafts and disperse raft proteins into the liquid-disordered matrix of the cell membrane[59]. Treatment of cells with cholesterol-sequestering agents (e.g., filipin or nystatin) or inhibition of cholesterol biosynthesis (e.g., lovastatin) as well as addition of exogenous cholesterol into cell membranes also disrupts raft structure leading to an impact on the functions of raft proteins[59]. As a consequence of these lipid raft-related perturbations, neutrophil functions (e.g., chemokine-induced calcium signaling, extracellular regulated kinase activity, cell polarization, shape change, adhesion, migration, integrin expression, and actin polymerization) are altered[64-68].

4.3. Effects of elevated cholesterol environments on neutrophil activity

Up to this point, we have described how perturbations in extracellular cholesterol levels modify the membrane physicochemistry and the mode by which these modifications may influence membrane protein-related signaling in the neutrophil. The altered cell signaling capacity of membrane-bound proteins is followed by changes in cell behavior that contribute to the principal role of the neutrophil as the first responder to tissue damage and infection. In this way, the influence of membrane cholesterol on the ability of the neutrophil to sense its environment extends to basic cell functions including cell adhesion and migration, phagocytosis, ROS production, and degranulation. We will discuss the effects of cholesterol on these cell functions in order to illustrate the link between the lipid bilayer properties and the control of neutrophil activation.

4.3.1. Expression of membrane adhesion molecules

Upon agonist stimulation, neutrophils exhibit upregulated expression of adhesion molecules that facilitate their recruitment to sites of inflammation by enabling their binding to other cells (e.g., leukocytes, platelets, endothelium)[33]. Two classes of adhesion molecules govern leukocyte interactions with other cells: selectins and integrins. In addition to the ligands for platelet (P)- or endothelial (E)- selectins, neutrophils constitutively express leukocyte (L)-selectins and β_2 (i.e., CD18) integrins, which participate in their initial capture and firm adhesion to other cells, respectively[33]. Currently, the impact of hypercholesterolemia on expression of the selectin family of adhesion molecules is unclear since neutrophils in a cholesterol-rich environment have been reported to exhibit both elevated surface expression[69] and cleavage of L-selectins[70]. In the case of the integrins, surface levels of CD18, particularly Mac-1 (CD11b/CD18), are elevated on neutrophils exposed to a hypercholesterolemic environment both in vitro and in vivo[29, 69, 70]. Notably, surface expression of Mac-1 by neutrophils in hypercholesterolemic patients positively correlates with serum cholesterol levels[29]. But, cholesterol enrichment does not appear to alter the expression of LFA-1 (CD11a/CD18)[69], another CD18 subtype. Thus, the influence of extracellular cholesterol levels on neutrophil adhesion molecule expression is receptor-specific.

Moreover, neutrophils exposed to elevated cholesterol levels undergo increased adhesive interactions with other cells. For example, neutrophils with increased membrane cholesterol exhibit enhanced tethering and firm arrest on activated endothelial cell monolayers[41, 71, 72]. Moreover, neutrophils exposed to hypercholesterolemia display increased heterotypic adhesion to platelets[73] as well as increased homotypic aggregation in response to 10 µM N-formyl-Met-Leu-Phe (fMLP)[74]. These studies confirm that cholesterol-dependent modulation of adhesion molecule expression has an impact on neutrophil adhesion to other leukocytes, platelets, or the endothelium lining the blood vessel lumen.

4.3.2. ROS production

Neutrophil-derived ROS includes superoxide (O_2^-), hydrogen peroxide (H_2O_2), hydroxyl radicals ($^{\cdot}OH$), and NO-related oxidants. Notably, total production of ROS by neutrophils in a hyperlipidemic environment positively correlates with levels of triglycerides and LDL but not with total amount of cholesterol in the plasma[75]. Recent studies, however, did demonstrate a positive correlation between O_2^- release rate and plasma cholesterol levels[29]. Interestingly, enhanced O_2^- release by neutrophils was detected in other clinical states associated with cardiovascular complications, namely hypertension and diabetes[76,

77], which are usually accompanied by hyperlipidemia. In fact, elevations in extracellular cholesterol levels have been shown to enhance neutrophil respiratory burst in response to agonist stimulation. Moreover, plasma activity of superoxide dismutase (SOD), which scavenges ROS, decreases with increases in total cholesterol[75].

4.3.3. Degranulation

Neutrophils contain four main types of granules: primary, secondary, and tertiary granules as well as secretory vesicles. These granules contain a multitude of cytokines (e.g., interleukins, tumor necrosis factor-α, etc.), enzymes (e.g., MPO, etc.), and proteases (e.g., cathepsins, MMPs, etc.). Upon activation, neutrophils degranulate and release these bioactive mediators into the extracellular milieu. Interestingly, although neutrophils from hyperlipidemic patients contain significantly lower levels of intracellular MPO, sera from these patients exhibit significantly higher levels of MPO[29]. These results point to a degranulation process that further links the activation state of neutrophils with the cholesterol levels in the blood environment.

5. Membrane cholesterol and the neutrophil mechanosensitivity to shear stress

In addition to the presence of inflammatory stimuli (e.g., oxLDL), elevated neutrophil activity in hypercholesterolemia may result from defects in their mechanotransduction of fluid shear stress, a control mechanism to prevent spontaneous neutrophil activity under physiological conditions[78, 79]. In this regard, the mechanosensitivity of neutrophils may serve as a key regulator of the inflammatory status of the circulation. We will first define the leukocyte mechanosensitivity to shear followed by a brief discussion of cellular mechanisms that link the extracellular flow environment to downstream neutrophil functions. Interestingly, such mechanotransduction processes occur across the plasma membrane that plays a critical role in regulating the activity of membrane proteins as well as the transmembrane movement of bioactive molecules. The direct contact of cell membrane with the extracellular flow environment makes it a likely target of local environmental factors (e.g., enhanced cholesterol abundance) that influence the neutrophil responsiveness to mechanical stimuli.

5.1. Regulation of neutrophil activity by fluid flow-derived shear stress

Neutrophils, either freely suspended in the bloodstream or adhered to/migrating on vascular endothelium, sense and respond to fluid shear stress[80-82]. Fluid shear stress (ranging from approximately 1 to 10 dyn/cm²) minimizes neutrophil activity levels[78]. The most obvious manifestation of the cell-inactivating effects of shear exposure on cell activity is the retraction of existing pseudopodia by non-cytokine-stimulated human neutrophils adhered to a surface and subjected to a non-uniform flow field imposed by a micropipette with a tip of diameter in the range of 4 – 8 μm[82] (Figure 3A). This situation models brief and spontaneous periods of blood stasis followed by reperfusion, a typical scenario in the microvessels. Under this condition, neutrophils sediment, attach, extend pseudopods, and migrate on the vascular endothelium. Upon reintroduction of fluid flow, these cells retract pseudopods and detach into the flow field in a mechanobiological fashion. Such a scenario has been documented using intravital microscopy of microvascular networks of rodents (e.g., mesentery, spinotrapezius muscle, cremaster muscle)[81-83]. The ability of shear stress to minimize pseudopod activity has been further confirmed for non-adherent heterogeneous leukocyte populations[84] exposed to a constant shear field (5 dyn/cm^2) in a cone-plate viscometer (Figure 3B).

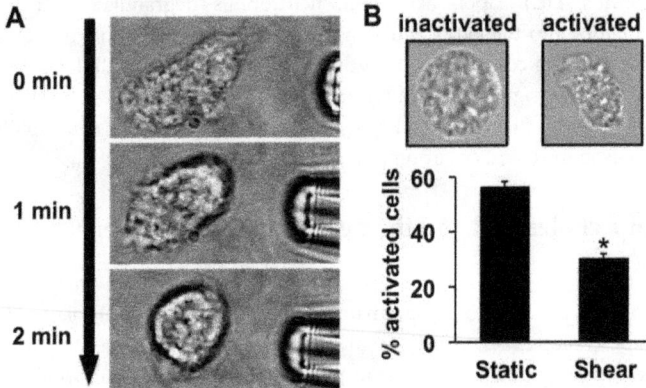

Figure 3. Deactivation of neutrophils under flow stimulation. A: A migrating/adherent neutrophil exposed to a micropipette flow (~ 2 dyn/cm^2) for 2 min. B: Non-adherent neutrophils in suspension exposed to cone-plate shear (5 dyn/cm^2) for 10 min. Bars are mean percentage of activated cells with pseudopods (see image insets) in each population tested ± SEM; *p < 0.05 compared to static condition using paired Student's t-test.

Notably, impairment of shear-induced pseudopod retraction by treating neutrophils with cell agonists above threshold concentrations, e.g. fMLP (>10^{-8} M), commits these cells to an activated (inflamed) phenotype and leads to their microvascular entrapment due to increases in adhesivity, size, and stiffness[35, 81, 82]. Thus, during inflammation, the biochemical milieu of the neutrophil overrides mechanobiological deactivation. Exposure to shear of magnitudes typically found in the macro- and micro- circulations is also associated with other attributes of neutrophil deactivation such as decreased surface expression of integrin receptors (i.e., CD18), depolymerization of the F-actin cytoskeleton, cell detachment, and attenuated phagocytic activity[81, 82]. Moreover, shear stress exposure enhances caspase 3-dependent apoptosis[85], in line with the relatively short lifespan (18 to 24 hrs) of these cells when they are passively circulating in the physiologic bloodstream. These observations support the key role of fluid flow-related shear stress as a biophysical stimulus that promotes neutrophil inactivation when cell activity is below a threshold level. As such, the mechanical influence of fluid flow serves an anti-inflammatory role.

5.2. Shear stress mechanotransduction at the neutrophil surface

An understanding of the fluid flow mechanoregulation of neutrophil activity in the circulation reveals clues regarding how impaired mechanosensitivity to flow may be a mitigating factor for hypercholesterolemic disorders. Membrane detachment during pseudopod retraction by migrating neutrophils in response to fluid shear stress points to two fundamental requirements that must be fulfilled by the cell signaling apparatus: 1) depolymerization of the F-actin cytoskeleton that serves as a structural and a signaling scaffold for neutrophil motility and 2) rapid disengagement of adhesion receptors that anchor the pseudopod to the underlying substrates. For suspended neutrophils, similar events are needed but, in this case, mechanisms must be in place to prevent the expression of adhesive proteins or interfere with engagement of adhesion molecule with substrates (e.g. foreign surfaces, other cells) presenting counter-receptors. These fundamental requirements point to the neutrophil surface components as critical players in mechanotransduction since the cell must sense the extracellular flow environment and remediate its interactions with the cellular microenvironment (e.g., the surrounding matrix and cells).

5.2.1. GPCRs and shear stress control of neutrophil pseudopod activity

Shear stress-induced pseudopod retraction by neutrophils occurs in parallel with a rapid decrease in F-actin content[86, 87]. Typically, remodeling of the F-actin cytoskeleton in leukocytes is controlled by the Ras superfamily of small guanine triphosphate (GTP)-binding proteins, particularly the small GTP-binding phosphatases (GTPases) including Rac1, Rac2, cdc42 and members of the Rho family (as reviewed in the literature[88-90]). Rather than stimulating the activity of molecules that coordinate pseudopod retraction (e.g., RhoA, MLCK), fluid shear stress appears to either inhibit (e.g., possibly through release of an inhibitor) or interfere with the ability of neutrophils to form and sustain pseudopod projections via reducing cytosolic activity of the key small GTPases (e.g., Rac1, Rac2) involved in actin polymerization[83]. These reported effects point to the actions of fluid shear stress on G protein signaling downstream of GPCRs that regulate neutrophil chemotaxis, such as the formyl peptide receptor (FPR).

Notably, fMLP, a ligand for FPR, dose-dependently impairs neutrophil pseudopod retraction responses to shear stimulation[81]. Along this line, HL-60-derived neutrophils subjected to shear stress exhibit reduced activity of $G_{\alpha i}$ downstream of FPR[91]. A critical piece of evidence pointing to FPR as a mechanosensory regulator of pseudopod retraction is the observation that transfection of FPR expression plasmid in undifferentiated HL-60 cells not only confers expression of this receptor but imparts on these cells the ability to form pseudopods that retract under the influence of fluid shear stress[91]. Furthermore, HL-60 promyelocytes differentiated into neutrophils and subsequently transfected with siRNA to silence FPR expression exhibit an attenuated pseudopod retraction response to shear exposure, despite the fact that these cells retain the ability to project pseudopods because of the presence of other cytokine-related GPCRs[91]. Together, these observations point to a role of fluid flow in regulating the activity of membrane-associated receptors by establishing the importance of membrane-bound GPCRs, specifically FPR, in the neutrophil pseudopod retraction response to shear stress. In conjunction with the dependence of GPCR activity on the membrane cholesterol content, it is conceivable that the influence of shear stress on GPCR activity is impacted by perturbations in extracellular cholesterol abundance and their effects on the cell membrane properties.

5.2.2. Cell surface CD18 integrins and shear stress regulation of neutrophil adhesion

Pseudopod retraction by migrating neutrophils subjected to fluid flow depends on their expression levels of CD18 integrins[92], consistent with the requirement of these receptors for cyclical pseudopod projection and retraction[93]. In addition to modulating CD18 interactions with their ligands (e.g. ICAM-1) during inflammation[94, 95], fluid shear stress appears to regulate integrin dynamics on the neutrophil surface under conditions that mimic low activation states by redistributing these receptors from areas of maximal shear stress to regions where shear is minimal, i.e. at focal adhesions. Moreover, shear exposure reduces CD18 levels on the surfaces of migrating, and also non-adherent, neutrophils even in the presence of inflammatory mediators, e.g. fMLP[81, 96]. Considering the role of CD18 in strengthening neutrophil attachment to the vascular wall, shear-mediated reductions in CD18 likely diminish the ability of cells to maintain adhesive attachments[97]. In this way, shear-mediated reductions in CD18 serve an anti-inflammatory role that ensures neutrophils in a non-inflamed environment remain in a non-adhesive state.

The mechanism underlying shear-induced reductions in CD18 surface levels involves proteolysis that occurs on the surfaces of migrating and suspended neutrophils. Proteolysis modulates the levels of a wide variety of transmembrane receptors on the neutrophil surface including L-selectin (involved in rolling interactions with endothelium)[98] and CD43, an anti-adhesive mucin-like molecule[99]. CD18 integrins also undergo cleavage of the intracellular domain by calpain to promote detachment of the cell uropod during neutrophil migration[100]. But shear-induced truncation of CD18 integrins differs from calpain-mediated cleavage in that the former involves lysosomal cysteine proteases (e.g., cathepsin B) that exert extracellular activity[96, 97]. Notably, the cell membrane is critically positioned between the intracellular levels, and the extracellular actions, of these proteases.

Additionally, cleavage of CD18 integrins under fluid flow also requires conformational changes in their extracellular domains[96]. Conformational activity of CD18 integrins involves shifts in the protein tertiary structure from a closed-bent to an open-extended configuration[96]. In the case of cytokine stimulation, this conformational change exposes ligand binding sites[101] that promote cell capture onto the vessel wall[95, 102]. Another consequence of CD18 conformational changes, which occur upon shear stress exposure, is to expose proteolytic cleavage sites[96]. With this evidence in mind, it is apparent that the physicochemical state of the cell membrane is a key factor in neutrophil mechanosensitivity that directly or indirectly affects the ability of shear stress to unfold the CD18 ectodomain.

5.2.3. RNS and ROS in shear mechanotransduction

Reactive nitrogen species (RNS; e.g., NO) and ROS are multi-functional free radical mediators of acute inflammation serving not only as anti-microbial agents but also as biological second messengers that influence leukocyte functions (e.g., chemotaxis, phagocytosis, etc.)[103, 104]. NO from exogenous and endogenous sources (such as membrane-associated NO synthase) inhibits neutrophil recruitment out of the microvasculature during acute inflammation[105, 106]. Interestingly, NO also enhances neutrophil pseudopod retraction in response to shear stress and counteracts the blocking effects of cell agonists (e.g. fMLP and platelet-activating factor)[81]. In contrast, ROS, particularly O^{2-}, interferes with the neutrophil shear response and is thought to contribute to the blocking effects of cell agonist, e.g. fMLP, on flow-induced pseudopod retraction[84].

Notably, the fact that inhibition of NO synthase activity in neutrophils has no effect on shear-induced pseudopod retraction[81] points to an exogenous source and an extracellular role for NO. This finding leaves open the possibility that the facultative effects of NO on the neutrophil shear response (i.e., pseudopod retraction) result from its ability to scavenge O^{2-} [103] and, in this way, mediate cell pseudopod activity[107, 108]. In support of this, SOD (an O^{2-} scavenger) also enhances the shear responses of fMLP-stimulated neutrophils[84]. Thus, O^{2-} is a critical mediator for neutrophil shear response. Since the cell membrane, particularly cholesterol-enriched lipid rafts, plays an important role in regulating the production/release of O^{2-}[109], its state may indirectly influence neutrophil mechanosensitivity to shear stress.

5.3. Neutrophil mechanosensitivity and cardiovascular disease

The accumulated evidence reported in the vascular mechanotransduction literature (see reviews[23, 78]) points to the following general paradigm. Exposure of vascular cells to physiological flows under normal (i.e., non-diseased, non-inflamed) conditions correlates with quiescence (i.e., baseline activity). This paradigm resulted from a multitude of studies that selectively examined the activity of various signaling pathways and putative force sensors in response to applied mechanical stresses. They, however, overlooked a subtle, but equally important, factor: mechanosensitivity or the degree to which cells respond to mechanical stresses. Just as biochemical perturbations (e.g. pathogens, inflammatory agonists) temporally and dose-dependently alter vascular cell activity leading to pathogenesis, so must changes in cell mechanosensitivity impact circulatory health.

Neutrophils experience wide variations in fluid stresses as they pass through the circulation and, thus function "normally" under a diverse array of mechanical stress distributions and magnitudes. In other words, aberrant mechanical stresses are unlikely to be a cause of cell dysfunction. What may change and contribute to "abnormal" behavior is their sensitivity to the surrounding fluid flow mechanoenvironment with a negative impact on the ability of fluid shear stress to deactivate the neutrophils. Along this line, the work of Geert Schmid-Schönbein at the University of California, San Diego has demonstrated that attenuated neutrophil shear responses contribute to the microvascular pathobiology observed in spontaneously hypertensive rats (SHRs)[110] and, in doing so, illustrated the potential impact of impaired shear stress mechanotransduction on cardiovascular health.

5.3.1. Impaired fluid shear responses and downstream effects on vascular pathophysiology

Significant features of the blood from SHRs are elevated numbers of circulating neutrophils, suppressed expression of adhesion molecules (e.g., selectins, CD18), and an activated phenotype[111-113]. Although the increased activity of neutrophils is not associated with increased adhesion to microvascular endothelium[78], their increased numbers raise peripheral vascular resistance[110]. One possible explanation is that circulating activated neutrophils in SHRs release vasoactive substances that constrict the small arteries and arterioles; this has been documented for atherosclerosis[114-116]. Extensive evidence, however, points to a hemorheological effect of leukocyte activation on microvascular resistance[1, 34, 36]. Specifically, the disturbed motion of white blood cells, due to pseudopod projection, significantly reduces erythrocyte velocities in the microcirculation increasing hemodynamic resistance and upstream blood pressures[36, 110] (see Figure 1).

The key evidence for the involvement of fluid flow mechanotransduction in microvascular abnormalities due to hypertension is that neutrophils from SHRs lack the ability to retract pseudopods in response to shear stress; in some cases, cells extend cellular projections under flow stimulation[110]. The underlying mechanism associated with the blockade and possible reversal of the pseudopod retraction response to shear stress reportedly involves the dependence of blood pressure in SHRs on the plasma level of glucocorticoid-related steroid hormones and the density of glucocorticoid receptors on the neutrophil surface[117, 118]. In line with this, glucocorticoid-treated[119] rats, like SHRs, exhibit elevated peripheral resistance in parallel with elevated numbers of neutrophils that lack a pseudopod retraction response to shear stress. Taken together, leukocyte shear mechanotransduction appears to be critical for the maintenance of a healthy circulation, particularly the microcirculation. Failure of this regulatory mechanism, e.g., due to impaired cell mechanosensitivity resulting from a pathological blood environment, may not only lead to sustained neutrophil activation but also result in disturbed blood flow. In this way, aberrant neutrophil mechanotransduction may contribute to microvascular damage that exacerbates ischemia-reperfusion injury or leads to peripheral vascular disease and downstream organ/tissue injury.

Studies on spontaneous hypertension also reveal a key point. Factors that drive phenotypic changes in neutrophils (e.g., from an inactivated to an activated state) dramatically alter their ability to sense the surrounding flow environment (i.e., mechanosensitivity) leading to the development of pathological behavior, including immune suppression. Intuitively, cell mechanosensitivity depends on the number and activity of proteins "moonlighting" as putative mechanosensors embedded in the cell membrane positioned at the interface between the intra- and extra- cellular milieu. These studies further strengthen the argument that the plasma membrane is a critical determinant of neutrophil mechanosensitivity.

5.3.2. The plasma membrane and shear stress mechanosensitivity

The fact that shear stress-induced neutrophil deactivation (e.g., FPR deactivation, G protein signaling, CD18 cleavage, pseudopod retraction, etc.) occurs in the absence of any passive cell deformation due to flow[120] substantiates the presence of a cell surface component(s) that transduces flow stimulation. Interestingly, neutrophils retract pseudopods independently of the fluid shear stress distribution imposed on the cell surface[82]. Thus, membrane properties appear to outweigh the location of mechanosensors on the cell surface. Moreover, non-adherent neutrophils respond to shear stress further emphasizing the importance of cell membrane-mediated over cell deformation-based (e.g., cytoskeleton-related, cell adhesion-dependent) neutrophil mechanotransduction.

The membrane itself may act as a mechanotransducer either via stress-induced changes in its fluidity[121-123] or through lipid rafts[62, 63, 124]. However, the concept that the membrane serves as a fluid stress sensor lacks the specificity that explains the diversity of cell type-specific responses to shear. An alternative, more plausible, viewpoint is that the cell membrane serves as mechanotransduction center for the cell. Along this line, the specificity associated with mechanotransduction depends on the specific mechanoreceptor(s) expressed by the cell. In this regard, a multitude of cell transmembrane proteins including various GPCRs[58, 91, 125], tyrosine kinase receptors[126-130], ion channels[131], NO synthases, and integrin-associated focal adhesions[132, 133] have been implicated as fluid shear stress transducers for a variety of cells (e.g., endothelial cells, osteoblasts, neutrophils) and microorganisms (e.g. dino-flagella)[134].

One potential action of fluid shear stress on transmembrane mechanosensors (e.g., FPR) is to alter their surface levels. In the case of GPCRs, exposing migrating neutrophil-like cells to parallel plate flow redistributes surface-associated FPRs to a perinuclear compartment in the cytosol[135]. These results suggest that internalization of FPRs under fluid shear stimulation leads to pseudopod retraction by counteracting their constitutive activity which drives pseudopod extension. It should be noted, however, that intact FPR must be present since cleavage of FPR is linked to an impaired ability of fluid shear stress to promote retraction of neutrophil pseudopods[136]. Since receptor internalization occurs across the lipid bilayer, shear-induced changes in mechanoreceptor surface levels may thus be a mechanosensitive neutrophil response influenced by properties of the cell membrane.

It is also feasible that the ability of shear stress to alter protein tertiary structure is a function of membrane properties. In addition to evidence regarding the influence of shear stress on the conformation of FPR and CD18 integrins, fluid flow also alters the structure of other membrane-bound GPCRs in other cell types including the bradykinin B_2 receptor for endothelial cells and the type I parathyroid hormone receptor for osteoblasts[58, 125]. Interestingly, physiologically relevant magnitudes of mechanical stresses are capable of physically altering the conformation of proteins[132, 133, 137]. Since these proteins are embedded in the cell membrane, it is possible that membrane properties influence flow-related perturbations of protein structure.

In the end, the physicochemical properties (e.g., fluidity, lipid rafts) of the cell membrane, with their influence on the ability of surface mechanosensors to adopt structural shifts under shear, come to the forefront in terms of how hypercholesterolemia modifies neutrophil mechanosensitivity. This is the topic of the next section.

5.3.3. Membrane cholesterol versus membrane fluidity in hypercholesterolemic impairment of neutrophil mechanosensitivity

Hypercholesterolemia is associated with chronic neutrophil activation and elevated blood cholesterol as well as cholesterol enrichment in the plasma membranes of blood cells. Based on the intimate relationship between protein dynamics (e.g., surface expression, conformational activity) and the cell membrane (as described in the previous section), the chemical and mechanical properties of the lipid bilayer may be critical determinants of the ability of neutrophils to sense fluid shear stress. Along this line, hypercholesterolemia-related membrane perturbations may reduce the neutrophil responsiveness to shear stress by interfering with critical mechanotransduction events, e.g. GPCR and CD18 conformational activity, protease release, and/or production of ROS, that must bidirectionally transmit biological activity across the cell membrane (Figure 4).

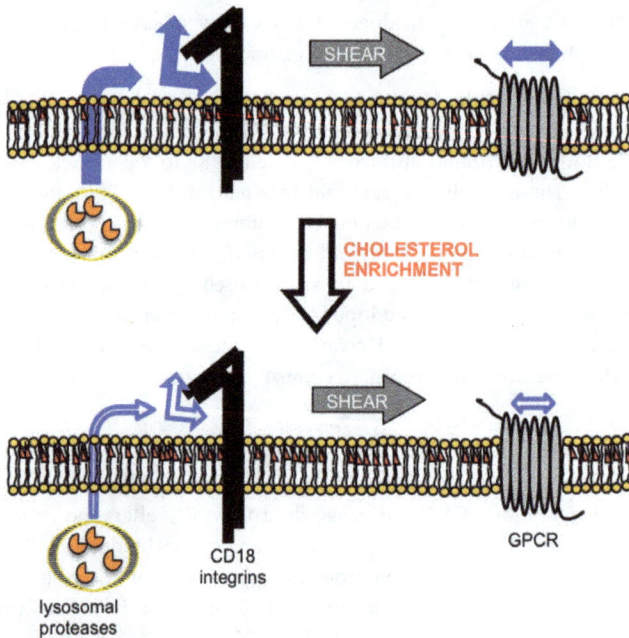

Figure 4. Effects of cholesterol abundance on neutrophil mechanotransduction. Elevations in extracellular cholesterol lead to membrane cholesterol enrichment which may alter cell mechanosensitivity either by influencing shear-induced structural changes of surface sensors, or by interfering with shear-induced release of lysosomal proteases. The cell membrane may also influence contributions from ROS/RNS (not shown).

(A)

(B)

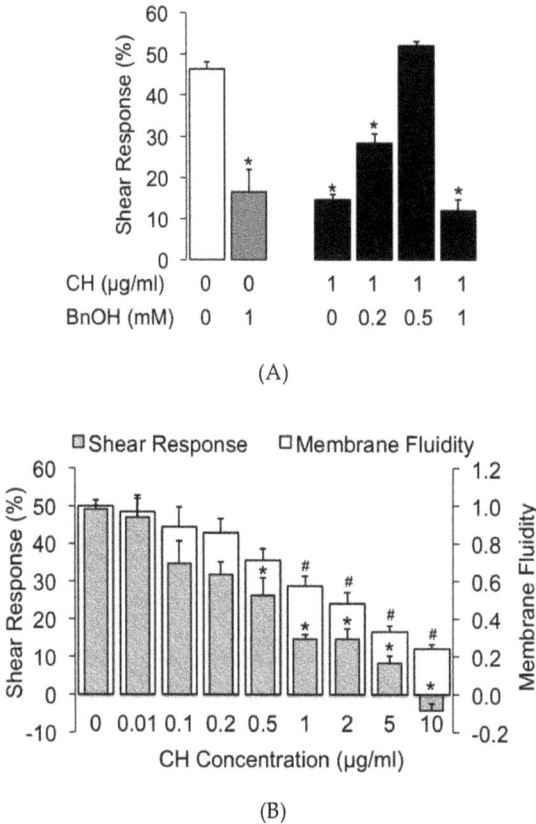

Figure 5. Relationship between membrane cholesterol-dependent fluidity and neutrophil shear responses. A: Recovery effects of benzyl alcohol (BnOH; a membrane fluidizer) on the shear response by neutrophils treated with cholesterol-enhancing agents (CH). B: Does-dependent effects of cholesterol enrichment on neutrophil shear response and membrane fluidity. Cone-plate shear: 5 dyn/cm² for 10 min. Bars are mean percentage of reductions in activated cells by shear ± SEM. *, #p < 0.001 compared to untreated cells using Student's t-test with Bonferrroni's adjustment.

Recently, we reported that neutrophil deactivation by shear stress depends on the cholesterol-dependent physicochemical properties (i.e., fluidity) of the cell membrane[40]. Fundamentally, we showed that the deactivating actions of fluid shear require a cell membrane containing an optimal level of cholesterol. Shear stress mechanotransduction is impaired if there is too much or too little cholesterol. Moreover, the membrane must be capable of supporting the formation of lipid rafts. But, the critical evidence from this work are our observations[40] that membrane fluidizer, benzyl alcohol, was capable of counteracting the rigidifying effects of membrane cholesterol enhancement (with cyclodextrin-cholesterol conjugates) and that the concentration of benzyl alcohol to achieve this depended on the amount of cholesterol loaded into the neutrophil membranes (Figure 5A). Thus, there is also an optimal membrane fluidity level permissive for shear-induced neutrophil deactivation. This was confirmed by regression analysis[40] which revealed a linear relationship (Figure 5B) between membrane cholesterol-related fluidity and the degree to which neutrophils within a population are inactivated by fluid flow. Membrane cholesterol enrichment therefore impairs neutrophil mechanosensitivity, at least in part, through its impact on membrane fluidity.

Interestingly, neutrophils from LDLr-/- mice fed a HFD exhibit a reduced and even reversed shear stress response relative to cells from similar mice maintained on a regular chow (i.e., normal) diet (ND)[40]. These observations were consistent with our in vitro data correlating membrane cholesterol levels with neutrophil mechanosensitivity[40]. In fact, the shear sensitivity of neutrophils from hypercholesterolemic mice tracks negatively with time-dependent increases in blood levels of cholesterol, particularly of the free form (Figure 6). Presumably, the gradual loading of cholesterol into the neutrophil membrane resulting from the progressive increases in the cholesterol concentration gradient across the outer leaflet of the cell membrane is responsible for the time-dependent decrease in shear mechanosensitivity. Impairment of neutrophil shear responses by membrane cholesterol enrichment may thus underlie the pathogenesis of hypercholesterolemic disorders via an effect on cell membrane fluidity which governs the ability of protein sensors to initiate a sufficient degree of mechanotransduction at the cell surface. As such, a chronic inflammatory state may develop.

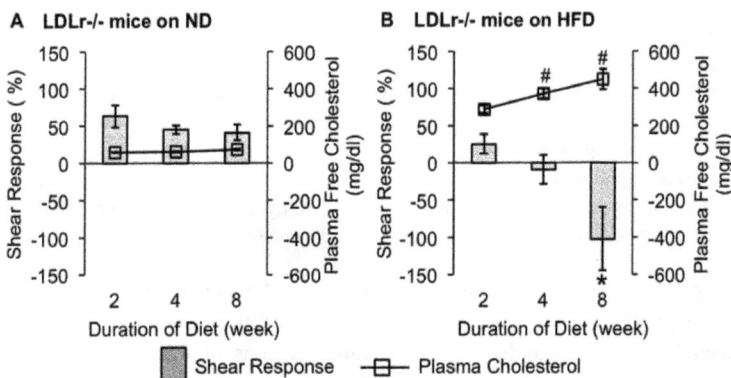

Figure 6. Correlation between neutrophil shear responses and serum levels of free cholesterol. A: LDLr-/- mice on normal diet (ND); B: LDLr-/- mice on high fat diet (HFD). Cone-plate shear: 5 dyn/cm^2 for 10 min. Bars and square dots are mean ± SEM. *, #p < 0.02 compared to 2-week using Student's t-test with Bonferrroni's adjustment.

6. Future directions

To date, the accumulated evidence strongly points to shear stress mechanotransduction as an important negative control mechanism for neutrophils flowing in blood under non-inflamed conditions and, thus, an important mediator of circulatory homeostasis. For the most part, the pathobiology of hypercholesterolemia is a process that takes decades to develop into a serious, life-threatening condition and tracks with gradual elevations in blood cholesterol levels. In addition, hypercholesterolemia is characterized by a chronic inflammatory phenotype associated with elevated levels of neutrophil activity in the blood. The question is how these two factors may be related or linked?

Based on the evidence presented in this chapter, the possibility that elevations in blood cholesterol levels impair the neutrophil-deactivating effects of fluid shear stress further suggests that vascular mechanotransduction is an important aspect of cardiovascular physiology and that the pathobiology of hypercholesterolemia may result, at least in part, from a putative disruption of this mechanotransducing function. This statement applies not only to neutrophils, but also to other cells in the circulation including the other white cells and the endothelium. Moreover, the presented evidence hints at the need to shift focus on the study of vascular mechanobiology from characterizing mechanotransduction (i.e., identifying mechanobiological signaling) in disease to actively investigating the influence of mechanosensitivity (i.e. the degree to which cells transduce fluid stresses) on vascular pathogenesis. In our case, we linked altered neutrophil mechanosensitivity with the gradual changes in blood cholesterol levels and leukocyte membranes during the development of hypercholesterolemia in LDLr-/- mice fed a fat-enriched diet. In light of our own evidence and those of others[3, 4, 8] showing that shear stress is anti-inflammatory for neutrophils, it is possible that a putative source of vascular dysfunction causal for hypercholesterolemic pathobiology is the aberrant neutrophil mechanosensitivity.

Despite recognition that vascular mechanotransduction is critical for circulatory homeostasis, there are no markers currently in use or, to our knowledge, in development that account for mechanosensitivity to predict vascular inflammatory status. Current indicators of inflammation include C-reactive protein (CRP; >3 mg/L is at cardiovascular risk) and serum amyloid protein A (SAA; >10 mg/L is at cardiovascular risk). But even though these two biochemical markers are the gold standard measures of inflammatory activity for blood[138, 139], they are upregulated when leukocyte activity levels are already elevated. It is thus not clear whether these molecules are viable "predictors" or just indicators of chronic inflammatory disorders. As such, understanding, characterizing, and formulating measures of neutrophil mechanosensitivity may prove useful in revealing earlier clues regarding the state of inflammation in blood.

In the end, the likelihood that a cholesterol-dependent loss of neutrophil sensitivity to fluid flow stimuli leads to pathological situations implicates a wide range of cardiovascular (and non-cardiovascular) diseases that correlate with both chronic inflammation and an altered cholesterol environment, e.g. hypercholesterolemia and diabetes[74, 140]. The critical issues are to increase efforts to define the link between chronic inflammation and impaired neutrophil mechanotransduction and to determine if chronic inflammation precedes or results from an impairment of vascular mechanotransduction. Further work is, therefore, needed to determine mechanistic-level connections between the cell surface, the flow sensors, the extracellular flow environment, and the influence of a hypercholesterolemic environment on these. The hope is that by fully defining the role of fluid mechanics in the physiological regulation of leukocytes, particularly the neutrophils, one may gain a better understanding of their role in the pathogenesis of cardiovascular disease.

Author details

Xiaoyan Zhang and Hainsworth Y. Shin*
Center for Biomedical Engineering, University of Kentucky, Lexington, KY, USA

7. References

[1] Mazzoni, M.C. and G.W. Schmid-Schönbein, *Mechanisms and consequences of cell activation in the microcirculation.* Cardiovasc Res, 1996. 32(4): p. 709-19.

[2] Gorman C, P.K.D.A., and Cray D, *The Fires Within.* Time, 2004. 163: p. 30-46.

[3] Underwood, A., *Quieting a Body's Defenses.* Newsweek, 2005.

[4] Libby, P., P.M. Ridker, and A. Maseri, *Inflammation and atherosclerosis.* Circulation, 2002. 105(9): p. 1135-43.

[5] Steinberg, D., *Atherogenesis in perspective: hypercholesterolemia and inflammation as partners in crime.* Nat Med, 2002. 8(11): p. 1211-7.

[6] Boullier, A., et al., *Scavenger receptors, oxidized LDL, and atherosclerosis.* Ann N Y Acad Sci, 2001. 947: p. 214-22; discussion 222-3.

[7] Stokes, K.Y., et al., *Hypercholesterolemia promotes inflammation and microvascular dysfunction: role of nitric oxide and superoxide.* Free Radic Biol Med, 2002. 33(8): p. 1026-36.

[8] Scalia, R., J.Z. Appel, 3rd, and A.M. Lefer, *Leukocyte-endothelium interaction during the early stages of hypercholesterolemia in the rabbit: role of P-selectin, ICAM-1, and VCAM-1.* Arterioscler Thromb Vasc Biol, 1998. 18(7): p. 1093-100.

[9] Stokes, K.Y., et al., *NAD(P)H oxidase-derived superoxide mediates hypercholesterolemia-induced leukocyte-endothelial cell adhesion.* Circ Res, 2001. 88(5): p. 499-505.

[10] Mori, N., et al., *Ischemia-reperfusion induced microvascular responses in LDL-receptor -/- mice.* Am J Physiol, 1999. 276(5 Pt 2): p. H1647-54.

[11] Lehr, H.A., et al., *P-selectin mediates the interaction of circulating leukocytes with platelets and microvascular endothelium in response to oxidized lipoprotein in vivo.* Lab Invest, 1994. 71(3): p. 380-6.

[12] Vink, H., A.A. Constantinescu, and J.A. Spaan, *Oxidized lipoproteins degrade the endothelial surface layer : implications for platelet-endothelial cell adhesion.* Circulation, 2000. 101(13): p. 1500-2.

---

* Corresponding Author

[13] Parthasarathy, S., D. Steinberg, and J.L. Witztum, *The role of oxidized low-density lipoproteins in the pathogenesis of atherosclerosis*. Annu Rev Med, 1992. 43: p. 219-25.

[14] Cunningham, K.S. and A.I. Gotlieb, *The role of shear stress in the pathogenesis of atherosclerosis*. Lab Invest, 2005. 85(1): p. 9-23.

[15] De Caterina, R., et al., *Nitric oxide decreases cytokine-induced endothelial activation. Nitric oxide selectively reduces endothelial expression of adhesion molecules and proinflammatory cytokines*. J Clin Invest, 1995. 96(1): p. 60-8.

[16] De Keulenaer, G.W., et al., *Oscillatory and steady laminar shear stress differentially affect human endothelial redox state: role of a superoxide-producing NADH oxidase*. Circ Res, 1998. 82(10): p. 1094-101.

[17] Nagel, T., et al., *Shear stress selectively upregulates intercellular adhesion molecule-1 expression in cultured human vascular endothelial cells*. J Clin Invest, 1994. 94(2): p. 885-91.

[18] Drechsler, M., et al., *Hyperlipidemia-triggered neutrophilia promotes early atherosclerosis*. Circulation, 2010. 122(18): p. 1837-45.

[19] Rotzius, P., et al., *Distinct infiltration of neutrophils in lesion shoulders in ApoE-/- mice*. Am J Pathol, 2010. 177(1): p. 493-500.

[20] van Leeuwen, M., et al., *Accumulation of myeloperoxidase-positive neutrophils in atherosclerotic lesions in LDLR-/- mice*. Arterioscler Thromb Vasc Biol, 2008. 28(1): p. 84-9.

[21] Ionita, M.G., et al., *High neutrophil numbers in human carotid atherosclerotic plaques are associated with characteristics of rupture-prone lesions*. Arterioscler Thromb Vasc Biol, 2010. 30(9): p. 1842-8.

[22] Naruko, T., et al., *Neutrophil infiltration of culprit lesions in acute coronary syndromes*. Circulation, 2002. 106(23): p. 2894-900.

[23] Davies, P.F., *Hemodynamic shear stress and the endothelium in cardiovascular pathophysiology*. Nat Clin Pract Cardiovasc Med, 2009. 6(1): p. 16-26.

[24] Burns, M.P. and N. DePaola, *Flow-conditioned HUVECs support clustered leukocyte adhesion by coexpressing ICAM-1 and E-selectin*. Am J Physiol Heart Circ Physiol, 2005. 288(1): p. H194-204.

[25] Maiellaro, K. and W.R. Taylor, *The role of the adventitia in vascular inflammation*. Cardiovasc Res, 2007. 75(4): p. 640-8.

[26] Mulligan-Kehoe, M.J., *The vasa vasorum in diseased and nondiseased arteries*. Am J Physiol Heart Circ Physiol, 2010. 298(2): p. H295-305.

[27] Ritman, E.L. and A. Lerman, *The dynamic vasa vasorum*. Cardiovasc Res, 2007. 75(4): p. 649-58.

[28] Soehnlein, O., *Multiple roles for neutrophils in atherosclerosis*. Circ Res, 2012. 110(6): p. 875-88.

[29] Mazor, R., et al., *Primed polymorphonuclear leukocytes constitute a possible link between inflammation and oxidative stress in hyperlipidemic patients*. Atherosclerosis, 2008. 197(2): p. 937-43.

[30] Nicholls, S.J. and S.L. Hazen, *Myeloperoxidase and cardiovascular disease*. Arterioscler Thromb Vasc Biol, 2005. 25(6): p. 1102-11.

[31] Chertov, O., et al., *Identification of human neutrophil-derived cathepsin G and azurocidin/CAP37 as chemoattractants for mononuclear cells and neutrophils.* J Exp Med, 1997. 186(5): p. 739-47.

[32] Soehnlein, O., et al., *Neutrophil secretion products pave the way for inflammatory monocytes.* Blood, 2008. 112(4): p. 1461-71.

[33] Schmid-Schönbein, G.W., *Analysis of inflammation.* Annu Rev Biomed Eng, 2006. 8: p. 93-131.

[34] Eppihimer, M.J. and H.H. Lipowsky, *Effects of leukocyte-capillary plugging on the resistance to flow in the microvasculature of cremaster muscle for normal and activated leukocytes.* Microvasc Res, 1996. 51(2): p. 187-201.

[35] Worthen, G.S., et al., *Mechanics of stimulated neutrophils: cell stiffening induces retention in capillaries.* Science, 1989. 245(4914): p. 183-6.

[36] Helmke, B.P., et al., *A mechanism for erythrocyte-mediated elevation of apparent viscosity by leukocytes in vivo without adhesion to the endothelium.* Biorheology, 1998. 35(6): p. 437-48.

[37] Lipowsky, H.H., *Microvascular rheology and hemodynamics.* Microcirculation, 2005. 12(1): p. 5-15.

[38] Lange, Y., et al., *Plasma membranes contain half the phospholipid and 90% of the cholesterol and sphingomyelin in cultured human fibroblasts.* J Biol Chem, 1989. 264(7): p. 3786-93.

[39] Day, A.P., et al., *Effect of simvastatin therapy on cell membrane cholesterol content and membrane function as assessed by polymorphonuclear cell NADPH oxidase activity.* Ann Clin Biochem, 1997. 34 (Pt 3): p. 269-75.

[40] Zhang, X., et al., *Membrane cholesterol modulates the fluid shear stress response of polymorphonuclear leukocytes via its effects on membrane fluidity.* Am J Physiol Cell Physiol, 2011. 301(2): p. C451-60.

[41] Oh, H., et al., *Membrane cholesterol is a biomechanical regulator of neutrophil adhesion.* Arterioscler Thromb Vasc Biol, 2009. 29(9): p. 1290-7.

[42] Cooper, R.A., *Influence of increased membrane cholesterol on membrane fluidity and cell function in human red blood cells.* J Supramol Struct, 1978. 8(4): p. 413-30.

[43] Goldstein, J.L. and M.S. Brown, *The low-density lipoprotein pathway and its relation to atherosclerosis.* Annu Rev Biochem, 1977. 46: p. 897-930.

[44] Brasaemle, D.L. and A.D. Attie, *Rapid intracellular transport of LDL-derived cholesterol to the plasma membrane in cultured fibroblasts.* J Lipid Res, 1990. 31(1): p. 103-12.

[45] Lara, L.L., et al., *Low density lipoprotein receptor expression and function in human polymorphonuclear leucocytes.* Clin Exp Immunol, 1997. 107(1): p. 205-12.

[46] Goldstein, J.L. and M.S. Brown, *Binding and degradation of low density lipoproteins by cultured human fibroblasts. Comparison of cells from a normal subject and from a patient with homozygous familial hypercholesterolemia.* J Biol Chem, 1974. 249(16): p. 5153-62.

[47] Basu, S.K., et al., *Degradation of cationized low density lipoprotein and regulation of cholesterol metabolism in homozygous familial hypercholesterolemia fibroblasts.* Proc Natl Acad Sci U S A, 1976. 73(9): p. 3178-82.

[48] Phillips, M.C., W.J. Johnson, and G.H. Rothblat, *Mechanisms and consequences of cellular cholesterol exchange and transfer.* Biochim Biophys Acta, 1987. 906(2): p. 223-76.

[49] Brasaemle, D.L., A.D. Robertson, and A.D. Attie, *Transbilayer movement of cholesterol in the human erythrocyte membrane.* J Lipid Res, 1988. 29(4): p. 481-9.

[50] Ohvo-Rekila, H., et al., *Cholesterol interactions with phospholipids in membranes.* Prog Lipid Res, 2002. 41(1): p. 66-97.

[51] Rothman, J.E. and D.M. Engelman, *Molecular mechanism for the interaction of phospholipid with cholesterol.* Nat New Biol, 1972. 237(71): p. 42-4.

[52] Chabanel, A., et al., *Influence of cholesterol content on red cell membrane viscoelasticity and fluidity.* Biophys J, 1983. 44(2): p. 171-6.

[53] Coderch, L., et al., *Influence of cholesterol on liposome fluidity by EPR. Relationship with percutaneous absorption.* J Control Release, 2000. 68(1): p. 85-95.

[54] Lenaz, G., *Lipid fluidity and membrane protein dynamics.* Biosci Rep, 1987. 7(11): p. 823-37.

[55] Rimon, G., et al., *Mode of coupling between hormone receptors and adenylate cyclase elucidated by modulation of membrane fluidity.* Nature, 1978. 276(5686): p. 394-6.

[56] Schramm, M., *Transfer of glucagon receptor from liver membranes to a foreign adenylate cyclase by a membrane fusion procedure.* Proc Natl Acad Sci U S A, 1979. 76(3): p. 1174-8.

[57] Lenaz, G.a.P.C., G., *Structure and Properties of Cell Membranes,* G. Benga, Editor 1985, CRC Press: Boca Raton, FLA. p. 73-136.

[58] Chachisvilis, M., Y.L. Zhang, and J.A. Frangos, *G protein-coupled receptors sense fluid shear stress in endothelial cells.* Proc Natl Acad Sci U S A, 2006. 103(42): p. 15463-8.

[59] Simons, K. and D. Toomre, *Lipid rafts and signal transduction.* Nat Rev Mol Cell Biol, 2000. 1(1): p. 31-9.

[60] Brown, D.A. and E. London, *Functions of lipid rafts in biological membranes.* Annu Rev Cell Dev Biol, 1998. 14: p. 111-36.

[61] Radel, C., M. Carlile-Klusacek, and V. Rizzo, *Participation of caveolae in beta1 integrin-mediated mechanotransduction.* Biochem Biophys Res Commun, 2007. 358(2): p. 626-31.

[62] Rizzo, V., et al., *In situ flow activates endothelial nitric oxide synthase in luminal caveolae of endothelium with rapid caveolin dissociation and calmodulin association.* J Biol Chem, 1998. 273(52): p. 34724-9.

[63] Rizzo, V., et al., *Rapid mechanotransduction in situ at the luminal cell surface of vascular endothelium and its caveolae.* J Biol Chem, 1998. 273(41): p. 26323-9.

[64] Marwali, M.R., et al., *Membrane cholesterol regulates LFA-1 function and lipid raft heterogeneity.* Blood, 2003. 102(1): p. 215-22.

[65] Niggli, V., et al., *Impact of cholesterol depletion on shape changes, actin reorganization, and signal transduction in neutrophil-like HL-60 cells.* Exp Cell Res, 2004. 296(2): p. 358-68.

[66] Pierini, L.M., et al., *Membrane lipid organization is critical for human neutrophil polarization.* J Biol Chem, 2003. 278(12): p. 10831-41.

[67] Seely, A.J., J.L. Pascual, and N.V. Christou, *Science review: Cell membrane expression (connectivity) regulates neutrophil delivery, function and clearance.* Crit Care, 2003. 7(4): p. 291-307.

[68] Tuluc, F., J. Meshki, and S.P. Kunapuli, *Membrane lipid microdomains differentially regulate intracellular signaling events in human neutrophils.* Int Immunopharmacol, 2003. 3(13-14): p. 1775-90.

[69] Stulc, T., et al., *Leukocyte and endothelial adhesion molecules in patients with hypercholesterolemia: the effect of atorvastatin treatment.* Physiol Res, 2008. 57(2): p. 184-94.

[70] Lehr, H.A., et al., *In vitro effects of oxidized low density lipoprotein on CD11b/CD18 and L-selectin presentation on neutrophils and monocytes with relevance for the in vivo situation.* Am J Pathol, 1995. 146(1): p. 218-27.

[71] Sugano, R., et al., *Polymorphonuclear leukocytes may impair endothelial function: results of crossover randomized study of lipid-lowering therapies.* Arterioscler Thromb Vasc Biol, 2005. 25(6): p. 1262-7.

[72] Furlow, M. and S.L. Diamond, *Interplay between membrane cholesterol and ethanol differentially regulates neutrophil tether mechanics and rolling dynamics.* Biorheology, 2011. 48(1): p. 49-64.

[73] Tailor, A. and D.N. Granger, *Hypercholesterolemia promotes leukocyte-dependent platelet adhesion in murine postcapillary venules.* Microcirculation, 2004. 11(7): p. 597-603.

[74] Lechi, C., et al., *Increased leukocyte aggregation in patients with hypercholesterolaemia.* Clin Chim Acta, 1984. 144(1): p. 11-6.

[75] Araujo, F.B., et al., *Evaluation of oxidative stress in patients with hyperlipidemia.* Atherosclerosis, 1995. 117(1): p. 61-71.

[76] Kristal, B., et al., *Participation of peripheral polymorphonuclear leukocytes in the oxidative stress and inflammation in patients with essential hypertension.* Am J Hypertens, 1998. 11(8 Pt 1): p. 921-8.

[77] Shurtz-Swirski, R., et al., *Involvement of peripheral polymorphonuclear leukocytes in oxidative stress and inflammation in type 2 diabetic patients.* Diabetes Care, 2001. 24(1): p. 104-10.

[78] Makino, A., et al., *Mechanotransduction in leukocyte activation: a review.* Biorheology, 2007. 44(4): p. 221-49.

[79] Shin, H.Y., et al., *Mechanobiological Evidence for the Control of Neutrophil Activity by Fluid Shear Stress* in *Mechanobiology Handbook*, J. Nagatomi, Editor 2011, CRC Press: Boca Raton, FL, USA. p. 139-75.

[80] Fukuda, S. and G.W. Schmid-Schönbein, *Centrifugation attenuates the fluid shear response of circulating leukocytes.* J Leukoc Biol, 2002. 72(1): p. 133-9.

[81] Fukuda, S., et al., *Mechanisms for regulation of fluid shear stress response in circulating leukocytes.* Circ Res, 2000. 86(1): p. E13-8.

[82] Moazzam, F., et al., *The leukocyte response to fluid stress.* Proc Natl Acad Sci U S A, 1997. 94(10): p. 5338-43.

[83] Makino, A., et al., *Control of neutrophil pseudopods by fluid shear: role of Rho family GTPases.* Am J Physiol Cell Physiol, 2005. 288(4): p. C863-71.

[84] Komai, Y. and G.W. Schmid-Schönbein, *De-activation of neutrophils in suspension by fluid shear stress: a requirement for erythrocytes.* Ann Biomed Eng, 2005. 33(10): p. 1375-86.

[85] Shive, M.S., W.G. Brodbeck, and J.M. Anderson, *Activation of caspase 3 during shear stress-induced neutrophil apoptosis on biomaterials.* J Biomed Mater Res, 2002. 62(2): p. 163-8.

[86] Shive, M.S., M.L. Salloum, and J.M. Anderson, *Shear stress-induced apoptosis of adherent neutrophils: a mechanism for persistence of cardiovascular device infections.* Proc Natl Acad Sci U S A, 2000. 97(12): p. 6710-5.

[87] Chen, H.Q., et al., *Effect of steady and oscillatory shear stress on F-actin content and distribution in neutrophils.* Biorheology, 2004. 41(5): p. 655-64.

[88] Cicchetti, G., P.G. Allen, and M. Glogauer, *Chemotactic signaling pathways in neutrophils: from receptor to actin assembly.* Crit Rev Oral Biol Med, 2002. 13(3): p. 220-8.

[89] Niggli, V., *Signaling to migration in neutrophils: importance of localized pathways.* Int J Biochem Cell Biol, 2003. 35(12): p. 1619-38.

[90] Tybulewicz, V.L. and R.B. Henderson, *Rho family GTPases and their regulators in lymphocytes.* Nat Rev Immunol, 2009. 9(9): p. 630-44.

[91] Makino, A., et al., *G protein-coupled receptors serve as mechanosensors for fluid shear stress in neutrophils.* Am J Physiol Cell Physiol, 2006. 290(6): p. C1633-9.

[92] Marschel, P. and G.W. Schmid-Schönbein, *Control of fluid shear response in circulating leukocytes by integrins.* Ann Biomed Eng, 2002. 30(3): p. 333-43.

[93] Anderson, S.I., et al., *Linked regulation of motility and integrin function in activated migrating neutrophils revealed by interference in remodelling of the cytoskeleton.* Cell Motil Cytoskeleton, 2003. 54(2): p. 135-46.

[94] Simon, S.I. and C.E. Green, *Molecular mechanics and dynamics of leukocyte recruitment during inflammation.* Annu Rev Biomed Eng, 2005. 7: p. 151-85.

[95] Simon, S.I. and H.L. Goldsmith, *Leukocyte adhesion dynamics in shear flow.* Ann Biomed Eng, 2002. 30(3): p. 315-32.

[96] Shin, H.Y., S.I. Simon, and G.W. Schmid-Schönbein, *Fluid shear-induced activation and cleavage of CD18 during pseudopod retraction by human neutrophils.* J Cell Physiol, 2008. 214(2): p. 528-36.

[97] Fukuda, S. and G.W. Schmid-Schönbein, *Regulation of CD18 expression on neutrophils in response to fluid shear stress.* Proc Natl Acad Sci U S A, 2003. 100(23): p. 13152-7.

[98] Walcheck, B., et al., *Neutrophil rolling altered by inhibition of L-selectin shedding in vitro.* Nature, 1996. 380(6576): p. 720-3.

[99] Carney, D.F., et al., *Effect of serine proteinase inhibitors on neutrophil function: alpha-1-proteinase inhibitor, antichymotrypsin, and a recombinant hybrid mutant of antichymotrypsin (LEX032) modulate neutrophil adhesion interactions.* J Leukoc Biol, 1998. 63(1): p. 75-82.

[100] Pfaff, M., X. Du, and M.H. Ginsberg, *Calpain cleavage of integrin beta cytoplasmic domains.* FEBS Lett, 1999. 460(1): p. 17-22.

[101] Arnaout, M.A., *Structure and function of the leukocyte adhesion molecules CD11/CD18.* Blood, 1990. 75(5): p. 1037-50.

[102] Radi, Z.A., M.E. Kehrli, Jr., and M.R. Ackermann, *Cell adhesion molecules, leukocyte trafficking, and strategies to reduce leukocyte infiltration.* J Vet Intern Med, 2001. 15(6): p. 516-29.

[103] Fialkow, L., Y. Wang, and G.P. Downey, *Reactive oxygen and nitrogen species as signaling molecules regulating neutrophil function.* Free Radic Biol Med, 2007. 42(2): p. 153-64.

[104] Guzik, T.J., R. Korbut, and T. Adamek-Guzik, *Nitric oxide and superoxide in inflammation and immune regulation.* J Physiol Pharmacol, 2003. 54(4): p. 469-87.

[105] Dal Secco, D., et al., *Nitric oxide inhibits neutrophil migration by a mechanism dependent on ICAM-1: role of soluble guanylate cyclase.* Nitric Oxide, 2006. 15(1): p. 77-86.

[106] Kubes, P., M. Suzuki, and D.N. Granger, *Nitric oxide: an endogenous modulator of leukocyte adhesion.* Proc Natl Acad Sci U S A, 1991. 88(11): p. 4651-5.

[107] Kubes, P., et al., *Nitric oxide synthesis inhibition induces leukocyte adhesion via superoxide and mast cells.* Faseb J, 1993. 7(13): p. 1293-9.

[108] Gaboury, J., et al., *Nitric oxide prevents leukocyte adherence: role of superoxide.* Am J Physiol, 1993. 265(3 Pt 2): p. H862-7.

[109] Vilhardt, F. and B. van Deurs, *The phagocyte NADPH oxidase depends on cholesterol-enriched membrane microdomains for assembly.* EMBO J, 2004. 23(4): p. 739-48.

[110] Fukuda, S., et al., *Contribution of fluid shear response in leukocytes to hemodynamic resistance in the spontaneously hypertensive rat.* Circ Res, 2004. 95(1): p. 100-8.

[111] Suzuki, H., et al., *Impaired leukocyte-endothelial cell interaction in spontaneously hypertensive rats.* Hypertension, 1994. 24(6): p. 719-27.

[112] Arndt, H., C.W. Smith, and D.N. Granger, *Leukocyte-endothelial cell adhesion in spontaneously hypertensive and normotensive rats.* Hypertension, 1993. 21(5): p. 667-73.

[113] Suematsu, M., et al., *The inflammatory aspect of the microcirculation in hypertension: oxidative stress, leukocytes/endothelial interaction, apoptosis.* Microcirculation, 2002. 9(4): p. 259-76.

[114] Kaul, S., R.C. Padgett, and D.D. Heistad, *Role of platelets and leukocytes in modulation of vascular tone.* Ann N Y Acad Sci, 1994. 714: p. 122-35.

[115] Mugge, A., et al., *Activation of leukocytes with complement C5a is associated with prostanoid-dependent constriction of large arteries in atherosclerotic monkeys in vivo.* Atherosclerosis, 1992. 95(2-3): p. 211-22.

[116] Faraci, F.M., et al., *Effect of atherosclerosis on cerebral vascular responses to activation of leukocytes and platelets in monkeys.* Stroke, 1991. 22(6): p. 790-6.

[117] DeLano, F.A. and G.W. Schmid-Schönbein, *Enhancement of glucocorticoid and mineralocorticoid receptor density in the microcirculation of the spontaneously hypertensive rat.* Microcirculation, 2004. 11(1): p. 69-78.

[118] Sutanto, W., et al., *Corticosteroid receptor plasticity in the central nervous system of various rat models.* Endocr Regul, 1992. 26(3): p. 111-8.

[119] Fukuda, S., H. Mitsuoka, and G.W. Schmid-Schönbein, *Leukocyte fluid shear response in the presence of glucocorticoid.* J Leukoc Biol, 2004. 75(4): p. 664-70.

[120] Sugihara-Seki, M. and G.W. Schmid-Schönbein, *The fluid shear stress distribution on the membrane of leukocytes in the microcirculation.* J Biomech Eng, 2003. 125(5): p. 628-38.

[121] Haidekker, M.A., N. L'Heureux, and J.A. Frangos, *Fluid shear stress increases membrane fluidity in endothelial cells: a study with DCVJ fluorescence.* Am J Physiol Heart Circ Physiol, 2000. 278(4): p. H1401-6.

[122] Butler, P.J., et al., *Rate sensitivity of shear-induced changes in the lateral diffusion of endothelial cell membrane lipids: a role for membrane perturbation in shear-induced MAPK activation.* Faseb J, 2002. 16(2): p. 216-8.

[123] Butler, P.J., et al., *Shear stress induces a time- and position-dependent increase in endothelial cell membrane fluidity.* Am J Physiol Cell Physiol, 2001. 280(4): p. C962-9.

[124] Ferraro, J.T., et al., *Depletion of plasma membrane cholesterol dampens hydrostatic pressure and shear stress-induced mechanotransduction pathways in osteoblast cultures.* Am J Physiol Cell Physiol, 2004. 286(4): p. C831-9.

[125] Zhang, Y.L., J.A. Frangos, and M. Chachisvilis, *Mechanical stimulus alters conformation of type 1 parathyroid hormone receptor in bone cells.* Am J Physiol Cell Physiol, 2009. 296(6): p. C1391-9.

[126] Chen, K.D., et al., *Mechanotransduction in response to shear stress. Roles of receptor tyrosine kinases, integrins, and Shc.* J Biol Chem, 1999. 274(26): p. 18393-400.

[127] Jin, Z.G., et al., *Ligand-independent activation of vascular endothelial growth factor receptor 2 by fluid shear stress regulates activation of endothelial nitric oxide synthase.* Circ Res, 2003. 93(4): p. 354-63.

[128] Lee, H.J. and G.Y. Koh, *Shear stress activates Tie2 receptor tyrosine kinase in human endothelial cells.* Biochem Biophys Res Commun, 2003. 304(2): p. 399-404.

[129] Milkiewicz, M., et al., *HIF-1alpha and HIF-2alpha play a central role in stretch-induced but not shear-stress-induced angiogenesis in rat skeletal muscle.* J Physiol, 2007. 583(Pt 2): p. 753-66.

[130] Shay-Salit, A., et al., *VEGF receptor 2 and the adherens junction as a mechanical transducer in vascular endothelial cells.* Proc Natl Acad Sci U S A, 2002. 99(14): p. 9462-7.

[131] Tarbell, J.M., S. Weinbaum, and R.D. Kamm, *Cellular fluid mechanics and mechanotransduction.* Ann Biomed Eng, 2005. 33(12): p. 1719-23.

[132] Kamm, R.D. and M.R. Kaazempur-Mofrad, *On the molecular basis for mechanotransduction.* Mech Chem Biosyst, 2004. 1(3): p. 201-9.

[133] Lee, S.E., R.D. Kamm, and M.R. Mofrad, *Force-induced activation of talin and its possible role in focal adhesion mechanotransduction.* J Biomech, 2007. 40(9): p. 2096-106.

[134] Chen, A.K., et al., *Evidence for the role of G-proteins in flow stimulation of dinoflagellate bioluminescence.* Am J Physiol Regul Integr Comp Physiol, 2007. 292(5): p. R2020-7.

[135] Su, S.S. and G.W. Schmid-Schönbein, *Internalization of Formyl Peptide Receptor in Leukocytes Subject to Fluid Stresses.* Cell Mol Bioeng, 2010. 3(1): p. 20-29.

[136] Chen, A.Y., et al., *Receptor cleavage reduces the fluid shear response in neutrophils of the spontaneously hypertensive rat.* Am J Physiol Cell Physiol, 2010. 299(6): p. C1441-9.

[137] Mofrad, M.R., et al., *Force-induced unfolding of the focal adhesion targeting domain and the influence of paxillin binding.* Mech Chem Biosyst, 2004. 1(4): p. 253-65.

[138] Gillmore, J.D., et al., *Amyloid load and clinical outcome in AA amyloidosis in relation to circulating concentration of serum amyloid A protein.* Lancet, 2001. 358(9275): p. 24-9.

[139] Johnson, B.D., et al., *Serum amyloid A as a predictor of coronary artery disease and cardiovascular outcome in women: the National Heart, Lung, and Blood Institute-Sponsored Women's Ischemia Syndrome Evaluation (WISE).* Circulation, 2004. 109(6): p. 726-32.
[140] Tomida, K., et al., *Hypercholesterolemia induces leukocyte entrapment in the retinal microcirculation of rats.* Curr Eye Res, 2001. 23(1): p. 38-43.

Management of Hyper and Dyslipoproteinemias

Chapter 10

Endoscopic Treatment of Metabolic Syndrome

Eduardo Guimarães Hourneaux de Moura, Ivan Roberto Bonotto Orso,
Bruno da Costa Martins and Guilherme Sauniti Lopes

Additional information is available at the end of the chapter

1. Introduction

For a long time, obesity has been known as a risk factor for cardiovascular disease, which is one of the main causes of death in developed countries. The prevalence of obesity (defined as having a body mass index [BMI] of 30 kg/m2 or more) is increasing in both developing and developed countries. A 5-kg/m2 increase in body mass index (BMI) increases the risk of cardiac complications by 29% [1]. This risk is due to the coexistence of other factors associated with obesity, such as hypertension, dyslipidemia, nonalcoholic fatty liver disease and abnormalities in glycemic metabolism. Resistances to insulin and lipid abnormalities are commonly found among obese patients with type 2 diabetes mellitus (T2DM) and are strongly related to an increase of cardiovascular risk. Resistance to insulin and consequent compensatory hyperinsulinemia significantly increase the risk of death due to cardiovascular diseases [2–5].

To identify the patients with metabolic syndrome, insulin resistance, and greater cardiovascular risk, there are criteria, established by the Adult Treatment Panel III (NCEP ATP III), that include the presence of three of more of the following: central obesity (abdominal circumference above 102 cm in men and 88 cm in women), increased triglycerides (greater than or equal to 150 mg/dl) or use of a lipid-lowering agent, reduced HDL cholesterol (lower than 40 mg/dl among men and lower than 50mg/dl among women) or use of a lipid lowering agent, hypertension (systolic arterial pressure greater than or equal to 130 mmHg or diastolic pressure greater than or equal to 90mmHg) or use of an antihypertensive agent, andglucose levels greater than or equal to 100 mg/dl or use of an oral hypoglycemic agent and/or insulin [6].

Despite this, recent studies propose that the use of the TG/HDL (Triglicerides/High Density Lipoprotein) ratio may be a more practical way to estimate insulin resistance. It is believed that the greater the ratio, the greater the insulin resistance of the patient. This ratio provides an estimate of the sensitivity to insulin and is as accurate as the criteria for the metabolic syndrome defined by the ATP III, the concentration of plasma insulin when fasting, or other estimates that measure the amount of glucose and the plasma concentration of insulin in order to evaluate its action [5, 7 -9]. Some studies suggest that the increase of the TG/HDL ratio may better predict the risk of cardiovascular diseases than do conventional risk factors such as hypertension, tobacco use, and physical activity [10].

To control the obesity and insulin resistance the initial steps are lifestyle changes aimed at controlling diet and increasing activity with the goal of reducing body weight, followed by the addition of orally active pharmacologic agents and insulin to the treatment regimen.

However, dietary modification and pharmaceutical therapy offer limited potential for sustained weight loss, effective in fewer than 5% of cases [2]. In a meta-analysis of pharmacotherapy for obesity, the percentages of patients achieving 5% and 10% weight loss thresholds by using anti obesity drugs were 54% and 18%, respectively, but a lack of adherence to treatment limited the efficacy and effectiveness [11].

Weight loss surgery, in contrast, has been shown to effect a more durable response . In addition, it can induce reversal of obesity-associated comorbidities [2,12].

Of interest is the observation that obese patients with diabetes who undergo certain gastric bypass procedures demonstrate improvement in glycemia, often within days of surgery and before significant weight loss. The exact mechanism responsible for this dramatic effect of surgical procedures for obesity on diabetes improvement is not fully understood; however, the surgical rearrangement of the anatomy of the gastrointestinal (GI) tract changes the location where partially digested nutrients first contact the intestine, suggesting that correction of dysfunctional homeostatic mechanisms may contribute to the glycemic improvement. Whether it is a pure effect of weight reduction or bypass of the hormonally active foregut has a primary effect remains a controversy. Hypothesis includes weight reduction, decreased caloric intake, and bypass of the hormonally active foregut [12-14].

However, weight loss surgery is associated with complications such as anastomotic leak and ulcer presenting a mortality rate estimates range between 0.1 and 2.0% [2].

Endoscopic weight loss therapies may provide some of the benefits of weight loss surgery while being reversible, with a lower risk profile, and being available to patients who do not qualify for surgery or are poor candidates for surgery. Those endoscopic solutions for weight loss are also applicable as metabolic procedures to address comorbidities as type 2 diabetes, dyslipidemia and nonalcoholic fatty liver disease.

2. Physiopathology of metabolic improvement after metabolic surgery

Several studies have shown a significant diabetes improvement in obese patients who undergo certain gastric bypass procedures. The improvement of glycemia is observed, often within days of surgery and before significant weight loss. The mechanism responsible for this improvement is not fully understood; however, the surgical rearrangement of the anatomy of the gastrointestinal tract may contribute to the glycemic improvement. If the glicemic improvement is a pure effect of weight reduction or bypass of the hormonally active foregut has a primary effect remains a controversy. Hypothesis to explain those effects includes weight reduction, decreased caloric intake, and bypass of the hormonally active.foregut. [12-14]

Two hypothesis have been proposed to explain the effect of duodenal jejunal bypass on type 2 diabetes. The "hindgut hypothesis" holds that diabetes control results from the expedited delivery of nutrient chyme to the distal intestine, enhancing a physiologic signal that improves glucose metabolism. A potential candidate mediator of this effect is glucagon-like peptide 1 (GLP-1). This incretin hormone is secreted by L cells of the distal bowel in response to intestinal nutrients. It stimulates insulin secretion and exerts proliferative and antiapoptotic effects on pancreatic *beta* cells.16 If proven true, the hindgut hypothesis would spur further research on methods to enhance signaling by GLP-1 (or other distal gut peptides) to treat type 2 diabetes [14-16]. An alternative hypothesis is that the effect of selected bariatric operations on diabetes depends on exclusion of the duodenum and proximal jejunum from the transit of nutrients, possibly preventing secretion of a putative signal that promotes insulin resistance and type 2 diabetes ("foregut hypothesis"). Although no obvious candidate molecules can be identified with current knowledge, if proven true, this hypothesis might open new avenues in the search for the cause and cures of diabetes [14,17,18].

Several reports supported that the duodeno jejunal exclusion (foregut hypothesis) owes to a direct effect of the bypass of the hormonally active foregut. Rubino and Marescaux showed in their study in animal model that bypassing a short segment of proximal intestine directly ameliorates type 2 diabetes, independently of effects on food intake, body weight, malabsorption, or nutrient delivery to the hindgut [14].

However,in previous studies, it was observed that a strict calorie restriction, as performed in the first weeks after bariatric surgery, could bring itself to a normalization of plasma glucose and insulin levels before body weight decrease [19,20].

Wei-Jei Lee in a study comparing the band (restrictive) with the gastric bypass (duodeno jejunal bypass), with a longer follow-up, showed that the gastric banding group had a similar improvement of insulin resistance to the bypass group while similar weight reduction was achieved. The gastric banding group had similar result at postoperative 6 months compared to the gastric bypass group at the first postoperative month. Also, 3 months to 1 year and 6 months to 2 years were compatible. Suggesting that for a long-term effect of resolution of insulin resistance, sustained weight reduction plays the key mechanism [12].

Improvement in the glycemic control, insulin resistance and metabolic syndrome after bariatric surgery is regulated by a complex mechanism and still there is no certainty whether it is a pure effect of calorie restriction and weight reduction or it is caused by the bypass of the hormonally active foregut .

3. Metabolic improvements with the intragastric baloon

The intragastric baloon is a spherical silicone elastomer balloon that is resistant to degradation by gastric acid for approximately 6 months. It can be placed endoscopicallyand filled with 400 to 700 ml of saline and methylene blue dye, which changes the color of the urine in the event of balloon rupture.

Balloon insertion and removal are performed under conscious sedation or general anesthesia. Before the insertion, an upper gastrointestinal endoscopy is performed to detect possible contraindications to the procedure. The baloon placement device is inserted through the mouth into the stomach. Then the balloon is positioned in the fundus under endoscopic control, and inflated by injecting saline mixed with 10-ml methylene blue into the catheter. Finally, once the desired volume has been injected, the balloon is released by a short pull on the catheter. The baloon should be removed after a maximum of 6 months because beyond this period, the risk of spontaneous balloon deflation significantly increases.

A meta-analysis by Imaz et al. [21] of 15 studies comprising 3698 patients estimated 14.7 kg weight loss, 32.1% excess weight loss (EWL), and 5.7 kg/m2 decrease in BMI after 6 months.

In a review including 22 studies with a total of 4371 patients implanted with the intragastric baloon, demonstrated a mean weight loss of 17.8 Kg, with extremes of the means of 4.9–28.5 kg and higher absolute values observed in higher BMI categories. [22].

A prospective study, evaluating the effect of the baloon on weight, insulin resistance, and liver steatosis in obese patients showed that 76% of the patients had a BMI decrease of 3.5 Kg/m^2 or more. The mean (SD) weight loss with respect to baseline values was 16.4 (8.2) kg with a corresponding mean (SD) BMI reduction of 6.4 (3.2) kg/m2. The absolute percentage of participants with glycemia levels of 100 mg/dL or higher decreased from 50% to 12%, those with triglyceridemia 150 mg/dL or higher from 58% to 19%, and those with abnormal ALT level from 38% to 7% [23].

Two studies (one randomized, one uncontrolled) totaling 143 patients have reported that, one year after BIB removal, patients had regained 41% and 28% (mean values, respectively) of the absolute weight loss observed at BIB removal [24,25]. Another study following 88 patients for a median of 22 month after baloon withdrawal, observed that (50%) regained some weight, 34 (39%) maintained their weight, and the remaining 10 (11%) continued to lose weight [23].

It is also important to consider that 20–40% of patients fail to achieve a significant weight loss (often defined as ≥10% baseline weight or ≥25% excess weight). Such failures may be related to the request of early baloon removal by patients who present a digestive or psychological intolerance to the baloon, to the early vanishing of anticipated effects on hunger and early satiety, or to patient's adaptation of food intake [23].

In conclusion, the BIB strategy may be an alternative to current management of obesity focused on lifestyle changes, drug therapy, and treating associated metabolic complications.

Although the baloon has not yet proved to be a convincing means of primary long term weight loss, it holds some promise for improving co-morbidities and quality of life in nonmorbidly obese patients or those who are unwilling to undergo bariatric surgery. New perspectives are also beginning to show its potential value in specific patient groups especially, for example, those preparing for surgery.

4. Metabolic improvements with duodeno jejunal bypass liner

A totally different concept, that of mimicking principles of bariatric surgery, has recently been applied in the development of the endoscopic duodenal-jejunal bypass liner. In addition to early satiety and delayed gastric emptying, the intraluminal sleeve aims at creating a duodenojejunal exclusion.

The DJBL is a sterilized, single-use endoscopic device, which is minimally invasive and employed under radioscopic control. It is composed of a nitinol anchoring with tiny lateral barbs for fixation and an impermeable plastic conduit made of a fluorine polymer 62 cm in length, which impedes contact of the chyme with bile–pancreatic secretions prior to the proximal segments of the jejunum. FIGURE 1

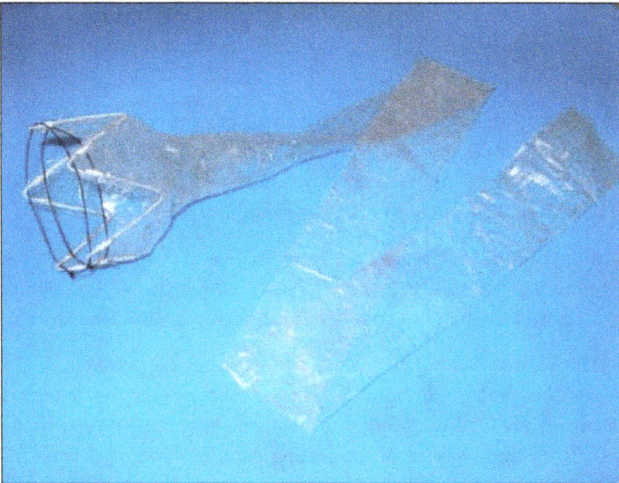

Figure 1. Impermeable plastic conduit and anchor system.

Endoscopic implantation is performed under general anesthesia. The device is introduced over a guidewire that has been previously positioned in the duodenal bulb with endoscopic assistance. The plastic conduit is stretched to overlay the duodenum and the proximal region of the jejunum. After the correct positioning of the plastic conduit, the anchoring system is freed, setting the device in the duodenal bulb. The infusion of a contrast agent is performed to verify the correct positioning of the prosthesis and the absence of obstructions within the plastic conduit. FIGURE 2

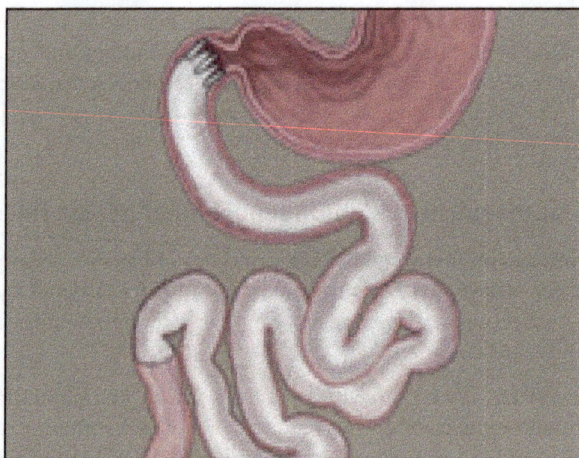

Figure 2. Implanted DJBL.

Previously studies with obese patients that used the duodenojejunal bypass liner (DJBL) demonstrated a significant weight loss. In addition, an improvement in the control of T2DM was observed, which was statistically greater than that of the group treated with a low-calorie diet [26-33].

Our group in the Gastrointestinal Endoscopy Unit of the University of São Paulo School of Medicine performed a prospective study to evaluates the effectiveness of this method over the control of hyperlipidemia, improvement of insulin resistance, metabolic syndrome, and in the potential benefit in the reduction of cardiovascular risk [34].

The inclusion criteria of the study were ages between 18 and 65 years, BMI≥35 kg/m2, T2DM with or without other comorbidities, and a triglyceride/high-density lipoprotein cholesterol ratio (TG/HDL) ratio greater than or equal to 3.5, indicating insulin resistance.

To identify patients with resistance to insulin and metabolic syndrome, TG/HDL ratio has proved to be an excellent practical indicator. This ratio estimates the resistance to insulin and is as accurate in terms of the clinical criteria for metabolic syndrome as specified by the Adult Treatment Panel III [6], the measure of the fasting concentration of plasma insulin,or other estimates that measure glycemia and the plasma concentration of insulin in order to identify individuals with insulin resistance [5, 7,8]. In addition, the TG/HDL ratio has proved to be an independent factor that was correlated with the risk of cardiovascular events [5].

A low TG/HDL ratio indicates large particles and a lower atherogenic potential of LDL cholesterol, while a high TG/ HDL ratio indicates a large population of small, dense, and pro-atherogenic particles of LDL cholesterol [7]. The lipid disorder consisting of the increase of plasma triglycerides and the reduction of HDL cholesterol, known as atherogenic dyslipidemia, is directly associated with insulin resistance and is also an independent risk factor for cardiovascular diseases [3–5, 7].

Eighty-one patients were selected for implantation of the device. Of these, 78 successfully received the implant and three were not given the implant due to anatomic factors (short bulb). Among the 78 patients, one was excluded from the analysis for not having performed the laboratory measures. Of the 77 remaining patients, we calculated the initial TG/ HDL ratio (at the time of the implant), identifying 54 patients (70%) with a ratio greater than or equal to 3.5, indicating the presence of insulin resistance and metabolic syndrome. These patients were included in the study and were monitored in order to evaluate whether or not an improvement occurred in this ratio during the period in which they had the implanted device.

We compared the TG/HDL ratio at the time of the implant with the ratio obtained after 6 months to evaluate whether there was an improvement in insulin resistance. We divided the patients into two groups: those who demonstrated a ratio below 3.5 in the end of the study, considered as control of the insulin resistance, and those that did not demonstrate values lower than 3.5. In the two groups, we evaluated whether control of T2DM and weight loss occurred during this period, and we correlated the influence of the control of diabetes and weight loss with the improvement of TG/HDL. We considered a significant weight loss to be a reduction of at least 10% of initial body weight and control of T2DM as an HbA1c level lower than 7%. The patients who presented a reduction of HbA1c levels greater than 1.5% yet did not obtain values lower than 7% were considered to have partial control over DM2.

The overall initial average of the TG/HDL ratio was 5.75 and presented a significant reduction down to 4.36 at the end of the 6 months (p=<0.001), indicating an improvement of insulin resistance (Table 1). Of these patients, 23 (42.6%) presented control of the TG/HDL ratio with values lower than 3.5 at the end of the study. This group presented a significant improvement in the ratio, which decreased from 5.15 to 2.85 (p<0.001). Thirty-one patients did not show a controlled TG/HDL ratio but rather a discrete improvement, with an initial average of 6.2 and a final of 5.47, with no statistical difference (p=0.1641).

	Patients N	Initial average TG/HDL ratio	Final average TG/HDL ratio	p
Baseline non controlled TG/HDL	54	5,75	4,36	0,001
Controlled TG/HDL at the end	23	5,15	2,85	<0,001
Not controlled TG/HDL at the end	31	6,2	5,47	0,1641

Table 1. Improvement on TG/HDL ratio.

In order to identify the differences between the group that presented an improvement in the ratio and the group that did not, we evaluated the control of T2DM (HbA1c improvement) and the success of weight loss (reduction >10% of initial weight).

In the evaluation of T2DM control (Table 2), we observed that all patients presented a significant improvement in the levels of HbA1c (p<0.001). In the group that controlled the

TG/HDL ratio, three patients already had diabetes controlled at the beginning of the study, with an initial average of HbA1c of 6.4% and a final of 5.83% (p=0.023). Fifteen patients did not have diabetes under control, presenting an initial average of 7.8%, and then developing control of T2DM, with a final average of 6.1% (p<0.001). Five patients did not have diabetes undercontrol and obtained partial control after the intervention. These patients presented higher initial levels of HbA1c than those of the group that controlled diabetes, with an initial average of 10.34%, and presented a significant reduction of the final average of 8.88% (p=0.03).

In the group that did not control the TG/HDL ratio, five patients already had T2DM under control, with an initial average of 6.6%mg/dl and a final average of 6.2% (p=0.037). Fifteen patients did not have diabetes under control and were able to bring it under control with initial and final averages of 8.5% and 6.4%, respectively (p=0.001). Eleven patients did not have diabetes under control and obtained partial control. These patients also presented a higher level of HbA1c than that of the patients who had controlled T2DM, with an initial average of 9.9%, reaching a significant reduction at the end of the study with an average of 7.7% (p=0.003), which is very close to the level required for T2DM control.

An association was not observed between the control of T2DM and an improvement in the TG/HDL ratio (Table 3).

Diabetes improvement on TG/HDL controlled patients				
Diabetes	Patients N	Inial HbA1c average	Final HbA1c average	p
Already controled and improved	3	6,4	5,38	0,023
Not controled who controlled	15	7,8	6,1	<0,001
Not controlled with partial control	11	10,34	8,88	0,03
HbA1c Worsening	0	-	-	.
Diabetes improvement on TG/HDL not controlled patients				
Diabetes	Patients	Inial HbA1c average	Final HbA1c average	p
Already controled and improved	5	6,6	6,2	0,037
Not controled who controlled	15	8,5	6,4	0,001
Not controlled with partial control	11	9,9	7,7	0,003
HbA1c Worsening	0	-	-	.

Legend: Diabetes evolution on TG/HDL controlled and not controlled patients. Glycemic improvement was statistically significant in all groups of patients.

Table 2. Diabetes improvement.

	Diabetes Improvement – HbA1c			HbA1c Worsening
	Already controled and improved	Not controlled who controlled	Not controlled with partial control	
TG/HDL controlled (n = 23)	3	15	5	0
TG/HDL not controlled (n= 31)	5	15	11	0

Legend: Relationship between improvement of diabetes and TG/HDL control. Patients improved diabetes regardless of having controlled the TG/HDL ratio (p 0,35).

Table 3. Relation between TG/HDL ratio control and HbA1C control.

In relation to weight loss (Table 4), the patients lost on average 12.6% of their initial weight. Among the 23 patients who controlled their TG/HDL ratio, 19 (82.6%) lost more than 10% of their initial weight. The average initial weight of these patients was 116.5 kg and the average final weight was 97 kg, constituting an average loss of 16.7% of initial weight. Four patients did not lose more than 10% of their weight. The average initial weight of these patients was 94.4 kg, and the average final weight was 87.47 kg, marking an average loss of 7.4%. In the group that did not control the TG/HDL ratio, 15 lost more than 10% of their initial weight (48%), with an average initial weight of 123.9 kg and an average final weight of 105.7 kg (loss of 14.6% in initial weight). Sixteen patients did not lose more than 10% of their weight, presenting an average initial weight of 111.9 kg and an average final weight of 103.5 kg, with an average loss of 7.5% in initial weight.

Weight loss on TG/HDL controlled patients				
Weight loss	Patients N (%)	Inicial average weight - Kg	Final average weight - Kg	Percentage of loss
Over than 10% of initial weight	19 (82,6)	116,5	97	16,7
Less than 10% of initial weight	4 (17,4)	94,4	87,47	7,4
Weight loss on TG/HDL not controlled patients				
Weight loss	Patients N (%)	Inicial weight average	Final weight average	Percentage of loss
Over than 10% of initial weight	15 (48)	123,9	105,7	14,6
Less than 10% of initial weight	16 (52)	111,9	103,5	7,5

Table 4. Weight loss

Comparing the patients who lost weight with the patients who controlled their TG/HDL ratio, an association can be observed between a weight loss greater than 10% of initial weight and control of the TG/HDL ratio (p<0.01), with an odds ratio of 5.06 (Table 5).

	Weight loss		
	Over than 10% of inicial weight	Less than 10% of inicial weight	Total
TG/HDL controlled - N (average weight loss %)	19 (16,7)	4 (7,4)	23
TG/HDL not controlled N (average weight loss %)	15 (14,6)	16 (7,5)	31

Legend: Relationship between weight loss and TG/HDL control. Control of TG/HDL ratio is related to weight loss greater than 10% of initial weight. (p 0,01- OR 5,06).

Table 5. Relation between TG/HDL ratio control and weight loss.

Of the 54 patients included in this 6 months study, 38 have completed (26 completed 24 weeks, 12 completed 20 weeks). Among the 16 patients left, 12 had the device removed at 16 weeks, 2 at 12 weeks, and 2 had the implant for just 4 weeks. The early implant removals occurred due to migration of the device in nine patients, the observation of a free device anchor during endoscopic exam in four patients, the presence of bleeding without migration in one patient, subject request in one case, and due to the decision of the researcher in one case.

In resume, all patients implanted with the device presented a statistically significant reduction of the levels of HbA1c, and the majority of these patients (70.3%) presented values lower than 7% at the end of the study and were therefore considered to be controlled diabetics. In addition, all patients presented a statistically significant reduction of initial weight, with an average general loss of 12.6% of initial weight. Regarding the improvement of insulin resistance and metabolic syndrome, there was a significant reduction of the TG/HDL ratio from 5.75 to 4.36 (p=0.0001). Of these patients, 42.6% controlled their insulin resistance, presenting a TG/HDL ratio value lower than 3.5 at the end of the study.

Among the patients who controlled the TG/HDL ratio, the reduction of the ratio went from 6.8 at the beginning of the study to 2.8 at the end (p<0.001).

In other study conducted by our group, twenty two implanted patients were followed during a period of 1 year [13]. In the full analysis population, the mean percentage excess weight loss was 35.5% (P < 0.0001). The reduction in excess body weight was reflected by reductions in BMI and waist circumference of 6.7 kg/m2 and 13.0 cm, respectively.

The improvement in glycemic control is convincingly demonstrated by the results with a percentage of subjects with HbA1c < 7% at baseline improved from 4.5% to 73.0% at final study assessment. Statistically significant reductions in fasting blood glucose (- 30.3 ± 10.2 mg/dL), fasting insulin (- 7.3 ± 2.6 lU/mL), and HbA1c (- 2.1 ± 0.3%) were observed.

Blood levels of total cholesterol, low-density lipoprotein cholesterol, and triglycerides also were significantly reduced during the study.

On this one year series, thirteen subjects completed the 52-week period, and 18 subjects completed at least 24 weeks. The mean duration of the implant period for all subjects was 41.9 ± 3.2 weeks. The reasons for early removal of the device were migration or rotation of the device (n = 3; 36, 36, and 48 weeks post implantation), GI bleeding (n = 1; 4 weeks post implantation), abdominal pain (n = 2; 21 and 30 weeks post-implantation), and principal investigator request due to subject's non compliance with study visits (n = 2; 20 and 32 weeks post implantation). The device was removed from one subject who presented an abdominal tumor not related to the device.

Sixteen subjects had HbA1c measured 3 and/or 6 months after explantation of the DJBL. These subjects demonstrated a mean decrease in HbA1c during the original 52-week study of - 2.3 ± 0.4%. Three and 6 months after removal of the device, their mean changes from baseline were - 2.3 ± 0.3% (n = 15) and - 1.7 – 0.7% (n = 11), respectively.

The DJBL offers a new non-surgical therapeutic possibility, positioned between pharmacological drugs and the various techniques employed in bariatric surgery. This technology platform may be employed prior to bariatric surgery to help control T2DM, in order to promote weight loss and a reduction of visceral fat, lipid control, a reduction of insulin resistance, and of cardiovascular risk,minimizing the risk of per operative clinical complications, accustoming the patient to a restricted diet that will be necessary in the post-operative period and can even be used as a substitution for bariatric surgery as a less invasive technique in selected cases.

5. Conclusion

As the prevalence of obesity increases, less invasive methods will be needed to obtain a sustained weight loss. Some new endoscopic tools and methods are being investigated and they could be applied as first-line therapy for obesity, to control the metabolic comorbidities, to reduce the operatory risk prior bariatric and metabolic surgery and as substitution of surgery in selected cases.

The intragastric baloon and the Duodeno Jejunal Bypass Liner are tools with promising results in the endoscopic treatment of obesity. They are still subject of research with a great potential for improvement. Although outcomes from the use of the intragastric baloon and the DJBL are not comparable to those of surgery with regard to weight loss and late results, these new techniques have showed an excelent result in ameliorating health status, in the control of the metabolic syndrome as well as improving the quality of life for a well selected group of patients.

Author details

Eduardo Guimarães Hourneaux de Moura*,
Bruno da Costa Martins, Guilherme Sauniti Lopes
Department of Gastroenterology, Gastrointestinal Endoscopy Unit,
Hospital das Clínicas - University of São Paulo School of Medicine, São Paulo, Brazil

Ivan Roberto Bonotto Orso
Department of Gastroenterology, Gastrointestinal Endoscopy Unit,
Hospital das Clínicas - University of São Paulo School of Medicine. São Paulo, Brazil
Department of Surgery, School of Medicine of the Assis Gurgacz Faculty,
Gastroclínica Cascavel, Brazil

6. References

[1] Bogers RP, Bemelmans WJ, Hoogenveen RT, et al. Association of overweight with increased risk of coronary heart disease partly independent of blood pressure and cholesterol levels: a metaanalysis of 21 cohort studies including more than 300 000 persons. Arch Intern Med. 2007;167(16):1720–8.

[2] Kumar N, Thompson CC. Endoscopic solutions for weight loss. Current Opinion in Gastroenterology 2011, 27:407–41

[3] McLaughlin T, Abbasi F, Cheal K, et al. Use of metabolic markers to identify overweight individuals who are insulin resistant. Ann Intern Med. 2003;139:802–9.

[4] Shishehbor MH, Hoogwef BJ, Lauer MS. Association of triglyceride to HDL cholesterol ratio with heart rate recovery. Diabetes Care. 2004;27(4):936–41.

[5] McLaughlin T, Reaven G, Abbasi F, et al. Is there a simple way to identify insulin resistant individuals at increased risk of cardiovascular disease? Am J Cardiol. 2005;96(3):399–404.

[6] Tong PC, Kong AP, SoWY, et al. The usefulness of the international diabetes federation and the national cholesterol education program's Adult Treatment Panel III definitions of the metabolic syndrome in predicting coronary heart disease in subjects with type 2 diabetes. Diabetes Care. 2007;30(5):1206–11.

[7] Hadaegh F, Dhalili D, Ghasemi A, et al. Triglyceride/HDLcholesterol ratio is an independent predictor for coronary heart disease in a population of Iranian men. Nutr Metab CardiovascDis. 2009;19(6):401–8.

[8] Vasques ACJ, Rosado LEFPM, Rosado GP, et al. Indicadores do perfil lipídico plasmático relacionados à resistência á insulina. Rev Assoc Méd Bras. 2009;55(3):342–6.

[9] Quizada Z, Paoli M, Zerpa Y, et al. The triglyceride/HDL cholesterol ratio as a marker of cardiovascular risk in obesechildren; association with traditional and emergent risk factors. Pediatr Diabetes. 2008;9(5):464–71.

[10] Kannel WB, Vasan RS, Keyes MJ, et al. Usefulness of the triglyceride high density lipoprotein versus the cholesterol high density lipoprotein ratio for predicting insulin

---

*Corresponding Author

resistance and cardiometabolic risk (from the Framingham Offspring Cohort). Am J Cardiol. 2008;101(4):497–501.

[11] Rucker D, Padwal R, Li SK, et al. Long term pharmacotherapy for obesity and overweight: update and meta-analysis. Br Med J 2007;335: 1194-9.)

[12] Lee WJ, Lee YC, Ser KH, Chen JC, Chen SC. Improvement of Insulin Resistance After Obesity Surgery: A Comparison of Gastric Banding and Bypass Procedures Obes Surg 2008; 18:1119–1125

[13] Moura,EGH, Martins BC, Lopes GS, Orso IR, Oliveira SL, Galva˜o Neto MP, et al. Metabolic Improvements in Obese Type 2 Diabetes Subjects Implanted for 1 Year with an Endoscopically Deployed Duodenal–Jejunal Bypass Liner. Diabetes Technol Ther 2012;14(2). DOI: 10.1089/dia.2011.0152

[14] Rubino F, Forgione A, Cummings DE, Vix M, Gnuli D, Mingrone G et al, The Mechanism of Diabetes Control After Gastrointestinal Bypass Surgery Reveals a Role of the Proximal Small Intestine in the Pathophysiology of Type 2. Ann Surg 2006;244: 741–749

[15] Mason EE. The mechanism of surgical treatment of type 2 diabetes. Obes Surg. 2005;15:459–461.

[16] Patriti A, Facchiano E, Sanna A, et al. The enteroinsular axis and the recovery from type 2 diabetes after bariatric surgery. Obes Surg. 2004; 14:840–848.

[17] Pories WJ, Albrecht RJ. Etiology of type II diabetes mellitus: role of the foregut. World J Surg. 2001;25:527–531.

[18] Rubino F, Gagner M, Gentileschi P, et al. The early effect of the Roux-en-Y gastric bypass on hormones involved in body weight regulation and glucose metabolism. Ann Surg. 2004;240:236 –242.

[19] Kelly DE, Wing R, Buonocore C, et al. Relative effect soft calorie restriction and weight loss in non-insulin-dependent diabetes mellitus. J Clin Endocrinol Metab. 1993;77:1287–93. 21.

[20] Ash S, Reeves MM, Yeo S, et al. Effect of intensive dietetic interventions on weight and glycemic control in overweight men with type II diabetes: a randomized trial. Int J Obes. 2003;27: 797–802.)

[21] Imaz I, Martı´nez-Cervell C, Garcı´a-Alvarez EE, et al. Safety and effectiveness of the intragastric balloon for obesity. A meta-analysis. Obes Surg 2008;7:841–846.

[22] Dumonceau JM. Evidence-based Review of the Bioenterics Intragastric Balloon for Weight Loss. Obes Surg (2008) 18:1611–1617

[23] Forlano R, Ippolito AM, Iacobellis A, et al. Effect of the BioEnterics intragastric balloon on weight, insulin resistance, and liver steatosis in obese patients. Gastrointest Endosc 2010; 71:927–933.

[24] Herve J, Wahlen CH, Schaeken A, et al. What becomes of patients one year after the intragastric balloon has been removed? Obes Surg. 2005;15:864–70.

[25] Mathus-Vliegen EM, Tytgat GN. Intragastric balloon for treatment- resistant obesity: safety, tolerance, and efficacy of 1-year balloon treatment followed by a 1-year balloon-free follow-up. Gastrointest Endosc. 2005;61:19–27.

[26] Gersin KS, Keller JE, Stefanidis D, et al. Duodenal–jejunal bypass sleeve: a totally endoscopic device for the treatment of morbid obesity. Surg Innov. 2007;14(4):275–8.

[27] Schauer P, Chand B, Brethauer S. New applications for endoscopy: the emerging field of endoluminal and transgastric bariatric surgery. Surg Endosc. 2007;21:347–56.

[28] Tarnoff M, Shikora S, Lembo A, et al. Chronic in-vivo experience with an endoscopically delivered and retrieved duodenal–jejunal bypass sleeve in a porcine model. SurG Endosc. 2008;22(4):1023–8.

[29] Tarnoff M, Shikora S, Lembo A. Acute technical feasibility of an endoscopic duodenal–jejunal bypass sleeve in a porcine model: a potentially novel treatment for obesity and type 2 diabetes. Surg Endosc. 2008;22(3):772–6.

[30] Tarnoff M, Rodriguez L, Escalona A, et al. Open label, prospective, randomized controlled trial of an endoscopic duodenal–jejunal bypass sleeve versus low calorie diet for pre-operative weight loss in bariatric surgery. Surg Endosc. 2009;23(3):650–6.

[31] Rodriguez-Grunert L, Galvao Neto MP, Alamo M, et al. First human experience with endoscopically delivered and retrieved duodenal– jejunal bypass sleeve. Surg Obes Relat Dis. 2008;4(1):55–9.

[32] Rodriguez L, Reyes E, Fagalde P, et al. Pilot clinical study of an endoscopic, removable duodenal–jejunal bypass liner for the treatment of type 2 diabetes. Diabetes Technol Ther. 2009;11 (11):725–32.

[33] Schouten R, Rijs CS, Bouvy ND, et al. A multicenter, randomized efficacy study of the EndoBarrier Gastrointestinal Liner for presurgical weight loss prior to bariatric surgery. Ann Surg. 2010;251(2):236–43

[34] Moura EGH, Orso IR, Martins BC, Lopes GS, Oliveira SL, et al. Improvement of Insulin Resistance and Reduction of Cardiovascular Risk Among Obese Patients with Type 2 Diabetes with the Duodenojejunal Bypass Liner. Obes Surg (2011) 21:941–947

The Confounding Factor of Apolipoprotein E on Response to Chemotherapy and Hormone Regulation Altering Long-Term Cognition Outcomes

Summer F. Acevedo

Additional information is available at the end of the chapter

1. Introduction

One player in health cognitive functioning is shown to be lipoproteins, essential in the metabolism and redistribution of lipids: cholesterol, phospholipids and triacylglycerol. There are several classes of lipoproteins which are used to transport lipids throughout the body and range in density (protein/lipid ratio); chylomicrons (contain dietary lipids), intermediate low density lipoproteins (IDL), very low density lipoproteins (VLDL, bad cholesterol), low density lipoproteins (LDLs) and the high density lipoproteins (HDLs, good cholesterol). The Apolipoprotein/Apoprotein gene family of proteins is part of the lipoprotein complexes that function as regulators of binding between lipoproteins and receptors. These proteins act as enzyme co-factors during lipid metabolism, helping to stabilize lipoproteins during transportation from cell or tissue to its destination [1].

Apolipoprotein E (ApoE), initially termed the "arginine-rich apoprotein", was first identified as a part of the VLDL complexes. ApoE is synthesized principally in the liver, but has also been found in other tissues such as the brain, ovaries, lungs, adrenals, spleen, muscle cells, and macrophages [2]. The three most common alleles of *ApoE* are *ApoE2*, *ApoE3*, *ApoE4* [3] found in the nervous system are primarily produced in astroglia and microglia. The three major isoforms differ at position 112 (*ApoE2/ApoE3* Cysteine, *ApoE4* Arginine) and 158 (*ApoE2* Cysteine, *ApoE3/ApoE4* Arginine), the amino acid substitutions at position 112 affect salt bridge formation within the protein, which ultimately impacts on lipoprotein preference, stability of the protein and on receptor binding activities of the isoforms [4]. Being an *ApoE4* carrier or having the *ApoE2/ApoE3* genotype is associated with higher triglyercide levels, higher VLDL levels, higher total cholesterol, higher total lipoproteins levels and elevated LDL or "bad cholesterol" levels all of which contribute to hypertension and diabetes confounding factors that have to be considered when designing a chemotherapy treatment regiment [5].

In rodent models, the lack of *ApoE* or ApoE4 protein expression leads to destabilization of cell membranes, increased apoptosis, and heightened sensitivity to neuronal trauma; whereas, ApoE3 and ApoE2 protein expression allow for healthy cell functioning and neuroprotection [4]. ApoE4 has detrimental effects in transgenic mice, including behavioral abnormalities, such as deficits in spatial learning and memory using Morris water maze (MWM) [6], as well as significant alterations in the hippocampus and cortex [4,7]. The studies in mice are consistent with clinical studies indicating reduced spatial learning and memory in those who carry the *ApoE4* allele [8]. Experiments have also demonstrated that the three isoforms of human *ApoE* gene have different effects on the development of neurodegenerative diseases. Those individuals that are *ApoE4* carriers have an increased risk of age-related mild cognitive impairments (MCI) and the development of Alzheimer's disease (AD) particularly in females [9].

In ovarian cancer, ApoE protein levels act as a potential tumor-associated marker as found in serous carcinomas, but not in serious borderline or normal ovarian surface epithelium cells [10]. Up-regulation of ApoE protein levels is also seen in breast carcinomas, pancreatic cancer, stomach carcinomas, colon carcinomas and prostate carcinomas [10]. Blockage of ApoE expression in the serous carcinoma cell lines leads to cell cycle arrest and apoptosis. Women infused with ApoE protein at the time of diagnosis showed significantly higher survival rates [10]. This data suggest that upregulation of ApoE expression may be a defense mechanism to help body fight carcinomas.

2. Chemotherapy, *ApoE* and memory

Over the last 20 years it has become apparent that chemotherapy drugs not only attack cancer cells, but also cross the blood brain barrier (BBB) leading to negative effects on cognitive processing which is known as Chemo-brain or Chem-fog [11,12]. Methotrexate, 5-Flourouracil (5-FU) the most common chemotherapy drugs used to treat breast, colorectal, head and neck cancers been shown in both rodent models and clinical studies to lead to neurocognitive deficits in a variety of domains including visual memory and visuospatial functioning [11,13,14]. In clinical studies, there is considerable variability between studies as to the extent and frequency of such impairments heretofore mentioned. However, rodent studies are clearly show that the drugs in the CMF (cyclophosphamide, methotrexate, 5-FU) regiment lead to decreased hippocampus cell proliferation and induce MWM memory impairments [15,16]. In addition, cytarabine (cytosine arabinoside) and ifosfamide among other chemotherapy have been shown to lead to memory impairments, hemiparesis, aphasia and progressive dementia [17]. However, there is significant lack of pre-clinical testing of most chemotherapeutic agents and their long-term effects on memory.

Currently, only one study has examined if *ApoE4* carrier status as a potential genetic risk factor in breast or lymphoma survivors. They have found that even after 8 years, *ApoE4* carriers displayed impairments, specifically in visual memory and spatial ability [18]. This suggests that the *ApoE* genotype may be a confounding factor to consider when examining post-chemotherapy neurocognitive status. More studies are necessary to confirm this result.

Studies in women treated with CMF regiment for breast cancer found increased total cholesterol, LDL, HDL cholesterol and Apolipoprotein A-1 (ApoA-1) in those who developed permanent amenorrhea (loss of menstrual cycle, induced menopause) [19]. Studies report that around 30% of patients who have gone through chemotherapy develop permanent amenorrhea [20]. There is evidence that *ApoE4* carriers have an earlier onset of natural menopause [21]. Age of menopause and being an *ApoE4* carrier are both risk factors for age related diseases including AD and coronary artery disease (CAD). There is also a direct connection between ApoE mRNA levels and estrogen in various tissues, including the brain [22] leading to regulation of neurite outgrowth [7]. The data suggests that ApoE is a critical intermediary in the estrogen related neuroplasticity [7]. Lack of estrogen along with expression of ApoE4 protein which has reduced ApoE functioning is one potential cause of impaired cognitive performance in some women that have undergone chemotherapy. Studies do not take this factor into account or narrowly examine menopausal status at the time of testing.

3. Apolipoproteins effects on secondary drug response

3.1. Tamoxifen

Other confounding factors that can affect cognitive status after chemotherapy include other medications patients are taking to control comorbid conditions or to treat the tumor itself. Tamoxifen is used as an estrogen receptor modulator (SERM) in estrogen receptor (ER) positive breast cancer carcinomas. Tamoxifen is a pure estrogen receptor blocker. As with other drugs, tamoxifen and its metabolites can cross the BBB affecting ER in various brain regions including the cerebral cortex, hippocampus and amygdala [23]. In combination with chemotherapy, tamoxifen appears to intensify the cognitive impairments, particularly in visual memory, verbal working memory and visuospatial ability [24]. The Anatrozole, Tamoxifen Combined (ATAC) trial, also found verbal memory and processing speed impairments post-chemotherapy treated only with tamoxifen compared to women only on a combined ATAC treatment (Table 1) [25]. This suggests that tamoxifen has confounding effects when given as part of the chemotherapy regiment. Tamoxifen also appears to have both agonist and antagonist properties in the brain with reported up-regulation of pro-inflammatory cytokines shown to be related to cognitive dysfunction [26]. Positron emission tomography (PET) imaging of survivors does show higher hypometabolism with dual chemotherapy and tamoxifen, not seen in women treated only with tamoxifen [27]. Animal models using repeated tamoxifen or combinations of methotrexate and 5-FU injections both produced deficits in acquisition and retention in an operant learning paradigm (Table 1)

[28]. These studies were conducted when women were still in treatment, leaving the question of potential long-term effects. There is a study that examined women who used tamoxifen for <4 years compared to >6 years of exposure. The study found that the current exposure led to greater memory deficits compared to non-users (Table 1) [29]. This suggests that while on therapy, patients may have acute memory impairments and that alternative drugs should be seriously considered.

In breast cancer survivors undergoing tamoxifen treatment, their total cholesterol, VLDL, high density lipoproteins (HDL) and Apolipoprotein B (ApoB) protein levels have been shown to decrease in both *ApoE4* carriers and non-*ApoE4* carriers (Table 1) [30,31]. Breast cancer patients who are *ApoE4* carriers have higher plasma triglyceride levels and altered ApoA-1/ApoB ratio. Both are risk factors for cardiovascular events, after tamoxifen treatment (Table 1) [30,31]. In non-*ApoE4* carriers, there were lower levels of lipoprotein (a) after treatment, but no effect on triglycerides or the ApoA-1/ApoB ratio (Table 1). This suggests that non-*ApoE4* carriers have a more positive response to tamoxifen with respect to lipid profiles and risk for cardiovascular complications [31]. Considering tamoxifen and *ApoE4* both have a deleterious effect on cognition and tamoxifen has an *ApoE* genotype dependent effect on lipid profiles, suggest further investigations are warranted to understand the mechanistic relationship.

3.2. Anatrozole

Anatrozole also known as arimidex is an aromatase inhibitor that lowers estrogen levels and is used as a treatment in estrogen positive breast cancer patients post-surgery. Results of the ATAC trial of 9399 women indicated that those only on arimidex had better clinical outcomes including vascular events and gynecological problems compared to the tamoxifen group with no differences seen in cognitive outcomes [25,32]. There data suggests that arimidex was the preferred initial treatment by women treated for breast cancer [32]. However, other studies indicate that women on arimidex treatment had greater cognitive decline than tamoxifen treatment in verbal and visual memory (Table 1) [33,34]. At this point no effects have been seen on cholesterol, lipoproteins or apolipoprotein levels in both animal models and clinical studies (Table 1) [35,36]. Although, there are not alternations in lipids levels, the cognitive side effects are of concern with this medication and should be examined with respect to *ApoE* genotype.

3.3. Letrozole

Letrozole, a potent aromatase interfering with adrenal steroid biosynthesis, has also been assessed as a replacement for tamoxifen or as secondary maintenance treatment after tamoxifen as part of the Breast International Group (BIG 1-98) trial [37,38]. In the BIG 1-98 study, better overall cognitive outcomes were seen in women on letrozole treatment compared to those on a tamoxifen treatment (Table 1) [37,38]. In the tamoxifen only group, increased endometrial cancer and vaginal bleeding where found [38]. Participants given letrozole only did displayed more incidences of skeletal and cardiac events and hypercholesterolemia compared to the tamoxifen group [37]. After letrozole treatment, unfavorable effects have been seen including increased serum total cholesterol, LDL and ApoB with atherogenic ratio risk of total cholesterol/HDL and LDL/HDL levels (Table 1) [39]. Therefore, the better cognitive outcomes may not out-weigh negative effects on lipoprotein levels, particularly in those with or at risk for hypertension and/or diabetes. The potential confounding factor of *ApoE* genotype has not been reported.

3.4. Exemestane

There is another alternative for estrogen suppression therapy exemestane, an aromatase inhibitor, general used after tamoxifen is not working in post-menopausal women. Exemestane has been examined as part of the randomized Tamoxifen and Exemestane Adjunctive Multinational (TEAM) trial [24]. The preliminary data from the TEAM trial found that tamoxifen is associated with lower verbal and executive function, while exemestane did not seem to alter cognitive performance levels (Table 1) [24]. Analysis of data from 72 patients as part of the EORTC trail 10958 indicates that treatment with exemestane resulted in reduced triglyceride levels and tamoxifen treatment increased triglyceride levels [40]. All other lipid parameters including HDL, ApoA-1, ApoB or Lip (a) levels at 8, 24 and 48 weeks were unchanged by either treatment [40]. Further studies will be needed to determine if stable cognitive performance and lipid levels are effect of *ApoE* genotype.

3.5. Raloxifene

Raloxifene, also a SERM and is used as a hormone replacement therapy, has the positive effects of estrogen on the skeletal system and is an antagonist of estrogen in breast or endometrial tissues [41]. Raloxifene treatment also appears to be less detrimental to cognitive function assessed by Modified Mini-Mental State (3MS) compared to tamoxifen [42]. Used to prevent osteoporosis, it has been shown that after three years of treatment raloxifene did not affect overall cognitive scores [43]. The Multiple Outcomes of Raloxifene Evaluation (MORE) study of 7478 women, reported finding that raloxifene treatment lowered the risk of cognitive decline in word list recall test and there was no overall effect on cognitive function (Table 1) [44]. A subset of the National Surgical Adjuvant Breast and Bowel Project (NSABP) Study of Tamoxifen and Raloxifene (STAR), the CoSTAR study for women at high risk for breast cancer did not find any cognitive effects of either drug (Table 1) [45]. Together the evidence supports that raloxifene treatment does not impair cognitive function the way that tamoxifen treatment and in fact may even lower the risk of cognitive decline [34].

With respect to lipid and lipoprotein levels post hoc analysis of 2659 women in the MORE study, found that raloxifene treatment in women with or without high triglycerides lead to reduced cholesterol levels with healthier lipoprotein parameters (Table 1) [46]. Studies in Greek women, found that LDL cholesterol levels were lower in women treated with raloxifene [47]. Raloxifene also appears to raise HDL levels and ApoA-1 while decreasing ApoB protein levels and improving ratios of total cholesterol to lipoproteins (Table 1). After one year of treatment with raloxifene women had reduced fat mass and trunk and central regions along with decreased adiposity in their truck and abdominal regions (Table 1) [48]. Overall, raloxifene treatment improves cholesterol health and alters fat distribution in a positive manner to help prevent obesity, making it a better candidate for overall health compared to tamoxifen. Its effects in relation to *ApoE* genotype have not been reported.

3.6. Estradiol

In post-menopausal estradiol (also known as 17β-estradiol or oestradiol) treatment is used for estrogen replacement therapy. Healthy post-menopausal women given estradiol display improved visuospatial abilities measured by a mental rotation task (Table 1) [49]. Other non-randomized studies in women with surgically induced amenorrhea or those with AD indicate that estrogen replacement treatment may help to improve or minimize cognitive deficits [50]. Even in men those given estradiol performed better on visual memory after treatment (Table 1) [51]. These results are consistent with improved memory in mice given other replacement estrogens treatments [52]. Over half of randomized clinical studies find significant improvements in cognition and attention after estrogen replacement therapy (Table 1) [53]. Estradiol has been shown to increase levels of ApoE in the brain, proposed to be beneficial for neuronal reorganization and repair [7]. In a health study of 3,393 women, results suggest that estrogen replacement reduces the risk of age-related cognitive decline in non-*ApoE4* women, but not in *ApoE4* carriers (Table 1) [54]. Another study with 181 post-menopausal women, also found the best learning and memory performance after estrogen replacement is seen in non-*ApoE4* carriers [55]. This suggests that knowing *ApoE* genotype may be helpful to assess potential response to estrogen replacement therapy.

Treatment	Function	Effects on Lipids/Apolipoproteins	Effects on Cognition
Tamoxifen	Estrogen receptor modulator (SERM) lowers estrogen function	Decreased total cholesterol, VLDL, HDL and ApoB protein levels.	Leads to impairments in visual memory, verbal working memory and visuospatial ability.
Anatrozole	Aromatase inhibitor lowers estrogen function	No effects seen in rodent or clinical studies.	Reduced verbal and visual memory compared to tamoxifen. Deficits

Treatment	Function	Effects on Lipids/Apolipoproteins	Effects on Cognition
			in rodent model operant learning paradigm.
Letrozole	Aromatase inhibitor lowers estrogen function.	Increased serum total cholesterol, LDL and ApoB increase risk of cardiovascular events.	Better overall cognitive outcomes compared to tamoxifen treatment.
Exemestane	Aromatase inhibitor lowers estrogen function.	Reduces triglyercide levels.	No effect on cognitive performance compared to Tamoxifen group that had lower verbal and executive functioning.
Raloxifene	SERM, lowers estrogen function in reproductive tissue and used as estrogen replacement therapy in non-reproductive tissues.	Lower cholesterol, LDL, ApoB protein levels and increases ApoA-1, HDL level leading to better cardiovascular health. Also shown to reduced adiposity and fat mass.	Lowered the risk of cognitive decline or has no effect.
Estradiol	Estrogen replacement therapy	Increase ApoE protein levels in brain. Reduces cognitive decline in only non-*ApoE4* carriers.	Treatment improves visuospatial abilities and visual memory.
Tibolone	Estrogen replacement therapy	Reduces total cholesterol, triglyceride levels, HDL and ApoA-1 levels.	Decreased anxiety, improved quality of like and semantic memory.
Cetrorelix	Used to reduce gonadotrophins and sex steroids	Increases ApoA-1 and HDL levels.	Anxiolytic, anti-depressive and improved beta-amyloid 25-35 associated memory consolidation impairments.

Table 1. Hormone treatments effects of lipids/apolipoproteins levels and cognition.

3.7. Tibolone

Tibolone is another drug used in hormonal replacement therapy having estrogenic, progestogenic, and androgenic effects. Long-term treatment does appear to decrease anxiety, improve semantic memory and overall quality of life; however, one study reported that those in treatment did score worse on attention task compared to women not on treatment (Table 1) [56,57]. Tibolone appears to be the most beneficial with respect to reducing total cholesterol, triglyercide, HDL and ApoA-1 levels compared to raloxifene and estradiol (Table 1) [47,58-60]. This suggests that hormone replacement therapy with medications such as tibolone in post-menopausal women are beneficial to cognitive health and lipid profiles.

3.8. Cetrorelix

Cetrorelix an antagonist of hypothalamic luteinizing hormone-releasing hormone (LHRH), is used in treatment of prostate carcinoma, benign prostatic hyperplasia, and ovarian cancer to reduce gonadotrophins and sex steroids [61]. In mice, a study suggests that it is anxiolytic, anti-depressive and able to correct beta-amyloid 25-35 associated memory consolidation impairments (Table 1) [61]. Injection of cetrorelix into ApoE deficient mice (ApoE$^{-/-}$) mice suggests that the associated suppression of testosterone leads to increased atherosclerosis despite lower cholesterol levels in the male mice [62]. In female ApoE$^{-/-}$ mice, the reduction in testosterone also leads to reduction in estradiol, insulin and HDL levels without effects on atherosclerosis [62]. In a pilot study conducted in men, treatment with cetrorelix resulted in increased ApoA-1, HDL, insulin and leptin consistently (Table 1) [63]. Therefore, when this drug is used within a chemotherapy treatment regiment it is important to carefully monitor lipid levels. Additional studies are needed to examine if ApoE genotype has any effect on response and potential long-term cognitive side effects of this drug.

4. Recommendations

1. Determine if an ApoE genotype can help assess what is most treatment useful including how to properly maintain lipid levels during chemotherapy.
2. Find out lipid levels, track and maintain determined treatment to reduce risk of post-chemotherapy cognitive impairments.
3. Examine ApoE genotype before selecting pharmacotherapy options pre or post-chemotherapy treatment.

5. Conclusion

Overall of the studies SERMs/aromatase inhibitors raloxifene or exemestane may be better alternatives to tamoxifen or letrozole treatment in terms of effects on cognitive deficits and overall health risk in women treated with chemotherapy. In addition, ApoE genotype and cholesterol levels need to be taken into account when examining efficacy of these drugs and in hormone replacement therapies as efficacy is dependent on *ApoE* genotype. Knowing these issues will help doctors to address them early for improving quality of life, reducing services used and saving millions of dollars in unneeded medical expenses. Realistically, the wonder drug that can cure all cancer and has no side effects will not be found. What is needed is to reduce the impact and intensity of cognitive side-effects as much as possible, taking into account an individual's physiology and genetics. The more we know about cognitive status across ages, ethnicity, lipid levels, and genetic status the better we can treat mind and body.

Author details

Summer F. Acevedo*
Department of Physiology, Pharmacology, and Toxicology, Psychology Program,
Ponce School of Medicine and Health Sciences, Ponce, Puerto Rico

Acknowledgement

We acknowledge the support of Tirtsa Porrata-Doria and the Ponce School of Medicine and Health Sciences (PSMHS) Research Center for Minority Institutions (RCMI), Molecular Biology Core Lab (G12 RR003050). Special thanks go to Robert Ritchie from the PSMHS/RCMI Publications Office (G12 RR003050/8G12MD007579-27). Additionally, would like to acknowledge Dr. Jacob Raber at Oregon Health Science University for mentoring and sparking my interest in apolipoproteins and cancer research.

6. References

[1] Han, X. (2004) The role of apolipoprotein E in lipid metabolism in the central nervous system. Cell Mol. Life Sci. 61: 1896-1906.
[2] Mahley, R. W., Y. Huang, and K. H. Weisgraber (2006) Putting cholesterol in its place: apoE and reverse cholesterol transport. J. Clin. Invest. 116: 1226-1229.
[3] Mahley, R. W. (1988) Apolipoprotein E: cholesterol transport protein with expanding role in cell biology. Science 240: 622-630.
[4] Mahley, R. W., K. H. Weisgraber, and Y. Huang (2009) Apolipoprotein E: structure determines function, from atherosclerosis to Alzheimer's disease to AIDS. J. Lipid Res. 50 Suppl: S183-188.
[5] Tso, T. K., J. T. Snook, R. A. Lozano, and W. B. Zipf (2001) Risk factors for coronary heart disease in type 1 diabetic children: the influence of apoE phenotype and glycemic regulation. Diabetes Res. Clin. Pract. 54: 165-171.
[6] Morris, R. (1984) Developments of a water-maze procedure for studying spatial learning in the rat. J. Neurosci. Method. 11: 47-60.

---

* Corresponding Author

[7] Struble, R. G., C. Cady, B. P. Nathan, and M. McAsey (2008) Apolipoprotein E may be a critical factor in hormone therapy neuroprotection. Front. Biosci. 13: 5387-5405.

[8] Berteau-Pavy, F., B. Park, and J. Raber (2007) Effects of sex and APOE epsilon4 on object recognition and spatial navigation in the elderly. Neuroscience 147: 6-17.

[9] Mahley, R. W., and Y. Huang (2006) Apolipoprotein (apo) E4 and Alzheimer's disease: unique conformational and biophysical properties of apoE4 can modulate neuropathology. Acta. Neurol. Scand. Suppl. 185: 8-14.

[10] Chen, Y. C., G. Pohl, T. L. Wang, P. J. Morin, B. Risberg, G. B. Kristensen, A. Yu, B. Davidson, and M. Shih Ie (2005) Apolipoprotein E is required for cell proliferation and survival in ovarian cancer. Cancer Res. 65: 331-337.

[11] Ahles, T. A., and A. Saykin (2001) Cognitive effects of standard-dose chemotherapy in patients with cancer. Cancer Invest. 19: 812-820.

[12] Meyers, C. A. (2008) How chemotherapy damages the central nervous system. J. Biol. 7: 11.

[13] Janelsins, M. C., S. Kohli, S. G. Mohile, K. Usuki, T. A. Ahles, and G. R. Morrow (2011) An update on cancer- and chemotherapy-related cognitive dysfunction: current status. Semin. Oncol. 38: 431-438.

[14] Nelson, C. J., N. Nandy, and A. J. Roth (2007) Chemotherapy and cognitive deficits: mechanisms, findings, and potential interventions. Palliat. Support. Care 5: 273-280.

[15] Seigers, R., S. B. Schagen, C. M. Coppens, P. J. van der Most, F. S. van Dam, J. M. Koolhaas, and B. Buwalda (2009) Methotrexate decreases hippocampal cell proliferation and induces memory deficits in rats. Behav. Brain Res. 201: 279-284.

[16] Winocur, G., J. Vardy, M. A. Binns, L. Kerr, and I. Tannock (2006) The effects of the anti-cancer drugs, methotrexate and 5-fluorouracil, on cognitive function in mice. Pharmacol. Biochem. Behav. 85: 66-75.

[17] Verstappen, C. C., J. J. Heimans, k. Hoekman, and T. J. Postma (2003) Neurotoxic complications of chemotherapy in patients with cancer: clinical signs and optimal management. Drugs 63: 1549-1563.

[18] Ahles, T. A., A. J. Saykin, W. W. Noll, C. T. Furstenberg, S. Guerin, B. Cole, and L. A. Mott (2003) The relationship of APOE genotype to neuropsychological performance in long-term cancer survivors treated with standard dose chemotherapy. Psycho-oncology 12: 612-619.

[19] Saarto, T., C. Blomqvist, C. Ehnholm, M. R. Taskinen, and I. Elomaa (1996) Effects of chemotherapy-induced castration on serum lipids and apoproteins in premenopausal women with node-positive breast cancer. J. Clin. Endocrinol. Metab. 81: 4453-4457.

[20] Phillips, K. A., and J. Bernhard (2003) Adjuvant breast cancer treatment and cognitive function: current knowledge and research directions. J. Nat. Cancer Inst. 95: 190-197.

[21] Koochmeshgi, J., S. M. Hosseini-Mazinani, S. Morteza Seifati, N. Hosein-Pur-Nobari, and L. Teimoori-Toolabi (2004) Apolipoprotein E genotype and age at menopause. Ann. N. Y. Acad. Sci. 1019: 564-567.

[22] Srivastava, R. A., N. Srivastava, M. Averna, R. C. Lin, K. S. Korach, D. B. Lubahn, and G. Schonfeld (1997) Estrogen up-regulates apolipoprotein E (ApoE) gene expression by

increasing ApoE mRNA in the translating pool via the estrogen receptor alpha-mediated pathway. J. Biol. Chem. 272: 33360-33366.

[23] Ciocca, D. R., and L. M. Roig (1995) Estrogen receptors in human nontarget tissues: biological and clinical implications. Endocr. Rev. 16: 35-62.

[24] Schilder, C. M., C. Seynaeve, L. V. Beex, W. Boogerd, S. C. Linn, C. M. Gundy, H. M. Huizenga, J. W. Nortier, C. J. van de Velde, F. S. van Dam, and S. B. Schagen (2010) Effects of tamoxifen and exemestane on cognitive functioning of postmenopausal patients with breast cancer: results from the neuropsychological side study of the tamoxifen and exemestane adjuvant multinational trial. J. Clin. Oncol. 28: 1294-1300.

[25] Shilling, V., V. Jenkins, L. Fallowfield, and T. Howell (2003) The effects of hormone therapy on cognition in breast cancer. J. Steroid Biochem. Mol. Biol. 86: 405-412.

[26] Wefel, J. S., A. E. Kayl, and C. A. Meyers (2004) Neuropsychological dysfunction associated with cancer and cancer therapies: a conceptual review of an emerging target. Br. J. Cancer 90: 1691-1696.

[27] Silverman, D. H., C. J. Dy, S. A. Castellon, J. Lai, B. S. Pio, L. Abraham, K. Waddell, L. Petersen, M. E. Phelps, and P. A. Ganz (2007) Altered frontocortical, cerebellar, and basal ganglia activity in adjuvant-treated breast cancer survivors 5-10 years after chemotherapy. Breast Cancer Res. Treat. 103: 303-311.

[28] Walker, E. A., J. J. Foley, R. Clark-Vetri, and R. B. Raffa (2011) Effects of repeated administration of chemotherapeutic agents tamoxifen, methotrexate, and 5-fluorouracil on the acquisition and retention of a learned response in mice. Psychopharmacology 217: 539-548.

[29] Paganini-Hill, A., and L. J. Clark (2000) Preliminary assessment of cognitive function in breast cancer patients treated with tamoxifen. Breast Cancer Res. Treat. 64: 165-176.

[30] Chang, N. W., F. N. Chen, C. T. Wu, C. F. Lin, and D. R. Chen (2009) Apolipoprotein E4 allele influences the response of plasma triglyceride levels to tamoxifen in breast cancer patients. Clin. Chim. Acta. 401: 144-147.

[31] Liberopoulos, E., S. A. Karabina, A. Tselepis, E. Bairaktari, C. Nicolaides, N. Pavlidis, and M. Elisaf (2002) Are the effects of tamoxifen on the serum lipid profile modified by apolipoprotein E phenotypes? Oncology 62: 115-120.

[32] Howell, A., J. Cuzick, M. Baum, A. Buzdar, M. Dowsett, J. F. Forbes, G. Hoctin-Boes, J. Houghton, G. Y. Locker, and J. S. Tobias (2005) Results of the ATAC (Arimidex, tamoxifen, Alone or in Combination) trial after completion of 5 years' adjuvant treatment for breast cancer. Lancet 365: 60-62.

[33] Collins, B., J. Mackenzie, A. Stewart, C. Bielajew, and S. Verma (2009) Cognitive effects of hormonal therapy in early stage breast cancer patients: a prospective study. Psycho-oncology 18: 811-821.

[34] Agrawal, K., S. Onami, J. E. Mortimer, and S. K. Pal (2010) Cognitive changes associated with endocrine therapy for breast cancer. Maturitas 67: 209-214.

[35] Lew, R., P. Komesaroff, M. Williams, T. Dawood, and K. Sudhir (2003) Endogenous estrogens influence endothelial function in young men. Circ. Res. 93: 1127-1133.

[36] Sadlonova, V., P. Kubatka, K. Kajo, D. Ostatnikova, G. Nosalova, K. Adamicova, and J. Sadlonova (2009) Side effects of anastrozole in the experimental pre-menopausal mammary carcinogenesis. Neoplasma 56: 124-129.

[37] Phillips, K. A., K. Ribi, Z. Sun, A. Stephens, A. Thompson, V. Harvey, B. Thurlimann, F. Cardoso, O. Pagani, A. S. Coates, A. Goldhirsch, K. N. Price, R. D. Gelber, and J. Bernhard (2010) Cognitive function in postmenopausal women receiving adjuvant letrozole or tamoxifen for breast cancer in the BIG 1-98 randomized trial. Breast 19: 388-395.

[38] Thurlimann, B., A. Keshaviah, A. S. Coates, H. Mouridsen, L. Mauriac, J. F. Forbes, R. Paridaens, M. Castiglione-Gertsch, R. D. Gelber, M. Rabaglio, I. Smith, A. Wardley, K. N. Price, and A. Goldhirsch (2005) A comparison of letrozole and tamoxifen in postmenopausal women with early breast cancer. N. Engl. J. Med. 353: 2747-2757.

[39] Elisaf, M. S., E. T. Bairaktari, C. Nicolaides, B. Kakaidi, C. S. Tzallas, A. Katsaraki, and N. A. Pavlidis (2001) Effect of letrozole on the lipid profile in postmenopausal women with breast cancer. Euro. J. Cancer 37: 1510-1513.

[40] Atalay, G., L. Dirix, L. Biganzoli, L. Beex, M. Nooij, D. Cameron, C. Lohrisch, T. Cufer, J. P. Lobelle, M. R. Mattiaci, M. Piccart, and R. Paridaens (2004) The effect of exemestane on serum lipid profile in postmenopausal women with metastatic breast cancer: a companion study to EORTC Trial 10951, 'Randomized phase II study in first line hormonal treatment for metastatic breast cancer with exemestane or tamoxifen in postmenopausal patients'. Ann. Oncol. 15: 211-217.

[41] Rey, J. R., E. V. Cervino, M. L. Rentero, E. C. Crespo, A. O. Alvaro, and M. Casillas (2009) Raloxifene: mechanism of action, effects on bone tissue, and applicability in clinical traumatology practice. Open. Orthop. J. 3: 14-21.

[42] Espeland, M. A., S. A. Shumaker, M. Limacher, S. R. Rapp, T. B. Bevers, D. H. Barad, L. H. Coker, S. A. Gaussoin, M. L. Stefanick, D. S. Lane, P. M. Maki, and S. M. Resnick (2010) Relative effects of tamoxifen, raloxifene, and conjugated equine estrogens on cognition. J. Women's Health (Larchmt). 19: 371-379.

[43] Yaffe, K., K. Krueger, S. Sarkar, D. Grady, E. Barrett-Connor, D. A. Cox, and T. Nickelsen (2001) Cognitive function in postmenopausal women treated with raloxifene. N. Engl. J. Med. 344: 1207-1213.

[44] Alejandre-Gomez, M., L. M. Garcia-Segura, and I. Gonzalez-Burgos (2007) Administration of an inhibitor of estrogen biosynthesis facilitates working memory acquisition in male rats. Neurosci. Res. 58: 272-277.

[45] Legault, C., P. M. Maki, S. M. Resnick, L. Coker, P. Hogan, T. B. Bevers, and S. A. Shumaker (2009) Effects of tamoxifen and raloxifene on memory and other cognitive abilities: cognition in the study of tamoxifen and raloxifene. J. Clin. Oncol. 27: 5144-5152.

[46] Dayspring, T., Y. Qu, and C. Keech (2006) Effects of raloxifene on lipid and lipoprotein levels in postmenopausal osteoporotic women with and without hypertriglyceridemia. Metabolism 55: 972-979.

[47] Christodoulakos, G. E., I. V. Lambrinoudaki, C. P. Panoulis, C. A. Papadias, E. E. Kouskouni, and G. C. Creatsas (2004) Effect of hormone replacement therapy, tibolone

and raloxifene on serum lipids, apolipoprotein A1, apolipoprotein B and lipoprotein(a) in Greek postmenopausal women. Gynecol. Endocrinol. 18: 244-257.

[48] Francucci, C. M., P. Daniele, N. Iori, A. Camilletti, F. Massi, and M. Boscaro (2005) Effects of raloxifene on body fat distribution and lipid profile in healthy post-menopausal women. J. Endocrinol. Invest. 28: 623-631.

[49] Duka, T., R. Tasker, and J. F. McGowan (2000) The effects of 3-week estrogen hormone replacement on cognition in elderly healthy females. Psychopharmacology 149: 129-139.

[50] Henderson, V. W., A. Paganini-Hill, C. K. Emanuel, M. E. Dunn, and J. G. Buckwalter (1994) Estrogen replacement therapy in older women. Comparisons between Alzheimer's disease cases and nondemented control subjects. Arch. Neurol. 51: 896-900.

[51] Kampen, D. L., and B. B. Sherwin (1996) Estradiol is related to visual memory in healthy young men. Behav. Neurosci. 110: 613-617.

[52] Liu, F., M. Day, L. C. Muniz, D. Bitran, R. Arias, R. Revilla-Sanchez, S. Grauer, G. Zhang, C. Kelley, V. Pulito, A. Sung, R. F. Mervis, R. Navarra, W. D. Hirst, P. H. Reinhart, K. L. Marquis, S. J. Moss, M. N. Pangalos, and N. J. Brandon (2008) Activation of estrogen receptor-beta regulates hippocampal synaptic plasticity and improves memory. Nature Neurosci. 11: 334-343.

[53] Vearncombe, K. J., and N. A. Pachana (2009) Is cognitive functioning detrimentally affected after early, induced menopause? Menopause 16: 188-198.

[54] Yaffe, K., M. Haan, A. Byers, C. Tangen, and L. Kuller (2000) Estrogen use, APOE, and cognitive decline: evidence of gene-environment interaction. Neurology 54: 1949-1954.

[55] Burkhardt, M. S., J. K. Foster, S. M. Laws, L. D. Baker, S. Craft, S. E. Gandy, B. G. Stuckey, R. Clarnette, D. Nolan, B. Hewson-Bower, and R. N. Martins (2004) Oestrogen replacement therapy may improve memory functioning in the absence of APOE epsilon4. J. Alzheimer's Dis. 6: 221-228.

[56] Fluck, E., S. E. File, and J. Rymer (2002) Cognitive effects of 10 years of hormone-replacement therapy with tibolone. J. Clin. Psychopharmacol. 22: 62-67.

[57] Gulseren, L., D. Kalafat, H. Mandaci, S. Gulseren, and L. Camli (2005) Effects of tibolone on the quality of life, anxiety-depression levels and cognitive functions in natural menopause: an observational follow-up study. Aust. N. Z. J. Obstet. Gynaecol. 45: 71-73.

[58] Creatsas, G., G. Christodoulakos, I. Lambrinoudaki, C. Panoulis, C. Chondros, and P. Patramanis (2003) Serum lipids and apolipoproteins in Greek postmenopausal women: association with estrogen, estrogen-progestin, tibolone and raloxifene therapy. J. Endocrinol. Invest. 26: 545-551.

[59] von Eckardstein, A., D. Crook, J. Elbers, J. Ragoobir, B. Ezeh, F. Helmond, N. Miller, H. Dieplinger, H. C. Bennink, and G. Assmann (2003) Tibolone lowers high density lipoprotein cholesterol by increasing hepatic lipase activity but does not impair cholesterol efflux. Clin. Endocrinol. (Oxf) 58: 49-58.

[60] Garefalakis, M., and M. Hickey (2008) Role of androgens, progestins and tibolone in the treatment of menopausal symptoms: a review of the clinical evidence. Clin. Interv. Aging 3: 1-8.

[61] Telegdy, G., M. Tanaka, and A. V. Schally (2009) Effects of the LHRH antagonist Cetrorelix on the brain function in mice. Neuropeptides 43: 229-234.

[62] von Dehn, G., O. von Dehn, W. Volker, C. Langer, G. F. Weinbauer, H. M. Behre, E. Nieschlag, G. Assmann, and A. von Eckardstein (2001) Atherosclerosis in apolipoprotein E-deficient mice is decreased by the suppression of endogenous sex hormones. Horm. Metab. Res. 33: 110-114.

[63] Buchter, D., H. M. Behre, S. Kliesch, A. Chirazi, E. Nieschlag, G. Assmann, and A. von Eckardstein (1999) Effects of testosterone suppression in young men by the gonadotropin releasing hormone antagonist cetrorelix on plasma lipids, lipolytic enzymes, lipid transfer proteins, insulin, and leptin. Exp. Clin. Endocrinol. Diabetes 107: 522-529.

Chapter 12

Anticholesterolemic and Antiatherogenic Effects of Taurine Supplementation is Model Dependent

Karl-Erik Eilertsen, Rune Larsen,
Hanne K. Mæhre, Ida-Johanne Jensen and Edel O. Elvevoll

Additional information is available at the end of the chapter

1. Introduction

Taurine (2-aminoethanesulfonic acid) is a sulphur-containing compound characterized as an amino acid. The presence of a sulfonic group, as opposed to a carboxyl group in other amino acids, gives taurine a pKa value of 1.5 and it is the most acidic amino acid. It is an exclusively free amino acid, i.e. it is not incorporated into proteins, but still widely distributed in most body tissues.

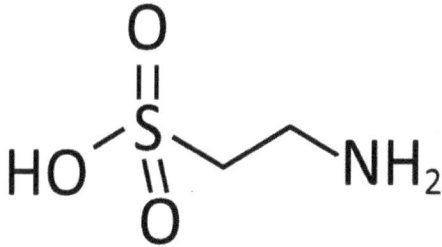

Figure 1. The structure of taurine

Taurine was identified almost two centuries ago and was named after the ox, *Bos taurus*, since it was first isolated from the bile of ox [1]. After its discovery, taurine was considered non-essential and biologically inert, however a multitude of functions have now been identified. Yet, all its physiological roles have not been fully elucidated. The phylogenetically oldest and best documented function of taurine is conjugation with bile acids in bile salt synthesis [2, 3]. In addition, taurine is involved in a variety of physiological processes as extensively reviewed [2], including neuromodulation in the central nervous system [4], energy production [5], protection against oxidation [6, 7] and immunomodulation

[8, 9]. An osmoregulatory role of taurine has also been established, playing a pivotal role in Central nervous system (CNS) cell volume regulation [10-12].

In felines taurine is considered indispensable and dietary deficiency leads to several clinical problems, including retinal degeneration and developmental abnormalities [13]. In humans it is regarded as a conditionally essential amino acid due to a limited ability to synthesize it [14, 15]. Taurine is now thought to play a more important role in human nutrition, and an increased dietary intake of taurine has been linked to several beneficial health outcomes in various diseases and medical conditions [16-18].

2. Taurine and nutrition

Estimates of dietary intake of taurine vary greatly. Although taurine content have been analysed in a variety of foods, it is usually excluded in food and nutrition data banks. Therefore, it is difficult to assess the dietary intake. A diet high in meat and especially seafood will provide a higher intake than a vegetarian diet which will provide very little taurine [19]. Mean ± SE dietary intake of taurine of 58 ± 19.5 mg/d was reported in omnivores [20], while it was not detected in a vegan diet. Laidlaw et al. [21] analysed taurine content in foods and calculated a taurine intake of less than 200 mg/d for individuals consuming a diet high in meat.

2.1. Taurine biosynthesis

Taurine is the most abundant intracellular free amino acid in the human body, the average amount being approximately 560 mmol (70g). The main organs of distribution are the retina, along with white blood cells, platelets, spleen, heart, muscle and brain [22].

As a product in the metabolism of sulphur-containing amino acids, taurine can be synthesised from its precursors methionine and cysteine, as shown in figure 2. The first step of the synthesis is methionine's reversible conversion to homocysteine by transmethylation and remethylation processes. Homocysteine can then be converted irreversibly to cysteine through the transsulfuration pathway catalyzed by cystathionine β-synthase and cystathionine γ-lyase [23]. Cysteine is, in turn, the origin of several biologically important molecules, including glutathione, inorganic sulphur and taurine [24]. Taurine can be synthesised from cysteine through several pathways, most commonly via cysteine sulfinic acid and hypotaurine, involving the enzymes cysteine dioxygenase (CDO) and cysteine sulfinic decarboxylase (CSAD) mainly present in the liver and brain. The activities of the enzymes involved, in particular the activity of CSAD, are both species and age dependent [25-27], being high in rodents and absent in cats. In addition, taurine synthesis is dependent on an adequate cysteine concentration, as production of glutathione is favoured when cysteine concentration is limited [28].

In humans the CSAD activity is low and the average daily synthesis of taurine ranges from 0.4 to 1.0 mmol (50-125mg). Excretion of taurine is very variable (0.22-1.85 mmol day-1) and affected by several factors such as genetics, age, gender, dietary intake, kidney function and health status [22]. The taurine body pool size is however regulated by the kidneys through renal absorption by the proximal tubule [14, 26, 29, 30].

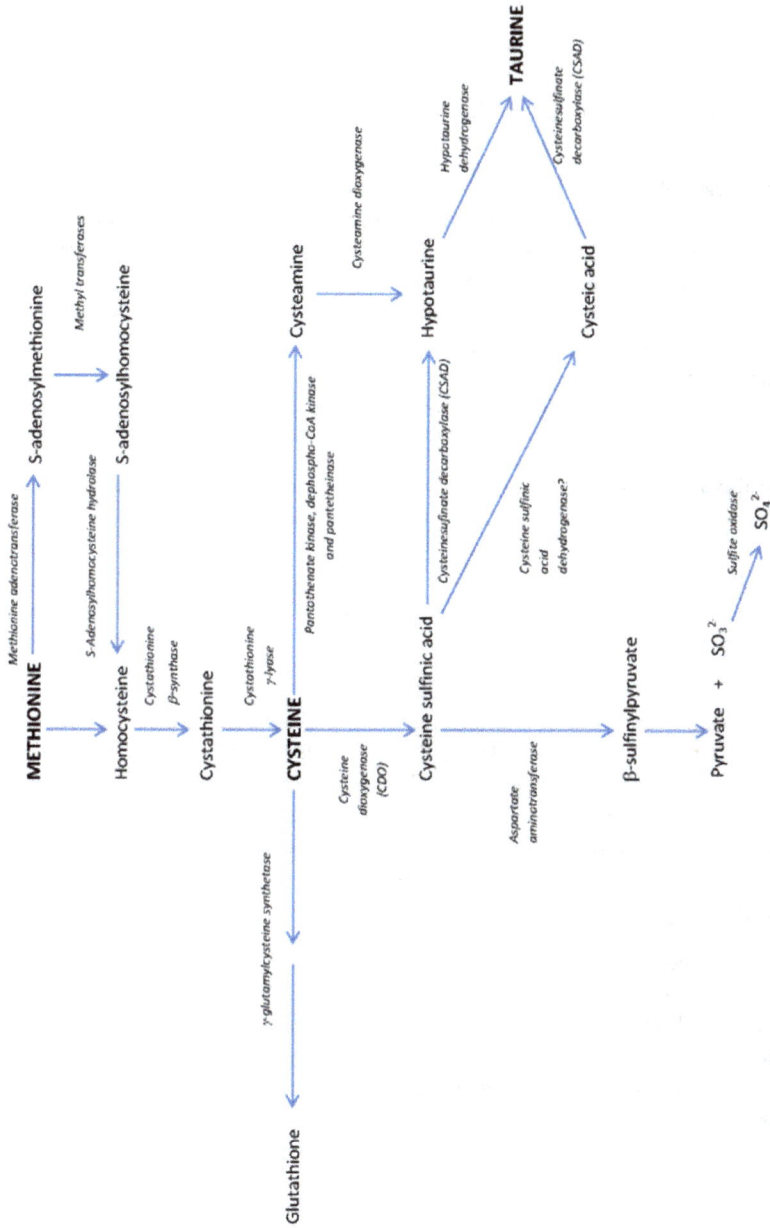

Figure 2. Taurine biosynthesis

2.2. Dietary sources

Taurine is found in most meats used for human consumption, whereas plants including grains, legumes, fruits and vegetables are devoid or contain only negligible amounts [21]. An exception is algae, mainly red algae (*Rhodophyta*), where notable amounts have been found [31, 32].

Taurine concentration has been investigated in a wide range of food products and it varies substantially between different marine and non-marine food items [17, 33]. A comparison of taurine concentrations in various foods is presented in table 1 [21, 34-43]. It is evident that seafood, and especially molluscs are high in taurine. Taurine is a key osmolyte in marine molluscs [44] and the highest taurine concentrations are found in marine bivalves and univalves [45]. Scallops and blue mussels are reported to have a respective taurine content of 827±15 and 510±12 mg per 100g raw muscle [21, 34]. In fact, the univalve abalone was already early in the last century 1918 exploited for preparation of taurine in large quantities [46].

There is also a tendency of taurine being more abundant in fish than in terrestrial animals. Taurine concentrations (mg per 100g raw fillets) of entire muscle of farmed Atlantic salmon (94 ± 16 mg), cod (120 ± 21 mg), saithe (162 ± 25 mg) and haddock (57 ± 6 mg) are reported to be intermediate [34]. Taurine content varies greatly between white and red muscle both in fish, poultry and mammals, with significantly higher levels being present in red muscle [21, 39, 40, 43], probably due to the increased vascularisation of these tissues.

Several studies investigating the retention and losses of taurine during food processing and preparation have been conducted [34, 36, 47-49]. Results indicate that taurine is susceptible to leaching losses similar to or even more than other free amino acids. Data on the oxidative and heat stability of taurine in foods is scarce. In milk, taurine losses seemed to proceed with the same degradation rate as lysine due to browning reactions [50].

2.3. Taurine supplementation

Taurine is maybe most famous for being an ingredient that is added to energy drinks, the concentration being approximately 4.0 g/L. Its physiological effect has been debated, with manufacturers, backed by studies, claiming that taurine in combination with other active ingredients may improve cognitive and muscular performance [51, 52]. The safety of taurine intake has also been investigated, especially in conjunction with its use in energy drinks. The European Food Safety Authority (EFSA) have concluded that taurine do not present any safety concerns with the levels currently used in energy drinks. The no observable adverse effect level (NOAEL) was at least 1000 mg/kg bw/day for pathological and behavioural changes, being much higher than an extreme consumer would be exposed to [53]. In their risk assessment, Shao and Hatchcock [54], found that absence of adverse effects was strong for taurine at supplemental intakes up to 3 g per day.

Food source	Taurine content (mg/100g wet weight) reported range ± SEM	References
Meat		
Beef, round	36	[38]
Beef (Bos taurus)	43 ± 8	[21]
Chicken, light meat	18 ± 3	[21]
Chicken, dark meat	169 ± 37	[21]
Turkey, light meat	30 ± 7	[21]
Turkey, dark meat	306 ± 69	[21]
Pork (loin)	61 ± 11	[21, 38]
Pork (loin)	50 ± 11	[21, 38]
Lamb, leg	45 ± 4	[38]
Veal	40 ± 13	[21]
Reindeer (loin)	62,4 ± 12	[42]
Red deer (loin)	28 ± 13	[37]
Fish		
Plaice	146 ± 5	[[35]
Cod fillet	120 ± 21	[34, 48]
Cod roe	365	[34]
Saithe fillet	162 ± 25	[34]
Haddock fillet	57 ± 6	[34]
African catfish	201 ± 32	[36]
Salmon fillet	94 ± 16	[34]
Mackerel	78	[35]
Bigeye tuna, white muscle	26	[41]
Bigeye tuna, dark muscle	270	[41]
Yellowfin tuna, white muscle	42	[41]
Yellowfin tuna, dark muscle	964	[41]
Bluefin tuna, white muscle	88	[41]
Bluefin tuna, dark muscle	195	[41]
Pacific saury, white muscle	223	[41]
Pacific saury, dark muscle	248	[41]
Milkfish, white muscle	95,7	[39]
Milkfish, dark muscle	309	[39]
Octopus	388 ± 13	[43]
Squid	356 ± 95	[21]
Fresh water fish		
Rainbow trout, white muscle	17	[40]
Rainbow trout, dark muscle	206	[40]

Coho salmon, white muscle	23	[40]
Coho salmon, dark muscle	275	[40]
Eel, white muscle	7	[40]
Eel, dark muscle	65	[40]
Catfish, white muscle	193	[40]
Catfish, dark muscle	465	[40]
Tilapia, white muscle	75	[40]
Tilapia, dark muscle	649	[40]
Carp, white muscle	129	[40]
Carp, dark muscle	579	[40]
Char, white muscle	15	[40]
Char, dark muscle	190	[40]
Sweet smelt, white muscle	137	[40]
Sweet smelt, dark muscle	294	[40]
Shellfish		
Peeled shrimps (Northern)	220 ± 2	[34]
Blue mussels	510 ± 12	[34]
Mussel	655 ± 72	[21]
Mussel	349	[43]
Clams	520 ± 97	[21, 43]
Scallops	827 ± 15	[21]
Scallop	332	[43]
Oysters	396 ± 29	[21]

Table 1. Taurine content in various food sources

Another food item where taurine is supplemented is in infant formulas. This practice started in the early 1980s after recognizing that preterm infants fed infant formulas had lower urine and plasma concentrations than infants fed pooled human milk [55]. The necessity of this supplementation remains disputed as clinical studies have not provided evidence of any clinical effects of growth and development in preterm or low birth weight infants [56]. High concentrations of taurine in the developing brain [57], as well as results from various animal studies clearly indicate the importance of taurine in neurodevelopment [58, 59].

2.4. Taurine and associated health benefits

An increased dietary intake of taurine has been associated with multiple beneficial health outcomes. Epidemiological data and animal studies suggests that dietary intake of taurine has beneficial effects on cardiovascular disease (CVD) [33, 60-62]. Perhaps the best characterized attribution of taurine is the antihypertensive effect although there are still questions about the exact mechanisms of action [63-66]. A long term effect of hypertension is the development of hypertrophy of the left ventricle, in which Angiotensin II (Ang II) plays an important role. Several studies have shown that taurine reverses these actions of Ang II [67, 68]. Animal studies have also indicated that taurine may reduce insulin resistance [69, 70], but most of the clinical studies have failed to prove the beneficial role of taurine in insulin resistance and diabetic complications [71, 72]. Taurine have also been found to ameliorate alcoholic steatohepatitis [73-75] in rats. In addition, some evidence have been brought forward of a potential therapeutic use of taurine in nonalcoholic fatty liver disease [76]. Despite taurine being linked to beneficial health outcomes in an increasing number of diseases and medical conditions, the number of studies is relatively small. The effects of taurine on cholesterol and CVD are most studied and documented.

3. Taurine and cholesterol metabolism

Perhaps the best studied function of taurine is its role in cholesterol metabolism. Cholesterol is metabolized and broken down to cholic acids, conjugated to taurine or glycine, and excreted in the bile [77].

3.1. Effects of taurine on circulating cholesterol levels

High blood cholesterol levels is the most pronounced risk factor for developing atherosclerosis, vascular inflammation and hardening of the arteries associated with excess cholesterol deposition in the vasculature. Taurine has generally been associated with a beneficial effect on blood cholesterol levels. Cholesterol is metabolized and broken down to cholic acids, conjugated to taurine or glycine, and excreted in the bile [77]. The conjugation pattern varies considerably across species. In dog and rat bile acids are entirely conjugated to taurine, whereas rabbits have all their bile acids conjugated to glycine. Species where glycine-conjugated bile acids dominate have higher blood cholesterol levels and are more susceptible to dietary induced hypercholesterolemia, and based on these observations it was hypothesized that dietary taurine might counteract dietary induced increase in blood cholesterol [2].

3.1.1. Effects of taurine on cholesterol levels in mice

Several studies have investigated the effect of dietary intake of taurine on lipids in different mice strains. Six months administration of 1% taurine (w/v) to the drinking water given to C57BL7/6J mice fed a high-fat diet resulted in reduced serum LDL and VLDL cholesterol and increased serum HDL cholesterol [78]. Similar results were obtained in a small study using the same mouse strain, where 1% taurine (w/w) added to a high cholesterol diet reduced serum triglycerides, total cholesterol and VLDL+LDL cholesterol levels already after 4 weeks treatment [79]. Cholesterol-fed and streptozotocin (STZ)-induced diabetic male ICR mice were given a diet enriched with 2% cholesterol (w/w) and 0.5% cholate (w/w) for 10 weeks [80]. In addition, mice received a daily dose of saline or taurine (50 or 100 mg/kg p.o.). Both taurine-treated groups had lower serum total and LDL cholesterol compared to the STZ/saline group.

In more extreme models such as apolipoprotein E-deficient (apoE[-/-]) mice fed a normal rodent chow supplemented with 2% taurine (w/w) for 12 weeks, serum VLDL, LDL and total cholesterol levels increased compared to mice without taurine supplementation [81]. Similar results has been reported for extreme spontaneously hyperlipidemic mice (SHL; KOR-*Apoe[shl]*), where 12 weeks treatment with 1% taurine (w/v) added to the drinking water increased serum HDL-cholesterol but did not affect serum total cholesterol or VLDL+LDL cholesterol levels [82].

3.1.2. Effects of taurine on cholesterol levels in rats

A large number of studies have investigated the effect of dietary taurine on dietary hypercholesterolemia in various rat models [65]. Rats fed high-fat diets seem to be the model with most consistent antihypercholesterolemic effects of dietary taurine. Rats have a relatively low taurine content in skeletal muscles (10-25 mg/100 g muscle) [83, 84].

Rats are generally not suitable for pharmacologically cholesterol and lipoprotein studies due to their substantially different lipid profile compared to humans.

Male wistar rats. Taurine supplementation does not alter plasma lipids in male wistar rats fed normal chow [85]. However, when these rats were fed high cholesterol diet containing 2% cholesterol and 1% cholic acid, dietary supplementation of 4% taurine significantly counteracted the observed increase in serum cholesterol by 44% [86]. This observation has been confirmed be several studies [87-90]. In Wistar male rats fed a cholesterol-containing diet (0.5% cholesterol w/w) for 40 days, serum cholesterol increased 5 fold compared to chow fed rats. Oral supplementation of 470 mg/kg/day taurine (0.5% w/v) in water lowered the increased serum cholesterol (54%) [91]. When these rats were fed a high-fat diet (11% coconut oil w/w) for 6 months, a daily oral supplementation of 1 mg taurine lowered serum cholesterol (37%), LDL cholesterol (34%), and triglycerides (95%), compared to the high-fat control diet [88]. Already after 14 days intervention, 5% dietary taurine (w/w) supplementation has been indicated to lower high cholesterol (1% cholesterol, 2.5% cholate) induced serum cholesterol (-42%) [90]. In wistar rats, the taurine effect has been indicated to be caused by an increased faecal bile acid excretion, increased hepatic cholesterol 7α-hydroxylase expression and activity [89, 90]. The rapid effect of taurine supplementation on serum cholesterol has been confirmed recently [92]. When fed a diet containing 60.7% sucrose, 9.0% lard, and 0.5% cholesterol for 14 days, serum cholesteryl ester and free cholesterol were reduced by 39% and 53% compared to rats fed control diet without taurine, respectively. Rats fed taurine also had smaller livers compared to control-fed rats. Hepatic cholesteryl esters were also reduced by approximately 20% in the taurine supplemented rats. This hypocholesterolemic effect was ascribed to a lower hepatic secretion of cholesteryl esters.

Streptozotocin-induced diabetic rats. Male Wistar rats injected with STZ are also used as a diabetic model. In these rats dietary taurine supplementation markedly reduced serum total cholesterol (-50%) induced by cholesterol-containing diet (1% cholesterol w/w) for 4 weeks [93].

Fructose-induced rat insulin resistance model. Male wistar rats fed a diet containing 60% fructose developed impaired glucose tolerance and insulin resistance [85]. Taurine administration (300 mg/kg/day *i.p.*) counteracted the fructose induced plasma total cholesterol, LDL cholesterol, and triglycerides by 11%, 21%, and 23%, respectively.

Spontaneously hypertensive rats (SHR). The effect of taurine supplementation on blood pressure in SHR rats was investigated already in the 1970ies [94]. How dietary administration of taurine affects cholesterol metabolism has not been reported in these studies. However, in a stroke-prone substrain of the SHR rats, taurine supplementation has been indicated to prevent high-fat/high cholesterol induced elevation of serum cholesterol in SHR rats [95].

Sprague-Dawley rats. Also in male Sprague-Dawley rats fed a high-fat, high cholesterol diet (HFCD; 10% corn oil, 1.5% cholesterol) supplemented with taurine (1.5% w/w) plasma cholesterol was lowered by 31% compared to HFCD control rats   [87]. LDL+VLDL cholesterol (-38%) and triglycerides (-43%) were also lower in rats supplemented with taurine compared to HFCD control rats. These results have been confirmed with an identical experimental setup for 5 weeks reporting a 20% and 25% reduction in serum total cholesterol and triglycerides, respectively [96]. In this model, plasma total cholesterol, LDL cholesterol and triglycerides were reduced in rats fed taurine supplemented cholesterol free diet compared the cholesterol free control diet [87].

3.1.3. Effects of taurine on cholesterol levels in rabbits

Different rabbit strains have been used to investigate the effects of dietary taurine supplementation on dietary induced hypercholesterolemia [86, 97, 98]. The results from the administration of taurine to rabbits have been ambiguous. In *Male New Zealand white rabbits*, fed a high cholesterol (1% w/w) diet, addition of 2.5% taurine (w/w) for 2.5 month reduced the serum total cholesterol and triglyceride levels by 22% and 38%, respectively, compared to high cholesterol diet alone [99]. In this study similar reductions were observed for hepatic and aorta lipid levels in these rabbits. However, when the same rabbit strain were given normal chow supplemented with 0.5% cholesterol (w/w) for 4 weeks no effect of dietary taurine (2.5% w/w) supplementation was observed [100]. Also when given a normal diet supplemented with 2% cholesterol (w/w), taurine added to the drinking water (0.1 or 0.5% w/v) for 14 weeks had no influence on serum cholesterol and triglycerides [97].

3.1.4. Effects of taurine on cholesterol levels in hamsters

Hamsters (Male Golden Syrian hamsters) have also been used as model for studying cholesterol metabolism. The rationale for this is that hamsters and humans have comparable blood cholesterol levels, hamsters use both taurine and glycine for bile acid conjugation and the lipoprotein profile in response to dietary cholesterol is comparable [101]. When Male Golden Syrian hamsters were fed a normal chow supplemented with 0.05% cholesterol or a 10% coconut oil, high-fat diet (0.05% cholesterol) for two weeks, taurine dissolved in drinking water (1% w/v) reduced serum total cholesterol in chow- (15% reduction) as well as high-fat diet-fed (42% reduction) hamsters [102]. A similar effect was observed for non-HDL (LDL+VLDL) cholesterol.

Recently, lipid metabolism has been closely studied in Male Golden Syrian Hamsters fed different diets with or without taurine for 4 weeks [103]. The groups received a high fat diet (chow mixed with 7% butter [w/w] and 0.2% cholesterol [w/w]) and drinking water without or supplemented with either 0.35% or 0.7% taurine (w/v). Hamsters given taurine was smaller, had less visceral fat and smaller livers after 4 weeks. Both taurine concentrations resulted in significant lower serum triglycerides, total cholesterol, and LDL+VLVL cholesterol. Up-regulated gene expression of the low-density lipoprotein receptor and CYP7A1 genes, paralleled by increased faecal cholesterol and bile acid concentrations in the taurine treated hamsters, indicated that the taurine effect on the cholesterol and lipid profiles is due to increased cholesterol metabolism.

3.1.5. Effects of taurine on cholesterol levels in humans

Historically, taurine has been believed to decrease blood cholesterol levels in adults. Only a limited number of studies have investigated the effect of oral taurine supplementation on blood cholesterol or lipoprotein levels in humans and ambiguous results have arisen from these. Early studies found no effect on serum cholesterol after incidental treatment of patients with 1.5 to 3 g taurine/day for up to 2 months [77, 104, 105]. To our knowledge there has been no well-designed random controlled clinical trial assessing the dose-response effect of oral taurine supplementation on blood lipids in healthy humans. However, the effect of taurine in relation to development of CVD has been documented through a human clinical trial. Results of a 7 week human intervention trial revealed that supplementation with 0.4 g taurine/day in combination with omega-3 fatty acids (1 g EPA+DHA/day) significantly improved the lipid profiles by reducing serum total and LDL cholesterol levels compared to supplementation with omega-3 fatty acids alone [106]. In another study the effects of oral supplementation with taurine (3 g/day) or placebo for 7 weeks was assessed in young obese healthy subjects [107]. In this study, taurine had no effect on serum cholesterol, but triglycerides and bodyweight was significantly reduced compared to placebo effect. Finally, a daily 6 g taurine supplementation to human healthy volunteers receiving a cholesterol-inducing diet for 3 weeks attenuated the expected increase of serum total cholesterol and LDL-cholesterol, whereas serum VLDL-cholesterol and triglyceride levels compared to the control group [108]. In insulin-dependent diabetes mellitus patients intake of taurine (1 g/day) reduced serum triglyceride levels, but no effect was observed on serum cholesterol [109]. Finally, in a randomized, double-blinded, crossover intervention, overweight non-diabetic men given a daily dose of 1.5 g taurine or placebo, no effect was reported on blood lipids [71]. In summary, results from oral taurine supplementation to humans are ambiguous, and further adequately designed interventions are warranted to further investigate the potential of taurine as a hypocholesterolemic agent.

4. Effects of taurine on atherogenesis/development of atherosclerosis

High blood cholesterol levels is the most pronounced risk factor for developing atherosclerosis, vascular inflammation and hardening of the arteries associated with excess cholesterol deposition in the vasculature. Apart from humans and monkeys, wild animals normally do not develop substantial atherosclerosis. There is however, an array of laboratory animal models in common use for studying the effects of pharmacological substances and dietary modifications on lesion formation. The effect of taurine has been investigated in several of these models.

4.1. Effects of taurine on atherosclerosis in mice

In the hyperlipidemic apoE$^{-/-}$ mice, taurine has been reported to delay atherogenesis by decreasing oxidized substances that cause inflammation, as well as increasing HDL-cholesterol [82]. Also in apoE$^{-/-}$ mice fed a normal rodent chow supplemented with 2% taurine (w/w) for 12 weeks, formation of atherosclerotic lesions were significantly reduced [81]. However this effect was independent of serum cholesterol as VLDL, LDL, and total cholesterol were increased.

Moreover, in spontaneously hyperlipidemic mice, taurine (1% w/v) provided through drinking water, was reported to suppress the development of lesion formation without affecting the levels of serum VLDL and LDL [82].

In our lab, apoE$^{-/-}$-mice were given Western diets (WD) containing 20% fat (w/w), 0.2% cholesterol (w/w) for 13 weeks [110]. The mice received WD, WD supplemented with 0.5% taurine (w/w) or WD supplemented with 0.5% taurine (w/w) in combination with a daily dose of marine long-chain omega-3 polyunsaturated fatty acids (n-3 PUFA) recommend in the dietary guidelines for humans. In these studies, taurine did not affect serum cholesterol or triglyceride levels alone or combination with n-3 PUFA. This may indicate that a larger supplementary dose of taurine is needed to prevent dietary induced hypercholesterolemia in apoE$^{-/-}$-mice.

4.2. Effects of taurine on lipid lesion formation in rats

Rats are generally not a suitable animal model for atherosclerosis as they do not develop lesion deposits resembling the early phase of human atherogenesis. However, in a rat model of balloon induced vascular neointima formation, supplementation with taurine (3% in drinking water) from 2 days before the surgical procedure and 14 days after, reduced vascular smooth muscle cell proliferation [111]. This key step in the initiation of atherogenesis was reduced by 28% compared to control fed rats. Taurine was located immunohistochemically mainly to the surface of the exposed media and adventitia of the injured carotid artery and higher levels were observed in the taurine treated rats. This corresponded to a lower vascular production of superoxide anion compared to the control animals. From these experiments it was concluded that the preventive effect of taurine towards neointima formation was attributable to anti-oxidative effects.

4.3. Effects of taurine on atherosclerosis in rabbits

The effect of dietary taurine on development of atherosclerosis has been investigated in different rabbit strains. Taurine has been indicated to prevent progression of atherosclerotic lesions in rabbits without affecting serum cholesterol in two different models. In New-Zealand white male rabbits given a diet containing 2% cholesterol (w/w), taurine added to the drinking water (0.1 or 0.5% w/v) for 14 weeks reduced the aortic deposition of fat [97]. This so-called anti-atherosclerotic effect was only significant for the highest taurine dose tested. Recently, it was indicated that the taurine antiatherosclerotic effect was evident in these rabbits after only 4 weeks on the atherogenic diet [100]. *Watanabe heritable hyperlipidemic (WHHL) rabbits* carries an inheritable mutation in the LDL receptor and is hence a typical genetically hyperlipidemic animal model. When WHHL rabbits were given drinking water containing 1% taurine (w/v) for 6 months they developed significantly less atherosclerotic lesion formation compared to rabbits not supplemented with taurine [112].

4.4. Effects of taurine on atherosclerosis in humans

Results on the effects of dietary taurine in humans are mainly from prospective studies. It is evident that individuals with high urinary excretion of taurine and high dietary intake of food high in taurine in general have fewer incidences of cardiovascular diseases compared to individuals with low dietary intake of taurine [33]. In addition, increased dietary intake of taurine either alone or in the combination with omega-3 fatty acids, has also been suggested to reduce MCP-1, an important risk factor of CVD [106]. No further randomised clinical trials on the effects of dietary supplementation of taurine on CVD disease markers has been reported.

5. Conclusion

Taurine appears to be able to prevent hypercholesterolemia and hepatic steatosis induced by high-fat and high-cholesterol diets in most animal models. The major mechanism by which taurine lowers serum cholesterol levels is by increased utilization of cholesterol for bile acid synthesis. In mice, rats, and hamsters, dietary intake of taurine cause reduction in diet-induced serum cholesterol accompanied by enhanced mRNA expression and enzymatic activity of 7a-hydroxylase, the rate-limiting enzyme of bile acid synthesis. In normal diets taurine does not appear to modify serum and liver cholesterol levels.

Dietary supplementation with taurine is indicated to have cardiovascular benefits. The effect on atherosclerosis appears to be highly dose- and model-dependent. In animal experiments using high-fat diets to induce increased levels of lipids, taurine has been demonstrated to significantly alleviate atherosclerotic lesions. The effects of taurine appear to be related to increased degradation and excretion of cholesterol as bile in the feces and the most common feature is that taurine increases expression and activity of cholesterol 7α-hydroxylase. Only a few studies have evaluated the effects of taurine in human subjects.

From the available data it is not possible to conclude about the proposed antihyperlipidemic and antiatherosclerotic, therefore more basic and clinical research on the effects of taurine supplementation on hypercholesterolemic and atherosclerotic effects are warranted. Randomized clinical trials of dietary taurine and taurine sources may provide further knowledge about the potential hypocholesterolemic and antiatherogenic effects of long-term dietary taurine supplementation in healthy volunteers and humans with hyperlipidemia, metabolic syndrome and cardiovascular diseases.

Author details

Karl-Erik Eilertsen, Rune Larsen,
Hanne K. Mæhre, Ida-Johanne Jensen and Edel O. Elvevoll
University of Tromsø, Norway

6. References

[1] Demarcay H. Ueber die Natur der Galle. Annalen der Pharmacie. 1938;27(3):270-291.

[2] Huxtable RJ. Physiological actions of taurine. Physiological Reviews. 1992;72(1):101-163.

[3] Jacobsen JG, Smith LH. Biochemistry and physiology of taurine and taurine derivatives. Physiological Reviews. 1968;48(2):424-.

[4] Belluzzi O, Puopolo M, Benedusi M, Kratskin I. Selective neuroinhibitory effects of taurine in slices of rat main olfactory bulb. Neuroscience. 2004;124(4):929-944.

[5] Giehl TJ, Qoronfleh MW, Wilkinson BJ. Transport, nutritional and metabolic studies of taurine in staphylococci. Journal of General Microbiology. 1987;133:849-856.

[6] Atmaca G. Antioxidant effects of sulfur-containing amino acids. Yonsei Medical Journal. 2004;45(5):776-788.

[7] Schaffer S, Azuma J, Takahashi K, Mozaffari M. Why is taurine cytoprotective? In: Lombardini JB, Schaffer SW, Azuma J, editors. Taurine 5: Beginning the 21st Century2003. p. 307-321.

[8] Schuller-Levis GB, Park E. Taurine and its chloramine: Modulators of immunity. Neurochemical Research. 2004;29(1):117-126.

[9] Wojtecka-Lukasik E, Czuprynska K, Maslinska D, Gajewski M, Gujski M, Maslinski S. Taurine - chloramine is a potent antiinflammatory substance. Inflammation Research. 2006;55:17-18.

[10] Lambert IH. Regulation of the cellular content of the organic osmolyte taurine in mammalian cells. Neurochemical Research. 2004;29(1):27-63.

[11] Oja SS, Saransaari P. Taurine as osmoregulator and neuromodulator in the brain. Metab Brain Dis. 1996;11(2):153-164.

[12] Olson JE, Martinho E. Regulation of taurine transport in rat hippocampal neurons by hypo-osmotic swelling. Journal of Neurochemistry. 2006;96(5):1375-89.

[13] Markwell PJ, Earle KE. Taurine - an essential nutrient for the cat - a brief review of the biochemistry of its requirement and the clinical consequences of deficiency. Nutrition Research. 1995;15(1):53-8.

[14] Schuller-Levis G, Park E. Is taurine a biomarker? Advances in Clinical Chemistry. 2006;41:1-21.

[15] Stapleton PP, Charles RP, Redmond HP, BouchierHayes DJ. Taurine and human nutrition. Clinical Nutrition. 1997;16(3):103-108.

[16] Bouckenooghe T, Remacle C, Reusens B. Is taurine a functional nutrient? Current Opinion in Clinical Nutrition and Metabolic Care. 2006;9(6):728-733.

[17] Wojcik OP, Koenig KL, Zeleniuch-Jacquotte A, Costa M, Chen Y. The potential protective effects of taurine on coronary heart disease. Atherosclerosis. 2010;208(1):19-25.

[18] Zulli A. Taurine in cardiovascular disease. Current Opinion in Clinical Nutrition and Metabolic Care. 2011;14(1):57-60.

[19] Laidlaw SA, Shultz TD, Cecchino JT, Kopple JD. Plasma and urine taurine levels in vegans. American Journal of Clinical Nutrition. 1988;47(4):660-663. Epub 1988/04/01.

[20] Rana SK, Sanders TA. Taurine concentrations in the diet, plasma, urine and breast milk of vegans compared with omnivores. British Journal of Nutrition. 1986;56(1):17-27. Epub 1986/07/01.

[21] Laidlaw SA, Grosvenor M, Kopple JD. The taurine content of common foodstuffs. Journal of Parenteral and Enteral Nutrition. 1990;14(2):183-188.

[22] Lourenco R, Camilo ME. Taurine: a conditionally essential amino acid in humans? An overview in health and disease. 2002;17:262-270.

[23] Brosnan JT, Brosnan ME. The sulfur-containing amino acids: An overview. Journal of Nutrition. 2006;136(6):1636S-1640S.

[24] Stipanuk MH, Dominy JE, Lee JI, Coloso RM. Mammalian cysteine metabolism: New insights into regulation of cysteine metabolism. Journal of Nutrition. 2006;136(6):1652S-1659S.

[25] Delarosa J, Stipanuk MH. Evidence for a rate-limiting role of cysteinesulfinate decarboxylase activity in taurine biosynthesis invivo. Comparative Biochemistry and Physiology B-Biochemistry & Molecular Biology. 1985;81(3):565-571.

[26] Jacobsen JG, Smith LH. Comparison of decarboxylation of cysteine sulphinic acid-1-14C and cysteic acid-1-14C by human, dot, and rat liver and brain. Nature. 1963;200(490):575-577.

[27] Worden JA, Stipanuk MH. A comparison by species, age and sex of cysteinesulfinate decarboxylase activity and taurine concentration in liver and brain of animals. Comparative Biochemistry and Physiology B-Biochemistry & Molecular Biology. 1985;82(2):233-239.

[28] Stipanuk MH. Role of the liver in regulation of body cysteine and taurine levels: A brief review. Neurochemical Research. 2004;29(1):105-110.

[29] Lourenco R, Camilo ME. Taurine: a conditionally essential amino acid in humans? An overview in health and disease. Nutricion hospitalaria : organo oficial de la Sociedad Espanola de Nutricion Parenteral y Enteral. 2002;17(6):262-270. Epub 2003/01/08.

[30] Tappaz ML. Taurine biosynthetic enzymes and taurine transporter: Molecular identification and regulations. Neurochemical Research. 2004;29(1):83-96.

[31] Dawczynski C, Schubert R, Jahreis G. Amino acids, fatty acids, and dietary fibre in edible seaweed products. Food Chemistry. 2007;103(3):891-839.

[32] Kataoka H, Ohnishi N. Occurrence of Taurine in Plants. Agricultural and Biological Chemistry. 1986;50(7):1887-1888.

[33] Yamori Y, Taguchi T, Hamada A, Kunimasa K, Mori H, Mori M. Taurine in health and diseases: consistent evidence from experimental and epidemiological studies. Journal of Biomedical Science. 2010;17.S6

[34] Dragnes BT, Larsen R, Ernstsen MH, Maehre H, Elvevoll EO. Impact of processing on the taurine content in processed seafood and their corresponding unprocessed raw materials. International Journal of Food Sciences and Nutrition. 2009;60(2):143-152.

[35] Gormley TR, Neumann T, Fagan JD, Brunton NP. Taurine content of raw and processed fish fillets/portions. European Food Research and Technology. 2007;225(5-6):837-842.

[36] Mierke-Klemeyer S, Larsen R, Oehlenschlager J, Maehre H, Elvevoll EO, Bandarra NM, et al. Retention of health-related beneficial components during household preparation of selenium-enriched African catfish (Clarias gariepinus) fillets. European Food Research and Technology. 2008;227(3):827-833.

[37] Purchas RW, Triumf EC, Egelandsdal B. Quality characteristics and composition of the longissimus muscle in the short-loin from male and female farmed red deer in New Zealand. Meat Science. 2010;86(2):505-510.

[38] Roe DA, Weston MO. Potential significance of free taurine in the diet. Nature. 1965;205:287-288.

[39] Shiau CY, Pong YJ, Chiou TK, Chai TY. Free amino acids and nucleotide-related compounds in milkfish (Chanos chanos) muscles and viscera. Journal of Agricultural and Food Chemistry. 1996;44(9):2650-2553.

[40] Suzuki T, Hirano T, Shirai T. Distribution of Extractive Nitrogenous Constituents in White and Dark Muscles of Fresh-Water Fish. Comparative Biochemistry and Physiology B-Biochemistry & Molecular Biology. 1990;96(1):107-111.

[41] Suzuki T, Hirano T, Suyama M. Free Imidazole Compounds in White and Dark Muscles of Migratory Marine Fish. Comparative Biochemistry and Physiology B-Biochemistry & Molecular Biology. 1987;87(3):615-619.

[42] Triumf EC, Purchas RW, Mielnik M, Maehre HK, Elvevoll E, Slinde E, et al. Composition and some quality characteristics of the longissimus muscle of reindeer in Norway compared to farmed New Zealand red deer. Meat Science. 2012;90(1):122-129.

[43] Zhao XH, Jia JB, Lin Y. Taurine content in Chinese food and daily taurine intake of Chinese men. Taurine 3. 1998;442:501-505.

[44] Lange R. Osmotic function of amino acids and taurine in mussel, mytilus edulis. Comparative Biochemistry and Physiology. 1963;10(2):173-179.

[45] Bradley HC. The occurence of taurin in invertebrate muscle. Science. 1904;20:25.

[46] Schmidt CLA, Watson T. A method for the preparation of taurin in large quantities. Journal of Biological Chemistry. 1918;33(3):499-500.

[47] Larsen R, Elvevoll EO. Water uptake, drip losses and retention of free amino acids and minerals in cod (Gadus morhua) fillet immersed in NaCl or KCl. Food Chemistry. 2008;107(1):369-376.

[48] Larsen R, Stormo SK, Dragnes BT, Elvevoll EO. Losses of taurine, creatine, glycine and alanine from cod (Gadus morhua L.) fillet during processing. Journal of Food Composition and Analysis. 2007;20(5):396-402.

[49] Purchas RW, Busboom JR, Wilkinson BHP. Changes in the forms of iron and in concentrations of taurine, carnosine, coenzyme Q(10), and creatine in beef longissimus muscle with cooking and simulated stomach and duodenal digestion. Meat Science. 2006;74(3):443-449.

[50] Saidi B, Warthesen JJ. Analysis and Heat-Stability of Taurine in Milk. Journal of Dairy Science. 1990;73(7):1700-1706.

[51] Geiss KR, Jester I, Falke W, Hamm M, Waag KL. The Effect of a Taurine-Containing Drink on Performance in 10 Endurance-Athletes. Amino Acids. 1994;7(1):45-56.

[52] Seidl R, Peyrl A, Nicham R, Hauser E. A taurine and caffeine-containing drink stimulates cognitive performance and well-being. Amino Acids. 2000;19(3-4):635-642.

[53] Authority EFS. Scientific Opinion of the Panel on Food Additives and Nutrient Sources added to Food on a request from the Commission on the use of taurine and D-glucurono-g-lactone as constituents of the so-called "energy" drinks. 2009;935: 1-31. The EFSA Journal. 2009;935:1-31.

[54] Shao A, Hathcock JN. Risk assessment for the amino acids taurine, l-glutamine and l-arginine. Regulatory Toxicology and Pharmacology. 2008;50(3):376-399.

[55] Chesney RW. Taurine - Is It Required for Infant Nutrition. Journal of Nutrition. 1988;118(1):6-10.

[56] Verner A, Craig S, McGuire W. Effect of taurine supplementation on growth and development in preterm or low birth weight infants. Cochrane Database of Systematic Reviews. 2007(4).

[57] Sturman JA, Gaull GE. Taurine in the Brain and Liver of Developing Human and Monkey. Journal of Neurochemistry. 1975;25(6):831-835.

[58] Chapman GE, Greenwood CE. Taurine in nutrition and brain development. Nutrition Research. 1988;8(8):955-968.

[59] Gaull GE. Taurine in pediatric nutrition: review and update. Pediatrics. 1989;83(3):433-442. Epub 1989/03/01.

[60] Yamauchitakihara K, Azuma J, Kishimoto S. Taurine protection against experimental arterial calcinosis in mice. Biochemical and Biophysical Research Communications. 1986;140(2):679-683.

[61] Yamori Y, Liu L, Mizushima S, Ikeda K, Nara Y, Group CS. Male cardiovascular mortality and dietary markers in 25 population samples of 16 countries. Journal of Hypertension. 2006;24(8):1499-1505.

[62] Yamori Y, Liu LJ, Ikeda K, Miura A, Mizushima S, Miki T, et al. Distribution of twenty-four hour urinary taurine excretion and association with ischemic heart disease mortality in 24 populations of 16 countries: Results from the WHO-CARDIAC Study. Hypertension Research. 2001;24(4):453-457.

[63] Hagar HH, El Etter E, Arafa M. Taurine attenuates hypertension and renal dysfunction induced by cyclosporine A in rats. Clinical and Experimental Pharmacology and Physiology. 2006;33(3):189-196.

[64] Hu JM, Xu XL, Yang JC, Wu GF, Sun CM, Lv QF. Antihypertensive Effect of Taurine in Rat. In: Azuma J, Schaffer SW, Ito T, editors. Advances in Experimental and Medical Biololgy. 2009. p. 75-84.

[65] Militante JD, Lombardini JB. Treatment of hypertension with oral taurine: experimental and clinical studies. Amino Acids. 2002;23(4):381-393.

[66] Nandhini AT, Thirunavukkarasu V, Anuradha CV. Potential role of kinins in the effects of taurine in high-fructose-fed rats. Canadian Journal of Physiology and Pharmacology. 2004;82(1):1-8.

[67] Azuma M, Takahashi K, Fukuda T, Ohyabu Y, Yamamoto I, Kim S, Iwao H, Schaffer SW, Azuma J. Taurine attenuates hypertrophy induced by angiotensin II in cultured neonatal rat cardiac myocytes. European Journal of Pharmacology. 2000;403(3):181-188.

[68] Rao MR, Tao L. Effects of taurine on signal transduction steps induced during hypertrophy of rat heart myocytes. In: Schaffer S, Lombardini JB, Huxtable RJ, editors. Taurine 3: Cellular and Regulatory Mechanisms1998. p. 137-143.

[69] Franconi F, Di Leo MAS, Bennardini F, Ghirlanda G. Is taurine beneficial in reducing risk factors for diabetes mellitus? Neurochemical Research. 2004;29(1):143-150.

[70] Franconi F, Loizzo A, Ghirlanda G, Seghieri G. Taurine supplementation and diabetes mellitus. Current Opinion in Clinical Nutrition and Metabolic Care. 2006;9(1):32-36.

[71] Brons C, Spohr C, Storgaard H, Dyerberg J, Vaag A. Effect of taurine treatment on insulin secretion and action, and on serum lipid levels in overweight men with a genetic predisposition for type II diabetes mellitus. European Journal of Clinical Nutrition. 2004;58(9):1239-1247.

[72] Ito T, Schaffer SW, Azuma J. The potential usefulness of taurine on diabetes mellitus and its complications. Amino Acids. 2012;42(5):1529-1539.

[73] Fang YJ, Chiu CH, Chang YY, Chou CH, Lin HW, Chen MF, Chen YC. Taurine ameliorates alcoholic steatohepatitis via enhancing self-antioxidant capacity and alcohol metabolism. Food Research International. 2011;44(9):3105-3110.

[74] Kerai MDJ, Waterfield CJ, Kenyon SH, Asker DS, Timbrell JA. Reversal of ethanol-induced hepatic steatosis and lipid peroxidation by taurine: A study in rats. Alcohol and Alcoholism. 1999;34(4):529-541.

[75] Wu GF, Yang JC, Sun CM, Luan XH, Shi J, Hu JM. Effect of Taurine on Alcoholic Liver Disease in Rats. Advances in Experimental and Medical Biology. 2009;643:313-22.

[76] Gentile CL, Nivala AM, Gonzales JC, Pfaffenbach KT, Wang D, Wei YR, Jiang H, Orlicky DJ, Petersen DR, Pagliassotti MJ, Maclean KN. Experimental evidence for therapeutic potential of taurine in the treatment of nonalcoholic fatty liver disease. American Journal of Physiology-Regulatory, Integrative andComparative Physiology. 2011;301(6):R1710-R722.

[77] Truswell AS, Mcveigh S, Mitchell WD, Brontest.B. Effect in Man of Feeding Taurine on Bile Acid Conjugation and Serum Cholesterol Levels. Journal of Atherosclerosis Research. 1965;5(5):526-529.

[78] Murakami S, Kondo-Ohta Y, Tomisawa K. Improvement in cholesterol metabolism in mice given chronic treatment of taurine and fed a high-fat diet. Life Sciences. 1998;64(1):83-91.

[79] Chen W, Matuda K, Nishimura N, Yokogoshi H. The effect of taurine on cholesterol degradation in mice fed a high-cholesterol diet. Life Sciences. 2004;74(15):1889-1898.

[80] Kamata K, Sugiura M, Kojima S, Kasuya Y. Restoration of endothelium-dependent relaxation in both hypercholesterolemia and diabetes by chronic taurine. European Journal of Pharmacology. 1996;303(1-2):47-53.

[81] Kondo Y, Toda Y, Kitajima H, Oda H, Nagate T, Kameo K, Murakami S. Taurine inhibits development of atherosclerotic lesions in apolipoprotein E-deficient mice. Clinical and Experimental Pharmacology and Physiology. 2001;28(10):809-815.

[82] Matsushima Y, Sekine T, Kondo Y, Sakurai T, Kameo K, Tachibana M, Murakami S. Effects of taurine on serum cholesterol levels and development of atherosclerosis in spontaneously hyperlipidaemic mice. Clinical and Experimental Pharmacology and Physiology. 2003;30(4):295-299.

[83] Matsuzaki Y, Miyazaki T, Miyakawa S, Bouscarel B, Ikegami T, Tanaka N. Decreased taurine concentration in skeletal muscles after exercise for various durations. Medicine and Science in Sports and Exercise. 2002;34(5):793-797.

[84] Yatabe Y, Miyakawa S, Miyazaki T, Matsuzaki Y, Ochiai N. Effects of taurine administration in rat skeletal muscles on exercise. Journal of Orthopaedic Science 2003;8(3):415-419.

[85] El Mesallamy HO, El-Demerdash E, Hammad LN, El Magdoub HM. Effect of taurine supplementation on hyperhomocysteinemia and markers of oxidative stress in high fructose diet induced insulin resistance. Diabetology & Metabolic Syndrome. 2010;2:46. Epub 2010/07/02.

[86] Herrmann RG. Effect of taurine, glycine and beta-sitosterols on serum and tissue cholesterol in the rat and rabbit. Circulation Research. 1959;7(2):224-227.

[87] Park T, Lee K, Um Y. Dietary taurine supplementation reduces plasma and liver cholesterol and triglyceride concentrations in rats fed a high-cholesterol diet. Nutrition Research. 1998;18(9):1559-1571.

[88] Sethupathy S, Elanchezhiyan C, Vasudevan K, Rajagopal G. Antiatherogenic effect of taurine in high fat diet fed rats. Indian Journal of Experimental Biology. Indian Journal of Experimental Biology. 2002;40(10):1169-1172.

[89] Sugiyama K, Ohishi A, Muramatsu K. Comparison between the plasma cholesterol-elevating effects of caffeine and methionine in rats on a high cholesterol diet. Agricultural and Biological Chemistry. 1989;53(11):3101-3103.

[90] Yokogoshi H, Mochizuki H, Nanami K, Hida Y, Miyachi F, Oda H. Dietary taurine enhances cholesterol degradation and reduces serum and liver cholesterol concentrations in rats fed a high-cholesterol diet. Journal of Nutrition. 1999;129(9):1705-1712.

[91] Masuda M, Horisaka K. Effect of taurine and homotaurine on bile acid metabolism in dietary hyperlipidemic rats. Journal of pharmacobio-dynamics. 1986;9(11):934-940. Epub 1986/11/01.

[92] Fukuda N, Yoshitama A, Sugita S, Fujita M, Murakami S. Dietary Taurine Reduces Hepatic Secretion of Cholesteryl Ester and Enhances Fatty Acid Oxidation in Rats Fed a High-Cholesterol Diet. Journal of Nutritional Science and Vitaminology. 2011;57(2):144-149.

[93] Mochizuki H, Takido J, Oda H, Yokogoshi H. Improving effect of dietary taurine on marked hypercholesterolemia induced by a high-cholesterol diet in streptozotocin-induced diabetic rats. Bioscience Biotechnology and Biochemistry. 1999;63(11):1984-1987.

[94] Nara Y, Yamori Y, Lovenberg W. Effect of Dietary Taurine on Blood-Pressure in Spontaneously Hypertensive Rats. Biochemical Pharmacology. 1978;27(23):2689-2692.

[95] Murakami S, Yamagishi I, Asami Y, Ohta Y, Toda Y, Nara Y, Yamori Y. Hypolipidemic effect of taurine in stroke-prone spontaneously hypertensive rats. Pharmacology. 1996;52(5):303-313.

[96] Choi MJ, Kim JH, Chang KJ. The effect of dietary taurine supplementation on plasma and liver lipid concentrations and free amino acid concentrations in rats fed a high-cholesterol diet. In: Oja SS, Saransaari P, editors. Taurine 62006. p. 235-242.

[97] Petty MA, Kintz J, Difrancesco GF. The effects of taurine on atherosclerosis development in cholesterol-fed rabbits. European Journal of Pharmacology. 1990;180(1):119-127.

[98] Takenaga T, Imada K, Otomo S. Hypolipidemic effect of taurine in golden Syrian hamsters. Taurine 4: Taurine and Excitable Tissues. 2000;483:187-192.

[99] Balkan J, Kanbagli O, Hatipoglu A, Kucuk M, Cevikbas U, Aykac-Toker G, Uysal M. Improving effect of dietary taurine supplementation on the oxidative stress and lipid levels in the plasma, liver and aorta of rabbits fed on a high-cholesterol diet. Bioscience Biotechnology and Biochemistry. 2002;66(8):1755-1758.

[100] Zulli A, Lau E, Wijaya BPP, Jin X, Sutarga K, Schwartz GD, Learmont J, Wookey PJ, Zinellu A, Carru C, Hare DL. High Dietary Taurine Reduces Apoptosis and Atherosclerosis in the Left Main Coronary Artery Association With Reduced CCAAT/Enhancer Binding Protein Homologous Protein and Total Plasma Homocysteine but not Lipidemia. Hypertension. 2009;53(6):1017-1022.

[101] Nistor A, Bulla A, Filip DA, Radu A. The hyperlipidemic hamster as a model of experimental atherosclerosis. Atherosclerosis. 1987;68(1-2):159-173.

[102] Murakami S, Kondo Y, Toda Y, Kitajima H, Kameo K, Sakono M, Fukuda N. Effect of taurine on cholesterol metabolism in hamsters: Up-regulation of low density lipoprotein (LDL) receptor by taurine. Life Sciences. 2002;70(20):2355-2366.

[103] Chang YY, Chou CH, Chiu CH, Yang KT, Lin YL, Weng WL, Chen YC. Preventive Effects of Taurine on Development of Hepatic Steatosis Induced by a High-Fat/Cholesterol Dietary Habit. Journal of Agricultural and Food Chemistry. 2011;59(1):450-457.

[104] Failey RB, Childress RH. Effect of Para-Aminobenzoic Acid on Serum Cholesterol Level in Man. American Journal of Clinical Nutrition. 1962;10(2):158-162.

[105] Hellstrom K, Sjovall J. Conjugation of Bile Acids in Patients with Hypothyroidism (Bile Acids and Steroids, 105). Journal of Atherosclerosis Research. 1961;1(3):205-210.

[106] Elvevoll EO, Eilertsen KE, Brox J, Dragnes BT, Falkenberg P, Olsen JO, Kirkhus B, Lamglait A, Østerud B. Seafood diets: Hypolipidemic and antiatherogenic effects of taurine and n-3 fatty acids. Atherosclerosis. 2008;200(2):396-402.

[107] Zhang M, Bi LF, Fang JH, Su XL, Da GL, Kuwamori T, Kagamimori S. Beneficial effects of taurine on serum lipids in overweight or obese non-diabetic subjects. Amino Acids. 2004;26(3):267-271.

[108] Mizushima S, Nara Y, Sawamura M, Yamori Y. Effects of oral taurine supplementation on lipids and sympathetic nerve tone. Taurine 2. 1996;403:615-622.

[109] Elizarova EP, Nedosugova LV. First experiments in taurine administration for diabetes mellitus - The effect on erythrocyte membranes. In: Huxtable RJ, Azuma J, Kuriyama K, Nakagawa M, Baba A, editors. Taurine 2: Basic and Clinical Aspects1996. p. 583-588.

[110] Eilertsen KE. Manuscript in preparation.

[111] Murakami S, Sakurai T, Toda Y, Morito A, Sakono M, Fukuda N. Prevention of neointima formation by taurine ingestion after carotid balloon injury. Vascular Pharmacology. 2010;53(3-4):177-184.

[112] Murakami S, Kondo Y, Sakurai T, Kitajima H, Nagate T. Taurine suppresses development of atherosclerosis in Watanabe heritable hyperlipidemic (WHHL) rabbits. Atherosclerosis. 2002;163(1):79-87.

Chapter 13

Cyclosporin A-Induced Hyperlipidemia

Maaike Kockx and Leonard Kritharides

Additional information is available at the end of the chapter

1. Introduction

Cyclosporin A (CsA) is an immunosuppressant drug widely used in organ transplant recipients and patients with auto-immune disorders. Long-term treatment with CsA is associated with hyperlipidemia and an increased risk of atherosclerosis. The mechanisms by which cyclosporin A causes hyperlipidemia are unclear. Cell and animal studies have pointed to various mechanisms that may mediate CsA-induced hyperlipidemia. In this review we will give an overview of CsA-induced hyperlipidemia, with a focus on the data available that might explain the underlying mechanism(s) and describe the available treatment regimes used to treat hyperlipidemia induced by immunosuppressant drugs.

2. Hyperlipidemia in humans after solid organ transplantation

Hyperlipidemia is observed in about 60% of kidney, liver, cardiac and bone marrow transplants after treatment with CsA (for review see [1,2]. There are multiple factors potentially contributing to hyperlipidemia in these patients, such as post-transplantation obesity, multiple drug therapy and diabetes. The concurrent use of steroids in particular, makes it hard to establish a direct contribution of CsA to dyslipidemia in humans, as corticosteroids are known to exacerbate hyperlipidemia in transplant recipients [3,4].

Studies investigating plasma lipids after CsA monotherapy are limited [4,5,6,7,8,9] and only a few studies have directly compared the combination of CsA therapy with low dose prednisolone with other immune suppressing strategies in combination with low dose steroids [10,11]. In general, these studies indicate that CsA treatment can independently lead to elevated plasma triglyceride and cholesterol levels in humans and that these effects are reversible upon cessation of immunosuppression therapy (Table 1). Animal studies (reviewed in [12]), where the effect of CsA can be studied in a more controlled background, indicate that CsA directly raises plasma lipid levels in rats, mice, guinea pigs and rabbits, and have proven that animals are valuable models to study mechanisms of CsA-induced hyperlipidemia.

Treatment	Patients	Patient number	Duration	Lipid effects	Reference
Monotherapy	Amyotrophic lateral sclerosis	36	2 mnths	TC ↑(21%) LDL-C ↑(31%) apoB ↑(12%) TG = HDL =	[5]
Monotherapy	Autologous bone marrow transplants	13	32 days	TC↑ (26%) LDL-C ↑ HDL-C ↓ TG = VLDL-C =	[13]
Monotherapy	Renal transplants	59	3-6 and 12 mnths	TC = LDL-C = apoB ↑ TG ↑ HDL-C ↓ apoA-I ↓	[8]
Monotherapy	Renal transplants	58	>1 yr	TC ↑ LDL-C ↑ apoB ↑ TG ↑ VLDL-C = HDL-C ↓ HDL2-C = HDL3-C ↓	[14]
Monotherapy and CsA/pred	Bone marrow transplants	180	100 days	TC ↑ LDL-C ↑ apoB ↑ TG ↑ VLDL-trig ↑ VLDL-C = HDL ↓ HDL2 ↓ HDL3 = apoA-I ↓	[4]
Monotherapy	Psoriasis	15	3 mnths	TC ↑ (22%) LDL-C ↑ (35%) TG = VLDL-C = HDL-C =	[9]

Treatment	Patients	Patient number	Duration	Lipid effects	Reference
ALG/aza/cort v CsA/ALG/aza/cort	Renal transplants	702	52 wks	TC ↑(20%) LDL-C ↑ TG ↑ HDL-C =	[7]
Aza/pred v CsA v CsA/pred	Renal transplants	9	3 mnths	TC ↑ LDL-C ↑ (45%) TG = VLDL-C = HDL-C =	[6]
Aza/pred v CsA/pred	Renal transplants	20	7.7 yrs	TC ↑ LDL-C ↑ apoB ↑ TG ↑ VLDL-C ↑ HDL-C ↓	[10]

ALG, Minnesota antilymphocyte globulin; aza, azathioprine; cort, corticosteroids; pred, prednisolone
TC, total cholesterol; TG, total triglyceride; LDL, low density lipoprotein; VLDL, very low density lipoprotein; HDL, high density lipoprotein; apo, apolipoprotein;

Table 1. Effect of CsA on plasma lipid parameters in humans

2.1. Plasma VLDL

Triglyceride-containing VLDL particles are produced in the liver via lipidation of apolipoprotein B (apoB) by microsomal triglyceride transfer protein (MTP), generating triglyceride-poor (VLDL2) as well as triglyceride-rich VLDL (VLDL1) particles, both of which can be secreted [15]. In plasma, VLDL is converted to intermediate-density lipoprotein (IDL) by lipoprotein lipase (LPL). IDL can be further hydrolyzed by lipases to low density lipoprotein (LDL). CsA increases plasma VLDL levels in transplant recipients and a concomitant increase in plasma apoB levels is observed [4,10,11]. It is unclear whether both plasma VLDL1 and VLDL2 levels are elevated. In contrast to LDL levels, plasma triglyceride and VLDL levels appear to increase only after long-term treatment with CsA (Table 1 and [8])

Hypertriglyceridemia in transplant patients is associated with increased plasma apolipoprotein CIII (apoCIII) levels [16,17,18] and decreased lipase activity (see below). As apoCIII inhibits LPL and hepatic lipase (HL) as well as uptake of triglyceride lipoprotein in liver, the increase of apoCIII may be an important contributor to hypertriglyceridemia found in transplant patients.

2.2. Plasma LDL

Plasma LDL levels appear to be consistently elevated by CsA [4,5,6,7,9,10,13,14] even in patients where plasma VLDL levels are not altered [5,6,9,13]. A correlation between CsA levels and plasma LDL-C has been described in some studies [19], but was not observed in others [5,20]. Regulation of plasma LDL levels is complex, depending on hepatic VLDL production, subsequent lipolysis of VLDL, clearance of LDL via the LDL receptor (LDLr) in the liver and conversion into bile. CsA may affect LDL metabolism at several levels (section 3.2).

2.3. Plasma HDL

Total plasma HDL levels are inversely correlated with the risk of cardiovascular disease [21]. HDL particles are however heterogeneous in size and composition, and occur as HDL2a, HDL2b, HDL3a, HDL3b and HDL3c which are progressively smaller in diameter and contain higher protein to lipid ratios. The precise contribution of various HDL subclasses to cardiovascular disease is currently unclear [21,22]. Plasma HDL cholesterol levels are determined by production of nascent HDL particles in the liver and intestine, by plasma transfer reactions of lipids between HDL and lipolysed triglyceride lipoproteins such as VLDL or chylomicrons, hepatic uptake of HDL lipids via the scavenger receptor class B1 (SRB1) HDL receptor in the liver, and renal clearance of small, lipid-poor apoA-I particles. Nascent HDL particles are formed by lipidation of apolipoprotein A-I (apoA-I) via the ATP-binding cassette transporter-1 (ABCA1) located in cellular membranes, although ABCA1-independent pathways of apoA-I lipidation also exist [23]. The formed lipid-poor HDL particles acquire more lipid after interaction with ABCG1 and mature by the subsequent esterification of cholesterol by lecithin-acyl transferase (LCAT). Further remodeling occurs by phospholipid transfer protein (PLTP) generating HDL2. HDL2 can be converted into HDL3 by hydrolysis via lipases and by transfer of cholesteryl esters to triglyceride-containing lipoproteins with the reciprocal exchange for triglycerides, which is mediated by cholesteryl ester transfer protein (CETP).

Immunosuppressive therapy has been reported to increase, decrease or leave HDL levels unaffected [5,10,11,24]. Parallel changes in plasma apoA-I levels are usually observed. Increased HDL levels are observed in most transplant patients, but this is most likely related to the concomitant treatment with steroids, which are known to increase plasma HDL [3]. CsA may affect particular subclasses of HDL more than others. Independently of steroids, plasma HDL levels, especially the HDL3 subpopulation, were found to inversely relate to plasma CsA levels [19]. In a study of bone marrow transplant recipients CsA decreased total plasma HDL, and in particular HDL2 [4]. In rats, a similar decrease in plasma HDL and HDL2 levels was observed after CsA treatment [25]. A recent study performed in pediatric renal transplant recipients showed that although total plasma HDL levels were not changed with CsA treatment, the relative proportion of HDL2b decreased while the relative proportion of HDL3a, HDL3b and HDL3c increased [26]. This is important as decreased HLD2b with increased HDL3b is associated with an atherogenic lipoprotein phenotype characterized by increased triglycerides and small dense LDL [27]. This result also emphases that simple monitoring of total HDL cholesterol may be insufficient to understand the consequences of CsA on HDL biology.

2.4. Plasma lipoprotein (a)

Lipoprotein (a) [Lp(a)] is a LDL-like lipoprotein consisting of LDL with one molecule of apoB covalently linked to a molecule of apolipoprotein (a). Plasma Lp(a) levels, and especially certain genetic Lp(a) variants, are independently associated with an increased risk for CVD [28,29]. Elevated Lp(a) plasma levels have been observed in renal transplant studies [14,30] this was however, not observed by others [31]. Although some studies suggested normalization of elevated Lp(a) levels after successful transplantation due to improved kidney function [31,32], CsA treatment has been indicated to independently increase Lp(a) levels in renal transplant recipients [8,14,33]. The mechanisms by which CsA affect plasma Lp(a) levels are unexplored, but may involve similar mechanisms to that of elevation of plasma LDL levels. As the LDLr does not play a major role in the clearance of Lp(a), the mechanism however, is unlikely mediated via effects of CsA on the LDLr (see section 3.2.1).

2.5. Qualitative differences in lipoproteins

2.5.1. Particle changes

Elevated plasma triglyceride levels are associated with the formation of triglyceride rich LDL particles that are more atherogenic [34]. A high prevalence of smaller denser LDL particles is observed in transplant recipients [35] and appears to be associated with CsA therapy [26,36]. Inhibition of lipoprotein lipase (LPL) activity is associated with the formation of small dense LDL subclasses. As apoCIII inhibits lipase activity, increased plasma apoCIII levels observed with CsA-treatment may explain inhibited lipase activity and subsequent increase in small dense LDL particles [17]. In addition decreased lipase activity could contribute to decreased HDL2 subclasses observed, while effects on CETP by CsA may help explain increases in HDL3 subfractions (see section 2.3 and 3.1.2).

2.5.2. Interaction of CsA with plasma lipoproteins

In whole blood CsA is primarily transported bound to lipoproteins (33%) and erythrocytes (58%) and whole blood CsA levels correlate with lipoprotein levels [37,38]. *In vitro* and *in vivo* studies show that in serum from healthy patients 50-60% of CsA is bound to HDL, 20-30% to LDL, 10-25% to VLDL with 10-15% bound to the non-lipoprotein proteins [39,40,41,42]. However, the proportion of CsA bound to the LDL and VLDL fractions increases in hyperlipidemic serum, without changing the amount bound to free protein [40,41], indicating that the distribution of CsA between the lipoprotein classes will change as plasma lipoprotein concentrations change. The binding of CsA to lipoprotein particles may also depend on lipoprotein composition. For example, Wasan et al. [41] showed that high triglyceride content of HDL was associated with a decreased percentage of CsA recovered in the HDL fraction and an increased percentage recovered in the VLDL fraction. Interestingly, treatment of patients with lipid lowering agents, such as statins have been reported to increase the unbound fraction of CsA and clearance of CsA in plasma [43].

Concerns have been raised about changes to the bioavailabilty and activity of CsA resulting from its binding to lipoproteins, especially as decreased CsA activity and increased toxicity have been observed in patients with hyperlipidemia [42,44]. CsA levels are higher in hyperlipidemic patients due to decreased clearance which was reversed after lipid-lowering with fibrates (reviewed in [37]). *In vitro* studies using skin fibroblasts indicate that CsA bound to LDL does not affect binding to cells via the LDLr, but uptake of CsA is inhibited [45]. These studies were confirmed in HepG2 and Jurkat Tcells which showed decreased uptake of CsA in the presence of LDL [40]. In line with these findings, uptake of CsA in tissues from rats was reduced when CsA was co-injected with lipoproteins [46].

3. Mechanisms of CsA-Induced hyperlipidemia – What we learn from cell and animal studies

As the effects of CsA in humans are confounded by many factors such as other medication, obesity, insulin resistance and nutritional status, cell and animal studies are useful to elucidate the mechanism(s) of CsA-induced hyperlipidemia. Figure 1 depicts the reported CsA-effects on VLDL, LDL and HDL metabolism.

3.1. VLDL

3.1.1. Effects of CsA on VLDL synthesis and secretion

CsA decreased apoB translocation over the endoplasmic reticulum (ER) membrane in the human liver cell line HepG2 [47]. It was suggested that this was due to a reduction in the efficiency of lipid transfer by inhibition of MTP, however whether MTP activity is inhibited by CsA was not investigated. These findings are in line with the report from Kaptein et al. [48], which showed that CsA inhibits VLDL and apoB secretion from HepG2 cells, by post-translational mechanisms. In contrast, in mice, CsA increased the rate of hepatic VLDL secretion *in vivo*, while total apoB secretion was unaffected [49]. No effect of CsA on levels of VLDL receptors in either adipose tissue or skeletal muscle were found [50] suggesting that VLDL uptake may not be affected by CsA. There are no studies that we are aware of studying the effect of CsA on *in vivo* VLDL synthesis in humans.

3.1.2. VLDL metabolism

Inhibition of lipolysis by CsA could contribute to increased plasma VLDL and reduced HDL concentrations. Various studies have investigated lipase activity in patients, but results may be confounded by co-treatment with steroids. HL activity was increased in cardiac transplant patients and correlated with CsA dose while lipoprotein lipase (LPL) activity was decreased in these patients [51]. Others have shown decreased HL as well as LPL activity in kidney transplant recipients [52]. More directly, Tory et al [53] showed suppression of LPL activity in plasma from normolipidemic subjects treated with CsA, while in rats, CsA dose- and time-dependent decreased plasma LPL activity [24]. In addition, LPL abundance in skeletal muscle and adipose tissue was decreased in rats [50]. These latter studies suggested CsA can inhibit LPL activity independently of steroids. Although the precise mechanism of CsA-inhibited LPL activity is unknown, it helps to explain increased triglyceride levels observed after CsA treatment.

Some studies show reduced cholesteryl ester transfer protein (CETP) activity in transplant recipients [54]. In contrast, CsA directly added to human plasma *ex vivo* increased CETP activity [53]. These apparently anomalous results may relate to differences between the direct effects of CsA on CETP itself and indirect effects secondary to changes in the concentrations of other lipoproteins, but remain unexplained. Since CETP transfers cholesteryl ester from HDL to apoB-containing lipoproteins with reciprocal transfer of triglycerides, any effect of CsA on CETP activity could be expected to have major effects on plasma lipoprotein profiles.

3.2. LDL

3.2.1. LDL synthesis and catabolism

We have recently reviewed this literature in detail [55]. There appear to be conflicting conclusions arising from *in vitro* and *in vivo* studies. One of the key discrepancies is the role of LDLr expression and LDL clearance by the liver in mediating CsA-hyperlipidemia. In general, *in vitro* studies are consistent with a role for decreased LDL receptor expression or activity in liver cells after exposure to CsA [48,56]. *In vivo* studies however, show mixed effects, with no effect or an increase in hepatic LDLr protein or mRNA levels [49,50]. Similarly 3-hydroxy-3-methyl-glutaryl-CoA reductase (HMG-CoAr), the rate limiting enzyme in cholesterol synthesis, mRNA levels were upregulated in HepG2 cells and mouse liver after CsA, but hepatic HMG-CoA reductase protein levels in rat liver were unaffected by CsA treatment [49,50,57]. In rats, CsA decreased the fractional catabolic rate of LDL [58]. One very important consideration is the difference in concentrations of CsA used in *in vitro* studies relative to those achieved *in vivo* under normal transplant immunosuppression. *In vitro* studies commonly use concentrations of 10 μg/ml whereas plasma levels of CsA in humans and in animal studies are typically in the order of 100 ng/ml. This apparent 10-fold difference in concentration may underestimate the difference in effective concentrations tested *in vivo* and *in vitro* studies because of the complicating effects of *in vivo* hyperlipidemia, which under some circumstances can lessen the effective concentration of CsA delivered to some tissues [46].

3.3. HDL

CsA effects on plasma HDL and HDL subclasses may be mediated by effects on the synthesis and/or formation of HDL as well as by effecting remodeling of HDL through changes in lipase and/or CETP activity (see 3.1.2)

3.3.1. Effect of CsA on HDL synthesis and formation

In vitro studies have indicated that CsA potently inhibits ABCA1 activity thereby inhibiting apoA-I lipidation, the first step in HDL formation [59,60,61]. This was associated with decreased ABCA1 turnover and an increase in total and cell-surface levels of ABCA1 [59]. Uptake, Internalization and re-secreton of apoA-I were however decreased by CsA, suggesting that ABCA1 trapped at the plasma membrane is dysfunctional [59,60]. *In vivo* studies using wild type C57Bl6 mice corroborated these *in vitro* findings. CsA lowered plasma HDL levels after 6 days of treatment [59]. A lowering in plasma HDL in mice was however not observed by others after long-term treatment of mice with CsA combined with a high fat diet [62]. As many aspects of lipid metabolism can be affected by CsA, it may be difficult to determine a causal effect on HDL levels via ABCA1 inhibition in an *in vivo* whole body system NB.

Direct effects of CsA on the expression of ABCA1 and apoA-I have also been reported and may contribute to the changes in HDL formation. The target of immunosuppression by CsA, Nuclear Factor of activated T-cells, cytoplasmic 2 (NFATc2), was found to bind the mouse ABCA1 promoter and mediate CsA-inhibition of ABCA1 expression by inflammatory stimuli [63]. In addition CsA has been found to inhibit apoA-I gene expression in human HepG2 cells and rats [64]. A recent proteomic study in HepG2 cells showed that CsA decreased secretion levels of apoA-I suggesting that the transcriptional effects of CsA on apoA-I expression may lead to decreased amounts of secreted apoA-I [65].

3.3.2. Effects on HDL metabolism

As mentioned above (section 3.1.2), CsA directly suppresses LPL activity and increases CETP activity in human plasma and animals (section 3.1.2). LPL activity is strongly associated with plasma HDL2 concentrations [66], and decreased LPL levels in CsA treatment may therefore contribute to decreased HDL2 levels [4,25]. On the other hand, increased CETP activity will generate triglyceride-rich HDL, which is converted to smaller HDL3 particles by HL [66].

3.4. Effects on bile acid synthesis and secretion

3.4.1. Effects on bile synthesis

In liver, cholesterol is converted to bile acids by 7α-hydroxylase (CYP7α) or 27-hydroxylase (CYP27A1) [67]. In healthy humans, CYP7α is considered the predominantly pathway while CYP27A1 accounts for 10% of bile acid synthesis and subsequent formation of chenodeoxycholate. However inhibition of Cyp7α can increase the contribution of the CYP27A1 pathway [68]. *In vitro* studies show that CsA inhibits both CYP27A1 activity and subsequent formation of chenodeoxycholate in human and animal liver extracts and in primary hepatocyte cultures [57,69,70,71]. A CsA responsive element has been mapped on the CYP27A1 promoter [72], indicating that CsA affects transcription of the CYP27A1 gene directly. In most of the *in vitro* studies, CYP7α activity was not affected by CsA [69,70]. *In vivo*, in rat however, CsA decreased CYP7α protein levels [50], indicating that the predominant bile acid synthesis pathway may also be affected by CsA. The inhibitory effect of CsA on bile synthesis is suggested to contribute to increased plasma lipid concentrations in transplant recipients. Radioisotope studies performed in children after liver transplantation demonstrated that CsA treatment significantly inhibits bile salts synthesis rates, especially that of chenodeoxycholate and that bile acid synthesis rate inversely correlates with plasma cholesterol and triglyceride levels [73].

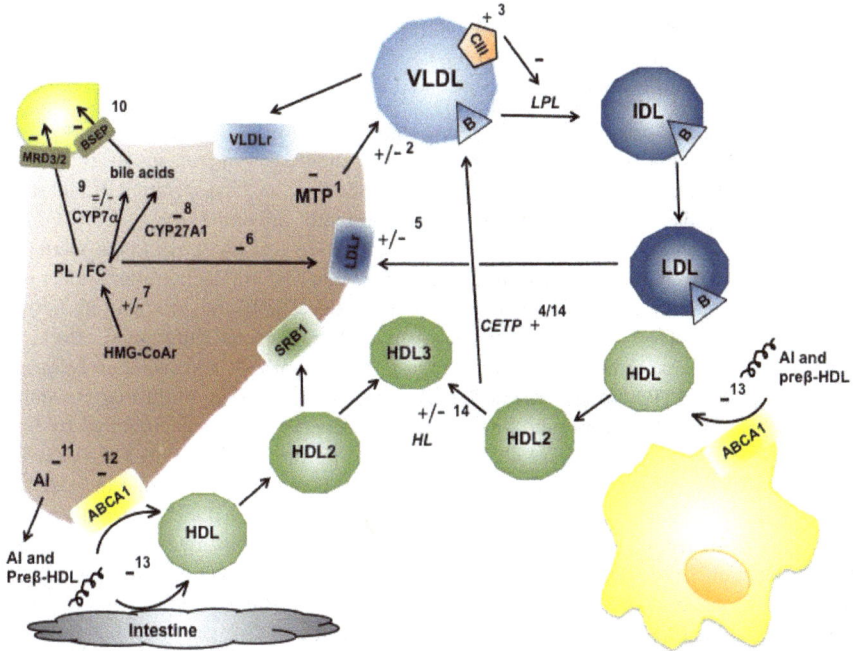

Figure 1. Mechanisms of CsA-mediated hyperlipidemia. Figure only displays pathways that are reported to be affected by CsA. 1) Inhibition of VLDL formation via inhibition of MTP, 2) Increased and decreased secretion of VLDL particles have been reported, 3) Decreased lipolysis of VLDL due to increased apoCIII and subsequent inhibition of LPL, 4) hypertriglyceridemia by increased CETP activity, 5) Increased LDL due to decreased LDLr expression as well as activity, 6) Increased liver FC content leading to decreased LDLr levels, 7) Increased and decreased levels of HMG-CoAr affecting cholesterol synthesis, 8/9) Inhibition of bile acid conversion via CYP27A1 or CYP7α leading to increased liver FC levels, however in most studies Cyp7α is not affected by CsA. NB: decreased CYP27A1 activity can increase HMG-CoAr levels via negative feedback, 10) Decreased flow of bile salts, cholesterol and phospholipids into bile, 11) Decreased expression and secretion of apoA-I, 12) Inhibition of ABCA1 expression, 13) inhibition of apoA-I lipidation via inhibition of ABCA1 activity 14) Stimulation of HL and CETP leads to increased formation of HDL2 to HDL3, however decreased HL activity has also been reported. VLDL, very low density lipoprotein; IDL, intermediate density lipoprotein; LDL, low density lipoprotein; HDL, high density lipoprotein; AI, apolipoprotein A-I, B, apolipoprotein B; CIII, apolipoprotein CIII; MTP, microsomal triglyceride transfer protein; LPL, lipoprotein lipase; HL, hepatic lipase; CETP, cholesteryl ester transfer protein; ABCA1, ATP-binding cassette transporter-1; SRB1, scavenger receptor class B1; LDLr, LDLreceptor; VLDLr, VLDLreceptor; PL, phospholipid; FC, free cholesterol; HMG-CoAr, 3-hydroxy-3-methyl-glutaryl-CoA reductase; CYP7α, 7α-hydroxylase; CYP27A1, 27-hydroxylase; MRD, multidrug resistance protein; BSEP, bile salt export protein.

The effects of CsA on CYP27A1 may relate to effects of CsA on cholesterol metabolism. 27-hydroxycholesterol is a potent negative feedback regulator of HMG-CoA reductase [74] and decreased CYP27A1 activity may therefore explain increased HMG-CoA reductase mRNA and cholesterol levels [57]. Although important in macrophages, it should be noted however that it is not clear whether such a feedback loop exists in liver cells [75]. Increased cholesterol synthesis could subsequently lead to downregulation of LDLr levels as observed in some CsA studies, also contributing to increased plasma cholesterol levels (see section 3.2.1).

Besides effects on bile acid synthesis CsA may affect bile flow. CsA treatment is associated with increased plasma bile acid concentrations and cholestasis in humans as well as in animal models [9,52,76]. Studies in rat indicate that bile flow and the secretion of bile salts, proteins and lipids into the bile are dose-dependently inhibited by CsA [52,76,77]. Interestingly, the changes in serum levels of bile acids are consistent with CsA-mediated inhibition of hepatocellular uptake of individual bile acids [78,79]. The inhibitory effect was greater for phosholipid secretion than that for cholesterol [80] and in some studies no inhibition of cholesterol excretion was observed [81], suggesting differential effects on transport mechanisms. Transport pumps involved in bile synthesis and secretion belong to the family of the ATP-binding cassette transporters which include, multidrug resistance proteins (MDR) and P-glycoprotein, and most of which are effectively inhibited by CsA [79,82]. Interestingly, comparison of the bile salt export pump (BSEP) activity from different species, showed that CsA inhibits bile salt transport with species and bile salt specific variation [83]. Rat BSEP was for example more effectively inhibited than mouse BSEP. Biliary cholesterol secretion is mediated via ABCG5 and ABCG8 [84]. Although both members of the ATP-binding cassette family, it has not been investigated whether CsA inhibits ABCG5/8 activity. As phospholipids are transported via MDR3, it is likely that differences in efficacy of CsA between inhibition of MRD3 and ABCG5/8 exist. It is clear that CsA can affect bile flow and secretion in cultured cells and animal models. It should be noted however, that in humans no inhibitory effect of CsA on secretion of bile acids and lipids or on bile composition after liver transplantation was observed [85]. Others have shown that although cholate synthesis was reduced by CsA, compensatory increased intestinal absorption counteracted this decrease [86]. It remains therefore unclear to what extent inhibition of bile flow and secretion by CsA are contributing to hypercholesteremia *in vivo*.

4. Therapies to address hyperlipidemia

Hyperlipidemia is associated with significant morbidity and mortality rates in transplant recipients [87]. Many strategies have been investigated to target dyslipidemia in transplant patients. A number of excellent comprehensive reviews have been published on the clinical management of hyperlipidemia and its risks (eg [88,89]). We will therefore restrict our comments to a very brief summary of this area.

4.1. Statins

Statins inhibit HMG-CoA reductase, the rate limiting enzyme in the cholesterol synthesis pathway and are world-wide the drug of choice to lower plasma LDL-C levels. Various statins have been tested in transplant patients and all show significant lowering of plasma cholesterol, LDL-C and apoB levels with some indicating improved survival rates (for review see [88,89,90]). A randomized trial, investigating the safety and efficacy of statins in renal transplant patients, the Assessment of LEscol in Renal Transplantation (ALERT) study, showed that fluvastatin effectively lowered LDL-C by 32% and reduced cardiac death and non-fatal myocardial infarction incidence significantly [91]. Importantly, statins may provide beneficial effects other then their lipid-lowering properties [92]. Wissing et al [93] reported improved flow mediated brachial artery vasodilatation by atorvastatin in kidney transplant patients and significant reductions in acute rejections have been observed in cardiac transplant patients [94].

Rhabdomyolysis, one of the few serious side effects of statins, is more common with high dose statin treatment. The risk is elevated in patients with renal disease and in patients taking drugs affecting statin metabolism, especially in those taking CsA [88,89]. All statins have the potential to interact with CsA, as CsA substantially increases plasma levels of all statins. Although this is most notable for those metabolized via the Cyp3A4 pathway, statins not metabolized via the Cyp3A4 pathway [95] such as pravastatin and fluvastatin are also affected [95], suggesting that the interaction of CsA and statins may involve other mechanisms such as inhibition of drug transporters. Simvastatin poses the highest risk of myopathy, and particular care must be taken with higher doses of this agent, with recommendations that doses of 10mg/d are not exceeded in transplant patients [89]. Because statin therapy has been associated with mortality benefit after transplantation, correction of hyperlipidemia using lower doses of statins is mandatory after transplantation. Therefore careful clinical monitoring of patients as well as measurement of creatine kinase levels to detect muscle injury is advised, and the use of statins that are not metabolized via CYP3A4, such as fluvastatin or pravastatin may be preferential [95].

4.2. Fibrates

Fibrates lower plasma triglyceride levels via activation of the Peroxisome Proliferator Activated Receptor alpha (PPARα) and may be useful in transplant patients with elevated plasma triglycerides especially in combination with statin treatment to lower plasma cholesterol levels. Gemfibrozil was found to significantly lower plasma triglyceride levels in heart transplant patients and increase long term survival [96,97]. Fenofibrate is less well studied in transplant patients and may be associated with increased nephrotoxicity [88,98]. Care must be taken administering fibrates with CsA, particularly in combination with statins as drug-drug interactions exist via CYP3A4 as well as the hepatic uptake transporter the organic anion transporting polypeptide 1B1 (OAT1B1).

4.3. Ezetimibe

Inhibition of intestinal cholesterol absorption to lower high plasma cholesterol levels may be used when statins or fibrates are ineffective or are not tolerated. Ezetimibe proved to be an effective drug lowering plasma LDL-C levels significantly by blocking cholesterol absorption in the small intestine [99]. To that point though, various studies showed effective LDL-C lowering in liver, cardiac and renal transplant recipients [99]. Although, drug-drug interaction between CsA and ezetimibe were suggested (See [88]), CsA levels in studied transplant patients were not affected by combined ezetimibe use (reviewed in [99]). Co-administration of ezetimibe with (low-dose) statins has been found to effectively reduce high plasma cholesterol levels in transplant recipients and may be useful in patients that resistant to high-dose statin or where target plasma lipid levels can not be achieved by statin therapy alone [100,101].

5. Conclusions

CsA-induced hyperlipidemia is well established and remains a significant clinical issue. CsA potentially affects many aspects of lipid and lipoprotein metabolism and the precise underlying mechanism(s) causing dyslipidemia are still unclear. Further mechanistic studies may lead to the generation immunosuppressants that do not cause hyperlipidemia or may help to develop strategies to effectively target CsA-induced hyperlipidemia.

Author details

Maaike Kockx
Macrophage Biology Group, Centre for Vascular Research, University of New South Wales, Sydney, Australia

Leonard Kritharides*
Macrophage Biology Group, Centre for Vascular Research, University of New South Wales, Sydney, Australia
Department of Cardiology, Concord Repatriation General Hospital, University of Sydney, Sydney, Australia

6. References

[1] Kobashigawa JA, Kasiske BL (1997) Hyperlipidemia in solid organ transplantation. Transplantation 63: 331-338.
[2] Miller LW (2002) Cardiovascular toxicities of immunosuppressive agents. Am J Transplant 2: 807-818.
[3] Strohmayer EA, Krakoff LR (2011) Glucocorticoids and cardiovascular risk factors. Endocrinol Metab Clin North Am 40: 409-417, ix.
[4] Lopez-Miranda J, Perez-Jimenez F, Torres A, Espino-Montoro A, Gomez P, et al. (1992) Effect of cyclosporin on plasma lipoproteins in bone marrow transplantation patients. Clin Biochem 25: 379-386.
[5] Ballantyne CM, Podet EJ, Patsch WP, Harati Y, Appel V, et al. (1989) Effects of cyclosporine therapy on plasma lipoprotein levels. JAMA 262: 53-56.

* Corresponding Author

[6] Raine AE, Carter R, Mann JI, Morris PJ (1988) Adverse effect of cyclosporin on plasma cholesterol in renal transplant recipients. Nephrol Dial Transplant 3: 458-463.

[7] Kasiske BL, Tortorice KL, Heim-Duthoy KL, Awni WM, Rao KV (1991) The adverse impact of cyclosporine on serum lipids in renal transplant recipients. Am J Kidney Dis 17: 700-707.

[8] Hilbrands LB, Demacker PN, Hoitsma AJ, Stalenhoef AF, Koene RA (1995) The effects of cyclosporine and prednisone on serum lipid and (apo)lipoprotein levels in renal transplant recipients. J Am Soc Nephrol 5: 2073-2081.

[9] Edwards BD, Bhatnagar D, Mackness MI, Gokal R, Ballardie FW, et al. (1995) Effect of low-dose cyclosporin on plasma lipoproteins and markers of cholestasis in patients with psoriasis. QJM 88: 109-113.

[10] Schorn TF, Kliem V, Bojanovski M, Bojanovski D, Repp H, et al. (1991) Impact of long-term immunosuppression with cyclosporin A on serum lipids in stable renal transplant recipients. Transpl Int 4: 92-95.

[11] Ichimaru N, Takahara S, Kokado Y, Wang JD, Hatori M, et al. (2001) Changes in lipid metabolism and effect of simvastatin in renal transplant recipients induced by cyclosporine or tacrolimus. Atherosclerosis 158: 417-423.

[12] Kockx M, Guo DL, Traini M, Gaus K, Kay J, et al. (2009) Cyclosporin A decreases apolipoprotein E secretion from human macrophages via a protein phosphatase 2B-dependent and ATP-binding cassette transporter A1 (ABCA1)-independent pathway. J Biol Chem 284: 24144-24154.

[13] Luke DR, Beck JE, Vadiei K, Yousefpour M, LeMaistre CF, et al. (1990) Longitudinal study of cyclosporine and lipids in patients undergoing bone marrow transplantation. J Clin Pharmacol 30: 163-169.

[14] Brown JH, Anwar N, Short CD, Bhatnager D, Mackness MI, et al. (1993) Serum lipoprotein (a) in renal transplant recipients receiving cyclosporin monotherapy. Nephrol Dial Transplant 8: 863-867.

[15] Adiels M, Olofsson SO, Taskinen MR, Boren J (2008) Overproduction of very low-density lipoproteins is the hallmark of the dyslipidemia in the metabolic syndrome. Arterioscler Thromb Vasc Biol 28: 1225-1236.

[16] Kimak E, Solski J, Baranowicz-Gaszczyk I, Ksiazek A (2006) A long-term study of dyslipidemia and dyslipoproteinemia in stable post-renal transplant patients. Ren Fail 28: 483-486.

[17] Badiou S, Garrigue V, Dupuy AM, Chong G, Cristol JP, et al. (2006) Small dense low-density lipoprotein in renal transplant recipients: a potential target for prevention of cardiovascular complications? Transplant Proc 38: 2314-2316.

[18] Tur MD, Garrigue V, Vela C, Dupuy AM, Descomps B, et al. (2000) Apolipoprotein CIII is upregulated by anticalcineurins and rapamycin: implications in transplantation-induced dyslipidemia. Transplant Proc 32: 2783-2784.

[19] Kuster GM, Drexel H, Bleisch JA, Rentsch K, Pei P, et al. (1994) Relation of cyclosporine blood levels to adverse effects on lipoproteins. Transplantation 57: 1479-1483.

[20] Ramezani M, Einollahi B, Ahmadzad-Asl M, Nafar M, Pourfarziani V, et al. (2007) Hyperlipidemia after renal transplantation and its relation to graft and patient survival. Transplant Proc 39: 1044-1047.

[21] Kontush A, Chapman MJ (2006) Functionally defective high-density lipoprotein: a new therapeutic target at the crossroads of dyslipidemia, inflammation, and atherosclerosis. Pharmacol Rev 58: 342-374.

[22] Rosenson RS, Brewer HB, Jr., Chapman MJ, Fazio S, Hussain MM, et al. (2011) HDL measures, particle heterogeneity, proposed nomenclature, and relation to atherosclerotic cardiovascular events. Clin Chem 57: 392-410.

[23] Zheng H, Kiss RS, Franklin V, Wang MD, Haidar B, et al. (2005) ApoA-I lipidation in primary mouse hepatocytes. Separate controls for phospholipid and cholesterol transfers. J Biol Chem 280: 21612-21621.

[24] Lopez-Miranda J, Perez-Jimenez F, Gomez-Gerique JA, Espino-Montoro A, Hidalgo-Rojas L, et al. (1992) Effect of cyclosporin on plasma lipoprotein lipase activity in rats. Clin Biochem 25: 387-394.

[25] Espino A, Lopez-Miranda J, Blanco-Cerrada J, Zambrana JL, Aumente MA, et al. (1995) The effect of cyclosporine and methylprednisolone on plasma lipoprotein levels in rats. J Lab Clin Med 125: 222-227.

[26] Zeljkovic A, Vekic J, Spasojevic-Kalimanovska V, Jelic-Ivanovic Z, Peco-Antic A, et al. (2011) Characteristics of low-density and high-density lipoprotein subclasses in pediatric renal transplant recipients. Transpl Int 24: 1094-1102.

[27] Berneis KK, Krauss RM (2002) Metabolic origins and clinical significance of LDL heterogeneity. J Lipid Res 43: 1363-1379.

[28] Erqou S, Kaptoge S, Perry PL, Di Angelantonio E, Thompson A, et al. (2009) Lipoprotein(a) concentration and the risk of coronary heart disease, stroke, and nonvascular mortality. JAMA 302: 412-423.

[29] Clarke R, Peden JF, Hopewell JC, Kyriakou T, Goel A, et al. (2009) Genetic variants associated with Lp(a) lipoprotein level and coronary disease. N Engl J Med 361: 2518-2528.

[30] Fonseca I, Queiros J, Costa S, Santos MJ, Henriques AC, et al. (2002) Lipoprotein(A) in renal transplant recipients. Transplant Proc 34: 370-372.

[31] Innocenti M, Lorenzetti M, Naldi F, Paleologo G, Pasquariello A, et al. (1998) Evaluation of lipoprotein A in renal transplant recipients. Transplant Proc 30: 2048.

[32] Black IW, Wilcken DE (1992) Decreases in apolipoprotein(a) after renal transplantation: implications for lipoprotein(a) metabolism. Clin Chem 38: 353-357.

[33] Webb AT, Reaveley DA, O'Donnell M, O'Connor B, Seed M, et al. (1993) Does cyclosporin increase lipoprotein(a) concentrations in renal transplant recipients? Lancet 341: 268-270.

[34] Packard CJ, Shepherd J (1997) Lipoprotein heterogeneity and apolipoprotein B metabolism. Arterioscler Thromb Vasc Biol 17: 3542-3556.

[35] Rajman I, Harper L, McPake D, Kendall MJ, Wheeler DC (1998) Low-density lipoprotein subfraction profiles in chronic renal failure. Nephrol Dial Transplant 13: 2281-2287.

[36] Quaschning T, Mainka T, Nauck M, Rump LC, Wanner C, et al. (1999) Immunosuppression enhances atherogenicity of lipid profile after transplantation. Kidney Int Suppl 71: S235-237.

[37] Akhlaghi F, Trull AK (2002) Distribution of cyclosporin in organ transplant recipients. Clin Pharmacokinet 41: 615-637.

[38] Gardier AM, Mathe D, Guedeney X, Barre J, Benvenutti C, et al. (1993) Effects of plasma lipid levels on blood distribution and pharmacokinetics of cyclosporin A. Ther Drug Monit 15: 274-280.

[39] Sgoutas D, MacMahon W, Love A, Jerkunica I (1986) Interaction of cyclosporin A with human lipoproteins. J Pharm Pharmacol 38: 583-588.

[40] Rifai N, Chao FF, Pham Q, Thiessen J, Soldin SJ (1996) The role of lipoproteins in the transport and uptake of cyclosporine and dihydro-tacrolimus into HepG2 and JURKAT cell lines. Clin Biochem 29: 149-155.

[41] Wasan KM, Pritchard PH, Ramaswamy M, Wong W, Donnachie EM, et al. (1997) Differences in lipoprotein lipid concentration and composition modify the plasma distribution of cyclosporine. Pharm Res 14: 1613-1620.

[42] De Klippel N, Sennesael J, Lamote J, Ebinger G, de Keyser J (1992) Cyclosporin leukoencephalopathy induced by intravenous lipid solution. Lancet 339: 1114.

[43] Akhlaghi F, McLachlan AJ, Keogh AM, Brown KF (1997) Effect of simvastatin on cyclosporine unbound fraction and apparent blood clearance in heart transplant recipients. Br J Clin Pharmacol 44: 537-542.

[44] de Groen PC, Aksamit AJ, Rakela J, Forbes GS, Krom RA (1987) Central nervous system toxicity after liver transplantation. The role of cyclosporine and cholesterol. N Engl J Med 317: 861-866.

[45] Wasan KM, Ramaswamy M, Kwong M, Boulanger KD (2002) Role of plasma lipoproteins in modifying the toxic effects of water-insoluble drugs: studies with cyclosporine A. AAPS PharmSci 4: E30.

[46] Lemaire M, Pardridge WM, Chaudhuri G (1988) Influence of blood components on the tissue uptake indices of cyclosporin in rats. J Pharmacol Exp Ther 244: 740-743.

[47] Macri J, Adeli K (1997) Studies on intracellular translocation of apolipoprotein B in a permeabilized HepG2 system. J Biol Chem 272: 7328-7337.

[48] Kaptein A, de Wit EC, Princen HM (1994) Cotranslational inhibition of apoB-100 synthesis by cyclosporin A in the human hepatoma cell line HepG2. Arterioscler Thromb 14: 780-789.

[49] Wu J, Zhu YH, Patel SB (1999) Cyclosporin-induced dyslipoproteinemia is associated with selective activation of SREBP-2. Am J Physiol 277: E1087-1094.

[50] Vaziri ND, Liang K, Azad H (2000) Effect of cyclosporine on HMG-CoA reductase, cholesterol 7alpha-hydroxylase, LDL receptor, HDL receptor, VLDL receptor, and lipoprotein lipase expressions. J Pharmacol Exp Ther 294: 778-783.

[51] Superko HR, Haskell WL, Di Ricco CD (1990) Lipoprotein and hepatic lipase activity and high-density lipoprotein subclasses after cardiac transplantation. Am J Cardiol 66: 1131-1134.

[52] Deters M, Kirchner G, Koal T, Resch K, Kaever V (2004) Everolimus/cyclosporine interactions on bile flow and biliary excretion of bile salts and cholesterol in rats. Dig Dis Sci 49: 30-37.

[53] Tory R, Sachs-Barrable K, Hill JS, Wasan KM (2008) Cyclosporine A and Rapamycin induce in vitro cholesteryl ester transfer protein activity, and suppress lipoprotein lipase activity in human plasma. Int J Pharm 358: 219-223.

[54] Atger V, Leclerc T, Cambillau M, Guillemain R, Marti C, et al. (1993) Elevated high density lipoprotein concentrations in heart transplant recipients are related to impaired plasma cholesteryl ester transfer and hepatic lipase activity. Atherosclerosis 103: 29-41.

[55] Kockx M, Jessup W, Kritharides L (2010) Cyclosporin A and atherosclerosis--cellular pathways in atherogenesis. Pharmacol Ther 128: 106-118.

[56] Rayyes OA, Wallmark A, Floren CH (1996) Cyclosporine inhibits catabolism of low-density lipoproteins in HepG2 cells by about 25%. Hepatology 24: 613-619.

[57] Gueguen Y, Ferrari L, Souidi M, Batt AM, Lutton C, et al. (2007) Compared effect of immunosuppressive drugs cyclosporine A and rapamycin on cholesterol homeostasis key enzymes CYP27A1 and HMG-CoA reductase. Basic Clin Pharmacol Toxicol 100: 392-397.

[58] Lopez-Miranda J, Vilella E, Perez-Jimenez F, Espino A, Jimenez-Pereperez JA, et al. (1993) Low-density lipoprotein metabolism in rats treated with cyclosporine. Metabolism 42: 678-683.

[59] Le Goff W, Peng DQ, Settle M, Brubaker G, Morton RE, et al. (2004) Cyclosporin A traps ABCA1 at the plasma membrane and inhibits ABCA1-mediated lipid efflux to apolipoprotein A-I. Arterioscler Thromb Vasc Biol 24: 2155-2161.

[60] Lorenzi I, von Eckardstein A, Cavelier C, Radosavljevic S, Rohrer L (2008) Apolipoprotein A-I but not high-density lipoproteins are internalised by RAW macrophages: roles of ATP-binding cassette transporter A1 and scavenger receptor BI. J Mol Med 86: 171-183.

[61] Karwatsky J, Ma L, Dong F, Zha X (2009) Cholesterol efflux to apoA-I in ABCA1-expressing cells is regulated by Ca2+ dependent-calcineurin signaling. J Lipid Res.

[62] Emeson EE, Shen ML (1993) Accelerated atherosclerosis in hyperlipidemic C57BL/6 mice treated with cyclosporin A. Am J Pathol 142: 1906-1915.

[63] Maitra U, Parks JS, Li L (2009) An innate immunity signaling process suppresses macrophage ABCA1 expression through IRAK-1-mediated downregulation of retinoic acid receptor alpha and NFATc2. Mol Cell Biol 29: 5989-5997.

[64] Zheng XL, Wong NC (2006) Cyclosporin A inhibits apolipoprotein AI gene expression. J Mol Endocrinol 37: 367-373.

[65] Van Summeren A, Renes J, Bouwman FG, Noben JP, van Delft JH, et al. (2011) Proteomics investigations of drug-induced hepatotoxicity in HepG2 cells. Toxicol Sci 120: 109-122.

[66] von Eckardstein A, Huang Y, Assmann G (1994) Physiological role and clinical relevance of high-density lipoprotein subclasses. Curr Opin Lipidol 5: 404-416.

[67] Anderson KE, Kok E, Javitt NB (1972) Bile acid synthesis in man: metabolism of 7 - hydroxycholesterol- 14 C and 26-hydroxycholesterol- 3 H. J Clin Invest 51: 112-117.

[68] Duane WC, Javitt NB (1999) 27-hydroxycholesterol: production rates in normal human subjects. J Lipid Res 40: 1194-1199.

[69] Souidi M, Parquet M, Ferezou J, Lutton C (1999) Modulation of cholesterol 7alpha-hydroxylase and sterol 27-hydroxylase activities by steroids and physiological conditions in hamster. Life Sci 64: 1585-1593.

[70] Princen HM, Meijer P, Wolthers BG, Vonk RJ, Kuipers F (1991) Cyclosporin A blocks bile acid synthesis in cultured hepatocytes by specific inhibition of chenodeoxycholic acid synthesis. Biochem J 275 (Pt 2): 501-505.

[71] Winegar DA, Salisbury JA, Sundseth SS, Hawke RL (1996) Effects of cyclosporin on cholesterol 27-hydroxylation and LDL receptor activity in HepG2 cells. J Lipid Res 37: 179-191.

[72] Segev H, Honigman A, Rosen H, Leitersdorf E (2001) Transcriptional regulation of the human sterol 27-hydroxylase gene (CYP27) and promoter mapping. Atherosclerosis 156: 339-347.

[73] Hulzebos CV, Bijleveld CM, Stellaard F, Kuipers F, Fidler V, et al. (2004) Cyclosporine A-induced reduction of bile salt synthesis associated with increased plasma lipids in children after liver transplantation. Liver Transpl 10: 872-880.

[74] Esterman AL, Baum H, Javitt NB, Darlington GJ (1983) 26-hydroxycholesterol: regulation of hydroxymethylglutaryl-CoA reductase activity in Chinese hamster ovary cell culture. J Lipid Res 24: 1304-1309.

[75] Javitt NB (2002) 25R,26-Hydroxycholesterol revisited: synthesis, metabolism, and biologic roles. J Lipid Res 43: 665-670.

[76] Stone BG, Udani M, Sanghvi A, Warty V, Plocki K, et al. (1987) Cyclosporin A-induced cholestasis. The mechanism in a rat model. Gastroenterology 93: 344-351.

[77] Roman ID, Monte MJ, Gonzalez-Buitrago JM, Esteller A, Jimenez R (1990) Inhibition of hepatocytary vesicular transport by cyclosporin A in the rat: relationship with cholestasis and hyperbilirubinemia. Hepatology 12: 83-91.

[78] Azer SA, Stacey NH (1994) Differential effects of cyclosporin A on transport of bile acids by rat hepatocytes: relationship to individual serum bile acid levels. Toxicol Appl Pharmacol 124: 302-309.

[79] Stieger B, Fattinger K, Madon J, Kullak-Ublick GA, Meier PJ (2000) Drug- and estrogen-induced cholestasis through inhibition of the hepatocellular bile salt export pump (Bsep) of rat liver. Gastroenterology 118: 422-430.

[80] Galan AI, Roman ID, Munoz ME, Cava F, Gonzalez-Buitrago JM, et al. (1992) Inhibition of biliary lipid and protein secretion by cyclosporine A in the rat. Biochem Pharmacol 44: 1105-1113.

[81] Chan FK, Shaffer EA (1997) Cholestatic effects of cyclosporine in the rat. Transplantation 63: 1574-1578.

[82] Bohme M, Jedlitschky G, Leier I, Buchler M, Keppler D (1994) ATP-dependent export pumps and their inhibition by cyclosporins. Adv Enzyme Regul 34: 371-380.

[83] Kis E, Ioja E, Nagy T, Szente L, Heredi-Szabo K, et al. (2009) Effect of membrane cholesterol on BSEP/Bsep activity: species specificity studies for substrates and inhibitors. Drug Metab Dispos 37: 1878-1886.

[84] Yu L, Gupta S, Xu F, Liverman AD, Moschetta A, et al. (2005) Expression of ABCG5 and ABCG8 is required for regulation of biliary cholesterol secretion. J Biol Chem 280: 8742-8747.

[85] Baiocchi L, Angelico M, De Luca L, Ombres D, Anselmo A, et al. (2006) Cyclosporine A versus tacrolimus monotherapy. Comparison on bile lipids in the first 3 months after liver transplant in humans. Transpl Int 19: 389-395.

[86] Hulzebos CV, Wolters H, Plosch T, Kramer W, Stengelin S, et al. (2003) Cyclosporin a and enterohepatic circulation of bile salts in rats: decreased cholate synthesis but increased intestinal reabsorption. J Pharmacol Exp Ther 304: 356-363.

[87] Toussaint C, Kinnaert P, Vereerstraeten P (1988) Late mortality and morbidity five to eighteen years after kidney transplantation. Transplantation 45: 554-558.

[88] Bilchick KC, Henrikson CA, Skojec D, Kasper EK, Blumenthal RS (2004) Treatment of hyperlipidemia in cardiac transplant recipients. Am Heart J 148: 200-210.

[89] Ballantyne CM, Corsini A, Davidson MH, Holdaas H, Jacobson TA, et al. (2003) Risk for myopathy with statin therapy in high-risk patients. Arch Intern Med 163: 553-564.

[90] Ojo AO (2006) Cardiovascular complications after renal transplantation and their prevention. Transplantation 82: 603-611.

[91] Holdaas H, Fellstrom B, Jardine AG, Holme I, Nyberg G, et al. (2003) Effect of fluvastatin on cardiac outcomes in renal transplant recipients: a multicentre, randomised, placebo-controlled trial. Lancet 361: 2024-2031.

[92] Blum A, Shamburek R (2009) The pleiotropic effects of statins on endothelial function, vascular inflammation, immunomodulation and thrombogenesis. Atherosclerosis 203: 325-330.

[93] Wissing KM, Unger P, Ghisdal L, Broeders N, Berkenboom G, et al. (2006) Effect of atorvastatin therapy and conversion to tacrolimus on hypercholesterolemia and endothelial dysfunction after renal transplantation. Transplantation 82: 771-778.

[94] Kobashigawa JA, Katznelson S, Laks H, Johnson JA, Yeatman L, et al. (1995) Effect of pravastatin on outcomes after cardiac transplantation. N Engl J Med 333: 621-627.

[95] Asberg A (2003) Interactions between cyclosporin and lipid-lowering drugs: implications for organ transplant recipients. Drugs 63: 367-378.

[96] Pflugfelder PW, Huff M, Oskalns R, Rudas L, Kostuk WJ (1995) Cholesterol-lowering therapy after heart transplantation: a 12-month randomized trial. J Heart Lung Transplant 14: 613-622.

[97] Stapleton DD, Mehra MR, Dumas D, Smart FW, Milani RV, et al. (1997) Lipid-lowering therapy and long-term survival in heart transplantation. Am J Cardiol 80: 802-805.

[98] Boissonnat P, Salen P, Guidollet J, Ferrera R, Dureau G, et al. (1994) The long-term effects of the lipid-lowering agent fenofibrate in hyperlipidemic heart transplant recipients. Transplantation 58: 245-247.

[99] Suchy D, Labuzek K, Stadnicki A, Okopien B Ezetimibe--a new approach in hypercholesterolemia management. Pharmacol Rep 63: 1335-1348.

[100] Yoon HE, Song JC, Hyoung BJ, Hwang HS, Lee SY, et al. (2009) The efficacy and safety of ezetimibe and low-dose simvastatin as a primary treatment for dyslipidemia in renal transplant recipients. Korean J Intern Med 24: 233-237.

[101] Lopez V, Gutierrez C, Gutierrez E, Sola E, Cabello M, et al. (2008) Treatment with ezetimibe in kidney transplant recipients with uncontrolled dyslipidemia. Transplant Proc 40: 2925-2926.

Chapter 14

Nutritional Management of Disturbances in Lipoprotein Concentrations

Somayeh Hosseinpour-Niazi, Parvin Mirmiran and Fereidoun Azizi

Additional information is available at the end of the chapter

1. Introduction

Atherogenic dyslipidemia includes increase in blood concentrations of LDL cholesterol, total cholesterol, triglycerides and decrease in high-density lipoprotein cholesterol, both of which are frequently associated with the development of cardiovascular diseases (CVDs) [1,2] .Treatment of dyslipidemia can reduce the risk of CVDs [3]. In both industrialized and non-industrialized countries, the prevalence of dyslipidemia is increasing (4-7), therefore management of dyslipidemia has become a mainstay of routine clinical practice for both public health and clinicians. Although the benefits of lipid-lowering therapy have been demonstrated most conclusively, the role of diet determinants in dyslipidemia needs to be further considered [8]. Diet plays an important role in the concentrations of lipoprotein and is the primary intervention for patients with dyslipidemia. Understanding the relationships between dietary determinants and dyslipidemia and the effect of diet on lipoprotein concentrations may help to identify the dietary changes needed to reduce health risks [2]. Dietary changes, including reduced intakes of saturated fat and cholesterol, increased intakes of polyunsaturated fatty acids, fish, fruits and vegetables, and reduced energy intakes may have beneficial effects on lipoprotein concentrations [9-12]. One important aspect of diet is dietary patterns that address the effect of the diet as a whole and thus may provide insight beyond the effects described for single nutrients or foods [13]. The effects of some dietary patterns including the Mediterranean diet, the dietary stop to hypertension (DASH) and traditional dietary patterns, on lipoprotein particles need to be discussed [14, 15, 16]. In addition, other aspects of diet, including herbal, phytochemical, and dietary supplement (plant stanols and sterols) also play important roles in the prevention and treatment of dyslipidemia and may improve lipoprotein concentrations [17]. We searched the medical literature for studies of the effects of diet and its component including macronutrient, dietary food groups, dietary patterns and herbal on disturbances of lipoprotein concentrations. The purpose of this chapter is to update current knowledge on the role of the following dietary determinants in lipoprotein concentrations and dyslipidemia: including 1) macronutrients (total fat, saturated fatty acids, trans fatty acids, n-6 polyunsaturated fatty acids, n-3 polyunsaturated fatty acids, dietary cholesterol, carbohydrate and protein), 2) food groups (grains and cereal, fruit and vegetables, dairy products, nuts, beans and legumes, and meat, fish, poultry and eggs), 3) dietary patterns (Mediterranean diet, Dietary to Stop Hypertension, western **diet** and healthy diet), and therapeutic life style change (TLC), 4) dietary **supplements**, (plant stanols and sterols), herbal and phytochemicals.

2. Diet and lipoprotein

Lipoprotein concentrations are affected by both genetic and environmental factors. Among environmental factors such as physical activity and smoking, diet is an important component in preventing and improving dyslipidemia. Diet intervention is recommended by the National Cholesterol Education Program (NCEP) guidelines as first-line therapy for the management of disturbances in lipoprotein concentrations. Also the Third report of the NCEP recommended that if dietary therapy do not improve disturbances in lipoprotein concentrations, non-pharmacologic therapeutic factors such as viscous fiber and plant stanols and sterols should be recommended prior to advancing to drug therapy[18].

3. Macronutrient and lipoprotein

3.1. Total fat

The Nutrition Committee of the American Heart Association (AHA) emphasises on that diets providing up to 40% of dietary energy as primarily unsaturated fat (20% MUFA, 10% SFA, 10% PUFA and 1% TFA) were as heart healthy as low-fat diets (<30% of dietary energy) [19]. The effects of different dietary fatty acids on lipid profiles should be considered in the evaluation of strategies for controlling of disturbances in lipoprotein concentrations. Changes in dietary fat composition are clearly associated with changes in lipoprotein concentrations. Types of dietary fatty acids include saturated fatty acids (SFAs), monounsaturated fatty acid (MUFAs), polyunsaturated fatty acid (PUFAs) and dietary cholesterol, the effects of which on lipoprotein concentrations will be discussed.

3.2. Dietary saturated fatty acids (SFAs)

Among the dietary fatty acids only dietary SFAs and *trans* fatty acids increase LDL cholesterol concentrations [18]. The major sources of dietary SFAs are fast foods, processed foods, high-fat dairy products (whole milk, cheese, butter, ice cream, and cream), high-fat red meats, tropical oils such as palm oil, coconut oil, and palm kernel oil, baked products and mixed dishes containing dairy fats, shortening, and tropical oils. Dietary SFAs increase LDL and total cholesterol concentrations, in comparison with all dietary fatty acids except *trans* fatty acids [20-21], by inhibiting LDL receptor activity and enhancing apolipoprotein (apo) β-containing lipoprotein production [22]. Every 1 percent increase of total energy from dietary SFAs raises the serum LDL cholesterol about 2 percent. Conversely, a 1 percent reduction in saturated fatty acids will reduce serum cholesterol by about 2 percent [23,24]. The LDL cholesterol–raising effect of dietary SFAs depends on the intake of dietary cholesterol and PUFAs. In high intakes of dietary cholesterol, dietary SFAs decreased LDL receptor activity and increased plasma LDL concentrations [25]. However, in the adequate of dietary PUFAs (5–10% of total energy), dietary SFAs have no effect on LDL clearance [22]. In addition different dietary SFAs have different effects on lipoprotein concentrations [29]. Short chain SFAs have been shown to have a stronger LDL cholesterol raising effect, such that lauric acid (12:0) raised LDL cholesterol the most, followed by myristic (14:0) and palmitic (16:0) acids. In contrast, stearic acid (18:0), as a long chain SFA, has no effect on LDL and HDL cholesterol or the TC: HDL cholesterol ratio, and even lowers serum cholesterol [27,28]. Finally, the effects of dietary SFAs can be modulated by the foods in which they are contained. Cheeses may have smaller effects on LDL cholesterol concentrations than butter, and fermented dairy foods, such as yogurt, have been associated with LDL reductions [29]. Reduced intakes of dietary SFAs and cholesterol are first steps for the purpose of achieving the LDL cholesterol goal (<100 mg/dl). To maximize LDL cholesterol lowering by reducing dietary SFAs, it will be necessary to lower intakes of dietary SFAs approximately to <7 percent of total energy [18]. However the replacement of dietary SFAs with other macronutrients is important. Although replacement of dietary SFAs with carbohydrate decrease total, LDL, and HDL cholesterol, it also increases triglycerides [20]; however replacement of dietary SFAs by PUFAs decreases concentrations of total, LDL, and the LDL/HDL cholesterol ratio by decreasing LDL cholesterol production and increasing LDL clearance [30]. Although replacement of dietary SFAs with PUFAs has been shown to decrease HDL cholesterol, it decreases LDL cholesterol even more substantially; thus, the HDL:LDL ratio is increased [23] and the TC:HDL cholesterol ratio is decreased [26]. Replacement of 5% of total energy from SFAs with PUFAs reduces CHD risk by 42% [31]. Replacement of dietary SFAs with MUFAs has also been associated with improving lipoprotein concentrations, although this effect is slightly less than when PUFAs are the replacement dietary fatty acid [23]. Replacement of dietary SFAs with both MUFAs and carbohydrate decrease LDL cholesterol; however replacement with MUFA was associated with lower reductions in HDL cholesterol and lower arises in triglyceride concentrations [32].

3.3. Trans fatty acids

Trans fatty acids contain at least one double bond in the *trans* configuration [40] and were the most harmful macronutrient that increase disturbances in lipoprotein concentrations [26,33,34]. Dietary *trans* fatty acids, produced during the hydrogenation of either vegetable or fish oils (industrial TFA), are found in manufacturing products such as cookies, pastries, and salad dressings; *trans* fatty acids are also formed during anaerobic bacterial fermentation of unsaturated fatty acids that occurs in the rumen of polygastric animals such as cattle, sheep, and goats (natural trans fatty acids), and hence found in dairy products derived from the animals' milk and meat [33,35]. Industrial and natural *trans* fatty acids contain similar types of these fatty acids, but in different proportions. Industrial *trans* fatty acids contain trans isomers of oleic acid, the major ones being C18:1 trans-9 (elaidic acid) and C18: 1 trans-10 [35]. Consumption of industrial *trans* fatty acids increases total, LDL cholesterol, and total to HDL cholesterol ratio and the LDL to HDL cholesterol ratio [33, 35-37] and decrease HDL cholesterol [40]. Data on the effects of natural *trans* fatty acids on plasma lipoproteins in humans are inconsistent. An equivalent of 1% natural *trans* fatty acids of daily energy, has no significant effect on total cholesterol, LDL cholesterol, apo B, triglyceride concentrations but may be associated with a reduction in plasma HDL cholesterol concentrations [38]. However high intakes of natural *trans* fatty acids, but not low intakes, have adverse effects [39]. Therefore both natural and industrial *trans* fatty acids have detrimental effects on lipoprotein concentrations and their intakes should be limited [40]. The effects of *trans* fatty acids on lipid profiles are also variable, depending on their chain length; long chain *trans* fatty acids may have more adverse effect on lipid profiles. Partially hydrogenated fish oil or *trans* alpha-linolenic acid had more detrimental effect on lipoprotein compared with isocaloric amount of partially hydrogenated soy bean oil [37,41]. Effect of *trans* fatty acids on lipoprotein concentrations is a current topic of debate. *Trans* fatty acid intake increases lipoprotein a and triglycerides when substituted for dietary SFAs [42,43]. Issues related to the potential change in lipoprotein a levels induced by *trans* fatty acid intake and risk for disease need to be clarified.

Dietary guidelines for American 2010 emphasize that consumption of *trans* fatty acids should be reduced as much as possible by limiting foods that contain sources of these fatty acids [43]. On the basis of these data, it should be attempts to substitute unhydrogenated oil for hydrogenated or SFAs in diet.

3.4. Monounsaturated fatty acids

Monounsaturated fatty acids have received increased attention as being potentially beneficial for their association with low rates of CHD in olive-oil consuming populations of the Mediterranean style diet [18]. The most common form of dietary MUFAs is oleic acid (18:1 n-9), which occurs in the cis form. Olive oil, canola oil, and sunflower oil are the main sources of dietary MUFAs. Oleic acid is an effective hypocholesterolemic factor when substituted for dietary SFAs. MUFA-rich oil consumption has been one of the strategies recommended for modulating the plasma lipid profile in humans. Diets containing high MUFA-rich foods reduce plasma total and LDL cholesterol levels and enrich LDL particles with cholesteryl oleate, a change in LDL particle composition that has been shown to confer atherogenicity [23, 45-48]. Also compared with diets rich in saturated fat, MUFA-rich diets lower apolipoprotein β concentrations along with declines in LDL cholesterol level [49,50]. Consumption of MUFA-rich diets also induces lower triglycerides and higher HDL cholesterol concentrations compared with low-fat, high-carbohydrate diets [51]. Long term MUFA-rich diets result in an earlier postprandial peak in plasma triglyceride and apo β-48 concentrations [52,53]; this mechanism is not clear, however oleic acid has been shown to be preferentially esterified into triglycerides in the enterocyte [54], which may be result a faster entry rate of chylomicrons into the circulation, reflecting accelerated rates of digestion and absorption or upregulation of chylomicron synthesis and secretion [55]. However MUFA-rich diets increase clearance of plasma triglycerides compared with isocaloric SFA-rich or high complex carbohydrate diets and therefore decrease triglyceride concentrations [51,56,57]. MUFA substitution for dietary SFAs suggest an effective dietary strategy for improving disturbances of lipoprotein concentrations, which currently recommended in most national and international dietary guidelines [18].

3.5. N-6 Polyunsaturated fatty acids

Dietary n-6 PUFAs such as linoleic acid (18:2) are widely found in a variety of vegetables and vegetables oils [58]. Conjugated linoleic acid (CLA), a group of naturally occurring fatty acids that are mainly present in foods from ruminant sources, is a collective term used to describe positional and geometric derivatives of linoleic acid containing conjugated double bonds [59].

CLA have beneficial effects on lipoprotein disturbances. CLA reduced total, LDL and VLDL cholesterol, especially atherogenic apolipoprotein β-rich lipoproteins and triglycerides concentrations [60,61]. CLA increases the excretion of sterols and consequently decreases serum cholesterol concentration [86].

3.6. N-3 Polyunsaturated fatty acids

Dietary sources of n-3 PUFAs are limited. The shorter chain n-3 PUFAs FA, α-linolenic acid (ALA), is found in many plants, but the longer chains eicosapentaenoic acid (EPA) and docosahexaenoic acid (DHA) are produced almost exclusively by cold water algae, which are, in turn, ingested by fish. Humans cannot synthesize the n-3 double bond, but they do have the elongase and desaturase enzymes to convert ALA to EPA and DHA, a conversion, which however is an inefficient process. The conversion of ALA to EPA may be further reduced as a result of large amounts of n-6 FA in the diet, which compete for the same enzymes. Some studies however have found that ALA, irrespective of n-6 PUFAs, has a beneficial effect of lipid profiles [63]. Mechanism of actions of the medium- and long chain n-3 fatty acids appears to be independent. ALA exerts most of its effects by modulating lipoproteins, while EPA and DHA may reduce triglyceride synthesis [64]. Experts currently recommend the consumption of EPA and DHA, rather than ALA, to meet dietary goals for dietary n-3 PUFA [65]. Long-chain n-3 PUFA reduce triglyceride concentrations. An intake of 4 g EPA and DHA per day results in a 25–30% decrease of fasting triglyceride concentrations in both normolipidaemic and hypertriacylglycerolaemic subjects (66). Compares EPA and DHA, EPA-ethyl ester shows no change in triglyceride concentrations, suggesting that DHA is the active agent in fish oil, that decreases triglyceride concentrations. Therefore among long chain n-3 PUFAs, EPA may produce favourable effects on triglyceride and HDL cholesterol concentrations [67,68]. The hypotriglyceridaemic effect of long chain n-3 PUFAs, mediated by several mechanisms such as enhanced hepatic fatty acid oxidation [69], inhibition of fatty acid and triglyceride synthesis, reduced assembly and secretion of VLDL triglyceride concentrations [70], facilitates triglyceride rich lipoprotein removal through enhanced LPL activity in plasma [71]. Significant increases in HDL have been observed after DHA supplementation [67,68,72,73]; it may be related to decreased cholesteryl ester transfer protein activity that reduces the exchange from HDL cholesterol ester and VLDL, resulting in larger, more cholesterol-rich HDL cholesterol particles [74,75].

Inconsistent effects of DHA on total and LDL cholesterol levels have been shown; some investigators found a LDL cholesterol-raising effect [68,76] or no significant changes in total cholesterol or LDL cholesterol concentration [77,78]. After supplementation with n-3 Long chain PUFA, limited amounts of triglycerides are available for packaging into VLDL, which results in VLDL particles with low triglycerides that are readily converted to LDL, increases LDL cholesterol concentrations [79]. N-3 PUFAs could increase production of LDL via conversion of VLDL to LDL by increased lipolysis of VLDL and/or increased lipolytic activity or decreased clearance of LDL, by decreases in LDL receptor binding activity or reduced LDL receptor expression [80]. ALA, an n-3 polyunsaturated fatty acid found mainly in plant sources, including flaxseed oil, canola oil, and walnuts, is a metabolic precursor of DHA and EPA and any risk reduction may be mediated through conversion to these fatty acids; ALA cannot be synthesized by humans, and therefore, it is an essential fatty acid in diet [58]. Although evidence indicates that consumption of long chain n-3 PUFAs from seafood reduces the risk factors of cardiovascular disease, the effect of ALA intake in these risk factors is less well established. Daily supplementation with ALA-rich flaxseed is reported to reduce total cholesterol, LDL-cholesterol [81,82]. Weight of the evidence favors recommendations for modest dietary consumption of ALA (2 to 3 g per day) for primary and secondary prevention of CHD [58]. The relationship between ALA intake and CHD risk was seen among participants who consumed very little seafood; among men with limited seafood intake, each 1 g per day ALA intake was associated with 50% lower risk of CVDs; in contrast among subjects with some seafood intake, ALA intake was not associated with CHD risk. If benefits of ALA are greatest when EPA and DHA intakes are very low, the consumption of plant sources of n-3 fatty acids may be particularly important for CHD prevention among individuals who do not regularly consume fish [58].

3.7. Dietary cholesterol

The main source of dietary cholesterol is eggs, which contribute about one-third of the cholesterol in the diet; intake of dietary cholesterol has increased in recent year. Other sources of dietary cholesterol include animal products, dairy, meats, poultry, and shellfish [83]. High cholesterol intakes increase LDL cholesterol and the degree of rise varies from person to person. On average, the response of serum cholesterol to dietary cholesterol as revealed is approximately 10 mg/dL per 100 mg dietary cholesterol per 1000 kcal [84,85]. A recent meta-analysis showed that dietary cholesterol raises the ratio of total to HDL cholesterol, adversely affecting the serum cholesterol profile [86].

3.8. Carbohydrate

Recommendations to decrease fat and increase carbohydrate intake have come under scrutiny. Diets low in fat necessarily has a high proportion of carbohydrates, and high carbohydrate diet increase triglycerides, reduce HDL cholesterol concentrations, and increase LDL cholesterol concentrations [87]. In addition to carbohydrate intake, the type of carbohydrate, according to glycemic index, most likely influences lipid profiles [88]. Glycemic index refers to the value obtained by feeding a carbohydrate load and measuring the level of blood glucose. Using the glycemic index, carbohydrates with a low glycemic index may decrease triglyceride concentrations and increase HDL cholesterol [89]. Also substituting low-GI foods for high-GI foods lowers triglyceride concentrations by 15 to 25% [138]. High-carbohydrate diets increase triglyceride concentrations, compared to high-fat diets [91] via enhance hepatic lipogenesis [92] and decrease the synthesis of lipoprotein lipase [93]. A high carbohydrate diet also increases glucose and insulin concentrations, the latter increasing lipogenesis, leading to increases in triglyceride concentrations, triglyceride-enriched VLDL particles, and increases the LDL cholesterol concentrations [94]. Therefore reductions in dietary carbohydrate have been associated with reduced concentrations of LDL cholesterol [95] and increase means LDL particle size [96].

High carbohydrate diets (>60 percent of total energy) are associated with lipoprotein disturbances; reduction in the content of carbohydrate have beneficial effects on lipid profiles. However substitution of carbohydrate with other macronutrients is important. When carbohydrates are substituted for SFAs, the fall in LDL cholesterol levels equals that with monounsaturated fatty acids, and however, compared with MUFAs, this substitution frequently causes a fall in HDL cholesterol and a rise in triglycerides [23,97]. When dietary carbohydrate is consumed along with high-fiber diets, however, the rise in triglycerides or fall in HDL cholesterol has been reported to be reduced [98,99]. Addition of n-3 PUFA to low-fat, high-carbohydrate diets decreases the adverse effects of carbohydrate on blood lipids [51,100]. Also refined- and whole grains, as sources of carbohydrate, have an essential role in the metabolism of lipid profiles, that will be discussed in the section on food groups. In a relatively short period of time, dietary consumption of fructose has increased several fold above the amount present in natural foods, because of the use of high fructose corn sweeteners and sucrose in manufactured foods [101]. In human diets approximately one-third of dietary fructose comes from fruit, vegetables, and other natural sources and two-thirds is added to beverages and food in the diet (e.g. soft drinks, fruit-flavored drinks, candies, jams, syrups, and bakery products). Although there is little evidence that modest amounts of fructose have detrimental effects on carbohydrate and lipid metabolism, larger doses have been associated with numerous metabolic abnormalities, suggesting that high fructose consumption adversely affects health. High levels of plasma triacylglycerols are a well-established consequence of dietary fructose intake [101]. Numerous mechanisms have been suggested to explain this phenomenon [102,103], e.g. enhanced hepatic lipogenesis, and therefore overproduction of VLDL [102,104].

3.9. Protein

Plant sources of protein are predominantly legumes, dry beans, nuts, and, to a lesser extent, grain products and vegetables, which are low in saturated fats and cholesterol. Animal sources of protein include dairy products, egg whites, fish, poultry, and meats. Dietary protein in general has little effect on lipoprotein profiles. However, substituting plant protein including wheat gluten, soy proteins for animal protein decrease serum cholesterol [104,105]. Advice on the use of soy foods to displace animal products is consistent with the AHA advisory on soy [107], which states that 50 g/d soy protein consumption reduces approximate 3% LDL-C with no apparent dose-response effect [108]. Maximum reduction in LDL cholesterol was achieved when ~50 g of soy protein when was replaced meat or dairy protein [109]. Soy is a complex protein with a globulin fraction to which its cholesterol-lowering effect has been attributed; this fraction digested to peptides with inhibitory effects on cholesterol synthesis [110]. Isoflavones or the saponins found in soy, are also responsible for the cholesterol-lowering effect of soy [111,112]. Soy and other vegetable proteins also reduce oxidized LDL due to antioxidant activity [112,113].

4. Dietary food groups and lipoprotein

4.1. Grains and cereal

Based on evidence from both population and intervention studies, the recommended intake of whole grains of the 2005 Dietary Guidelines for Americans, is at least three ounces per day [114]. The Dietary Guidelines Advisory Committee (DGAC) 2010 Report emphasizes fiber-rich carbohydrate foods such as whole grains and vegetables, fruits, and cooked dry beans and peas, it specifically recommends that half of the grains consumed be whole grains, hence some whole grains should replace refined grains [115]. Similar recommendations are made by the American Heart Association [116] and the American Diabetes Association [117]. Whole grains are referred to as "complex" or "high-quality" carbohydrates, mainly due to their dietary fiber content [118], which has a beneficial effect on body weight, and lipid profiles because they are usually less energy-dense and more satiating than refined-grain foods [119] may be due to their high fiber content. Among whole grains, oat and barley have an advantage over wheat and brown rice in lowering serum lipids [120,121,122], contain viscous fibres, including β-glucacon [118] that lower serum cholesterol; 3.5 g of β-glucan from oats reduces LDL-C by 5% [123,124]. β-glucan interferes with reabsorption of bile acids and cholesterol by binding to bile acids, leading to increase bile acid excretion and lowering the bile acid levels in the liver and thereby increasing the conversion rate of cholesterol to bile acids. A viscous fiber intake of 10–25 g/d is recommended by the National Cholesterol Education Program's Adult Treatment Panel III as an additional diet option to decrease LDL cholesterol; an intake of 5–10 g/d lowers LDL-C by about 5% [126].

4.2. Fruit and vegetables

The 2010 Dietary Guidelines for Americans, recommend consuming sufficient amounts (5-13 servings, depending on energy needs) and a varieties of fruits and vegetables to reduce the risk of developing chronic diseases [115]; fruits, vegetables, or both should be emphasized at each meal, being major sources of vitamins C, E, and A, beta-carotene, other vitamins, fiber, flavonoids, and some minerals. Snacks and desserts that contain fruits and/or vegetables can be low in saturated fat, total fat, and cholesterol, and are very nutritious [18]. Fruits and vegetable intakes do not significantly change HDL cholesterol concentrations, but do decrease total and LDL cholesterol [9,127-132]. The protective effect of fruit and vegetables against CVDs is from their water-soluble and also viscous fibers (e.g. pectins) [133]. Viscous fiber increases fecal bile acid losses [134] and chenodeoxycholic acid synthesis [135].

4.3. Dairy products

Dairy products are important sources of protein, calcium, phosphorus, and vitamin D. The recommendation for intakes of dairy products is 2-3 serving per day; fat-free milk or 1 percent fat milk, fat-free or low-fat cheese (e.g., ≤3g per 1 oz serving), 1 percent fat cottage cheese or imitation cheeses made from vegetable oils, and fat-free or low-fat yogurt are good choices. Fat-free milk and other fat-free or low-fat dairy products provide as much or more calcium and protein than whole milk dairy products, with little or no saturated fat [18].

Recent studies confirm that milk products were associated with lower small dense LDL, and triglyceride concentrations, and higher HDL cholesterol [136]. In the CARDIA study, obese subjects with more frequent consumption of dairy products showed a trend towards lower risk of dyslipidaemia [137]. Minerals (calcium, magnesium), protein (casein and whey) and vitamins (riboflavin and vitamin B-12) have the hypocholesterolaemic effect of dairy product. The possible hypolipidaemic mechanism of calcium includes decreased intestinal absorption of cholesterol, bile acids, or fat [138], decreased fatty acid synthesis, increasing lipolysis, all of which lead to decreased triacylglycerol stores [139]. Milk proteins (whey) [140] or peptides [141] may also play a role. Whey may act independently or synergistically with the calcium; attenuate lipogenesis, and accelerate lipolysis [142]. Dairy products contain SFAs that could affect the blood lipid profile. A recent meta-analysis of 21 prospective cohort studies showed that the harmful effects of SFAs on CHD are still controversial [143]. An inverse association was shown between milk-specific fatty acids in serum cholesterol esters with serum cholesterol and apolipoprotein β levels [144]. Consumption of fat-free dairy products might decrease plasma cholesterol levels, while whole milk has neither a hypo- nor hypercholesterolaemic effect [139]. SFAs in dairy products can adversely influence CHD, although the effect of SFAs on CHD risk depends on the source of calories by which it is substituted to maintain energy balance [145]. Different dairy products have different effects on the lipid profiles. The LDL-C-raising effect of cheese was less than that of butter at comparable intakes of total fat and saturated fat [146,147]. Butter fat may increase total and LDL cholesterol by down-regulation of LDL removal from the circulation [148]. Fermented dairy products may have a favourite effect on lipid profiles. The protective effect of yogurt [139,149], a fermented dairy product, was shown to reduce absorption of cholesterol and therefore prevent dyslipidemia; it is thought to increase calcium bioavailability through its high acidity [149]. Fermented milk products may decrease cholesterol levels more than non-fermented products [149-151]. Probiotic yogurt decreased total cholesterol by 4% and LDL cholesterol by 5% [149]. A meta-analysis of fermented dairy products has shown a possible cholesterol lowering property, through the high content of probiotic bacteria [152].

4.4. Nuts

Although nuts are high in fat, in most nuts the predominant fats are unsaturated. Studies over the last decade have demonstrated favourable effects of nuts in modifying lipid risk factors for CHD [153]. However, their use is not yet part of standard advice for patients with hyperlipidemia, despite recognized health benefits for the general population. Intake of nuts fits well with current American Heart Association guidelines [19] to replace dietary SFAs with unsaturated fats and with the National Cholesterol Education Program (NCEP) guidelines to increase intake of dietary MUFAs [153]. Less atherogenic plasma lipid profiles associated with long-term consumption of nuts [154,155]. Addition of nuts to the habitual diet of both normocholesterolemic and hypercholesterolemic subjects results in a significant reduction in plasma total and LDL cholesterol, whereas HDL remains unchanged or increases [155-158]. One-percent reductions in LDL cholesterol would be achieved with daily intakes of 4-11 g of walnuts, pecans, peanuts, macadamias, and pistachios [50,155,157-161]. There are several components in nuts i.e. high MUFA, high PUFAs : SFAs ratio, proteins (specially high arginin), plant sterols, fiber, and associated phenolic substances, which may all contribute to the cardioprotective effect of nuts [154,162]. Also replacement of dietary SFAs with MUFAs due to the high MUFA content of nuts and high content of vitamin E in nuts reduce susceptibility of LDL to oxidation, a key event in the development of CVDs [233]. Consumption of almonds, either as the whole nut or the oil, lower total and LDL cholesterol concentrations. Addition of 100 g of almonds to the diets reduces total cholesterol by 9-16% and LDL cholesterol by 12-19 % in hypercholesterolemic subjects [164]; in one study almond consumption also reduced fasting triglyceride concentrations by 14%, compared with baseline [165]. Macadamia is another nut that improve lipid disturbances, and its inclusion as part of a healthy diet favourably altered the plasma lipid profile, despite the nuts being high in fat; their consumption reduced plasma total and LDL cholesterol concentrations and increase HDL cholesterol without any change in the triglyceride concentrations [166]. These changes could contribute to high MUFA intake and lower intake of PUFA and SFA consumption of macadamia nuts. Of nuts, walnuts are unique in improving dyslipidemia because they are a rich source of PUFAs, especially α-linolenic acid and linoleic acid; 100 g of walnuts contain 65.2 g fat; mainly from PUFAs (47.2 g) including α-linolenic acid (9.1 g) and linoleic acid (38.1 g) [167]. In a meta-analysis, consumption of walnuts resulted in decrease in total and LDL cholesterol concentrations, whereas HDL cholesterol and triglycerides were not affected [168]. Despite favourable effects of nuts on dyslipidemia, the intake of nuts should fit within the calorie and fat goal [18].

4.5. Beans and legumes

Legumes include a variety of beans such as navy, pinto, kidney, garbanzo, lima beans and peas such as split green peas or lentils. The Dietary Guidelines for Americans suggest consuming 3 cups of legumes per week [18, 169]. Legumes are a rich source of soluble dietary fiber and vegetable protein and have long been known to be hypercholesterolaemic foods [170,171]. One-half cup of cooked beans or peas can provide a range of dietary fiber from 4.6 g in fava beans up to 9.6 g fiber in navy beans, with a half cup of chick peas providing 6.2 g of total fiber, and 1.3 grams soluble dietary fiber [169]. In a meta-analysis both total and LDL cholesterol decreased, while HDL cholesterol did not change significantly, when diets uses supplemented with non-soy legumes [169]. The hypocholesterolaemic property of legumes is associated with the water-soluble fibre. Dietary fiber in legumes is not digested in the small intestine but be fermented in the colon and produces short chain fatty acids such as acetate, propionate and butyrate [172,173]; that inhibits hydroxy-3-methylglutaryl-CoA reductase, the limiting enzyme for cholesterol synthesis. Dietary fiber also decrease LDL cholesterol concentration by partially interrupting the enterohepatic circulation of bile acids via binding to bile acids in the intestines and preventing their re-absorption [174]. Consequently, an increase in the production of bile acids decreases the liver pool of cholesterol and increases uptake of serum cholesterol by the liver, decreasing thereby circulating cholesterol in the blood [175]. Another hypercholesterolemic component of legume is phytochemicals, which has been shown to reduce blood cholesterol levels and is present in small to moderate amounts in many types of legumes, such as chickpeas [176]. Dietary modification strategies that target the reduction of risk factors for CVDs should include an increase in legume consumption in addition to other strategies which have been of proven benefit [169].

4.6. Meat, fish, poultry and eggs

Recommendation for intakes of meat, fish and poultry are up to 5 oz per day from lean meats (beef, pork, and lamb), poultry, and fish [18]. To achieve NCEP dietary goals, individuals are often counselled to reduce the amount and frequency of red meat consumption because of its hypercholestromia effects [177-179]. Cholesterol raising effects of red meats appears to result from high contents of SFAs [177,179]. Therefore, lean red meats that provide small amounts of these fatty acids do not adversely influence the blood lipid profile, compared with lean white meats. In isoenergetic low-fat diets, lean meat, fish and, poultry had similar effects on blood lipid response in both hypercholesterolemic and normocholesterolemic subjects [178,180,181]. Data available suggest that meat protein, per se, is not hypercholesterolemic [177,181,182]. The blood cholesterol-raising potential of meat products appears to be a function of their SFA fat and cholesterol contents. Therefore, substituting lean for higher fat red meat should favourably influence serum total cholesterol and LDL-C levels. Incorporating lean beef, fish, or poultry into the AHA diet can be beneficial in lower disturbances of lipid profile in patients with hypercholesterolemia [178,183]. Therefore the hypercholesterolemic subjects known to be at high risk for CVDs, could be advised to include lean fish as well as lean beef or poultry without skin in an AHA diet to reduce their lipoprotein disturbances [184,185]; normolipidemic subjects can also incorporate lean fish in an AHA diet [184], although it is not necessary to eliminate or drastically reduce intake of lean red meat consumption because it is a rich source of iron, zinc and vitamin B12. One of the dietary recommendations in the prevention of CVDs is to limit egg consumption, because they have been shown to be a major source of dietary cholesterol (One egg contains 200 mg/cholesterol) that increases both serum total and LDL-cholesterol concentrations [21,86,186]. Several epidemiologic studies however found no relation between egg consumption and risk of coronary heart disease [187,188], may be because dietary cholesterol increases not only concentrations of total and LDL cholesterol but also concentrations of HDL cholesterol [21,186,189,190]. Egg intake has been also shown to promote the formation of large LDL particles, which is less atherogenic [191]. Therefore dietary recommendations aimed at restricting egg consumption should not be generalized to include all individuals [191].

4.7. Dietary pattern

Using single nutrients or dietary food groups have some limitations in assessing their effect on lipid profiles separately because nutrients and foods are consumed in combination. To date, dietary patterns consider how foods are consumed in combination, and are used to evaluate the effects of overall nutritional habits on health status. There are two dietary patterns that demonstrate the beneficial effect on disturbances of lipoprotein concentrations; there include the dietary to stop hypertension (DASH) and the Mediterranean diet. The DASH dietary pattern, rich in fruits, vegetables, and low-fat dairy foods, emphasizes fish, poultry, and whole grains, and is reduced in total fat, SFAs and cholesterol, red meat, sweets, and sweetened beverages [192,193]; it lowers total, LDL and HDL cholesterols, without any adverse effects on triglyceride concentrations [194]; all of these coupled with decrease in blood pressure, reduce 10-year coronary heart disease risk of approximately 12% [194]. The Mediterranean dietary pattern consists of: (a) daily consumption: of non refined cereals and products (whole grain bread, pasta, brown rice, etc), vegetables (2 – 3 servings/ day), fruits (6 servings/day), olive oil (as the main added lipid) and dairy products (1 – 2 servings/day), (b) weekly consumption: of fish (4–5 servings/week), poultry (3 – 4 servings/week), olives, pulses, and nuts (3 servings/ week), potatoes, eggs and sweets (3 – 4 servings/week) and monthly consumption: of red meat and meat products (4 – 5 servings/month). It is also characterized by moderate consumption of wine (1 – 2 wineglasses/day). Mediterranean diet is a diet poor in SFAs and PUFAs but rich in MUFA (oleic acid) provided by the olive oil. The ratio of MUFAs : SFAs fat ratio is high > 2 [195]. This diet pattern is associated with reduction in total and LDL-cholesterol, and also a significant effect on triglycerides and VLDL concentrations, and a small positive or no effect on HDL-cholesterol [196-199] and improves dyslipidemia in dislipidemic patients [200]. This diet also includes antioxidant vitamins and phenolic compounds, and therefore reduces levels of circulating oxidized LDL and increases total antioxidant capacity [201]. Beside these two dietary patterns, other dietary pattern such as the western, and healthy dietary patterns affect lipoprotein profiles. The western pattern is characterized by high consumption of food such as refined grains, french fries, and red meats that have detrimental effects on lipid profiles. The healthy pattern included non-hydrogenated fat, vegetables, eggs, and fish and was negatively associated with lipoprotein disturbances [202-205]. In addition of dietary patterns, therapeutic lifestyle change is another dietary approach that ATP III recommends to reduce risks for CHD. This dietary approach includes the following: 1) Reduced intakes of dietary SFAs (<7% of total calories) and cholesterol (<200 mg/d), 2) weight reduction, 3) increased physical activity, and 4) therapeutic options for enhancing LDL lowering such as plant stanols/ sterols (2 g/d) and increased viscous (soluble) fiber (10-25 g/d) [18].

5. Dietary supplement

5.1. Plant stanols and sterols

Dyslipidemia may be treated with dietary interventions, including the daily consumption of foods with added plant stanols or plant sterols. Plant sterols are isolated from soybean and tall pine-tree oils. Also some foods such as macadamia nuts are a rich source (1.28 mg/g lipid) of plant sterols. Plant sterols can be esterified to unsaturated fatty acids, creating sterol esters, to increase lipid solubility. Hydrogenating of sterols produces plant stanols. Plant stanols and sterols are available in commercial margarines. Daily consumption of 2 g plant stanols or plant sterols, expressed as free plant stanol or plant sterol equivalents improves dyslipidemia [18]. FDA confirms a daily dose of plant sterols and stanols of 2 g per day as safe, a dose which reduces LDL cholesterol by 10% [206], with little or no change in HDL cholesterol or triglyceride levels. There were no apparent added benefits at higher doses of palnts stanols and sterols. Plant stanols and sterols compete with absorption of dietary cholesterol and bile acid [8]. The consumption of plant stanols and sterols is an effective LDL cholesterol lowering strategy for patients who are undergoing statin therapy. The lipid-lowering response to combined plant stanols and sterols/statin therapy target both intestinal and hepatic cholesterol metabolism. Consumption of plant stanols and sterols reduces intestinal cholesterol absorption and reduces hepatic cholesterol synthesis. Consumption of statins simultaneously with plant stanols and sterols inhibit hepatic cholesterol synthesis and therefore reduce in LDL cholesterol concentrations [8]. Plant sterols/stanols reduce absorption of dietary carotenoids, and decrease levels of plasma betacarotene; therefore increased intakes of fruits and vegetables are recommended with consumption of plant stanols/sterols[18].

5.2. Herbal

There is a need to identify additional non-pharmacologic therapeutic options for cholesterol lowering. There is also a need to find products that are more practical for the consumer than viscous fiber and plant stanols and sterols to permit widespread adoption.

5.2.1. Flavonoid

Flavonoids have 2 aromatic rings that are bound by an oxygenated heterocyclic ring. On the basis of their chemical structure, they are divided into several subclasses: flavones, flavonols, flavanones, flavan-3-ols, anthocyanins and isoflavones. Flavones and flavonols are found in leaf vegetables and onion. Flavanones are mainly found in grapefruits and citrus fruits. Tea and cocoa are the richest sources of flavan-3-ols. Soy and soy products such as tofu, and miso are the main sources of isoflavones [207,208]. Although increased resistance of LDL to oxidation was observed after treatment with various synthetic pharmaceutical agents, an effort is made to identify natural food products which can offer antioxidant defense against LDL oxidation. Polyphenolic flavonoids are powerful antioxidants and their antioxidative capacity is related to their chemical structure [209].

Incubation of LDL with flavonoids protects the lipoprotein against oxidation [210]. Certain flavonoids such as querectin could have a potentially protective role in suppression of LDL oxidation, regardless of the effect of antioxidant vitamins [211] via scavenging radicals and reduce total and LDL cholesterol concentrations, by reducing the hepatic lipogenesis [212].

5.2.2. Tea

The effect of tea on lipid profiles is uncertain. Although some studies have found no lipid-lowering effects from green or black tea consumption, most showed hypolipidemic effects for tea [213-218]. The association between tea drinking and lipid profile concentrations was linear for up to 10 cups per day, beyond which the association disappeared [219]. Daily consumption of 10 cups of green tea was associated with a reduction of approximately 2% in serum total cholesterol [219]. Tea also is a major source of flavonoids, the predominant ones in green tea being catechins. Theaflavins are polyphenol pigments present in black tea, formed from the polymerization of catechins during fermentation of green tea [220]. Catechins reduce intestinal cholesterol absorption [221], reduce hepatic cholesterol content [222] and increase fecal excretion of total fatty acids, neutral sterols, and acidic sterols [223] and up-regulate the LDL receptor in liver cells [224]. Polyphenol in black tea also increases fecal excretion of total lipids and cholesterol [225].

5.2.3. Chocolate

The beneficial effects of chocolate on healthy humans have been widely addressed in recent years. Supplementation of cocoa products affects lipid profiles in subjects with cardiovascular-related diseases such as hypercholesterolemia, glucose intolerance, and hypertension as well as healthy individuals [226-228]. Consumption of cocoa and dark chocolate increase the concentration of HDL cholesterol [229] and plasma antioxidant capacity, decrease the formation of lipid oxidation products, and inhibit the oxidation of LDL [230]. In a meta-analysis study, cocoa was associated with small decreases in total and LDL cholesterol, but not HDL cholesterol concentrations [231]. Cocoa products contain more polyphenols than teas. A particular group of flavonoids, namely, the flavan-3-ols was found in chocolate (flavanols) [232]. Moderate consumption of cocoa or dark chocolate, have potential health benefits [231], however, a high dose of polyphenols has been shown to exert cytotoxic effects on liver cells [233] and higher polyphenol supplementation may counteract its beneficial biological effects on lipid metabolism [234].

5.2.4. Fenugreek

Fenugreek (Trigonella foenum-graecum), an annual medicinal plant of the Fabaceae family is well documented for its pharmacological properties. Fenugreek seeds have been historically used for the treatment of various chronic diseases such as diabetes, dyslipidemia, and obesity [235,236]. The seeds of Fenugreek contain many nutrients including protein, carbohydrates, fat, vitamin, and minerals, fiber, saponins, choline and trigonelline, polyphenolic flavonoids, steroid saponins,   polysaccharides mainly galactomannans and 4-hydroxyisoleucine [237,239], the fiber and saponin components of the seeds have been shown to have hypocholesterolemic effect [240], and the beneficial effect of raw fenugreek seeds on elevated serum cholesterol levels has been well established [241]. Raw fenugreek seeds reduce serum total cholesterol, LDL cholesterol, VLDL cholesterol and triglyceride concentrations, without altering the HDL fraction [242]; intakes of 20–25 g in three divided doses yielded maximum benefit in the control of cholesterol concentrations [243]. Its use as a dietary adjunct however is limited because of its bitterness. Soaking and washing of fenugreek seeds in water overnight removes the bitterness to a certain extent and makes then edible [243,244].

5.2.5. Ginseng

The beneficial metabolic effects of ginseng on lipid profiles as a hypolipidemic agent were reported over 20 years ago [245-247]. Ginseng leads to reduction of cholesterol and triglyceride concentrations in liver and serum. Administration of red ginseng powder and extract reduces plasma total cholesterol, triglycerides, FFA, and increased HDL-C [248,249]. Ginseng saponins may decrease blood cholesterol concentrations by increasing cholesterol excretion through bile acid formation [249,250]. Ginsenoside, one of active components of ginseng saponins, may accelerate serum cholesterol turnover by increased cholesterol degradation and excretion in the feaces notwithstanding increased hepatic cholesterogenesis [250,251]. Ginseng saponins as ginsenosides increase LDL receptors by promoting the synthesis of LDL receptors[252].

5.2.6. Ginger

Ginger has been listed in the "Generally Recognized as Safe" by FDA [338]; fresh ginger rhizome contains polyphenolic compounds such as gingerols; zingerone, which is the major active component and gingerol, is one of the most abundant constituents in the gingerol series and also responsible for its characteristic pungent taste [253,254]. Ginger oleo-resin and dried ginger rhizome reduce hypercholesterolaemia. The speculated mechanism for these compounds is by disrupting cholesterol absorption from the gastro-intestinal tract [255], which may be due to the presence of niacin in ginger, and it causes increased clearance of VLDL, lowers triglyceride levels, increases hepatic uptake of LDL and inhibition of cholesterogenesis [256]. Ginger powder significantly reduces the extent of lipid peroxidation and improves plasma antioxidant capacity, which decreases plasma-free radicals [257]. Moreover, polyphenolic flavonoids present in ginger may prevent coronary artery disease by reducing plasma cholesterol levels or by inhibiting LDL oxidation [258]. Reduction in serum triglycerides is dose dependent; doses of 200 and 400 mg/kg of ginger are more effective as antihypercholesterolaemics than atorvastatin when given for 4 weeks and are equivalent to it when given for shorter period under the same conditions of diet and life style for the treatment of the same pathologic condition. The triglyceride lowering effect of ginger may be due to ginger's ability to enhance lipase activity [255].

5.2.7. Licorice

Licorice root, derived from the plant Glycyrrhiza glabra is used widely in Asia as a sweetener or a spice, contains flavonoids from the flavan and chalcone subclasses, and has a antioxidative properties [259]. Licorice-derived glabridin binds to the LDL particle and protects it from oxidation by its capacity to scavenge free radicals and its property to reduce the LDL aggregation [260,261].

6. Conclusion

Diet therapy is the initial recommended intervention for prevention of and managing disturbances of lipoprotein concentrations, prior to advancing to drug therapy. Further research on the association between dietary components and lipoprotein disturbances is recommended.

Author details

Parvin Mirmiran[*]
Department of Clinical Nutrition and Dietetics, Faculty of Nutrition Sciences and Food Technology, National Nutrition and Food Technology Research Institute, Shahid Beheshti University of Medical Sciences, Tehran, Iran

Somayeh Hosseinpour-Niazi
Nutrition Related Non-Communicable Disease, Research Institute for Endocrine Sciences, Shahid Beheshti University of Medical Sciences, Tehran, Iran

Fereidoun Azizi
Endocrine Research Center, Research Institute for Endocrine Sciences, Shahid Beheshti University of Medical Sciences, Tehran, Iran

Acknowledgement

The authors would like to acknowledge Ms N. Shiva for language editing of the manuscript and to express their appreciation to Emad Yuzbashian for his valuable help. This study was funded by a grant from the Research Institute of Endocrine Sciences, Shadid Beheshti University of Medical Sciences, Tehran, Iran. None of the authors had any personal or financial conflicts of interest.

7. References

[1] Robins SJ, Lyass A, Zachariah JP, Massaro JM, Vasan RS (2011) Insulin resistance and the relationship of a dyslipidemia to coronary heart disease: the Framingham Heart Study. Arterioscler Thromb Vasc Biol. 31: 1208-1214.

* Corresponding Author

[2] Musunuru K (2010) Atherogenic dyslipidemia: cardiovascular risk and dietary intervention. Lipids. 45:907-914.

[3] Goff DC Jr, Bertoni AG, Kramer H, Bonds D, Blumenthal RS, Tsai MY, eat al (2006) Dyslipidemia prevalence, treatment, and control in the Multi-Ethnic Study of Atherosclerosis (MESA): gender, ethnicity, and coronary artery calcium. Circulation. 113:647-656.

[4] Hosseini-Esfahani F, Mousavi Nasl Khameneh A, Mirmiran P, Ghanbarian A, Azizi F (2011) Trends in risk factors for cardiovascular disease among Iranian adolescents: the tehran lipid and glucose study, 1999-2008. J Epidemiol. 21:319-328.

[5] Esteghamati A, Meysamie A, Khalilzadeh O, Rashidi A, Haghazali M, Asgari F, et al (2009) Third national Surveillance of Risk Factors of Non-Communicable Diseases (SuRFNCD-2007) in Iran: methods and results on prevalence of diabetes, hypertension, obesity, central obesity, and dyslipidemia. BMC Public Health.9:167.

[6] Goodman SG, Langer A, Bastien NR, McPherson R, Francis GA, Genest JJ Jr, et all (2010) DYSIS Canadian Investigators. Prevalence of dyslipidemia in statin-treated patients in Canada: results of the DYSlipidemia International Study (DYSIS). Can J Cardiol. 26:e330-335.

[7] Aguilar-Salinas CA, Gómez-Pérez FJ, Rull J, Villalpando S, Barquera S, Rojas R (2010) Prevalence of dyslipidemias in the Mexican National Health and Nutrition Survey 2006. Salud Publica Mex. 52:44-53.

[8] Rideout TC, Harding SV, Marinangeli CP, Jones PJ (2010) Combination drug-diet therapies for dyslipidemia. Transl Res. 155:220-227.

[9] Mirmiran P, Noori N, Zavareh MB, Azizi F (2009) Fruit and vegetable consumption and risk factors for cardiovascular disease. Metabolism. 58:460-8.

[10] Mirmiran P, Ramezankhani A, Azizi F (2009) Combined effects of saturated fat and cholesterol intakes on serum lipids: Tehran Lipid and Glucose Study. Nutrition. 25:526-531

[11] Mirmiran P, Mirbolooki M, Heydarian P, Salehi P, Azizi F (2008) Intrafamilial associations of lipid profiles and the role of nutrition: the Tehran lipid and glucose study. Ann Nutr Metab. 52: 68-73.

[12] Azadbakht L, Mirmiran P, Hedayati M, Esmaillzadeh A, Shiva N, Azizi F(2007) Particle size of LDL is affected by the National Cholesterol Education Program (NCEP) step II diet in dyslipidaemic adolescents. Br J Nutr. 98: 134-139.

[13] Hu FB (2002) Dietary pattern analysis: a new direction in nutritional epidemiology. Curr Opin Lipidol. 13:3–9.

[14] Azadbakht L, Mirmiran P, Esmaillzadeh A, Azizi T, Azizi F (2005) Beneficial effects of a Dietary Approaches to Stop Hypertension eating plan on features of the metabolic syndrome. Diabetes Care. 28: 2823-31.

[15] Azadbakht L, Mirmiran P, Esmaillzadeh A, Azizi F (2006) Dietary diversity score and cardiovascular risk factors in Tehranian adults. Public Health Nutr. 9: 728-736.

[16] Hosseini-Esfahani F, Jessri M, Mirmiran P, Bastan S, Azizi F (2010) Adherence to dietary recommendations and risk of metabolic syndrome: Tehran Lipid and Glucose Study. Metabolism. 59: 1833-1842.

[17] Bahadoran Z, Mirmiran P, Hosseinpanah F, Hedayati M, Hosseinpour-Niazi S, Azizi F (2011) Broccoli sprouts reduce oxidative stress in type 2 diabetes: a randomized double-blind clinical trial. ur J Clin Nutr. 65: 972-977.

[18] National Cholesterol Education Program (NCEP) Expert Panel on Detection, Evaluation, and Treatment of High Blood Cholesterol in Adults (Adult Treatment Panel III) (2002) Third Report of the National Cholesterol Education Program (NCEP) Expert Panel on Detection, Evaluation, and Treatment of High Blood Cholesterol in Adults (Adult Treatment Panel III) final report. Circulation.106: 3143-3421.

[19] Krauss RM, Eckel RH, Howard B, et al. (2000) AHA Dietary Guidelines: revision 2000: A statement for healthcare professionals from the Nutrition Committee of the American Heart Association. Circulation 102, 2284–2299.

[20] Siri-Tarino PW, Sun Q, Hu FB, Krauss RM (2010) Saturated fat, carbohydrate, and cardiovascular disease. Am J Clin Nutr. 91:502–9.

[21] Clarke R, Frost C, Collins R, Appleby P, Peto R (1997) Dietary lipids and blood cholesterol: quantitative meta-analysis of metabolic ward studies. BMJ. 314:112–117.

[22] Dietschy JM (1998) Dietary fatty acids and the regulation of plasma low density lipoprotein cholesterol concentrations. J Nutr. 128: 444–448.

[23] Mensink RP, Katan MB (1992) Effect of dietary fatty acids on serum lipids and lipoproteins. A meta-analysis of 27 trials. Arterioscler Thromb. 12:911–919.

[24] Kris-Etherton PM, Yu S (1997) Individual fatty acid effects on plasma lipids and lipoproteins: human studies. Am J Clin Nutr;65: 1628-1644.

[25] Hayes KC, Khosla P, Hajri T, Pronczuk A (1997) Saturated fatty acids and LDL receptor modulation in humans and monkeys. Prostaglandins Leukot Essent Fatty Acids. 57: 411–418.

[26] Mensink RP, Zock PL, Kester AD, Katan MB (2003) Effects of dietary fatty acids and carbohydrates on the ratio of serum total to HDL cholesterol and on serum lipids and apolipoproteins: a metaanalysis of 60 controlled trials. Am J Clin Nutr. 77:1146–1155.

[27] Bonanome A & Grundy SM (1988) Effect of dietary stearic acid on plasma cholesterol and lipoprotein levels. N Engl J Med. 318, 1244–1248.

[28] Grande F, Anderson JT & Keys A (1970) Comparison of effects of palmitic and stearic acids in the diet on serum cholesterol in man. Am J Clin Nutr. 23, 1184–1193.

[29] German JB, Gibson RA, Krauss RM, et al (2009) A reappraisal of the impact of dairy foods and milk fat on cardiovascular disease risk. Eur J Nutr, 48:191–203.

[30] National Cholesterol Education Program (1994) Second report of the expert panel on detection, evaluation, and treatment of high blood cholesterol in adults (Adult Treatment Panel II). Circulation. 89:1333-445.

[31] Siri-Tarino PW, Sun Q, Hu FB, Krauss RM (2010) Saturated fatty acids and risk of coronary heart disease: modulation by replacement nutrients. Curr Atheroscler Rep. 12: 384-390.

[32] Berglund L, Lefevre M, Ginsberg HN, et al (2007) Comparison of monounsaturated fat with carbohydrates as a replacement for saturated fat in subjects with a high metabolic risk profile: studies in the fasting and postprandial states. Am J Clin Nutr. 86:1611–1620

[33] Lichtenstein AH (1997) Trans fatty acids, plasma lipid levels, and risk of developing cardiovascular disease. A statement for healthcare professionals from the American Heart Association. Circulation. 95: 2588-2590.

[34] Sun Q, Ma J, Campos H, et al (2007) A prospective study of trans fatty acids in erythrocytes and risk of coronary heart disease. Circulation. 115: 1858–1865

[35] Brouwer IA, Wanders AJ, Katan MB (2010) Effect of animal and industrial trans fatty acids on HDL and LDL cholesterol levels in humans--a quantitative review. PLoS One. 5: 9434.

[36] Almendingen K, Jordal O, Kierulf P, Sandstad B, Pedersen JI (1995) Effects of partially hydrogenated fish oil, partially hydrogenated soybean oil, and butter on serum lipoproteins and Lp[a] in men. J Lipid Res. 36: 1370–1384.

[37] Vermunt SH, Beaufrere B, Riemersma RA, Sebedio JL, Chardigny JM, et al (2001) Dietary trans alpha-linolenic acid from deodorised rapeseed oil and plasma lipids and lipoproteins in healthy men: the TransLinE Study. Br J Nutr. 85: 387–392.

[38] Lacroix E, Charest A, Cyr A, Baril-Gravel L, Lebeuf Y, Paquin P, et al (2012) Randomized controlled study of the effect of a butter naturally enriched in trans fatty acids on blood lipids in healthy women. Am J Clin Nutr. 95: 318-25.

[39] Motard-Belanger A, Charest A, Grenier G, Paquin P, Chouinard Y, et al (2008) Study of the effect of trans fatty acids from ruminants on blood lipids and other risk factors for cardiovascular disease. Am J Clin Nutr. 87: 593–599.

[40] Mozaffarian D, Katan MB, Ascherio A, Stampfer MJ, Willett WC (2006) Trans fatty acids and cardiovascular disease. N Engl J Med. 354: 1601–1613.

[41] Almendingen K, Jordal O, Kierulf P, Sandstad B, Pedersen JI (1995) Effects of partially hydrogenated fish oil, partially hydrogenated soybean oil, and butter on serum lipoproteins and Lp[a] in men. J Lipid Res. 36: 1370–1384.

[42] Zock PL, Mensink RP (1996) Dietary trans-fatty acids and serum lipoproteins in humans. Curr Opin Lipidol. 7: 34-37

[43] United States Department of Agriculture, United States Department of Health and Human Services: Report of the Dietary Guidelines Advisory Committee on the Dietary Guidelines for Americans. Washington, DC: Government Printing Office; 2010

[44] Heyden, S (1994) Polyunsaturated and monounsaturated fatty acids in the diet to prevent coronary heart disease via cholesterol reduction. Ann. Nutr. Metab. 38: 117–122.

[45] Gustafsson I-B, Vessby B, Ohrvall M, Nydahl M (1994) A diet rich in monounsaturated rapeseed oil reduces the lipoprotein cholesterol concentration and increases the relative content of n23 fatty acids in serum in hyperlipemic subjects. Am J Clin Nutr. 59:667–74

[46] Mensink RP (1994) Dietary monounsaturated fatty acids and serum lipoprotein levels in healthy subjects. Atherosclerosis. 110. 65–68.

[47] Roche, H. M., Zampelas, A. & Knapper, J.M.E (1998) Effect of longterm olive oil dietary intervention on postprandial triacylglycerol and factor VII metabolism. Am. J. Clin. Nutr. 68: 552–560.

[48] Ginsberg, H. N., Barr, S. L., Gilbert, A., Karmally,W., Deckelbaum, R., Kaplan, K., Ramakrishnan, R., Holleran, S. & Dell, R. B. (1990) Reduction of plasma cholesterol

levels in normal men on an American Heart Association Step 1 diet or a Step 1 diet with added monounsaturated fat. N. Engl. J. Med. 322: 574–579.

[49] Allman-Farinelli MA, Gomes K, Favaloro EJ, Petocz P (2005) A diet rich in high-oleic-acid sunflower oil favorably alters low-density lipoprotein cholesterol, triglycerides, and factor VII coagulant activity. J Am Diet Assoc. 105: 1071-1079.

[50] Rajaram S, Burke K, Connell B, Myint T, Sabate J (2001) A monounsaturated fatty-acid pecan-enriched diet favourably alters the serum lipid profile of healthy men and women. J Nutr. 131:2275-2279.

[51] Jiménez-Gómez Y, Marín C, Peérez-Martínez P, Hartwich J, Malczewska-Malec M, Golabek I, et al (2010) A low-fat, high-complex carbohydrate diet supplemented with long-chain (n-3) fatty acids alters the postprandial lipoprotein profile in patients with metabolic syndrome. J Nutr. 140: 1595-1601.

[52] Roche HM, Zampelas A, Knapper JM,Webb D, Brooks C, Jackson KG, et al (1998) Effect of long-term olive oil dietary intervention on postprandial triacylglycerol and factor VII metabolism. Am J Clin Nutr. 68: 552–560.

[53] Roche, H. M., Zampelas, A., Jackson, K. G., Williams, C. M., Gibney, M. J. (1998) The effect of test meal monounsaturated fatty acid:saturated fatty acid ratio on postprandial lipid metabolism. Br. J. Nutr. 79: 419–424.

[54] Dashti N, Smith EA, Alaupovic P (1990) Increased production of apolipoprotein B and its lipoproteins by oleic acid in Caco-2 cells. J Lipid Res. 31:113–23.

[55] Silva KD, Kelly CN, Jones AE, Smith RD, Wootton SA, Miller GJ, et al (2003). Chylomicron particle size and number, factor VII activation and dietary monounsaturated fatty acids. Atherosclerosis. 166: 73–84.

[56] Rajaram S, Burke K, Connell B, et al (2001) A monounsaturated fatty acid-rich pecan-enriched diet favorably alters the serum lipid profile of healthy men and women. J Nutr. 131:2275–2279

[57] Curb DJ, Wergowske G, Dobbs JC, Abbot RD, Huang B (2000) Serum lipid effects of a high-monounsaturated fat diet based on macadamia nuts. Arch Int Med. 160: 1154-1158.

[58] Mozaffarian D (2005)Does alpha-linolenic acid intake reduce the risk of coronary heart disease? A review of the evidence. Altern Ther Health Med. 11: 24-30.

[59] Wandders AJ, Brouwer IA, Siebelink E, Katan MB (2010) Effect of a high intake of conjugated linoleic acid on lipoprotein levels in healthy human subjects. PLoS One. 5: 9000.

[60] Kostogrys RB, Pisulewski PM (2010) Effect of conjugated linoleic acid (CLA) on lipid profile and liver histology in laboratory rats fed high-fructose diet. Environ Toxicol Pharmacol. 30: 245-250.

[61] Nicolosi RJ, Rogers EJ, Kritchevsky D, Scimeca JA, Huth PJ (1997) Dietary conjugated linoleic acid reduces plasma lipoproteins and early aortic atherosclerosis in hypercholesterolemic hamsters. Artery. 22: 26.

[62] Yang L, Yeung SY, Huang Y, Wang HQ, Chen ZY. referential incorporation of trans, trans-conjugated linoleic acid isomers into the liver of suckling rats. Br J Nutr. 2002 Mar;87(3):253-60.

[63] Mirmiran P, Hosseinpour-Niazi S, Naderi Z, Bahadoran Z, Sadeghi M, Azizi F (2012) Association between interaction and ratio of ω-3 and ω-6 polyunsaturated fatty acid and the metabolic syndrome in adults. Nutrition. 27.

[64] Poudyal H, Panchal SK, Diwan V, Brown L (2011) Omega-3 fatty acids and metabolic syndrome: effects and emerging mechanisms of action. Prog Lipid Res. 50: 372-387.

[65] Kris-Etherton PM, Grieger JA, Etherton TD (2009) Dietary reference intakes for DHA and EPA. Prostaglandins Leukot Essent Fatty Acids. 81:99–104.

[66] Harris WS, Miller M, Tighe AP, Davidson MH, Schaefer EJ (2008) Omega-3 fatty acids and coronary heart disease risk: clinical and mechanistic perspectives. Atherosclerosis. 197:12–24.

[67] Egert S, Kannenberg F, Somoza V, Erbersdobler H,Wahrburg U (2009) Dietary alpha-linolenic acid, EPA, and DHA have differential effects on LDL fatty acid composition but similar effects on serum lipid profiles in normolipidemic humans. J Nutr. 139:861–8

[68] Geppert J, Kraft V, Demmelmair H, Koletzko B (2006) Microalgal docosahexaenoic acid decreases plasma triacylglycerol in normolipidaemic vegetarians: a randomized trial. Br J Nutr. 95: 779–86.

[69] Clarke SD, Jump D (1997) Polyunsaturated fatty acids regulate lipogenic and peroxisomal gene expression by independent mechanisms. Prostaglandins Leukot Essent Fatty Acids. 57:65–9.

[70] Nestel PJ (2000) Fish oil and cardiovascular disease: lipids and arterial function. Am J Clin Nutr. 71, 228–231.

[71] Harris WS, Lu G, Rambjor GS,Walen AI, Ontko JA, Cheng Q, Windsor SL (1997) Influence of n-3 fatty acid supplementation on the endogenous activities of plasma lipases. Am J Clin Nutr. 66:254–260.

[72] Maki KC, McKenney JM, Reeves MS, Lubin BC, Dicklin MR (2008) Effects of adding prescription omega-3 acid ethyl esters to simvastatin (20 mg/day) on lipids and lipoprotein particles in men and women with mixed dyslipidemia. Am J Cardiol.102:429–33.

[73] Neff LM, Culiner J, Cunningham-Rundles S, Seidman C, Meehan D, Maturi J, Wittkowski KM, et al (2011) Algal docosahexaenoic acid affects plasma lipoprotein particle size distribution in overweight and obese adults. J Nutr. 141: 207-213.

[74] Abbey M, Clifton P, Kestin M, Belling B & Nestel P (1990) Effect of fish oil on lipoproteins, lecithin:cholesterol acyltransferase, and lipid transfer protein activity in humans. Arteriosclerosis. 10: 85–94.

[75] Buckley R, Shewring B, Turner R, Yaqoob P, Minihane AM (2004) Circulating triacylglycerol and apoE levels in response to EPA and docosahexaenoic acid supplementation in adult human subjects. Br J Nutr. 92: 477–483.

[76] Theobald HE, Chowienczyk PJ,Whittall R, Humphries SE, Sanders TA (2004) LDL cholesterol-raising effect of low-dose docosahexaenoic acid in middle-aged men and women. Am J Clin Nutr. 79: 558–563.

[77] Conquer JA, Holub BJ (1998) Effect of supplementation with different doses of DHA on the levels of circulating DHA as non-esterified fatty acid in subjects of Asian Indian background. J Lipid Res. 39: 286–292.

[78] Nestel P, Shige H, Pomeroy M, Cehun M, Abbey M, Raederstorff D (2002) The n-3 fatty acids eicosapentaenoic acid and docosahexaenoic acid increase systemic arterial compliance in humans. Am J Clin Nutr 76: 326–330

[79] Griffin BA (2001) The effect of n-3 fatty acids on low density lipoprotein subfractions. Lipids. 36, 91–97.

[80] Lu G, Windsor SL, Harris WS (1999) Omega-3 fatty acids alter lipoprotein subfraction distributions and the in vitro conversion of very low density lipoproteins to low density lipoproteins. J Nutr Biochem. 10:151–158.

[81] Cunnane SC, Hamadeh MJ, Liede AC, Thompson LU, Wolever TM, Jenkins DJ (1995) Nutritional attributes of traditional flaxseed in healthy young adults. Am J Clin Nutr. 61:62–8.

[82] Lucas EA, Wild RD, Hammond LJ, Khalil DA, Juma S, Daggy BP, et al (2002) Flaxseed improves lipid profile without altering biomarkers of bone metabolism in postmenopausal women. J Clin Endocrinol Metab. 87:1527–1532.

[83] Putnam J, Gerrior S (1999) Trends in the U.S. food supply, 1970-97. In: America's eating habits: changes and consequences. Washington, D.C.: United States Department of Agriculture, Economic Research Service. 133-60.

[84] Grundy SM, Barrett-Connor E, Rudel LL, Miettinen T, Spector AA (1988) Workshop on the impact of dietary cholesterol on plasma lipoproteins and atherogenesis. Arteriosclerosis. 8:95-101

[85] National Research Council. Diet and health: implications for reducing chronic disease risk. Washington, D.C.: National Academy Press, 1989: 171-201

[86] Weggemans RM, Zock PL, Katan MB (2001) Dietary cholesterol from eggs increases the ratio of total cholesterol to high-density lipoprotein cholesterol in humans: a meta-analysis. Am J Clin Nutr. 73: 885-91.

[87] Schaefer EJ, Gleason Jam (2009) Dietary fructose and glucose differentially affect lipid and glucose homeostasis J Nutr, 139:1257S–1262S

[88] Barclay AW, Petocz P, McMillan-Price J, et al (2008) Glycemic index, glycemic load, and chronic disease risk—a meta—analysis of observational studies. Am J Clin Nutr. 87:627–637.

[89] Barclay AW, Petocz P, McMillan-Price J, et al (2008) Glycemic index, glycemic load, and chronic disease risk–a meta-analysis of observational studies. Am J Clin Nutr. 87:627–37.

[90] Pelkman CL (2001) Effects of the glycemic index of foods on serum concentrations of high-density lipoprotein cholesterol and triglycerides. Curr Atheroscler Rep. 3:456–461.

[91] Sacks FM, Katan M (2002) Randomized clinical trials on the effects of dietary fat and carbohydrate on plasma lipoproteins and cardiovascular disease. Am J Med.;113 Suppl 9 : 13–24.

[92] Aarsland A, Chinkes D, Wolfe RR (1997) Hepatic and whole-body fat synthesis in humans during carbohydrate overfeeding. Am J Clin Nutr. 65:1774–82

[93] McNeel RL, Mersmann HJ (2005) Low- and high-carbohydrate diets: body composition differences in rats. Obes Res. 13:1651–60.

[94] Volek JS, Phinney SD, Forsythe CE, Quann EE, Wood RJ, Puglisi MJ, et al (2009) Carbohydrate restriction has a more favorable impact on the metabolic syndrome than a low fat diet. Lipids. 44: 297-309.

[95] Krauss RM, Blanche PJ, Rawlings RS, et al (2006) Separate effects of reduced carbohydrate intake and weight loss on atherogenic dyslipidemia. Am J Clin Nutr. 83:1025–1031

[96] Sharman MJ, Kraemer WJ, Love DM, Avery NG, Gomez AL, Scheett TP, Volek JS (2002) A ketogenic diet favorably affects serum biomarkers for cardiovascular disease in normal-weight men. J Nutr. 132: 1879–1885.

[97] Turley ML, Skeaff CM, Mann JI, Cox B (1998) The effect of a low-fat, high-carbohydrate diet on serum high density lipoprotein cholesterol and triglyceride. Eur J Clin Nutr. 52:728-32.

[98] Vuksan V, Sievenpiper JL, Owen R, Swilley JA, Spadafora P, Jenkins DJA (2000) Beneficial effects of viscous dietary fiber from Konjac-mannan in subjects with the insulin resistance syndrome: results of a controlled metabolic trial. Diabetes Care.23:9-14.

[99] Grundy SM, Florentin L, Nix D, et al. (1988) Comparison of monounsaturated fatty acids and carbohydrates for reducing raised levels of plasma cholesterol in man. Am J Clin Nutr 47: 965–969.

[100] Jiménez-Gómez Y, Marín C, Peérez-Martínez P, Hartwich J, Malczewska-Malec M (2010) A low-fat, high-complex carbohydrate diet supplemented with long-chain (n-3) fatty acids alters the postprandial lipoprotein profile in patients with metabolic syndrome. J Nutr. 140: 1595-1601.

[101] Basciano H, Federico L, Adeli K (2005) Fructose, insulin resistance, and metabolic dyslipidemia. Nutr Metab (Lond). 2: 5.

[102] Busserolles J, Zimowska W, Rock E, Rayssiguier Y, Mazur A (2002) ts fed a high sucrose diet have altered heart antioxidant enzyme activity and gene expression Life Sci. 71: 1303-1312.

[103] Girard A, Madani S, Boukortt F, Cherkaoui-Malki M, Belleville J, Prost J (2006) Fructose-enriched diet modifies antioxidant status and lipid metabolism in spontaneously hypertensive rats. Nutrition. 22: 758-766.

[104] Elliott SS, Keim NL, Stern JS, Teff K, Havel PJ (2002) Fructose, weight gain, and the insulin resistance syndrome. Am J Clin Nutr.76: 911-922.

[105] Rukmini C, Raghurm TC (1991) Nutritional and biochemical aspects of the hypolipidemic action of rice bran oil: a review. J Am Coll Nutr. 10: 593–601.

[106] Turnbull WH, Leeds AR, Edwards DG (1992) Mycoprotein reduces blood lipids in free-living subjects. Am J Clin Nutr. 55: 415–419.

[107] Erdman JW Jr (2000) AHA Science Advisory: soy protein and cardiovascular disease: a statement for healthcare professionals from the Nutrition Committee of the AHA. Circulation. 102: 2555–2559.

[108] Sacks FM, Lichtenstein A, Van Horn L, Harris W, Kris-Etherton P, Winston M (2006) Soy protein, isoflavones, and cardiovascular health: an American Heart Association Science Advisory for professionals from the Nutrition Committee. Circulation. 113: 1034–1044.

[109] Jenkins DJ, Mirrahimi A, Srichaikul K, Berryman CE, Wang L, Carleton A, et al (2010) Soy protein reduces serum cholesterol by both intrinsic and food displacement mechanisms. J Nutr. 140: 2302-2311.

[110] Lovati MR, Manzoni C, Gianazza E, Arnoldi A, Kurowska E, Carroll KK, Sirtori CR (2000) Soy protein peptides regulate cholesterol homeostasis in Hep G2 cells. J Nutr. 130: 2543–2549.

[111] Zhuo XG, Melby MK, Watanabe S (2004) Soy isoflavone intake lowers serum LDL cholesterol: a meta-analysis of 8 randomized controlled trials in humans. J Nutr. 134: 2395–2400.

[112] Wiseman H, O'Reilly JD, Adlercreutz H, Mallet AI, Bowey EA, Rowland IR, et al (2000) Isoflavone phytoestrogens consumed in soy decrease F(2)-isoprostane concentrations and increase resistance of low-density lipoprotein to oxidation in humans. Am J Clin Nutr. 72: 395–400

[113] Jenkins DJ, Kendall CW, Vidgen E, Augustin LS, van Erk M, Geelen A, et al (2001) High-protein diets in hyperlipidemia: effect of wheat gluten on serum lipids, uric acid, and renal function. Am J Clin Nutr. 74:57–63.

[114] United States Department of Agriculture, United States Department of Health and Human Services: Dietary Guidelines for Americans, 6th edn. Washington, DC: Government Printing Office; 2005.

[115] United States Department of Agriculture, United States Department of Health and Human Services: Report of the Dietary Guidelines Advisory Committee on the Dietary Guidelines for Americans. Washington, DC: Government Printing Office; 2010

[116] Lichtenstein AH, Appel LJ, Brands M, Carnethon M, Daniels S, Franch HA, et al (2006) Summary of American Heart Association Diet and Lifestyle Recommendations revision 2006. Arterioscler Thromb Vasc Biol. 26:2186–2191

[117] American Diabetes Association (2007) Nutrition recommendations and interventions for diabetes: a position statement of the American Diabetes Association. Diabetes Care 30: 48–65.

[118] Harris KA, Kris-Etherton PM (2010) Effects of whole grains on coronary heart disease risk. Curr Atheroscler Rep. 12: 368-376.

[119] Koh-Banerjee P, Rimm EB (2003) Whole grain consumption and weight gain: a review of the epidemiological evidence, potential mechanisms and opportunities for future research. Proc Nutr Soc. 62: 25–29

[120] Kashtan H, Stern HS, Jenkins DJ, Jenkins AL, Hay K, Marcon N, et al (1992) Wheat-bran and oat-bran supplements' effects on blood lipids and lipoproteins. Am J Clin Nutr.55: 976-980.

[121] Shimizu C, Kihara M, Aoe S, Araki S, Ito K, Hayashi K, et al (2008) Effect of high beta-glucan barley on serum cholesterol concentrations and visceral fat area in Japanese men—a randomized, double-blinded, placebo-controlled trial. Plant Foods Hum Nutr. 63: 21–25

[122] Kelly SA, Summerbell CD, Brynes A, Whittaker V, Frost G (2007) Wholegrain cereals for coronary heart disease. Cochrane Database Syst Rev, 2:CD005051.

[123] US FDA. Food labeling: health claims: soluble fiber from whole oats and risk of coronary heart disease. Docket 95P–0197. Washington, DC: US FDA; 2001. p. 15343–4.

[124] Brown L, Rosner B, Willett WW, Sacks FM (1999) Cholesterol-lowering effects of dietary fiber: a meta-analysis. Am J Clin Nutr. 69:30–42.

[125] Papathanasopoulos A, Camilleri M (2010) Dietary fiber supplements: effects in obesity and metabolic syndrome and relationship to gastrointestinal functions. Gastroenterology. 138:65–72.

[126] National Heart Lung and Blood Institute: Third Report of the Expert Panel on Detection, Evaluation, and Treatment of High Blood Cholesterol in Adults (Adult Treatment Panel III). Edited by National Cholesterol Education Program. Bethesda, MD: National Institutes of Health; 2002.

[127] Dragsted LO, Krath B, Ravn-Haren G, Vogel UB, Vinggaard AM, Bo Jensen P, et al (2006) Biological effects of fruit and vegetables. Proc Nutr Soc. ;65: 61-7.

[128] Jenkins DJ, Kendall CW, Popovich DG, Vidgen E, Mehling CC, Vuksan V, et al (2001) Effect of a very-high-fiber vegetable, fruit, and nut diet on serum lipids and colonic function. Metabolism. 50:494–503

[129] Djousse L, Arnett DK, Coon H, Province MA, Moore LL, Ellison RC (2004) Fruit and vegetable consumption and LDL cholesterol: the national heart, lung, and blood institute family heart study. Am J Clin Nutr.79: 213-217

[130] Ballesteros MN, Cabrera RM, Saucedo MS, Yepiz-Plascencia GM, Ortega MI, Valencia ME (2001) Dietary fiber and lifestyle influence serum lipids in free living adult men. J Am Coll Nutr. 20:649–655.

[131] Stone NJ (2001) Lowering low-density cholesterol with diet: the important role of functional foods as adjuncts. Coron Artery Dis. 12:547–552.

[132] Fornes NS, Martins IS, Hernan M, Velasquez-Melendez G, Ascherio A (2000) Frequency of food consumption and lipoprotein serum levels in the population of an urban area, Brazil. Rev Saude Publica. 34: 380–387

[133] Pereira MA, O'Reilly E, Augustsson K, Fraser GE, Goldbourt U, Heitmann BL, Hallmans G, et al (2004) Dietary fiber and risk of coronary heart disease: a pooled analysis of cohort studies. Arch Intern Med. 164: 370–376.

[134] Kritchevsky D, Story JA (1974) Binding of bile salts in vitro by nonnutritive fiber. J Nutr. 104:458-462.

[135] Everson GT, Daggy BP, McKinley C, Story JA (1992) Effects of psyllium hydrophilic mucilloid on LDL-cholesterol and bile acid synthesis in hypercholesterolemic men. J Lipid Res. 33:1183-1192.

[136] Sjogren P, Rosell M, Skoglund-Andersson C, Zdravkovic S, Vessby B, de Faire U, et al (2004) Milkderived fatty acids are associated with a more favorable LDL particle size distribution in healthy men. J Nutr. 134: 1729– 1735

[137] Pereira MA, Jacobs DR Jr, Van Horn L, Slattery ML, Kartashov AI, Ludwig DS (2002) Dairy consumption, obesity, and the insulin resistance syndrome in young adults: the CARDIA Study. JAMA. 287: 2081–2089.

[138] Shahkhalili Y, Murset C, Meirim I, Duruz E, Guinchard S, Cavadini C, et al (2001) Calcium supplementation of chocolate: effect on cocoa butter digestibility and blood lipids in humans. Am J Clin Nutr. 73: 246–252.

[139] Pfeuffer M, Schrezenmeir J (2007) Milk and the metabolic syndrome. Obes Rev.8:109 – 118.

[140] Takeuchi T, Shimizu H, Ando K, Harada E (2004) Bovine lactoferrin reduces plasma triacylglycerol and NEFA accompanied by decreased hepatic cholesterol and triacylglycerol contents in rodents. Br J Nutr. 91: 533–538.

[141] Nagaoka S, Futamura Y, Miwa K, Awano T, Yamauchi K, Kanamaru Y, et al (2001) Identification of novel hypocholesterolemic peptides derived from bovine milk beta lactoglobulin. Biochem Biophys Res Commun. 281: 11–17.

[142] Shah, H. (2000) Effects of milk-derived bioactives: an overview. Br. J. Nutr. 84: 3–10.

[143] Siri-Tarino PW, Sun Q, Hu FB, Krauss RM (2010) Meta-analysis of prospective cohort studies evaluating the association of saturated fat with cardiovascular disease. Am J Clin Nutr. 91: 535-46

[144] Samuelson G, Bratteby LE, Mohsen R, Vessby B (2001) Dietary intake in healthy adolescents: inverse relationship between the estimated intake of saturated fatty acids and serum cholesterol. Br J Nutr. 85: 333–341.

[145] Jakobsen MU, O'Reilly EJ, Heitmann BL, Pereira MA, Bälter K, Fraser GE, et al (2009) Major types of dietary fat and risk of coronary heart disease: a pooled analysis of 11 cohort studies. Am J Clin Nutr. 89:1425–1432.

[146] Nestel PJ, Chronopulos A, Cehun M (2005) dairy fat in cheese raises LDL cholesterol less than that in butter in mildly hypercholesterolaemic subjects. Eur J Clin Nutr.59: 1059-1063.

[147] Biong AS, Müller H, Seljeflot I, Veierød MB, Pedersen JI, et al (2004) A comparison of the effects of cheese and butter on serum lipids, haemostatic variables and homocysteine. Br J Nutr. 92: 791–797

[148] Matthan NR, Welty FK, Barrett PH, Harausz C, Dolnikowski GG, Parks JS, et al (2004) Dietary hydrogenated fat increases high-density lipoprotein apoA-I catabolism and decreases low-density lipoprotein apoB-100 catabolism in hypercholesterolemic women. Arterioscler Thromb Vasc Biol. 24:1092–1097

[149] Pfeuffer M, Schrezenmeir J (2000) Bioactive substances in milk with properties decreasing risk of cardiovascular diseases. Br J Nutr. 84: 155–159

[150] Agerhol-Larsen L, Bell ML, Grunwald GK, Astrup A (2000) The effect of a probiotic milk product on plasma cholesterol: a metaanalysis of short-term intervention studies. Eur J Clin Nutr. 54: 856–860.

[151] Xiao JZ, Kondo S, Takahashi N, Miyaji K, Oshida K, Hiramatsu A, et al (2003) Effects of milk products fermented by Bifidobacterium longum on blood lipids in rats and healthy adult male volunteers. J Dairy Sci. 86: 2452– 2461

[152] Agerholm-Larsen L, Bell ML, Grunwald GK, Astrup A (2000) The effect of probiotic milk product on plasma cholesterol: a metaanalysis of short-term intervention studies. Eur. J. Clin. Nutr. 54: 856–860.

[153] Expert Panel on Detection, Evaluation, and Treatment of High Blood Cholesterol in Adults. Executive Summary of the Third Report of the National Cholesterol Education Program (NCEP) Expert Panel on Detection, Evaluation, and Treatment of High Blood Cholesterol in Adults (Adult Treatment Panel III). JAMA. 2001;285:2486–2497

[154] Fraser GE (1999) Nut consumption, lipids, and risk of a coronary event. Clin Cardiol. 22:1–5.

[155] Kris-Etherton PM, Pearson TA, Wan Y, Hargrove RL, Moriarty K, Fishell V, et al (1999) High-monounsaturated fatty acid diets lower both plasma cholesterol and triglyceride concentrations. Am J Clin Nutr. 70:1009–1015

[156] Abbey M, Noakes M, Belling GB, Nestel PJ (1994) Partial replacement of saturated fatty acids with almonds or walnuts lowers total plasma cholesterol and low-density-lipoprotein cholesterol. Am J Clin Nutr.. 59: 995–999

[157] Morgan WA, Clayshulte BJ (2000) Pecans lower low-density lipoprotein cholesterol in people with normal lipid levels. J Am Diet Assoc. 100: 312–318.

[158] Zambón D, Sabaté J, Muñoz S, Campero B, Casals E, Merlos M, et al (2000) Substituting walnuts for monounsaturated fat improves the serum lipid profile of hypercholesterolemic men and women: a randomized crossover trial. Ann Intern Med. 132: 538–546

[159] Sabaté J, Fraser GE, Burke K, Knutsen SF, Bennett H, Lindsted KD (1993) Effects of walnuts on serum lipid levels and blood pressure in normal men. N Engl J Med. 328:603- 607.

[160] Edwards K, Kwaw I, Matud J, Kurtz I (1999) Effect of pistachio nuts on serum lipid levels in patients with moderate hypercholesterolemia. J Am Coll Nutr.18:229–232

[161] O'Byrne DJ, Knauft DA, Shireman RB (1997) Low fat-monounsaturated rich diets containing high-oleic peanuts improve serum lipoprotein profiles. Lipids. 32:687–695

[162] Schaefer EJ, Lichtenstein AH, Lamon-Fava S, Contois JH, Li Z, Rasmussen H, McNamara JR, et al (1995) Efficacy of a National Cholesterol Education Program Step 2 diet in normolipidemic and hypercholesterolemic middle-aged and elderly men and women. Arterioscler Thromb Vasc Biol. 15:1079–1085.

[163] Reaven P, Parthasarathy S, Grasse BJ, Miller E, Steinberg D, Witztum JL (1993) Effects of oleate-rich and linoleate-rich diets on the susceptibility of low density lipoprotein to oxidative modification in mildly hypercholesterolemic subjects. J Clin Invest. 91:668–676

[164] Spiller, G. A., Jenkins, D.A.J., Bosello, O., Gates, J. E., Cragen, L. N., Bruce B (1998) Nuts and plasma lipids: an almond-based diet lowers LDL-C while preserving HDL-C. J. Am. Coll. Nutr. 17: 285–290.

[165] Hyson DA, Schneeman BO, Davis PA (2002) Almonds and almond oil have similar effects on plasma lipids and LDL oxidation in healthy men and women. Almonds and almond oil have similar effects on plasma lipids and LDL oxidation in healthy men and women. J Nutr. 132: 703-707

[166] Garg ML, Blake RJ, Wills RB (2003) Macadamia nut consumption lowers plasma total and LDL cholesterol levels in hypercholesterolemic men. J Nutr. 133:1060-1063.

[167] Li L, Tsao R, Yang R, Kramer JK, Hernandez M (2007) Fatty acid profiles, tocopherol contents, and antioxidant activities of heartnut (Juglans ailanthifolia var. cordiformis) and Persian walnut (Juglans regia L.). J Agric Food Chem. 55:1164-1169.

[168] Banel DK, Hu FB (2009) Effects of walnut consumption on blood lipids and other cardiovascular risk factors: a meta-analysis and systematic review. Am J Clin Nutr. 90: 56-63

[169] Bazzano LA, Thompson AM, Tees MT, Nguyen CH, Winham DM (2011) Non-soy legume consumption lowers cholesterol levels: a meta-analysis of randomized controlled trials. Nutr Metab Cardiovasc Dis. 21: 94-103.

[170] Anderson JW, Major AW (2002) Pulses and lipaemia, short- and long term effect: potential in the prevention of cardiovascular disease. Br J Nutr. 88:263-271

[171] Duranti M (2006) Grain legume proteins and nutraceutical properties. Fitoterapia. 77: 67-82

[172] Mallillin AC, Trinidad TP, Raterta R, Dagbay K, Loyola AS (2008) Dietary fiber and fermentability characteristics of root crops and legumes. Br J Nutr. 100: 485–488.

[173] Roberfroid M (1997) Health benefits of non-digestible oligosaccharides. In Dietary Fiber in Health and Disease (Advances in Experimental Biology), p. 427 [D Kritchevsky and C Bonfield, editors]. New York: Plenum Press.

[174] Duane WC (1997) Effects of legume consumption on serum cholesterol, biliary lipids, and sterol metabolism in humans. J Lipid Res 38: 1120–1128

[175] Galisteo M, Duarte J, Zarzuelo A (2008) Effects of dietary fibers on disturbances clustered in the metabolic syndrome. J Nutr Biochem. 19: 71-84

[176] Rochfort S, Panozzo J (2007) Phytochemicals for health, the role of pulses. J Agric Food Chem. 55:7981-7994

[177] Denke M (1994) Role of beef tallow, an enriched source of stearic acid, in a cholesterol lowering diet. Am J Clin Nutr. 60: 1044S-1049S

[178] Scott LW, Dunn JK, Pownall HJ, Brauchi DJ, McMann MC, Herd JA, et al (1994) Effects of beef and chicken consumption on plasma lipid levels in hypercholesterolemic men. Arch Intern Med. 154: 1261-1267.

[179] Wolmarans P, Benadé AJS, Kotze TJvW, Daubitzer AK, Marais MP, Laubscher R (1991) Plasma lipoprotein response to substituting fish for red meat in the diet. Am J Clin Nutr. 53:1171–1176

[180] Davidson MH, Hunninghake D, Maki KC, Kwiterovich PO, Kafonek S (1999) Comparison of the effects of lean red meat vs lean white meat on serum lipid levels among free-living persons with hypercholesterolemia. Arch Intern Med. 159:1331–1338

[181] Lankinen M, Schwab U, Erkkilä A, Seppänen-Laakso T, Hannila ML, Mussalo H, et al (1991) Effects of a lean beef diet and of a chicken and fish diet on lipoprotein profiles. Nutr Metab Cardiovasc Dis. 1: 25-30.

[182] Morgan S, Sinclair A, O'Dea K (1993) Effect on serum lipids of addition of safflower oil or olive oil to very-low-fat diets rich in lean beef. J Am Diet Assoc. 93: 644- 648.

[183] Grundy SM. Cholesterol and atherosclerosis. Diagnosis and treatment. New York: Gower Medical Publishing, 1990

[184] Lacaille B, Julien P, Deshaies Y, Lavigne C, Brun L-D, Jacques H (2000) Responses of plasma lipoproteins and sex hormones to the consumption of lean fish incorporated in a prudent-type diet in normolipidemic men. J Am Coll Nutr. 19: 745–753

[185] Gascon A, Jacques H, Moorjani S, Deshaies Y, Brun L-D, Julien P (1996) Plasma lipoprotein profile and lipolytic activities in response to the substitution of lean white fish for other animal protein sources in premenopausal women. Am J Clin Nutr. 63: 315–321

[186] Howell WH, McNamara DJ, Tosca MA, Smith BT, Gaines JA (1997) Plasma lipid and lipoprotein responses to dietary fat and cholesterol: a meta-analysis. Am J Clin Nutr.65:1747–1764.

[187] Dawber TR, Nickerson RJ, Brand FN, Pool J (1982) Eggs, serum cholesterol, and coronary heart disease. Am J Clin Nutr. 36:617–625

[188] Hu FB, Stampfer MJ, Rimm EB, Manson JE, Ascherio A, Colditz GA, et al (1999) A prospective study of egg consumption and risk of cardiovascular disease in men and women. JAMA. 281:1387–1394.

[189] Mayurasakorn K, Srisura W, Sitphahul P, Hongto PO (2008) High-density lipoprotein cholesterol changes after continuous egg consumption in healthy adults. J Med Assoc Thai. Mar. 91: 400-407.

[190] Mutungi G, Ratliff J, Puglisi M, Torres-Gonzalez M, Vaishnav U, Leite JO (2008) Dietary cholesterol from eggs increases plasma HDL cholesterol in overweight men consuming a carbohydrate-restricted diet. J Nutr. 138:272-176

[191] Fernandez ML (2006) Dietary cholesterol provided by eggs and plasma lipoproteins in healthy populations. Curr Opin Clin Nutr Metab Care. 9: 8-12.

[192] Harsha DW, Sacks FM, Obarzanek E, Svetkey LP, Lin PH, Bray GA (2004) Effect of dietary sodium intake on blood lipids: results from the DASH-sodium trial. Hypertension. 43: 393-398.

[193] Miller ER 3rd, Erlinger TP, Appel LJ (2006) The effects of macronutrients on blood pressure and lipids: an overview of the DASH and OmniHeart trials. Curr Atheroscler Rep. 8: 460-465.

[194] Obarzanek E, Sacks FM, Vollmer WM, Bray GA, Miller ER, Lin PH (2001) Effects on blood lipids of a blood pressure lowering diet: the Dietary Approaches to Stop Hypertension (DASH) Trial. Am J Clin Nutr. 74:80–89.

[195] Willett WC, Sacks F, Trichopoulou A, Drescher G, Ferro-Luzzi A, Helsing E, et al (1995) Mediterranean diet pyramid: a cultural model for healthy eating. Am J Clin Nutr. 6: 1402-1406

[196] Tzima N, Pitsavos C, Panagiotakos DB, Skoumas J, Zampelas A, Chrysohoou C, et al (2007) Mediterranean diet and insulin sensitivity, lipid profile and blood pressure levels, in overweight and obese people; the Attica study. Lipids Health Dis. 6:22.

[197] Salen P, de Lorgeril M (1999) [Hyperlipidemias. Concern with the Mediterranean diet]. Presse Med.28:2018-2024.

[198] Demarin V, Lisak M, Morović S (2011) Mediterranean diet in healthy lifestyle and prevention of stroke. Acta Clin Croat. 50: 67-77.

[199] Willett WC. Public Health Nutr 2006;9:105

[200] Mekki K, Bouzidi-bekada N, Kaddous A, Bouchenak M (2010) Mediterranean diet improves dyslipidemia and biomarkers in chronic renal failure patients. Food Funct.1: 10-5.

[201] Pitsavos, C., Panagiotakos, D.B., Tzima, N., Chrysohoou, C., Economou, M., Zampelas, A, et al (2005) Adherence to the Mediterranean diet is associated with total antioxidant capacity in healthy adults: the ATTICA study. Am. J. Clin. Nutr. 82: 694–699.

[202] Bouchard-Mercier A, Paradis AM, Godin G, Lamarche B, Pérusse L, Vohl MC (2010) Associations between dietary patterns and LDL peak particle diameter: a cross-sectional study. J Am Coll Nutr.29: 630-67.

[203] Ganguli D, Das N, Saha I, Biswas P, Datta S, Mukhopadhyay B (2011) Major dietary patterns and their associations with cardiovascular risk factors among women in West Bengal, India. Br J Nutr. 105:1520-1529

[204] Lim JH, Lee YS, Chang HC, Moon MK, Song Y (2011) Association between dietary patterns and blood lipid profiles in Korean adults with type 2 diabetes. J Korean Med Sci. 26: 1201-1208.

[205] Ambrosini GL, Huang RC, Mori TA, Hands BP, O'Sullivan TA, de Klerk NH, et al (2010) Dietary patterns and markers for the metabolic syndrome in Australian adolescents. Nutr Metab Cardiovasc Dis.20:274-283.

[206] FDA Talk Paper. FDA authorizes new coronary heart disease health claim for plant sterol and plant stanol esters. 5 September 2000. Washington (DC). Available from: http://www.fda.gov/Food/Labeling.Nutrition/LabelClaims/HealthClaimsMeetingSignifi cantScientificAgreementSSA/ucm074747.htm

[207] Manach C, Scalbert A, Morand C, Remesy C, Jimenez L (2004) Polyphenols: food sources and bioavailability. Am J Clin Nutr. 79:727–747

[208] Otaki N, Kimira M, Katsumata S, Uehara M, Watanabe S, Suzuki K (2009) Distribution and major sources of flavonoid intakes in the middle-aged Japanese women. J Clin Biochem Nutr. 44:231-238.

[209] Higgins JPT, Green S, eds. Highly sensitive search strategies for identifying reports of randomized controlled trials in MEDLINE. Cochrane handbook for systematic reviews of interventions 4.2.5 (updated May 2005); Appendix 5b. The Cochrane Library, Issue 3, 2005. Chichester, UK: Wiley & Sons, Ltd. 2005.

[210] Frankel, E.N., Kanner, J., German, J.B., Parks, E., Kinsella, J.E (1993) Inhibition of oxidation of human low-density lipoprotein by phenolic substances in red wine. Lancet. 341:454-457.

[211] Otaki N, Kimira M, Katsumata S, Uehara M, Watanabe S, Suzuki K (2009) Distribution and major sources of flavonoid intakes in the middle-aged Japanese women. J Clin Biochem Nutr. 44: 231-8.

[212] Odbayar TO, Badamhand D, Kimura T, Takashi Y, Tsushida T, Ide T (2006) Comparative studies of some phenolic compounds quercetin, rutin, and ferulic acid) affecting hepatic fatty acid synthesis in mice. J Agric Food Chem. 54:8261–8265

[213] Princen HM, van Duyvenvoorde W, Buytenhek R, Blonk C, Tijburg LB, Langius JA, et al (1998) No effect of consumption of green and black tea on plasma lipid and

antioxidant levels and on LDL oxidation in smokers. Arterioscler Thromb Vasc Biol. 18:833–841

[214] Bingham, S. A., Vorster, H., Jerling, J. C., Magee, E., Mulligan, A., Runswick, S. A., Cummings, J. H. (1997) Effect of black tea drinking on blood lipids, blood pressure and aspects of bowel habit. Brit. J. Nutr. 78: 41–55

[215] van het Hof KH, de Boer HS, Wiseman SA, Lien N, Westrate JA, Tijburg LB (1997) Consumption of green or black tea does not increase resistance of low-density lipoprotein to oxidation in humans. Am J Clin Nutr. 66:1125-1132

[216] McAnlis GT, McEneny J, Pearce J, Young IS (1998) Black tea consumption does not protect low density lipoprotein from oxidative modification. Eur J Clin Nutr. 52:202-206

[217] Duffy SJ, Vita JA, Holbrook M, Swerdloff PL, Keaney JF (2001) Effect of acute and chronic tea consumption on platelet aggregation in patients with coronary artery disease. Arterioscler Thromb Vasc Biol. 21: 1084-1089

[218] Maron DJ, Lu GP, Cai NS, Wu ZG, Li YH, Chen H, Zhu JQ, et al (2003) Cholesterol-lowering effect of a theaflavin-enriched green tea extract: a randomized controlled trial. Arch Intern Med. 163:1448-1453.

[219] Tokunaga S, White IR, Frost C, Tanaka K, Kono S, Tokudome S (2002) Green tea consumption and serum lipids and lipoproteins in a population of healthy workers in Japan. Ann Epidemiol.12: 157-165.

[220] Kris-Etherton PM, Keen CL (2002) Evidence that the antioxidant flavonoids in tea and cocoa are beneficial for cardiovascular health. Curr Opin Lipidol. 13:41-49.

[221] Ikeda I, Imasato Y, Sasaki E, Nakayama M, Nagao H, Takeo T, et al (1992) Tea catechins decrease micellar solubility and intestinal absorption of cholesterol in rats. Biochim Biophys Acta. 1127:141– 146.

[222] Yang TT, Koo MW (2000) Chinese green tea lowers cholesterol level through an increase in fecal lipid excretion. Life Sci. 66:411-423

[223] Chan PT, Fong WP, Cheung YL, Huang Y, Ho WK, Chen ZY (1999) Jasmine green tea epicatechins are hypolipidemic in hamsters (Mesocricetus auratus) fed a high fat diet. J Nutr. 129:1094-1101.

[224] Bursill C, Roach PD, Bottema CD, Pal S (2001) Green tea upregulates the low-density lipoprotein receptor through the sterol-regulated element binding protein in HepG2 liver cells. J Agric Food Chem. 49: 5639-5645

[225] Matsumoto N, Okushio K, Hara Y (1998) Effect of black tea polyphenols on plasma lipids in cholesterol-fed rats. J Nutr Sci Vitaminol. 44:337-342

[226] Crews WD Jr, Harrison DW, Wright JW (2008) A double-blind, placebocontrolled, randomized trial of the effects of dark chocolate and cocoa on variables associated with neuropsychological functioning and cardiovascular health: Clinical findings from a sample of healthy, cognitively intact older adults. Am J Clin Nutr. 87:872–880.

[227] Farouque HM, Leung M, Hope SA, Baldi M, Schechter C, Cameron JD, Meredith IT (2006) Acute and chronic effects of flavanol-rich cocoa on vascular function in subjects with coronary artery disease: a randomized double-blind placebo-controlled study. Clin Sci. 111: 71–80.

[228] Lecumberri E, Goya L, Mateos R, Alía M, Ramos S, Izquierdo-Pulido M, et al (2007) A diet rich in dietary fiber from cocoa improves lipid profile and reduces malondialdehyde in hypercholesterolemic rats. Nutrition. 23:332–41.

[229] Rein D, Lotito S, Holt RR, Keen CL, Schmitz HH, Fraga CG (2000) Epicatechin in human plasma. In vivo determination and effect of chocolate consumption on plasma oxidation status. J Nutr. 130: 2109–2114.

[230] Mathur S, Devaraj S, Grundy SM, Jialal I (2002) Cocoa products decrease low density lipoprotein oxidative susceptibility but do not affect biomarkers of inflammation in humans. J Nutr. 132:3663-3667.

[231] Jia L, Liu X, Bai YY, Li SH, Sun K, He C, et al (2010) Short-term effect of cocoa product consumption on lipid profile: a meta-analysis of randomized controlled trials. Am J Clin Nutr. 92:218–225.

[232] Keen CL, Holt RR, Oteiza PI, Fraga CG, Schmitz HH (2005) Cocoa antioxidants and cardiovascular health. Am J Clin Nutr. 81:298-303.

[233] Schmidt M, Schmitz HJ, Baumgart A, Guédon D, Netsch MI, Kreuter MH, Schmidlin CB, et al (2005) Toxicity of green tea extracts and their constituents in rat hepatocytes in primary culture. Food Chem Toxicol. 43: 307-14.

[234] Vinson JA, Proch J, Bose P, Muchler S, Taffera P, Shuta D, et al (2006) Chocolate is a powerful ex vivo and in vivo antioxidant, an antiatherosclerotic agent in an animal model, and a significant contributor to antioxidants in the European and American Diets. J Agric Food Chem. 54:8071–6.

[235] Basch E, Ulbricht C, Kuo G, Szapary P, Smith M (2003) Therapeutic applications of fenugreek. Altern Med Rev. 8:20–27.

[236] Handa T, Yamaguchi K, Sono Y, Yazawa K (2005) Effects of fenugreek seed extract in obese mice fed a high-fat diet. Biosci Biotechnol Biochem. 69: 1186–1188.

[237] Gupta R, Nair S (1999) Antioxidant flavonoids in common Indian diet. South Asian J Prev Cardio. 3:83-94.

[238] Petit PR, Sauvaire YD, Hillaire-Buys DM, Leconte OM, Baissac YG, Ponsin GR (1995) Steroid saponins from fenugreek seeds: extraction, purification, and pharmacological investigation on feeding behavior and plasma cholesterol. Steroids. 60:674-80

[239] Broca C, Breil V, Cruciani-Guglielmacci C, Manteghetti M, Rouault C, Derouet M, et al (2004) Insulinotropic agent ID-1101 (4-hydroxyisoleucine) activates insulin signaling in rat. Am J Physiol Endocrinol Metab. 287:463-71

[240] Dixit PP, Misar A, Mujumdar AM, Ghaskadbi S (2010) Pre-treatment of Syndrex protects mice from becoming diabetic after streptozotocin injection. Fitoterapia. 81(5):403-12

[241] Udayasekhara Rao P, Sesikeran B, Srinivasa Rao P, Nadamnui A, Vikas Rao V, Ramachandra RP (1996) Short term Nutritional and safety evaluation of Fenugreek. Nutr Res 16(9): 1495–1505

[242] Praveen KB, Dasgupta DJ, Prashar BS, Kaushal SS (1987) Preliminary Report: Effective reduction of LDL cholesterol by indigenous plant products. Current Science. 56(12): 80–81

[243] Saibaba A, Raghuram TC (1997) Fenugreek – The wonder Seed. Nutrition 31(2): 21–25.

[244] Neeraja A, Rajyalakshmi P (1996) Hypoglycemic effect of processed fenugreek seeds in humans. J Food Sci Technol. 33: 427–430.

[245] Muwalla MM, Abuirmmeileh NM (1991) Suppression of avian hepatic cholesterogenesis by dietary ginseng. J Nutr Biochem. 1:518– 521

[246] Zheng X, Yan Y (1991) The effect of ginsenosides of ginseng stem and leaf (GSL) on the lipid regulation and lipid peroxidation in chronic hyperlipidemic rabbits. Zhongguo Yaolixue Tonbao. 7:110–116

[247] Kim SH, Park KS (2003) Effects of Panax ginseng extract on lipid metabolism in humans. Pharmacol Res. 48:511-513.

[248] Kim SH, Park KS (2003) Effects of Panax ginseng extract on lipid metabolism in humans. Pharmacol Res. 48:511–513.

[249] Yamamoto M, Kumagai A (1984) Long term ginseng effects on hyperlipidemia in man with further study of its actions on atherogenesis and fatty liver rats. In: Proceedings of the 4th International Ginseng Symposium at Korea Ginseng & Tobaco Research Institute. p. 13–20.

[250] Joo CN. The preventive effect of Korean ginseng saponins on aortic atheroma formation in prolonged cholesterol fed rabbits. In: Proceedings of the 3rd International Ginseng Symposium at Korea Ginseng & Tobaco Research Institute; 1980. p. 27–36

[251] Kang DG, Yun YG, Ryoo JH, Lee HS (2002) Anti-hypertensive effect of water extract of danshen on renovascular hypertension through inhibition of the renin angiotensin system. Am J Chin Med. 30:87–93

[252] Yokozawa T, Kobayashi T, Kawai A, Oura H, Kawashima Y (1985) Hyperlipidemia-improving effects of ginseoside-Rb2in cholesterolfed rats. Chem Pharm Bull. 33:722–729.

[253] Langner E, Greifenberg S, Gruenwald J (1998) Ginger: history and use. Adv Ther 15:25–44

[254] Gruenwald J, Brendler T, Jaenicke C (2000) PDR for herbal medicines, 2nd edn. Medical Economics Company, Inc, Montvale, NJ

[255] Bhandari U, Kanojiah R, Pillai KK (2005) Effect of ethanol extract of Zingiber officinale on dyslipidaemia in diabetic rats. J Ethnopharmacol 97: 227–230

[256] Mary JM, John PK (2000) Agents used in hyperlipidaemia. In: Katzung BG (ed) Basic and clinical pharmacology, 8th edn. McGraw Hill Comp, New York, pp 581–595

[257] Afshari AT, Shirpoor A, Farshid A, Saadatian R, Rasmi Y, Saboory E et al (2007) The effect of ginger on diabetic nephropathy, plasma antioxidant capacity and lipid peroxidation in rats. Food Chem 101:148–153

[258] Belinky PA, Aviram M, Fuhrman B, Rosenblat M, Vaya J (1998) The antioxidative effects of the isoflavan glabridin on endogenous constituents of LDL during its oxidation. Atherosclerosis. 137:49–61

[259] Demizu S, Kajiyama K, Takahashi K, Hiraga Y, Yamamoto S, Tamura Y, et al (1988) Antioxidant and antimicrobial constituents of licorice: isolation and structure elucidation of a new benzofuran derivative. Chem Pharm Bull (Tokyo). 36: 3474-3479.

[260] Belinky PA, Aviram M, Fuhrman B, Rosenblat M, Vaya J (1998) The antioxidative effects of the isoflavan glabridin on endogenous constituents of LDL during its oxidation. Atherosclerosis. 137: 49

[261] Fuhrman B, Volkova N, Kaplan M, Presser D, Attias J, Hayek T, et al (2002) Antiatherosclerotic effects of licorice extract supplementation on hypercholesterolemic patients: increased resistance of LDL to atherogenic modifications, reduced plasma lipid levels, and decreased systolic blood pressure. Nutrition. 18: 268-273.

Chapter 15

Lipoproteins and Apolipoproteins of the Ageing Eye

Edward Loane

Additional information is available at the end of the chapter

1. Introduction

In this chapter, we outline the structure of the retina and the aetiopathogenesis of the major age-related eye disease: age-related macular degeneration (AMD). We then discuss the role that lipoproteins and apolipoproteins play in the ageing eye and in the development of AMD.

2. The macula and retina

The macula is the central part of the retina, the neurosensory portion of the eye, and it is responsible for detailed central and colour vision due to its high concentration of cone photoreceptors. Anatomically, the macula is centred on the foveola, and has a ganglion cell layer of more than one cell in thickness. The macula has a diameter of approximately 5.5 mm. The macula is characterised by a yellowish colour (hence the term *macula lutea*, which is Latin for 'yellow spot'), attributable to the presence of macular pigment (MP).[1] The concentration of MP peaks at the centre of the macula, where the appearance of the 'yellow spot' may be clearly evident on clinical examination or fundus photography [Figure 1]. MP is optically undetectable outside the macula.[2] Within the layer structure of the retina, the highest concentration of MP is seen in the receptor axon layer and the inner plexiform layer.[1]

The retina consists of a neurosensory portion comprised of nine individual layers, and an external retinal pigment epithelium (RPE). The RPE plays an important physiological role in the maintenance of neurosensory retinal health, through functions including Vitamin A metabolism, phagocytosis of photoreceptor outer segments, maintenance of the outer blood-retina barrier, heat exchange, and the active transport of substances in and out of the RPE.[3] The blood supply of the retina is derived from the inner retinal vasculature and the outer choriocapillaris. Non-pathological changes that occur in the RPE with age include an increase in cellular pleomorphism and a decrease in cell number, with migration of peripheral RPE cells towards the macula, reduced melanin composition, and an accumulation of the age-pigment lipofuscin.[4;5] These changes may lead to a reduction in the metabolic activity of the RPE, with subsequent apoptosis, which pre-dates pathological change.[5;6] The RPE is separated from the choriocapillaris by Bruch's membrane (BrM). BrM is a semipermeable filtration barrier, comprised of five individual layers.[7;8] Disruption of BrM may result in alteration of its filtration properties, impacting on the function of the RPE and the neurosensory retina.[9] Changes that occur in BrM with age include an increase in its overall thickness, with a reconfiguration of associated lipids and proteins and the accumulation of debris.[10;11] When this debris accumulates between BrM and the RPE, it is referred to as a basal laminar deposit (BlamD) and is not specifically pathological in nature.[12] However, when deposits accumulate within the inner collagenous layer of BrM, they are referred to as basal linear deposits (BlinDs) and are a histopathological hallmark of AMD.[13] These deposits (BlamDs and BlinDs) contain a wide range of constituents including collagen, inflammatory proteins and lipoproteins. When sufficient debris accumulates in BlinDs, they are visible clinically as drusen.[14;15]

3. Age-related macular degeneration

Age-related macular degeneration (AMD) is the leading cause of blindness in people over 50 years of age in the developed world, and it results in loss of central and colour vision if not treated, or if not amenable to treatment.[16-18] The loss of central vision impacts greatly on the individual, as their ability to perform simple daily tasks, such as reading, watching television, driving and recognizing people's faces becomes increasingly difficult. Thus, their quality of life and their ability to lead an independent life diminish significantly as the disease progresses. The peripheral retina is not affected in individuals with AMD, regardless of stage, such that, in the absence of other ocular pathology, peripheral (navigational) vision remains unchanged.

It is currently estimated that late AMD affects 513,000 people in the United Kingdom (2.4% of those over the age of 50), and that this number will increase to 679,000 by the year 2020.[19] Prevalence data from the United States in 2004 estimated that more than 1.75 million individuals were affected by the disease, with this latter figure expected to rise to almost 3 million by the year 2020.[20] The prevalence of this condition is likely to increase dramatically in the future, as a result of increasing life-expectancy and the resultant increasing senescence of society.[21] Data from the National Eye Institute in the United States in 2004 indicated that the prevalence of advanced AMD in people over 40 years of age was 1.47%, rising to 15% in white females aged over 80 years. Beyond its impact on the individual sufferer,[22] the predicted increase in longevity (Figure 2), coupled with the predicted growth in world population (Figure 3) will significantly increase the socio-economic burden that AMD places on countries and their health-care systems.[23-26]

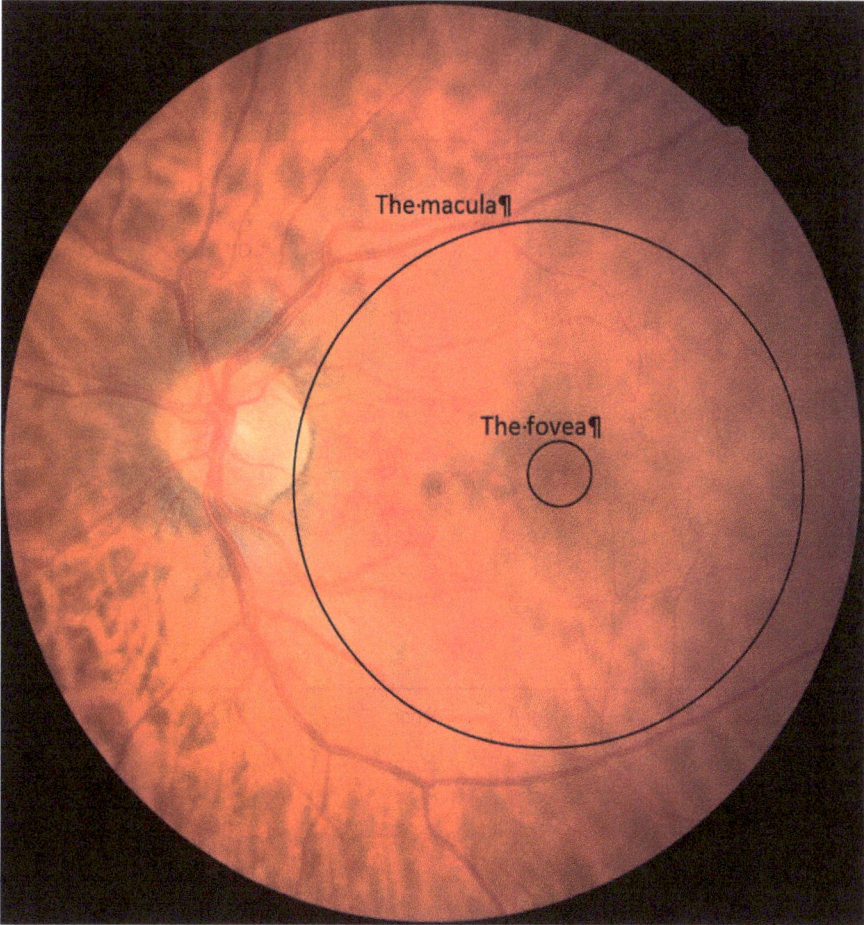

Figure 1. Colour fundus photograph showing the macula, surrounding the fovea, which is centred on the foveola (not marked, but evident as the 'yellow spot') of a left eye.

Male and Female Life Expectancy 1950-2050

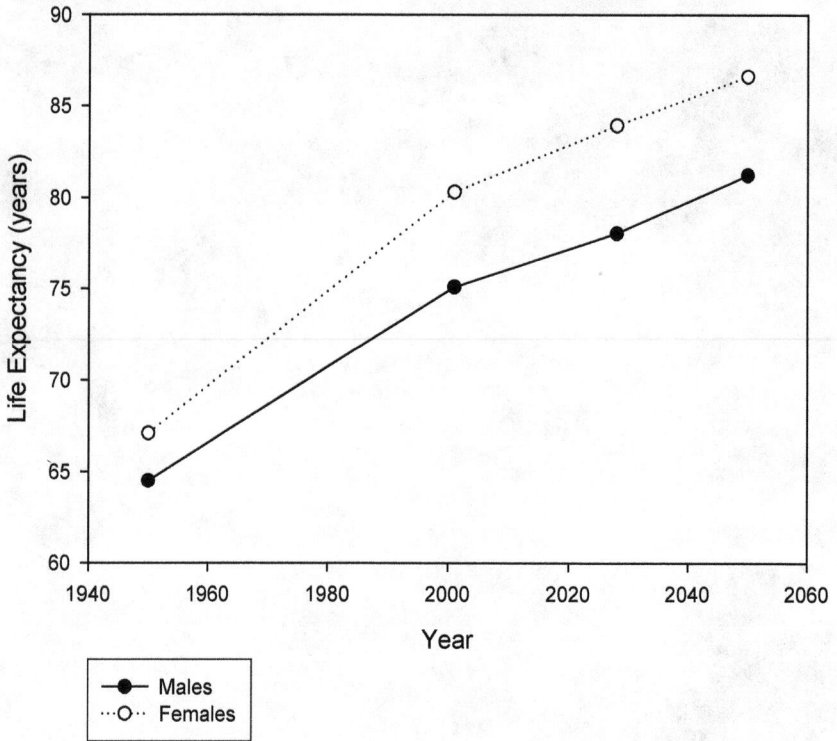

Figure 2. Male and female life expectancy 1950-2050.
* Figures from 1950 and 2001: Irish Department of Health and Children data;
  Projected figures for 2028 and 2050: USA data.

World Population 1950-2050

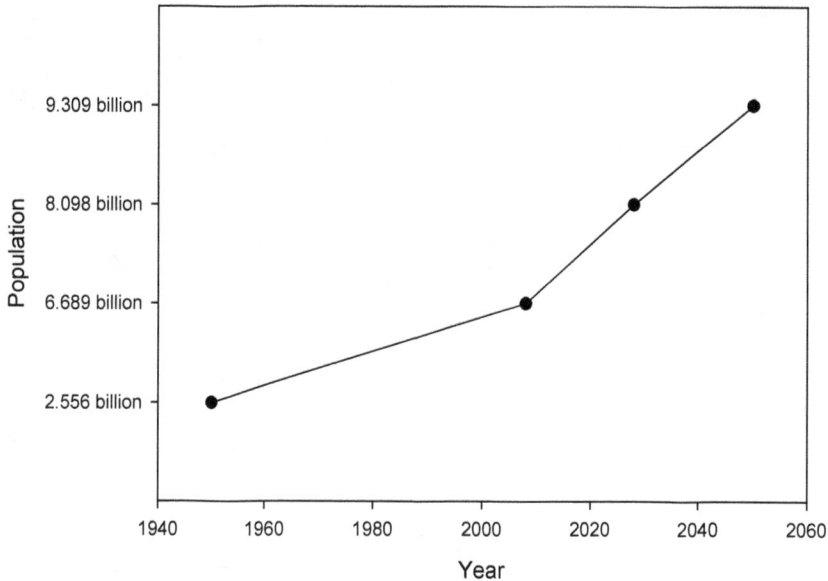

Figure 3. World population 1950-2050 (predicted).

4. Classification of AMD

In 1995, the International Age-Related Maculopathy Epidemiological Study Group clarified the definition and core grading system used to detect and define AMD.[27] This was done to homogenize the systems used to identify and classify this disease in all future clinical and epidemiological studies. This current classification system defines AMD primarily on the basis of morphological changes, without reference to visual acuity.

AMD is defined as a disorder of the macular area, most often clinically apparent after 50 years of age, and characterised by any of the following findings, which are not patently due to another disorder:

1. Soft drusen ≥ 63 μm in diameter. Drusen are whitish-yellow spots that lie external to the neurosensory retina or the RPE (Figure 4). Drusen may be soft and confluent, soft distinct, or soft indistinct. Hard drusen do not, of themselves, characterize AMD.
2. Hyperpigmentation in the outer retina or choroid associated with drusen.
3. Hypopigmentation of the RPE, most often more sharply demarcated than drusen, without any visible choroidal vessels associated with drusen.

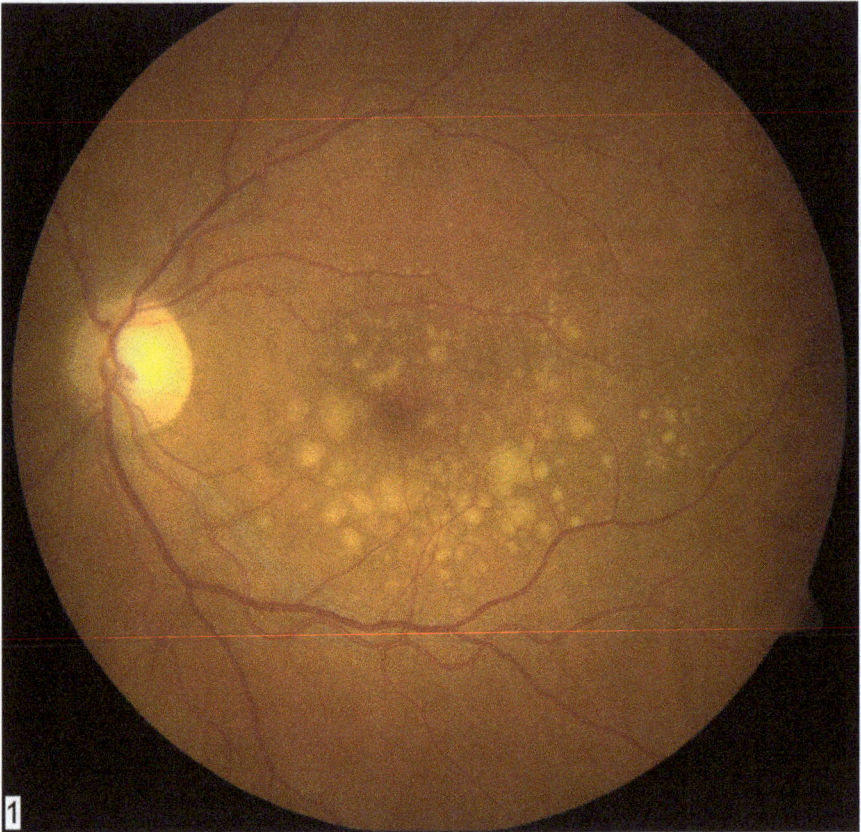

Figure 4. Macular soft drusen of a left eye.

These age-related pathological changes, which are associated with progressive accumulation of debris under the retina, predispose to the late stage of AMD.[28;29] Late AMD is classified as either geographic atrophy (atrophic AMD) or neovascular AMD (choroidal neovascularisation, also referred to as 'exudative AMD' or 'disciform AMD').

Geographic atrophy (GA) is characterised by the following, which is not patently due to another disorder:

1. Any sharply delineated area of hypopigmentation, or depigmentation, or apparent absence of the RPE, in which the choroidal vasculature is more visible than in the surrounding area. The area of atrophy must be ≥ 175 μm in diameter (Figure 5).

Neovascular AMD is characterised by any of the following, which are not patently due to another disorder:

1. RPE detachment(s), which may be associated with neurosensory retinal detachment.
2. Subretinal or sub-RPE neovascularisation.
3. Epiretinal, intraretinal, subretinal, or sub-RPE glial tissue or fibrin-like deposits.
4. Subretinal haemorrhage (Figure 6).
5. Hard exudates (lipids) within the macular area, related to any of the above, in the absence of other retinal vascular disease.

Rarely, neovascular AMD may develop in an area of GA. If this happens, the affected eye is re-classified as having neovascular AMD.

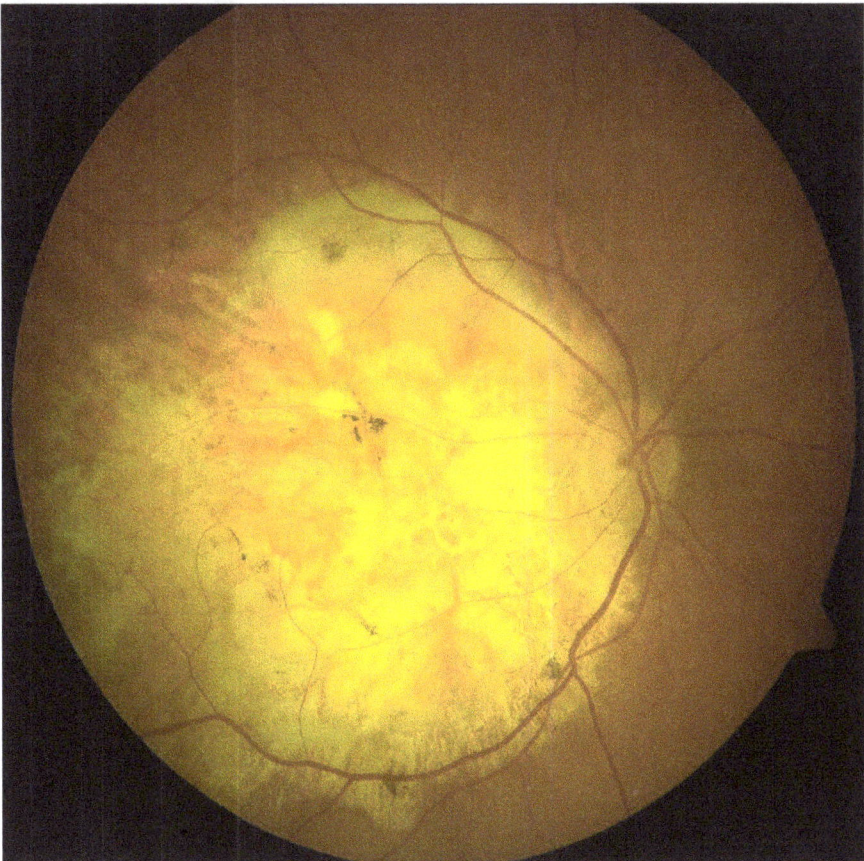

Figure 5. Geographic atrophy, affecting the entire macula of a right eye.

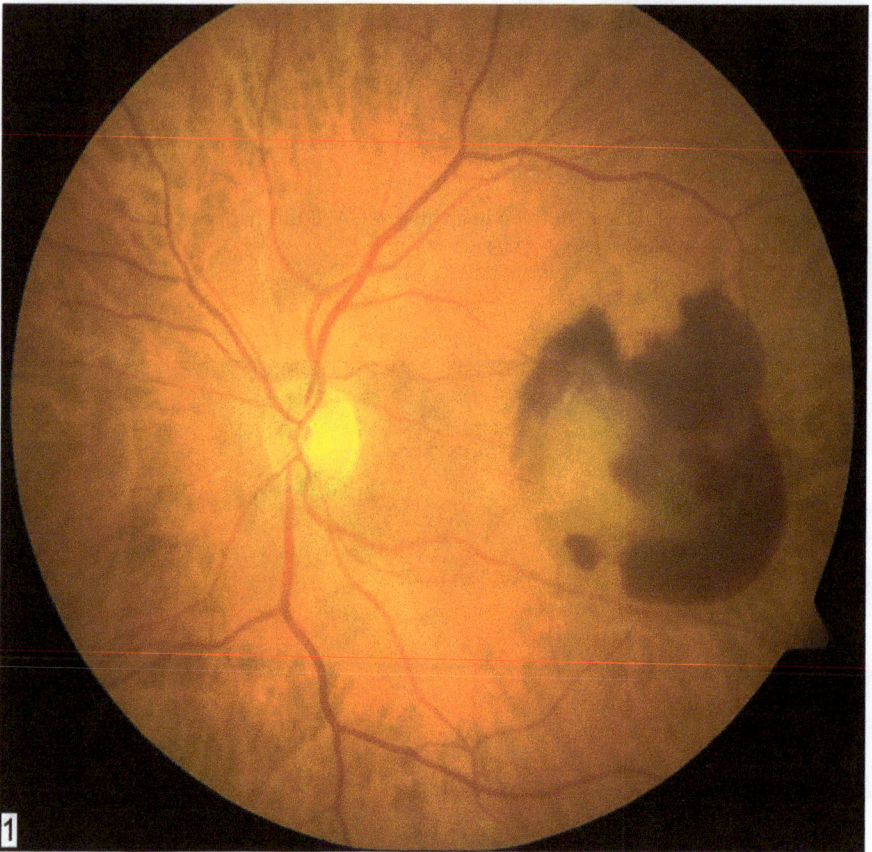

Figure 6. Neovascular AMD, showing sub-retinal haemorrhage in a left eye.

5. Pathogenesis of AMD

AMD has a multi-factorial pathogenesis.[30;31] Therefore, the development of AMD is dependent on a complex interaction between an individual's genetic composition (genotype) and lifestyle (or environmental) factors. This interaction is complex and incompletely understood; however, certain factors have been well established as representing risk for this condition, whereas others are known as putative risk factors, according to our current understanding of this disease. The well-established risk factors for the development of AMD are: increasing age, a positive family history of AMD (including specific genotypes), and tobacco smoking.[30;32;33] Therefore, tobacco smoking is the only proven environmental/lifestyle risk factor for this disease.[34;35] Putative risk factors include: obesity,[36;37] hypertension,[38] light iris colour,[39] cumulative sunlight exposure,[40] and a diet low in anti-oxidant fruits and vegetables,[41] particularly those containing the hydroxy-carotenoids: lutein and zeaxanthin.[42] Although the pathogenesis of AMD remains incompletely understood, there is a growing consensus that one or more of the following processes contribute to this condition: inflammation; oxidative stress; cumulative blue light damage; RPE cell and BrM dysfunction; reduced foveolar choroidal circulation.

6. Macular pigment

Macular pigment (MP) is composed of the hydroxy-carotenoids lutein (L), zeaxanthin (Z), and *meso*-zeaxanthin (*meso*-Z). L and Z are of dietary origin and are not synthesized *de novo* in humans, whereas *meso*-Z is not found in a conventional western diet, but is understood to be primarily formed in the retina following conversion from L.[43;44] Interestingly, it has been shown that L is the dominant carotenoid in the diet,[45] whereas Z/*meso*-Z have been shown to be the dominant carotenoids at the central macula.[46;47] MP is found in highest concentration at the central macula, where it functions as a powerful antioxidant and acts as a filter of actinic short wavelength blue light, thus limiting (photo-)oxidative damage to retinal cells.[48] These properties of MP are believed to be the mechanism whereby it may protect against the development, and/or progression, of AMD.

Although MP is entirely of dietary origin, it is also subject to heritability, as reported in 2005 by Liew *et al.* in a classic twin study.[49] In that study of 76 monozygotic and 74 dizygotic female twin pairs, they estimated that heritability accounted for between 67% and 85% of an individual's MP level. However, to date a direct significant association between MP levels and the major risk genes for AMD has not been shown.[50]

MP can be measured *in vivo* by non-invasive psychophysical means, resulting in an MP optical density measurement.[51;52]

7. Lipoproteins

Circulating lipoproteins consist of a complex of triglycerides, phospholipids and cholesterol, and one or more specific proteins, referred to as apolipoproteins. The association of lipoproteins with high affinity receptors on cell surfaces regulates lipid metabolism and transport in the body.[53] Lipoproteins are classified into the following six groups: chylomicrons; chylomicron remnants; very low density lipoproteins (VLDL); intermediate density lipoproteins (IDL); low density lipoproteins (LDL); high density lipoproteins (HDL).[53]

Chylomicrons are synthesised by the intestine and deliver dietary triglycerides to muscle and adipose tissue, and dietary cholesterol to the liver. Lipoprotein lipase, located at capillary endothelial cell surfaces, hydrolyses the triglyceride core of the chylomicron, thus liberating fatty acids and glycerol, which are used as energy sources by various cells, or are taken up by adipocytes and stored as triglycerides. Chylomicron remnants, which are rich in cholesterol, result from chylomicron metabolism, and are rapidly cleared by the liver.[53]

Subsequently, the liver synthesises a second class of triglyceride-rich lipoprotein, referred to as VLDL, which, upon secretion, functions as a transporter of lipids and cholesterol. In the bloodstream, VLDL undergoes progressive removal of triglycerides from its core by lipoprotein lipase, in a similar way to chylomicrons. The VLDL particles thus become increasingly smaller, leading to the formation of IDL, and LDL. LDL are the final metabolic products of VLDL and are responsible for most of the cholesterol transport in serum.[53]

HDL are the smallest lipoproteins, arising from several sources including the intestine and liver. HDL are involved in a process known as 'reverse cholesterol transport', whereby HDL acquire cholesterol from cells and deliver it to the liver.[53] This is a particularly important mechanism in humans, as the quantities of cholesterol transported out of the gut and liver far exceed the quantities converted to steroid hormones, or those lost through the skin in sebum. Thus, unless the requirement for cell membrane repair or synthesis is high, excess cholesterol must be returned to the liver for excretion.[54]

8. Association of carotenoids with plasma lipoproteins

The majority of plasma carotenoids are transported on LDL, with 55% of total carotenoids associated with this lipoprotein, whereas HDL is associated with 33%, and VLDL is associated with 10-19%, of the total carotenoids.[55] However, in the case of the hydroxy-carotenoids, L and Z, some studies have reported that they are relatively equally distributed between LDL and HDL molecules, but other studies have reported that HDL is the preferential carrier of the MP carotenoids in plasma.[56;57]

MP is inversely related to percentage body fat.[58] Interestingly, Viroonudomphol *et al.* have demonstrated lower levels of HDL in overweight and obese subjects, consistent with the possibility that a relative lack of HDL may impair transport and/or retinal capture of the carotenoids.[59] Furthermore, Seddon and co-workers have demonstrated a significantly increased risk of AMD in association with obesity.[33] These findings have prompted the suggestion that an individual's lipoprotein, and apolipoprotein, profile may influence the transport and delivery of these carotenoids to the retina, with a consequential impact on MP.

A recent study, designed to investigate the respective relationships between lipoprotein profile, MP optical density and serum concentrations of L and Z, was conducted in 302 healthy adult subjects.[60] This study found that there was a statistically significant inverse association between serum triglyceride concentration and MP optical density, and an inverse association between serum triglyceride concentration and serum L concentration in subjects with a positive family history of AMD. There have been no previous reports on the association between serum triglyceride concentration and either MP optical density or serum concentrations of L and/or Z. Elevated serum triglyceride concentration is an element of an undesirable lipoprotein profile and represents risk for cardiovascular disease.[61;62] Since there is an inverse association between serum triglyceride concentration and serum HDL concentration,[62] one could expect an inverse association between serum triglyceride concentration and serum L, since HDL appears to be the most important lipoprotein involved in the transport of L in serum. This expected inverse association was observed in subjects with a positive family history of AMD. In this study sample there was a positive and significant association between serum HDL concentration (and serum cholesterol concentration) and serum L and Z concentrations. Of note, there was no significant association observed between MP optical density and either serum cholesterol concentration or serum HDL concentration. There was also no association between serum LDL concentration and MP optical density (or serum concentrations of its constituent carotenoids). These findings suggest that a desirable lipoprotein profile (higher serum HDL, lower serum LDL and lower serum triglyceride concentrations) is associated with greater serum L concentration. However, the impact of lipoprotein profile on the capture and/or stabilization of these carotenoids at the macula, where they comprise MP, is less clear from this data.

In this study, the lipoprotein particle-concentration of L and/or Z in serum was not directly measured, nor were lipoprotein subspecies measured, as performed by Goulinet et al.[57] In their study, they fractionated HDL and LDL subspecies on the basis of their hydrated density by gradient ultracentrifugation, and they found that serum L and Z (combined) were relatively equally distributed between HDL and LDL; but more importantly, they found that there was a progressive decrease in the concentration of these carotenoids with increasing density (and decreasing lipoprotein particle size) from light to dense LDL. They also found that the majority of macular carotenoid transport by LDL was accounted for by the most abundant subspecies, LDL3 (intermediate LDL) and LDL4 (dense LDL). This is highly relevant to the transport of L and Z in serum, as LDL3 and LDL4, despite being the most abundant subspecies of LDL in that study, had reduced particle-concentrations of these carotenoids compared to less dense LDL subspecies, making them more vulnerable to oxidation.[63] LDL is the primary component of total cholesterol,[62] and has previously been reported in various studies to transport between 22-44% of L and Z in serum.[55;57;64-66] Of note, it has been shown that there is no significant difference in the transport of L and Z by lipoproteins between subjects with and without AMD.[65]

The findings of Goulinet et al in relation to HDL were similar to that of LDL, in that there was a progressive and marked decrease in HDL particle concentration of L and Z, with maximal carotenoid concentration evident in the lightest, largest HDL subspecies (HDL2-1), and minimal concentration in the densest HDL. Certainly, the findings of Goulinet et al with respect to HDL, in concert with our findings, are consistent with the view that HDL plays an important role in the transport of L and Z in human serum, and are provocative given that AMD and cardiovascular disease share certain antecedants.[32;57;60;67-70] Furthermore, and again consistent with a shared pathogenesis between AMD and cardiovascular disease, the finding of an inverse association between serum triglyceride concentration and MP optical density (and between serum triglyceride concentration and serum L concentration) in subjects with a positive family history of AMD, is noteworthy.[60] Since AMD has been shown to be associated with low serum concentrations of L,[71] and given that risk factors for AMD are associated with a relative lack of MP,[31] our observations are yet another example of how AMD and cardiovascular disease share risk factors.[32;60-62;67-70]

In 2007, Connor *et al* reported on the role that HDL plays in the transport of L and Z in serum in a study involving WHAM chicks.[64] WHAM chickens have a recessive sex-linked mutation in the *ABCA1* transporter gene that results in very low circulating HDL concentration, with normal, or increased, concentrations of other plasma lipoproteins, particularly LDL. The analogous mutation in humans results in Tangier disease, which is characterized by a similar deficiency in circulating HDL concentration.[72] In their study, involving 24 WHAM chicks and 24 control chicks, Connor *et al* found that one-day old WHAM chicks had only 9% of the L concentration in plasma when compared with control chicks, and only 6% of the retinal concentration of controls (the corresponding concentrations of Z were 6% and 9%, respectively). Following a high-L diet for 28 days, there was a significant increase in the plasma and retinal concentrations of L in WHAM chicks and controls, but the increases were still greatly inferior in the WHAM chicks when compared with control chicks and, furthermore, still did not reach the concentrations observed in the one-day old control chicks. The observations of Connor *et al* suggest an important role for HDL in the transport of L and Z in serum and/or their incorporation into the retina, and are consistent with our findings.[60;64]

Interestingly, although all subjects in our study were healthy volunteers with no evidence of ocular pathology, it is notable that, on average, subjects with a positive family history of AMD had a higher serum concentration of L than subjects with a negative family history of AMD, yet MP optical density levels in both groups were comparable, as were serum concentrations of HDL.[60] As was shown in this study, and as has previously been documented,[73] serum concentrations of L and Z generally correlate positively with MP optical density. Therefore, it is plausible to suggest that in the subjects in this study with a positive family history of AMD, the delivery to, and/or uptake by, the retina of the macular carotenoids is defective when compared to subjects without such a family history.[60] Indeed, although MP optical density levels were comparable between subjects with and without a family history of AMD, subjects with a positive family history of this disease also had higher serum L concentrations. This is consistent with the observations of Nolan *et al*, where a relative lack of MP was seen in association with a positive family history of AMD in 828 healthy subjects, but where dietary and serum concentrations of L and Z were comparable for subjects with and without a family history of this condition, suggesting defective retinal capture of circulating L and/or Z in persons who are genetically predisposed to AMD.[31] Mechanisms governing the retinal capture and/or stabilization of L and/or Z may be subject to influence by HDL subspecies profile, by affecting receptor-mediated uptake of these carotenoids from serum. Indeed, apolipoprotein profile is probably a determinant of retinal uptake of the macular carotenoids from serum, reflected in our recently reported finding that individuals with at least one Apo ε4 allele exhibit significantly higher MP optical density than individuals without this protective allele, despite statistically comparable serum concentrations of L and Z.[74] Interestingly, the lack of an association between MP optical density and either serum cholesterol concentration or serum HDL concentration in our study would suggest that our observations are more likely due to impaired uptake and/or stabilization of circulating L and/or Z by the macula than due to any impact the HDL subspecies profile may have on the transport of the macular carotenoids in serum.

Another recent study has shown somewhat conflicting evidence regarding the association between circulating lipoprotein levels and MP levels in serum and in the macula.[75] These differences may be attributable to differences in the methods used to measure serum lipoproteins, although it should be noted that this study also found a positive association between serum L and serum HDL levels, underscoring the importance of HDL as a transporter of L in serum. However, it should be emphasised that a notable paucity of data still remains regarding the mechanism(s) whereby L and Z accumulate in the liver, are repackaged into lipoproteins, and transported via the circulatory system to specific target tissues such as the retina.

9. Apolipoproteins

Plasma lipoproteins include one or more protein constituents, known as apolipoproteins. Apolipoproteins have been classified into several subgroups, including apolipoprotein A (ApoA), apolipoprotein B (ApoB), apolipoprotein C (ApoC), and apolipoprotein E (ApoE). These subgroups are themselves further sub-classified, for example: ApoA-I, ApoA-II etc. Each lipoprotein class is associated with certain apolipoproteins, for example: chylomicrons and VLDL are associated with ApoB; chylomicrons, VLDL and HDL are associated with ApoE.[76] The primary role of apolipoproteins is the transport and redistribution of lipids amongst various tissues in the body. Specific apolipoproteins are recognised by cell surface receptors, and this facilitates the high affinity binding required for delivery to target tissues. Certain apolipoproteins also act as cofactors of enzymes involved in lipoprotein metabolic pathways, including those of lipoprotein lipase and lecithin-cholesterol acyl transferase (LCAT), which catalyse the formation of cholesterol esters. Another role of specific apolipoproteins is the maintenance of the structure of lipoproteins, by stabilizing their micellar structure, and by providing a hydrophilic surface in association with phospholipids.[53] The function of apolipoproteins has provoked interest in their possible role in a range of degenerative conditions. In particular, several investigators have suggested an association between ApoE and various diseases, including Alzheimer's disease, atherosclerosis and AMD.[77-80]

Abalain et al. investigated the association between AMD and serum levels of lipoproteins and lipoparticles.[78] They found that there was no difference in serum ApoA-I and ApoB levels between AMD patients and controls. However, they found that serum ApoE levels were higher, and that serum ApoC-III levels were lower, in AMD patients compared with controls. The higher level of serum ApoE in AMD patients is consistent with the findings of Boerwinkle and Utermann, who found that the Apo ε4 allele is associated with lower serum ApoE levels, and that the Apo ε2 allele is associated with higher serum levels of ApoE.[79] ApoC-III interferes with lipoprotein metabolism and, when associated with ApoB as a lipoparticle, it has been shown to be involved in atherogenesis.[80] Abalain et al. found no difference in the levels of this particular lipoparticle between AMD patients and controls.[78] The evidence to date suggests that, of the apolipoproteins, ApoE has the strongest association with AMD.

10. Apolipoprotein E

ApoE is a structural component of plasma chylomicrons, VLDL, and a subclass of HDL. It is a 299 amino-acid protein, and is synthesised in a large number of tissues including the spleen, kidneys, lungs, adrenal glands, liver, brain and retinal Müller cells.[81] ApoE is polymorphic, with three common isoforms: E2, E3 and E4, which are coded for by three separate alleles: Apo ε2, Apo ε3 and Apo ε4. These alleles are differentiated on the basis of cysteine-arginine residue interchanges at sites 112 and 158 in the amino acid sequence.[82] As a result of this polymorphism, six common phenotypes exist: three homozygous phenotypes (ε3ε3, ε2ε2, ε4ε4) and three heterozygous phenotypes (ε2ε3, ε2ε4, ε3ε4). ApoE is crucial to many processes, including: cholesterol transport and metabolism; receptor-mediated uptake of specific lipoproteins; heparin binding; formation of cholesteryl-ester-rich particles; lipolytic processing of type IIIβ-VLDL; inhibition of mitogenic stimulation of lymphocytes; transport of lipids within the brain.[53]

ApoE is an important regulator of cholesterol metabolism because of its affinity for ApoE-specific receptors in the liver, and its affinity for LDL receptors in the liver and other peripheral tissues requiring cholesterol.[53] ApoE-specific receptors are present on the membranes of hepatic parenchymal cells, and have a high binding affinity for chylomicron remnants, IDL and a sub-class of HDL. ApoE also regulates the activity of several lipid-metabolising enzymes, including lipoprotein lipase, and LCAT.

ApoE is found in greatest concentrations in the liver. However, it is also the predominant apolipoprotein in the brain, and is responsible for lipid transport and cholesterol regulation within the central nervous system (CNS). ApoE is a major component of plasma and cerebrospinal fluid, and plays a fundamental role after CNS injury, where it appears to regulate the transport of cholesterol and phospholipids during the early and intermediate phases of the reinnervation process.[83;84]

ApoE polymorphisms result in differences in the metabolism of ApoE-containing lipoprotein particles.[85] For example, it is possible that certain ApoE polymorphisms affect their ability to interact with lipoprotein lipase in the conversion of VLDL to LDL.[86] Indeed, ApoE polymorphism influences plasma lipid levels both in sedentary states and in their response to exercise, and it is therefore believed to be related to risk for coronary artery disease. In general, carriers of the Apo ε4 allele have higher levels of total cholesterol and LDL-cholesterol than those with the Apo ε3 allele. ApoE polymorphism also appears to play a role in the responsiveness of blood lipids to dietary and lipid-lowering drug interventions. Thus, the ApoE gene-environmental interactions contribute to population variance in blood lipid-lipoprotein levels.[87]

ApoE receptors also play an important role in lipoprotein metabolism. The primary physiological role of ApoE is to facilitate the binding of lipoproteins to LDL receptors, thereby regulating the uptake of cholesterol required by the cell. For instance, large amounts of lipids are released from degenerating cell membranes after nerve cell loss, thus stimulating astrocytes to synthesise ApoE, which binds these excess lipids and distributes them appropriately for reuse in cell membrane biosynthesis.[88] This observation prompted Klaver *et al.* to speculate that a high degree of ApoE biosynthesis is required to support the high rate of photoreceptor renewal at the macula.[88] Indeed, it has been demonstrated that mice which were fed a high-fat diet, or which were deficient in ApoE, exhibit an increase in the thickness of BrM, which is seen in association with ageing and with AMD.[89]

Ishida *et al.* identified the presence of ApoE and lipids at the inner aspect of the RPE, and proposed that both compounds may be secreted by the RPE.[90] The role of ApoE in reverse cholesterol transport prompted the authors to suggest that this apolipoprotein may also facilitate the efflux of lipids from the RPE into the adjacent BrM, and they proposed a possible pathway for RPE cell-secreted lipids to cross BrM, where partially digested or undigested photoreceptor outer segments are secreted across the basal surface in association with ApoE. Subsequent binding with HDL at BrM may then facilitate desorption of the lipid particles into the circulation.[90]

In the retina ApoE is synthesised in Müller cells and in the RPE, and the presence of ApoE has been demonstrated in drusen.[81;91;92] It has been suggested, therefore, that age and/or disease-related disruption of normal ApoE function may result in the accumulation of lipoproteins at the interface between the RPE and BrM, consistent with observations that lipid deposits in drusen are largely composed of cholesteryl esters and unsaturated fatty acids.

These findings are consistent with the view that ApoE plays an important physiological role in the maintenance of macular health, and that an impaired ApoE system may affect the functional integrity of BrM. Furthermore, there is a biologically plausible rationale whereby the ApoE profile might influence the transport, capture, and stabilization of key compounds, such as L and Z, at the macula.

11. Lipoproteins, apolipoproteins and the retina

As noted previously, the ageing retina features changes in the RPE and BrM, which include changes in the lipoprotein and apolipoprotein composition of both structures. These changes may progress to the disease state of AMD. In recent times, evidence accrued from light microscopy, ultrastructural studies, lipid histochemistry, isolated lipoprotein assays, and gene expression analysis had led to the identification of many of the constituents that deposit in the RPE and BrM with age and AMD.[93] One of the universal changes that occurs with age is the development of BlamDs between the RPE and BrM.[11;12] This process may progress to the development of a 'lipid wall', mainly composed of neutral lipid deposits, decreasing the permeability of BrM and hindering metabolic activity between the RPE and BrM, preceding pathological changes associated with AMD.[10;93;94] When these deposits accumulate within the inner collagenous layer of BrM, they are referred to as basal linear deposits (BlinDs) and are a histopathological hallmark of AMD, which, when sufficiently large, can be recognised clinically as drusen.[13-15;95]

Much of the debris that accumulates in BrM in the form of BlinDs is composed of lipoproteins and lipoprotein particles.[14] It has been found that almost 60% of the total cholesterol within these lipoproteins is esterified cholesterol.[96] Furthermore, the esterified cholesterol within BrM was enriched between 16 and 40-fold compared to plasma. If these extracellular lipid deposits had been derived from plasma, more than 90% of the phospholipid would be phosphatidylcholine, whereas in actual fact, these lipoproteins are comprised of less than 50% phosphatidylcholine.[96] Indeed, the composition of drusen, which are essentially large BlinDs, has been shown to include esterified and unesterified cholesterol, and multiple apolipoproteins, including apolipoproteins B, A-I, C-I, C-II, and E, appearing with frequencies ranging from 100% (ApoE) to approximately 60% (A-I).[88;91;97;98] Interestingly, ApoC-III, although abundant in plasma, is present in fewer drusen (16.6%) than ApoC-I (93.1%), which is not present in plasma in large quantities, indicating either a specific retention of plasma-derived apolipoproteins within drusen, or an intraocular source for these apolipoproteins.[93] It is now understood that the majority of lipoproteins in BrM have undergone intracellular processing within the RPE prior to secretion as neutral lipids, mainly esterified cholesterol.[99;100] The RPE origin has been definitively shown by two groups using metabolic labelling and immunoprecipitation in rat-derived and human-derived RPE cell lines that were shown to secrete full-length ApoB.[101;102] This evidence is further strengthened by the finding of microsomal triglyceride transfer protein within native human RPE, indicating that the RPE is capable of secreting lipoprotein particles.[102] The pattern of lipid deposition in BrM with age, in which debris appears firstly in the elastic layer and then fills in towards the RPE, is also consistent with this lipid being primarily of RPE origin.[103]

The hydrophobic nature of the age-related thickening of BrM has been implicated in the aetiopathogenesis of AMD. In the case of Apo E, it is noteworthy that ApoE4 presents a positive charge relative to both ApoE2 and ApoE3. ApoE4 possesses arginine at residue 112 of the amino acid sequence, whereas ApoE3 possesses cysteine at this position, and in the case of ApoE2, the most frequent variant has cysteine instead of the normally occurring arginine at residue 158. Thus, ApoE3 presents a neutral charge, and ApoE2 a negative charge, relative to ApoE4.[53] Souied *et al.* suggested that this difference in charges between the ApoE isoforms may also contribute to differences in the clearance of debris through BrM.[104]

It appears that Müller cells are the most prominent biosynthetic sources of ApoE in the neural retina, and RPE cells are the most prominent sources in the RPE/choroid.[91] However, it remains unclear whether the concentration of ApoE in the cytoplasm of some RPE cells, especially those in close proximity to drusen, is the result of biosynthesis or selective accumulation. It has been shown that, in both the central and peripheral nervous systems, ApoE expression by astrocytes is up-regulated in response to neuronal injury and neuro-degenerative disease.[84;105;106] Indeed, there is evidence for ApoE up-regulation by Müller cells in degenerating human retina, where increased ApoE immuno-reactivity is found in the sub-retinal space of detached retinas[107] and in the Müller cells of retinas affected by glaucoma or AMD.[108] Furthermore, the relatively high levels of ApoE mRNA detected in the retina, especially in the eyes of older donors and in an individual with documented AMD, support the view that up-regulation by retinal glia may be responsible for the observed increase in ApoE expression.[91]

12. Apo ε4 allele status and AMD

ApoE gene status is believed to be a determinant of AMD risk.[88;104;109-111] The *ApoE* gene has three separate alleles: Apo ε2, Apo ε3 and Apo ε4, resulting in six common phenotypes: three homozygous (ε3ε3, ε2ε2, ε4ε4) and three heterozygous (ε2ε3, ε2ε4, ε3ε4) phenotypes. The ε4 allele has been found to be associated with a reduced risk of AMD, whereas the ε2 allele has been associated with an increased risk of developing this disease.[88;104;109-113]

Due to the lack of cysteine residues at positions 112 and 158, preventing the formation of disulphide bridges with ApoA-II or other peptide components, the Apo ε4 allele has an inability to form dimers. It has been suggested that this inability of the Apo ε4 allele to form dimers, when compared with the Apo ε2 and Apo ε3 alleles, favours easier transport of lipids through BrM because of the smaller sized lipid particles, thus protecting against a loss of permeability of BrM.[104]

In the same way, it is possible that the neurosensory retina and the RPE respond to conditions of high oxidative injury by up-regulation of ApoE synthesis and/or accumulation, with implications for selective capture and stabilisation of L and Z in the retina.[91] It has been demonstrated that there is selective binding of certain receptors within the CNS to HDL particles enriched with ApoE, and that there is a lack of binding of these receptors to HDL particles deficient in ApoE.[114] Should this selectivity of the uptake mechanism be dependent on the ApoE polymorphism of the transporting lipoproteins, and given that the Apo ε4 allele is putatively protective for AMD, it is tempting to hypothesise that retinal capture of L and Z may be related to apolipoprotein profile. In other words, the apolipoprotein composition as well as the lipoprotein profile, may play an important role in the transport and delivery of L and Z, and their subsequent accumulation and stabilisation within the retina.[115] Therefore, it is possible that the putative protective effect of the Apo ε4 allele against AMD is attributable, at least in part, to the role its phenotypic expression (ApoE4) plays in the transport and delivery of the macular carotenoids to the retina, and to their stabilisation within the retina. Furthermore, recent research has shown an association between possession of at least one Apo ε4 allele and higher levels of MP across the macula, which is consistent with the view that apolipoprotein profile influences the transport and/or retinal capture of the macular carotenoids.[74]

13. Conclusion

In conclusion, the role that lipoproteins and apolipoproteins play in the ageing eye and in the aetiopathogenesis of AMD is complex and, as yet, incompletely understood. Lipoproteins and apolipoproteins play an important role in the delivery of potentially protective nutrients from the digestive tract to the eye. The local ocular metabolic activity, centred on the RPE and BrM, involves an exchange of nutrients from the choroidal circulatory system via BrM to the RPE and retina, with a reverse process whereby waste products are removed from the retina by the RPE through BrM in association with locally produced lipoproteins and apolipoproteins (particularly ApoB and ApoE). Unfortunately, over time it appears that these lipoproteins and apolipoproteins can accumulate between the RPE and BrM, and within BrM, leading to degradation in the metabolic efficiency between these two structures and the choroidal circulation. This deposition has been described as a 'lipid wall' and precedes the development of AMD.[93;94] Methods to detect and arrest or delay this process before it becomes clinically apparent and visually consequential to the patient have yet to be developed. Recent advances in our understanding of the lipoprotein and apolipoprotein molecular biology of the ageing and AMD-affected eye will help to direct future treatment strategies.[100]

Author details

Edward Loane

Department of Ophthalmology, Mater Misericordiae University Hospital, Dublin, Ireland

14. References

[1] Snodderly DM, Brown PK, Delori FC, Auran JD. The Macular Pigment .1. Absorbance Spectra, Localization, and Discrimination from Other Yellow Pigments in Primate Retinas. Investigative Ophthalmology & Visual Science 1984;25(6):660-73.

[2] Werner JS, Donnelly SK, Kliegl R. Aging and human macular pigment density. Appended with translations from the work of Max Schultze and Ewald Hering. Vision Research 1987;27:275-68.

[3] Bok D. The retinal pigment epithelium: a versatile partner in vision. J Cell Sci Suppl 1993;17:189-95.:189-95.

[4] Boulton M, yhaw-Barker P. The role of the retinal pigment epithelium: topographical variation and ageing changes. Eye (Lond) 2001 Jun;15(Pt 3):384-9.

[5] Del Priore LV, Kuo YH, Tezel TH. Age-related changes in human RPE cell density and apoptosis proportion in situ. Invest Ophthalmol Vis Sci 2002 Oct;43(10):3312-8.

[6] Dunaief JL, Dentchev T, Ying GS, Milam AH. The role of apoptosis in age-related macular degeneration. Arch Ophthalmol 2002 Nov;120(11):1435-42.

[7] American Academy of Ophthalmology. Basic and Clinical Science Course, Section 2: Fundamentals and Principles of Ophthalmology. 2011.

[8] Snell RS, Lemp MA. Clinical Anatomy of the Eye. Second ed. Wiley-Blackwell; 1998.

[9] Marshall J. The ageing retina: physiology or pathology. Eye (Lond) 1987;1(Pt 2):282-95.

[10] Zarbin MA. Current concepts in the pathogenesis of age-related macular degeneration. Arch Ophthalmol 2004;122(4):598-614.

[11] Pauleikhoff D, Harper CA, Marshall J, Bird AC. Aging changes in Bruch's membrane. A histochemical and morphologic study. Ophthalmology 1990 Feb;97(2):171-8.

[12] van der Schaft TL, de Bruijn WC, Mooy CM, de Jong PT. Basal laminar deposit in the aging peripheral human retina. Graefes Arch Clin Exp Ophthalmol 1993 Aug;231(8):470-5.

[13] Curcio CA, Millican CL. Basal linear deposit and large drusen are specific for early age-related maculopathy. Arch Ophthalmol 1999 Mar;117(3):329-39.

[14] Curcio CA, Presley JB, Millican CL, Medeiros NE. Basal deposits and drusen in eyes with age-related maculopathy: evidence for solid lipid particles. Exp Eye Res 2005 Jun;80(6):761-75.

[15] Lommatzsch A, Hermans P, Muller KD, Bornfeld N, Bird AC, Pauleikhoff D. Are low inflammatory reactions involved in exudative age-related macular degeneration? Morphological and immunhistochemical analysis of AMD associated with basal deposits. Graefes Arch Clin Exp Ophthalmol 2008 Jun;246(6):803-10.

[16] Bressler NM. Age-related macular degeneration is the leading cause of blindness. JAMA 2004 Apr 21;291(15):1900-1.

[17] Congdon NG, Friedman DS, Lietman T. Important causes of visual impairment in the world today. JAMA 2003 Oct 15;290(15):2057-60.

[18] Klein R, Wang Q, Klein BEK, Moss SE, Meuer SM. The Relationship of Age-Related Maculopathy, Cataract, and Glaucoma to Visual-Acuity. Investigative Ophthalmology & Visual Science 1995;36(1):182-91.

[19] Owen CG, Jarrar Z, Wormald R, Cook DG, Fletcher AE, Rudnicka AR. The estimated prevalence and incidence of late stage age related macular degeneration in the UK. Br J Ophthalmol 2012 Feb 13.

[20] Friedman DS, O'Colmain BJ, Munoz B, Tomany SC, McCarty C, de Jong PT, et al. Prevalence of age-related macular degeneration in the United States. Arch Ophthalmol 2004 Apr;122(4):564-72.

[21] van Leeuwen R, Klaver CC, Vingerling JR, Hofman A, de Jong PT. Epidemiology of age-related maculopathy: a review. Eur J Epidemiol 2003;18(9):845-54.

[22] Augustin A, Sahel JA, Bandello F, Dardennes R, Maurel F, Negrini C, et al. Anxiety and depression prevalence rates in age-related macular degeneration. Invest Ophthalmol Vis Sci 2007 Apr;48(4):1498-503.

[23] Gupta OP, Brown GC, Brown MM. Age-related macular degeneration: the costs to society and the patient. Curr Opin Ophthalmol 2007 May;18(3):201-5.

[24] Owen CG, Fletcher AE, Donoghue M, Rudnicka AR. How big is the burden of visual loss caused by age related macular degeneration in the United Kingdom? Br J Ophthalmol 2003 Mar 1;87(3):312-7.

[25] Bandello F, Lafuma A, Berdeaux G. Public health impact of neovascular age-related macular degeneration treatments extrapolated from visual acuity. Invest Ophthalmol Vis Sci 2007 Jan;48(1):96-103.

[26] Cruess AF, Zlateva G, Xu X, Soubrane G, Pauleikhoff D, Lotery A, et al. Economic burden of bilateral neovascular age-related macular degeneration: multi-country observational study. Pharmacoeconomics 2008;26(1):57-73.

[27] Bird AC, Bressler NM, Bressler SB, Chisholm IH, Coscas G, Davis DM, et al. An international classification and grading system for age-related maculopathy and age-

related macular degeneration. The International ARM Epidemiological Study Group. Survey of Ophthalmology 1995;39(5):367-74.

[28] Gass JD. Pathogenesis of disciform detachment of the neuroepithelium. Am J Ophthalmol 1967 Mar;63(3):Suppl-139.

[29] Sarks SH. Council Lecture. Drusen and their relationship to senile macular degeneration. Aust J Ophthalmol 1980 May;8(2):117-30.

[30] Tomany SC, Wang HJ, van Leeuwen R, Klein R, Mitchell P, Vingerling JR, et al. Risk factors for incident age-related macular degeneration - Pooled findings from 3 continents. Ophthalmology 2004;111(7):1280-7.

[31] Nolan JM, Stack J, O'Donovan O, Loane E, Beatty S. Risk factors for age-related maculopathy are associated with a relative lack of macular pigment. Exp Eye Res 2007 Jan;84(1):61-74.

[32] Delcourt C, Michel F, Colvez A, Lacroux A, Delage M, Vernet MH, et al. Associations of cardiovascular disease and its risk factors with age-related macular degeneration: the POLA study. Ophthalmic Epidemiology 2001 Sep;8(4):237-49.

[33] Seddon JM, Cote J, Davis N, Rosner B. Progression of age-related macular degeneration: associated with body mass index, waist circumference, and waist-hip ratio. Arch Ophthalmol 2003;121:785-92.

[34] Tomany SC, Cruickshanks KJ, Klein R, Klein BEK, Knudtson MD. Sunlight and the 10-year incidence of age-related maculopathy - The Beaver Dam eye study. Arch Ophthalmol 2004;122(5):750-7.

[35] SanGiovanni JP, Chew EY, Clemons TE, Ferris FL, III, Gensler G, Lindblad AS, et al. The relationship of dietary carotenoid and vitamin A, E, and C intake with age-related macular degeneration in a case-control study: AREDS Report No. 22. Arch Ophthalmol 2007 Sep;125(9):1225-32.

[36] Hammond BR, Johnson MA. The Age-related Eye Disease Study (AREDS). Nutrition Reviews 2002;60(9):283-8.

[37] Klein R, Klein BEK, Franke T. The Relationship of Cardiovascular-Disease and Its Risk-Factors to Age-Related Maculopathy - the Beaver Dam Eye Study. Ophthalmology 1993;100(3):406-14.

[38] Hyman L, Schachat AP, He QM, Leske MC. Hypertension, cardiovascular disease, and age-related macular degeneration. Arch Ophthalmol 2000;118(3):351-8.

[39] Tomany SC, Klein R, Klein BEK. The relationship between iris color, hair color, and skin sun sensitivity and the 10-year incidence of age-related maculopathy - The beaver dam eye study. Ophthalmology 2003;110(8):1526-33.

[40] Klein R, Tomany SC, Cruickshanks KJ, Klein BEK. Sunlight and the 10-year incidence of age-related maculopathy. The Beaver Dam Eye Study. Arch Ophthalmol 2004 May;122(5):750-7.

[41] Delcourt C, Carriere I, Delage M, Barberger-Gateau P, Schalch W. Plasma lutein and zeaxanthin and other carotenoids as modifiable risk factors for age-related maculopathy and cataract: the POLA Study. Invest Ophthalmol Vis Sci 2006 Jun;47(6):2329-35.

[42] Sommerburg O, Keunen JEE, Bird AC, van Kuijk FJGM. Fruits and vegetables that are sources for lutein and zeaxanthin: the macular pigment in human eyes. Br J Ophthalmol 1998;82(8):907-10.

[43] Bone RA, Landrum JT, Hime GW, Cains A, Zamor J. Stereochemistry of the Human Macular Carotenoids. Investigative Ophthalmology & Visual Science 1993;34(6):2033-40.

[44] Johnson EJ, Neuringer M, Russell RM, Schalch W, Snodderly DM. Nutritional manipulation of primate retinas, III: effects of lutein or zeaxanthin supplementation on adipose tissue and retina of xanthophyll-free monkeys. Investigative Ophthalmology Visual Science 2005 Feb 1;46(2):692-702.

[45] Bialostosky K, Wright JD, Kennedy-Stephenson J, McDowell M, Johnson CL. Dietary intake of macronutrients, micronutrients, and other dietary constituents: United States 1988-94. Vital Health Stat 11 2002 Jul;(245):1-158.

[46] Bone RA, Landrum JT, Fernandez L, Tarsis SL. Analysis of the macular pigment by HPLC - Retinal distribution and age study. Investigative Ophthalmology & Visual Science 1988;29(6):843-9.

[47] Snodderly DM, Handelman GJ, Adler AJ. Distribution of individual macular pigment carotenoids in central retina of macaque and squirrel monkeys. Investigative Ophthalmology & Visual Science 1991;32(2):268-79.

[48] Snodderly DM. Evidence for Protection Against Age-Related Macular Degeneration by Carotenoids and Antioxidant Vitamins. Am J Clin Nutr 1995;62(6):S1448-S1461.

[49] Liew SHM, Gilbert C, Spector TD, Mellerio J, Marshall J, van Kuijk FJGM, et al. Heritability of Macular Pigment: a Twin Study. Investigative Ophthalmology & Visual Science 2005;46(12):4430-6.

[50] Loane E, Nolan JM, McKay GJ, Beatty S. The association between macular pigment optical density and CFH, ARMS2, C2/BF, and C3 genotype. Exp Eye Res 2011 Nov;93(5):592-8.

[51] Loane E, Stack J, Beatty S, Nolan JM. Measurement of macular pigment optical density using two different heterochromatic flicker photometers. Curr Eye Res 2007 Jun;32(6):555-64.

[52] Wooten BR, Hammond BR, Land RI, Snodderly DM. A practical method for measuring macular pigment optical density. Investigative Ophthalmology & Visual Science 1999;40(11):2481-9.

[53] Mahley RW, Innerarity TL, Rall SC, Jr., Weisgraber KH. Plasma lipoproteins: apolipoprotein structure and function. J Lipid Res 1984 Dec 1;25(12):1277-94.

[54] Durrington PN. Lipoproteins and their metabolism. In: Durrington PN, editor. Hyperlipidaemia: Diagnosis and Management. Butterworth-Heinemann Ltd; 1989.

[55] Clevidence BA, Bieri JG. Association of carotenoids with human plasma lipoproteins. Methods in Enzymology 1993;214:33-46.

[56] Erdman JW, Jr., Bierer TL, Gugger ET. Absorption and transport of carotenoids. Ann N Y Acad Sci 1993 Dec 31;691:76-85.

[57] Goulinet S, Chapman MJ. Plasma LDL and HDL subspecies are heterogenous in particle content of tocopherols oxygenated and hydrocarbon carotenoids - Relevance to

oxidative resistance and atherogenesis. Arteriosclerosis Thrombosis and Vascular Biology 1997;17(4):786-96.

[58] Nolan J, O'Donovan O, Kavanagh H, Stack J, Harrison M, Muldoon A, et al. Macular pigment and percentage of body fat. Investigative Ophthalmology Visual Science 2004 Nov 1;45(11):3940-50.

[59] Viroonudomphol D, Pongpaew P, Tungtrongchitr R, Changbumrung S, Tungtrongchitr A, Phonrat B, et al. The relationships between anthropometric measurements, serum vitamin A and E concentrations and lipid profiles in overweight and obese subjects. Asia Pacific Journal of Clinical Nutrition 2003;12(1):73-9.

[60] Loane E, Nolan JM, Beatty S. The respective relationships between lipoprotein profile, macular pigment optical density, and serum concentrations of lutein and zeaxanthin. Invest Ophthalmol Vis Sci 2010 Nov;51(11):5897-905.

[61] Hokanson JE, Austin MA. Plasma triglyceride level is a risk factor for cardiovascular disease independent of high-density lipoprotein cholesterol level: a meta-analysis of population-based prospective studies. J Cardiovasc Risk 1996 Apr;3(2):213-9.

[62] Morrison A, Hokanson JE. The independent relationship between triglycerides and coronary heart disease. Vasc Health Risk Manag 2009;5(1):89-95.

[63] Dejager S, Bruckert E, Chapman MJ. Dense low density lipoprotein subspecies with diminished oxidative resistance predominate in combined hyperlipidemia. J Lipid Res 1993 Feb 1;34(2):295-308.

[64] Connor WE, Duell PB, Kean R, Wang Y. The Prime Role of HDL to Transport Lutein into the Retina: Evidence from HDL-Deficient WHAM Chicks Having a Mutant ABCA1 Transporter. Investigative Ophthalmology & Visual Science 2007 Sep 1;48(9):4226-31.

[65] Wang W, Connor SL, Johnson EJ, Klein ML, Hughes S, Connor WE. Effect of dietary lutein and zeaxanthin on plasma carotenoids and their transport in lipoproteins in age-related macular degeneration. Am J Clin Nutr 2007 Mar;85(3):762-9.

[66] Cardinault N, Abalain JH, Sairafi B, Coudray C, Grolier P, Rambeau M, et al. Lycopene but not lutein nor zeaxanthin decreases in serum and lipoproteins in age-related macular degeneration patients. Clin Chim Acta 2005 Jul 1;357(1):34-42.

[67] Klein R, Klein BEK, Tomany SC, Cruickshanks KJ. The association of cardiovascular disease with the long-term incidence of age-related maculopathy - The Beaver Dam Eye Study. Ophthalmology 2003;110(4):636-43.

[68] Rizzo M, Berneis K. Low-density lipoprotein size and cardiovascular risk assessment. QJM 2006 Jan;99(1):1-14.

[69] Snow KK, Seddon JM. Do age-related macular degeneration and cardiovascular disease share common antecedents? Ophthalmic Epidemiology 1999;6:125-43.

[70] Klein R, Deng Y, Klein BE, Hyman L, Seddon J, Frank RN, et al. Cardiovascular disease, its risk factors and treatment, and age-related macular degeneration: Women's Health Initiative Sight Exam ancillary study. Am J Ophthalmol 2007 Mar;143(3):473-83.

[71] Gale CR, Hall NF, Phillips DIW, Martyn CN. Lutein and zeaxanthin status and risk of age-related macular degeneration. Investigative Ophthalmology & Visual Science 2003;44(6):2461-5.

[72] Brooks-Wilson A, Marcil M, Clee SM, Zhang LH, Roomp K, van DM, et al. Mutations in ABC1 in Tangier disease and familial high-density lipoprotein deficiency. Nat Genet 1999 Aug;22(4):336-45.

[73] Beatty S, Nolan J, Kavanagh H, O'Donovan O. Macular pigment optical density and its relationship with serum and dietary levels of lutein and zeaxanthin. Archives of Biochemistry and Biophysics 2004;430(1):70-6.

[74] Loane E, McKay GJ, Nolan JM, Beatty S. Apolipoprotein E genotype is associated with macular pigment optical density. Invest Ophthalmol Vis Sci 2010 May;51(5):2636-43.

[75] Renzi LM, Hammond BR, Jr., Dengler M, Roberts R. The relation between serum lipids and lutein and zeaxanthin in the serum and retina: results from cross-sectional, case-control and case study designs. Lipids Health Dis 2012 Feb 29;11:33.:33.

[76] Mahley RW, Innerarity TL. Lipoprotein receptors and cholesterol homeostasis. Biochim Biophys Acta 1983 May 24;737(2):197-222.

[77] Corder EH, Saunders AM, Strittmatter WJ, Schmechel DE, Gaskell PC, Small GW, et al. Gene dose of apolipoprotein E type 4 allele and the risk of Alzheimer's disease in late onset families. Science 1993 Aug 13;261(5123):921-3.

[78] Abalain JH, Carre JL, Leglise D, Robinet A, Legall F, Meskar A, et al. Is age-related macular degeneration associated with serum lipoprotein and lipoparticle levels? Clinica Chimica Acta 2002 Dec;326(1-2):97-104.

[79] Boerwinkle E, Utermann G. Simultaneous effects of the apolipoprotein E polymorphism on apolipoprotein E, apolipoprotein B, and cholesterol metabolism. Am J Hum Genet 1988 Jan;42(1):104-12.

[80] Parra HJ, Arveiler D, Evans AE, Cambou JP, Amouyel P, Bingham A, et al. A case-control study of lipoprotein particles in two populations at contrasting risk for coronary heart disease. The ECTIM Study. Arterioscler Thromb 1992 Jun;12(6):701-7.

[81] Shanmugaratnam J, Berg E, Kimerer L, Johnson RJ, Amaratunga A, Schreiber BM, et al. Retinal muller glia secrete apolipoproteins E and J which are efficiently assembled into lipoprotein particles. Brain Res Mol Brain Res 1997 Oct 15;50(1-2):113-20.

[82] Utermann G, Langenbeck U, Beisiegel U, Weber W. Genetics of the apolipoprotein E system in man. Am J Hum Genet 1980 May;32(3):339-47.

[83] Boyles JK, Zoellner CD, Anderson LJ, Kosik LM, Pitas RE, Weisgraber KH, et al. A role for apolipoprotein E, apolipoprotein A-I, and low density lipoprotein receptors in cholesterol transport during regeneration and remyelination of the rat sciatic nerve. J Clin Invest 1989 Mar;83(3):1015-31.

[84] Poirier J. Apolipoprotein E in animal models of CNS injury and in Alzheimer's disease. Trends Neurosci 1994 Dec;17(12):525-30.

[85] Gregg RE, Zech LA, Schaefer EJ, Stark D, Wilson D, Brewer HB, Jr. Abnormal in vivo metabolism of apolipoprotein E4 in humans. J Clin Invest 1986 Sep;78(3):815-21.

[86] Ehnholm C, Mahley RW, Chappell DA, Weisgraber KH, Ludwig E, Witztum JL. Role of apolipoprotein E in the lipolytic conversion of β-very low density lipoproteins to low density lipoproteins in type III hyperlipoproteinemia. PNAS 1984 Sep 1;81(17):5566-70.

[87] Leon AS, Togashi K, Rankinen T, Despres JP, Rao DC, Skinner JS, et al. Association of apolipoprotein E polymorphism with blood lipids and maximal oxygen uptake in the sedentary state and after exercise training in the HERITAGE family study. Metabolism 2004 Jan;53(1):108-16.

[88] Klaver CC, Kliffen M, van Duijn CM, Hofman A, Cruts M, Grobbee DE, et al. Genetic association of apolipoprotein E with age-related macular degeneration. Am J Hum Genet 1998 Jul;63(1):200-6.

[89] Ong JM, Zorapapel NC, Rich KA, Wagstaff RE, Lambert RW, Rosenberg SE, et al. Effects of cholesterol and apolipoprotein E on retinal abnormalities in apoE-deficient mice. Investigative Ophthalmology & Visual Science 2001 Jul 1;42(8):1891-900.

[90] Ishida BY, Bailey KR, Duncan KG, Chalkley RJ, Burlingame AL, Kane JP, et al. Regulated expression of apolipoprotein E by human retinal pigment epithelial cells. J Lipid Res 2004 Feb 1;45(2):263-71.

[91] Anderson DH, Ozaki S, Nealon M, Neitz J, Mullins RF, Hageman GS, et al. Local cellular sources of apolipoprotein E in the human retina and retinal pigmented epithelium: implications for the process of drusen formation. American Journal of Ophthalmology 2001 Jun;131(6):767-81.

[92] Dentchev T, Milam AH, Lee VM, Trojanowski JQ, Dunaief JL. Amyloid-beta is found in drusen from some age-related macular degeneration retinas, but not in drusen from normal retinas. Mol Vis 2003 May 14;9:184-90.

[93] Curcio CA, Johnson M, Huang JD, Rudolf M. Apolipoprotein B-containing lipoproteins in retinal aging and age-related macular degeneration. J Lipid Res 2010 Mar;51(3):451-67.

[94] Ruberti JW, Curcio CA, Millican CL, Menco BP, Huang JD, Johnson M. Quick-freeze/deep-etch visualization of age-related lipid accumulation in Bruch's membrane. Invest Ophthalmol Vis Sci 2003 Apr;44(4):1753-9.

[95] Sarks S, Cherepanoff S, Killingsworth M, Sarks J. Relationship of Basal laminar deposit and membranous debris to the clinical presentation of early age-related macular degeneration. Invest Ophthalmol Vis Sci 2007 Mar;48(3):968-77.

[96] Curcio CA, Millican CL, Bailey T, Kruth HS. Accumulation of cholesterol with age in human Bruch's membrane. Invest Ophthalmol Vis Sci 2001 Jan;42(1):265-74.

[97] Malek G, Li CM, Guidry C, Medeiros NE, Curcio CA. Apolipoprotein B in Cholesterol-Containing Drusen and Basal Deposits of Human Eyes with Age-Related Maculopathy. Am J Pathol 2003 Feb 1;162(2):413-25.

[98] Li CM, Clark ME, Chimento MF, Curcio CA. Apolipoprotein localization in isolated drusen and retinal apolipoprotein gene expression. Invest Ophthalmol Vis Sci 2006 Jul;47(7):3119-28.

[99] Ebrahimi KB, Handa JT. Lipids, lipoproteins, and age-related macular degeneration. J Lipids 2011;2011:802059. Epub;%2011 Jul 28.:802059.

[100] Curcio CA, Johnson M, Rudolf M, Huang JD. The oil spill in ageing Bruch membrane. Br J Ophthalmol 2011 Dec;95(12):1638-45.

[101] Wu T, Fujihara M, Tian J, Jovanovic M, Grayson C, Cano M, et al. Apolipoprotein B100 secretion by cultured ARPE-19 cells is modulated by alteration of cholesterol levels. J Neurochem 2010 Sep;114(6):1734-44.

[102] Li CM, Presley JB, Zhang X, Dashti N, Chung BH, Medeiros NE, et al. Retina expresses microsomal triglyceride transfer protein: implications for age-related maculopathy. J Lipid Res 2005 Apr;46(4):628-40.

[103] Huang JD, Presley JB, Chimento MF, Curcio CA, Johnson M. Age-related changes in human macular Bruch's membrane as seen by quick-freeze/deep-etch. Exp Eye Res 2007 Aug;85(2):202-18.

[104] Souied EH, Benlian P, Amouyel P, Feingold J, Lagarde JP, Munnich A, et al. The epsilon4 allele of the apolipoprotein E gene as a potential protective factor for exudative age-related macular degeneration. Am J Ophthalmol 1998 Mar;125(3):353-9.

[105] Mouchel Y, Lefrancois T, Fages C, Tardy M. Apolipoprotein E gene expression in astrocytes: developmental pattern and regulation. Neuroreport 1995 Dec 29;7(1): 205-8.

[106] Snipes GJ, McGuire CB, Norden JJ, Freeman JA. Nerve injury stimulates the secretion of apolipoprotein E by nonneuronal cells. PNAS 1986 Feb 15;83(4):1130-4.

[107] Schneeberger SA, Iwahashi CK, Hjelmeland LM, Davis PA, Morse LS. Apolipoprotein E in the subretinal fluid of rhegmatogenous and exudative retinal detachments. Retina 1997;17(1):38-43.

[108] Kuhrt H, Hartig W, Grimm D, Faude F, Kasper M, Reichenbach A. Changes in CD44 and ApoE immunoreactivities due to retinal pathology of man and rat. J Hirnforsch 1997;38(2):223-9.

[109] Baird PN, Guida E, Chu DT, Vu HTV, Guymer RH. The ε 2 and ε4 alleles of the apolipoprotein gene are associated with age-related macular degeneration. Investigative Ophthalmology & Visual Science 2004 May 1;45(5):1311-5.

[110] Simonelli F, Margaglione M, Testa F, Cappucci G, Manitto MP, Brancato R, et al. Apolipoprotein E polymorphisms in age-related macular degeneration in an Italian population. Ophthalmic Res 2001 Nov;33(6):325-8.

[111] Zareparsi S, Reddick AC, Branham KEH, Moore KB, Jessup L, Thoms S, et al. Association of apolipoprotein E alleles with susceptibility to age-related macular degeneration in a large cohort from a single center. Investigative Ophthalmology & Visual Science 2004;45(5):1306-10.

[112] Bojanowski CM, Shen D, Chew EY, Ning B, Csaky KG, Green WR, et al. An apolipoprotein E variant may protect against age-related macular degeneration through cytokine regulation. Environ Mol Mutagen 2006 Oct;47(8):594-602.

[113] Fritsche LG, Freitag-Wolf S, Bettecken T, Meitinger T, Keilhauer CN, Krawczak M, et al. Age-related macular degeneration and functional promoter and coding variants of the apolipoprotein E gene. Hum Mutat 2008 Dec 18.

[114] Stewart JE, Skinner ER, Best PV. Receptor binding of an apolipoprotein E-rich subfraction of high density lipoprotein to rat and human brain membranes. The International Journal of Biochemistry & Cell Biology 1998 Mar 1;30(3):407-15.

[115] Schneider WJ, Kovanen PT, Brown MS, Goldstein JL, Utermann G, Weber W, et al. Familial dysbetalipoproteinemia. Abnormal binding of mutant apoprotein E to low density lipoprotein receptors of human fibroblasts and membranes from liver and adrenal of rats, rabbits, and cows. J Clin Invest 1981 Oct;68(4):1075-85.

Permissions

The contributors of this book come from diverse backgrounds, making this book a truly international effort. This book will bring forth new frontiers with its revolutionizing research information and detailed analysis of the nascent developments around the world.

We would like to thank Saša Frank and Gerhard Kostner, for lending their expertise to make the book truly unique. They have played a crucial role in the development of this book. Without their invaluable contribution this book wouldn't have been possible. They have made vital efforts to compile up to date information on the varied aspects of this subject to make this book a valuable addition to the collection of many professionals and students.

This book was conceptualized with the vision of imparting up-to-date information and advanced data in this field. To ensure the same, a matchless editorial board was set up. Every individual on the board went through rigorous rounds of assessment to prove their worth. After which they invested a large part of their time researching and compiling the most relevant data for our readers. Conferences and sessions were held from time to time between the editorial board and the contributing authors to present the data in the most comprehensible form. The editorial team has worked tirelessly to provide valuable and valid information to help people across the globe.

Every chapter published in this book has been scrutinized by our experts. Their significance has been extensively debated. The topics covered herein carry significant findings which will fuel the growth of the discipline. They may even be implemented as practical applications or may be referred to as a beginning point for another development. Chapters in this book were first published by InTech; hereby published with permission under the Creative Commons Attribution License or equivalent.

The editorial board has been involved in producing this book since its inception. They have spent rigorous hours researching and exploring the diverse topics which have resulted in the successful publishing of this book. They have passed on their knowledge of decades through this book. To expedite this challenging task, the publisher supported the team at every step. A small team of assistant editors was also appointed to further simplify the editing procedure and attain best results for the readers.

Our editorial team has been hand-picked from every corner of the world. Their multi-ethnicity adds dynamic inputs to the discussions which result in innovative outcomes. These outcomes are then further discussed with the researchers and contributors who give their valuable feedback and opinion regarding the same. The feedback is then collaborated with the researches and they are edited in a comprehensive manner to aid the understanding of the subject.

Apart from the editorial board, the designing team has also invested a significant amount of their time in understanding the subject and creating the most relevant covers. They scrutinized every image to scout for the most suitable representation of the subject and create an appropriate cover for the book.

The publishing team has been involved in this book since its early stages. They were actively engaged in every process, be it collecting the data, connecting with the contributors or procuring relevant information. The team has been an ardent support to the editorial, designing and production team. Their endless efforts to recruit the best for this project, has resulted in the accomplishment of this book. They are a veteran in the field of academics and their pool of knowledge is as vast as their experience in printing. Their expertise and guidance has proved useful at every step. Their uncompromising quality standards have made this book an exceptional effort. Their encouragement from time to time has been an inspiration for everyone.

The publisher and the editorial board hope that this book will prove to be a valuable piece of knowledge for researchers, students, practitioners and scholars across the globe.

List of Contributors

Ruth Prassl and Peter Laggner
Institute of Biophysics and Nano systems Research, Austrian Academy of Sciences, Graz, Austria

Benjamin Dieplinger
Department of Laboratory Medicine, Konvent hospital Barmherzige Brüder Linz, Austria

Hans Dieplinger
Division of Genetic Epidemiology, Department of Human Genetics and Molecular Pharmacology, Medical University of Innsbruck, Austria

Stanislav Oravec
2nd Department of Internal Medicine, Faculty of Medicine, Comenius University, Bratislava, Slovak Republic

Johannes Mikl
Department of Cardiology, Hietzing Hospital, Austria

Kristina Gruber
Department of Internal Medicine, Landesklinikum Thermenregion Baden, Austria

Elisabeth Dostal
Krankenanstalten Dr. Dostal, Vienna, Austria

D.S. Mshelia and A.A. Kullima
University of Maiduguri/University of Maiduguri Teaching Hospital, Nigeria

Göran Walldius
Department of Epidemiology, Institute of Environmental Medicine (IMM), Karolinska Institutet, Stockholm, Sweden

Assia Rharbi and Zohra Bakkoury
Equipe: AMIPS Ecole Mohammadia des Ingénieurs,
Université Mohammed V, Agdal, Rabat – Morocco

Afaf Mikou
Laboratoire GAIA, Spectroscopie Faculté des Sciences Ain Chock – Université Hassan II, Casablanca – Morocco

Khadija Amine
Laboratoire GAIA, Spectroscopie Faculté des Sciences Ain Chock - Université Hassan II, Casablanca – Morocco
Laboratoire de Recherche sur les Lipoprotéine et l'Athérosclérose, Unité de Recherche Associée au
CNRS-URAC 34-, Faculté des Sciences Ben Msik-Casablanca, Université Hassan II Mohammedia, Morocco

Anass Kettani
Laboratoire de Recherche sur les Lipoprotéine et l'Athérosclérose, Unité de Recherche Associée au CNRS-URAC 34-, Faculté des Sciences Ben Msik-Casablanca, Université Hassan II Mohammedia, Morocco

Abdelkader Betari
ENSA Oujda, Université Mohammed Premier Oujda, Morocco

Adebowale Saba and Olayinka Oridupa
University of Ibadan, Nigeria

Jelena Umbrasiene
Medical Academy, Lithuanian University of Health Sciences, Kaunas, Lithuania, Kaunas Region Cardiology Society, Kaunas, Lithuania

Ruta-Marija Babarskiene
Kaunas Region Cardiology Society, Kaunas, Lithuania, Department of Cardiology, Medical Academy, Lithuanian University of Health Sciences, Kaunas, Lithuania

Jone Vencloviene
Institute of Cardiology, Medical Academy, Lithuanian University of Health Sciences, Kaunas, Lithuania, Vytautas Magnus University, Kaunas, Lithuania

Xiaoyan Zhang and Hainsworth Y. Shin
Center for Biomedical Engineering, University of Kentucky, Lexington, KY, USA

Eduardo Guimarães Hourneaux de Moura, Bruno da Costa Martins, Guilherme Sauniti Lopes
Department of Gastroenterology, Gastrointestinal Endoscopy Unit, Hospital das Clínicas - University of São Paulo School of Medicine, São Paulo, Brazil

Ivan Roberto Bonotto Orso
Department of Gastroenterology, Gastrointestinal Endoscopy Unit, Hospital das Clínicas - University of São Paulo School of Medicine. São Paulo, Brazil Department of Surgery, School of Medicine of the Assis Gurgacz Faculty,Gastroclínica Cascavel, Brazil

Summer F. Acevedo
Department of Physiology, Pharmacology, and Toxicology, Psychology Program, Ponce
School of Medicine and Health Sciences, Ponce, Puerto Rico

**Karl-Erik Eilertsen, Rune Larsen, Hanne K. Mæhre, Ida-Johanne Jensen and Edel O.
Elvevoll**
University of Tromsø, Norway

Maaike Kockx
Macrophage Biology Group, Centre for Vascular Research, University of New South
Wales, Sydney,
Australia

Leonard Kritharides
Macrophage Biology Group, Centre for Vascular Research, University of New South
Wales, Sydney,
Australia
Department of Cardiology, Concord Repatriation General Hospital, University of Sydney,
Sydney, Australia

Parvin Mirmiran
Department of Clinical Nutrition and Dietetics, Faculty of Nutrition Sciences and Food
Technology, National Nutrition and Food Technology Research Institute, Shahid Beheshti
University of Medical Sciences, Tehran, Iran

Somayeh Hosseinpour-Niazi
Nutrition Related Non-Communicable Disease, Research Institute for Endocrine Sciences,
Shahid Beheshti University of Medical Sciences, Tehran, Iran

Fereidoun Azizi
Endocrine Research Center, Research Institute for Endocrine Sciences, Shahid Beheshti
University of Medical Sciences, Tehran, Iran

Edward Loane
Department of Ophthalmology, Mater Misericordiae University Hospital, Dublin, Ireland